B.B.A. LIBRARY – VOLUME 12

MEMBRANE-ACTIVE COMPLEXONES

B.B.A. LIBRARY

B.B.A. LIBRARY

VOLUME 12

MEMBRANE-ACTIVE COMPLEXONES

by

YU. A. OVCHINNIKOV

V.T. IVANOV

and

A.M. SHKROB

U.S.S.R. Academy of Sciences, Shemyakin Institute for Chemistry of Natural Products, Moscow (U.S.S.R.)

ELSEVIER SCIENTIFIC PUBLISHING COMPANY

AMSTERDAM – OXFORD – NEW YORK

1974

ELSEVIER SCIENTIFIC PUBLISHING COMPANY
335 JAN VAN GALENSTRAAT
P.O. BOX 211, AMSTERDAM, THE NETHERLANDS

AMERICAN ELSEVIER PUBLISHING COMPANY, INC.
52 VANDERBILT AVENUE
NEW YORK, NEW YORK 10017

This book is simultaneously published in the Russian language by Nauka (Publishing House of the U.S.S.R. Academy of Sciences) in 1974 under the title Мембрано-активные комплексоны, 456 pp. including 52 tables and 251 illustrations, 1130 references.

LIBRARY OF CONGRESS CARD NUMBER: 73-85226

ISBN 0-444-41159-3

PRINTED IN THE NETHERLANDS

Preface

Many of the important trends in the biology and chemistry of natural products are becoming ever more involved with studies of the structure and function of biological membranes. Such a tendency follows naturally from the part the membranes play in such vital processes as, for instance, respiration and photosynthesis. It will be no exaggeration to say that the progress achieved lately in mechanistic studies of these processes is in no small measure due to the extensive use of antibiotics selectively increasing the cation permeability of the membranes. Such compounds frequently called ionophores have at present become highly popular tools in the hands of biophysicists and biochemists investigating phenomena associated with transmembrane ion transport. In the course of study of the antibiotic ionophores it was found that their membrane activity is the direct outcome of their ability to form complexes with alkali metal ions. This discovery has given birth to and spurred on the tempestuous development of the chemistry of a heretofore unknown type of macrocyclic complexone. It is these complexones, in particular such complexones that bind alkali metal ions and promote their transport across artificial and biological membranes, which form the subject matter of this book.

The book is roughly divided into two parts. Chapters I—IV deal with the structure, conformations and properties of both the free complexone molecules and their complexes, so they have a more or less chemical imprint. The concluding Chapters V and VI are devoted to the action of these compounds on artificial and biological membranes. In Chapters I—V the authors have tried to make a comprehensive survey of the relevant problems, incorporating data published to the end of 1972. In the last Chapter, dealing with the effect of membrane-active complexones on biological objects, only a few trends which are considered to be of particular interest have been discussed. The authors feel that the topics selected will suffice for the reader to acquaint himself with the mechanisms of the most characteristic effects and with the directed use of complexones in studies of membrane processes.

It is hoped that this book will prove helpful to both organic chemists interested in complexones and their application in the pure and applied sciences, and to biologists striving to understand membrane function at the molecular level.

Naturally certain areas such as physical methods of conformational analysis and also many of the biological problems could be discussed here only in broad outline in order not to lose track of the main objectives and not to make the book too cumbersome for the wide group of workers for which it is intended. The authors are also aware of the controversiality of many of the problems discussed here and will be very grateful for all critical comments.

Deep acknowledgement is expressed to Drs. A. A. Lev, E. A. Liberman, V. S. Markin, Yu. G. Molotkovskii and V. P. Skulachev for their valuable advice and comments, which have been taken into account in the preparation of the book. The authors also wish to express their great indebtedness to Dr. G. Peck both for his knowledgeable translation of the Russian manuscript into English and for his many helpful remarks which have often made them look with fresh eyes on many a page of their writing. We are grateful to Dr. Yu. B. Shvetsov for expert assistance in the preparation of the manuscript. Finally, we are indebted to authors and publishers who cooperated fully in giving the permission to reproduce numerous figures.

YU. A. OVCHINNIKOV
V. T. IVANOV
A. M. SHKROB

Introduction

In the second half of the sixties it was found that various naturally occurring and synthetic neutral macrocyclic compounds — peptides, depsipeptides, depsides, polyethers — are capable of binding cations in solution, holding them in the intramolecular cavities by means of their polar (ether, amide, ester, etc.) groupings. The ion selectivity of such interaction depends not only on the number and nature of the ligands, but is also a function of the conformational characteristics of the whole molecule. It is to this specific property of the macrocyclic complexones that their outstanding ability to discriminate between alkali metal ions is due. Complexes formed by the neutral macrocyclic compounds carry a positive charge and are highly lipophilic, the metal ion being shielded from the solvent and counter-ions.

The considerable interest which is being manifested in such macrocyclic compounds is due not only to their unusual complexing properties, but to their unique ability to increase selectively the alkali ion permeability of biological membranes. This effect is responsible for the antibiotic activity of some of the microbial-produced macrocyclic complexones such as valinomycin and the enniatins.

Certain compounds formally of non-cyclic structure were found also to be capable of binding metal ions, the resulting complexes being folded in a pseudocyclic conformation. To such complexones belongs the group of nigericin antibiotics. The coordination sphere of the bound cation in these compounds, as in the macrocyclic complexones, is filled as a rule by non-ionic polar groups. However, the antibiotics also contain a dissociable carboxyl group so that their complexes are of a salt-like character with no net charge. The biological activity of the nigericin antibiotics is due to their ability to induce non-electrogenic transmembrane exchange of cations and protons. A unique position among the membrane-active substances is occupied by the gramicidins A, B, C and D. These antibiotics are linear peptides which in solution do not bind alkali metal ions to a noticeable degree. However, their membrane activity together with their conformational characteristics have led to the inference that these molecules are capable of accommodating the alkali metal ions in the act of their induced transfer across biological or model membranes.

The neutral macrocyclic complexones, the nigericin antibiotics and the gramicidins thus resemble each other in both the nature of their interaction with cations and their biological activity. Hence there are valid grounds for considering them together in this book, combining them into a single class of membrane-active complexones. We have also included macrocyclic complexones (antamanide, monamycins) which in both the nature of their action on biological objects and their behavior in model systems differ from the above antibiotics. Their further study will no doubt broaden our concepts of the modes of action of the complexones.

The rapid development of the chemistry of membrane-active complexones has led to elucidation of the structure and conformational states of both the free molecules and the cation complexes of valinomycin, the nactins, nigericin, etc. The data obtained lie at the basis of studies now in progress aimed to shed further light on the cause for the high cationic selectivity and on the mode of functioning of these compounds in membranes. One of the basic objectives in such studies is the synthesis of a wide spectrum of compounds required by biologists and medical workers for the directed and selective increase of the ion permeability of biomembranes. At the same time it becomes more and more evident that detailed study of the conformation-dependent binding of ions by the macrocyclic complexones lays the grounds for the structural elucidation of cation binding centers in such biological systems as, for instance, K,Na-dependent transport ATPase.

Up until recently macrocyclic complexones have been used mainly in biological studies; now new promising areas of application have arisen:

in analytical and inorganic chemistry — for the extractive and chromatographic separation of metal ions and the solubilization of sparingly soluble salts;

in organic chemistry — for affecting the course, rate and stereospecificity of reactions;

in ionometry — for the production of highly selective cation-sensitive membrane electrodes.

This highly incomplete enumeration shows that the chemistry of macrocyclic complexones, originating with a study of the membrane-active antibiotics, is today at the threshold of a new stage in its development connected with applications in multifarious scientific and technical fields.

Contents

Chapter I

GENERAL CHARACTERISTICS OF ALKALI METAL–BINDING MACROCYCLIC COMPOUNDS AND THEIR ANALOGS

I.A. Depsipeptides

Among the macrocyclic compounds forming specific alkali metal complexes, a prominent position is assumed by the cyclic depsipeptides, atypical peptides consisting of amino and hydroxy acid residues linked by amide, *N*-methylamide and ester bonds. Such compounds include, for instance, the antibiotics valinomycin, the enniatins and monamycin whose activity is connected with just this complexing ability. Quite possibly the biological action of several other naturally occurring cyclodepsipeptides (see review [900]) proceeds through some stage of metal-ion binding.

Structural studies on the naturally occurring cyclodepsipeptides carried out in Shemyakin's laboratory resulted in the accumulation of considerable numbers of their synthetic analogs, so that when they came to be recognized as inducers of transmembrane cation transport — an event of considerable importance to molecular biology — many types of these compounds were available in the laboratory, facilitating study of their structure—activity relationships. Moreover, elucidation of the principles governing the complexing reaction soon made possible the synthesis of new analogs with predetermined properties.

I.A.1. *Valinomycin*

The antibiotic valinomycin was first isolated by Brokmann and Schmidt-Kastner in 1955 from extracts of *Streptomyces fulvissimus* [98]. Soon after, MacDonald obtained this compound from the culture of an unidentified *Streptomyces* species, PRL 1649 [572]. In 1962 a substance which proved to be identical with valinomycin was discovered among the antifungal products of *Streptomyces* No. 5901 [100]; and finally, in 1964, Japanese workers described the isolation from a *Streptomyces* No. 329 culture (which they subsequently named *Streptomyces tsusimaensis*) of a compound displaying all the physical, chemical and antimicrobial properties of valinomycin [686]. Recently, methods for the semi-industrial production of valinomycin have been elaborated [573, 953].

In the studies mentioned above different methods for the isolation and

2

purification of the antibiotic were used, including crystallization from dibutyl ether [98], chromatography on alumina [572, 953], sublimation in vacuum [572] and fractional precipitation of impurities by different solvents [100] etc. Pure, crystalline valinomycin is in the form of colorless platelets, readily soluble in organic solvents, even in heptane (to the extent of $2.3 \cdot 10^{-3}$ mole·l^{-1} [521]) and sparingly soluble in water ($7.2 \cdot 10^{-7}$ mole·l^{-1} [521]). It is stable on storage.

Valinomycin has a fairly broad antimicrobial spectrum, being highly active against a number of Gram-positive and acid-fast bacteria, yeasts and phytopathogenic fungi (see Table 2). For example, the minimal in-vivo inhibiting concentration with respect to *Mycobacterium tuberculosis* H_{37} is 0.5 μg/ml [686]. According to Brown [100] the following concentrations are effective against various pathogenic yeasts: *Histoplasma capsulatum*, 3.13 μg/ml; *Blastomyces dermatitidis*, 1.56 μg/ml; *Candida albicans* 12.5 μg/ml. The interest of the Japanese workers in this antibiotic arose from its potent, specific in-vitro activity against the rice infecting phytopathogenic fungus *Piricularia oryzae*. Experiments on rice kernels inoculated with this fungus and then sprayed with valinomycin suspension led them to the conclusion that this antibiotic can be useful in rice farming.

At first, two possible formulae, namely the cyclic octadepsipeptides (10) and (13), were proposed for valinomycin, since these agreed with the results obtained at that time of elementary analysis, molecular weight determination, acid and alkaline hydrolysis and spectral measurements [97].

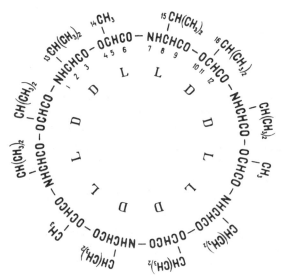

Fig. 1. Valinomycin.

However, when the compounds as formulated were synthesized, they were found to be biologically inactive and to differ considerably in other properties from the naturally occurring antibiotic [920]. Redetermination of the molecular weight [99, 895] showed that valinomycin must actually be a cyclododecadepsipeptide and this was unequivocally proved by Shemyakin and coworkers by means of total synthesis [895, 896].

As one can see from Fig. 1, valinomycin consists of three identical tetradepsipeptide fragments D-Val—L-Lac—L-Val—D-HyIv* with alternating amino and hydroxy acid residues. In view of this the synthesis (Scheme 1) was started by acylating the hydroxy groups of *tert*-butyl-L-lactate and D-hydroxyisovalerate with benzyloxycarbonyl-D- and L-valine; after removal of the protective groups the resultant didepsipeptides were condensed to the tetradepsipeptide Z—D-Val—L-Lac—L-Val—D-HyIv—OBu^t. Further build-up of the depsipeptide chain was also achieved by means of amide bonding. Since activation of the C-terminal hydroxy acid residues does not cause extensive racemization, the peptide bond was formed by means of the acid chloride method which, although highly efficient, is rarely used in peptide chemistry because it causes considerable racemization of C-terminal amino acids. This method was also used under high dilution conditions to cyclize the linear 12-membered depsipeptide, resulting in yields of valinomycin of up to 25%. It has recently been shown that Merrifield's solid-phase method [288, 289, 630] can be successfully applied to synthesis of linear peptide and depsipeptide analogs of the valinomycins.

The preparation of labeled valinomycins has been described in a number of papers. Thus, in the course of a biosynthetic study of this antibiotic, variously ^{14}C-labeled specimens were obtained. The biosynthesis of valinomycin in a medium containing D,L-[14C]valine according to an improved MacDonald procedure [572—574, 953] gave a preparation with a specific activity of 150 000 counts/min per mg. For assigning the NH signals in the valinomycin NMR spectra (see Part III.A.1.a) one of the three L-valine residues was replaced by an L-[15N]valine residue to give the

---

* The generally accepted IUPAC abbreviations [1111] are used to designate the amino acids and the protective groups. The hydroxy acid residues are designated as follows:

Glyc = O—CH$_2$—CO

Lac = O—CH—CO
$\quad\quad\quad$ |
$\quad\quad\quad$ CH$_3$

HyIv = O—CH—CO
$\quad\quad\quad$ |
$\quad\quad\quad$ CH(CH$_3$)$_2$

HyIc = O—CH—CO
$\quad\quad\quad$ |
$\quad\quad\quad$ CH$_2$CH(CH$_3$)$_2$

HyMeV = O—CH—CO
$\quad\quad\quad$ |
$\quad\quad\quad$ CH(CH$_3$)C$_2$H$_5$

HyDec = O—CH—CH$_2$CO
$\quad\quad\quad$ |
$\quad\quad\quad$ CH$_2$(CH$_2$)$_5$CH$_3$

($\Delta^{4,5}$)HyIv = O—CH—CO
$\quad\quad\quad$ |
$\quad\quad\quad$ C(CH$_3$)=CH$_2$

4

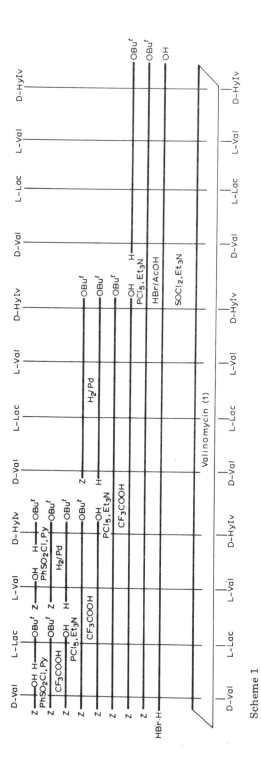

Scheme 1

[15]N-labeled analog (2) [433]. Finally, Pressman [791] has mentioned synthesis of a tritiated valinomycin using Wilzbach's method [1080]; however, the optical purity of the compound is uncertain because racemization could have occurred under the conditions of $H-^3H$ exchange.

Valinomycin forms very stable equimolar complexes with K^+, Rb^+ and Cs^+ ions in methanol and ethanol solutions (Table 1); complexes with Li^+, Na^+ and NH_4^+ and with the alkaline earth and transition metals are considerably less stable. The KCNS and K_2PtCl_6 complexes have been isolated in the crystalline state [751, 757]. The cation affinity of valinomycin decreases sharply in aqueous solutions [15, 904]. For instance, the stability constants of the complexes with K^+, Rb^+ and Cs^+ amount to only 2.3, 5.9 and 0.12 $l \cdot mole^{-1}$ and no formation of a Na^+ complex can be detected under these conditions [258]. In general, valinomycin excels all known metal-binding macrocyclic compounds in K/Na-selectivity and to this it owes its exceptionally widespread use in the studies of transmembrane ion transport.

I.A.2. Valinomycin analogs

Numerous valinomycin analogs (Table 2) were synthesized in the course of studies of the relation between the structure, physicochemical properties and biological activity of cyclodepsipeptides. These differed from the parent compound in ring size (compounds 10 and 82), in amino and hydroxy acid configurations (compounds 14, 19, 26, 36, 49, 54, 56, 63, 74—76), in the side chains (compounds 15—17, 20, 21, 28—34, 37, 38, 41—44, 47, 50—52, 55, 57, 58, 61, 62 and 79) and in the number and position of the ring amide and hydroxy acid groupings (compounds 18, 23, 25, 27, 35, 39, 40, 45, 46, 48, 53, 59 and 60). Compounds (3—9, 11—13, 22, 24, 64—73, 77, 78, 80, 81 and 83—85) differ from valinomycin in more than one of these characteristics. Bearing in mind the C_3 symmetry of the primary valinomycin structure, its equi-ring analogs can be classified according to whether the modifications are in only one (14—43), two (44—48) or in all three (49—81) of its tetradepsipeptide D-Val—L-Lac—L-Val—D-HyIv* units. Analogs with changed configuration of all asymmetric centers (74), with reversed direction of acylation (75) or with interchanged amide and ester groups (77), (76) and (78—80), or those obtained by combinations of such procedures are generally referred to as topochemical analogs [433, 902].

All compounds in Table 2 have been prepared in the Shemyakin Institute for Chemistry of Natural Products (U.S.S.R. Academy of Sciences, Moscow) [267, 433, 848, 921, 925, 1038] by the usual methods for solution

---

* In Table 2 this unit is designated as A.

TABLE 1

STABILITY CONSTANTS OF THE VALINOMYCIN COMPLEXES WITH DIFFERENT CATIONS*

Solvent	Li+	Na+	K+	Rb+	Cs+	NH4+	Ag+	Tl+	Mg2+	Ca2+	Sr2+	Ba2+
Methanol [326]	<5	4.7 12[1083] 14[384]	80 000 27 000[15] 8 000[1083]	180 000	26 000	47	8000	5400	<5	500	170	2200
Ethanol [15, 904]	—	<50	2600 000**	2900 000	650 000	—	—	—	—	—	—	—
99.5% Aqueous ethanol [668]			1200 000									

* The values for the bivalent cation complexes are too low owing to neglect of (metal·anion)+ pair formation.
** Previously given [700, 897, 903] stability constants for the valinomycin·K+ complex were later corrected [15, 904].

synthesis of depsipeptides [923, 924] excepting compounds (73), (75) and (81) which were synthesized by the solid-phase method [288, 630]. As with valinomycin, ester bonds were formed first, further chain build-up and cyclization being carried out by amide bonding. Compound (60) was prepared by methylation of the valinomycin amide bonds according to the method of Das *et al.* [187]. Yields of the 12-membered cyclotetra-depsipeptide analogs (3—7) were markedly lower than those of the larger ring analogs, their formation being accompanied by doubling, tripling and other side reactions [925]. For instance, in cyclizing the linear tetra-depsipeptide H—D-Val—L-Lac—D-Val—D-HyIv—OH, cyclododeca- and cyclo-hexadecadepsipeptides (54) and (83) were obtained in yields of 4 and 2.5%, respectively, whereas the expected cyclotetradepsipeptide could not be isolated at all. Doubling of linear octa- and dodecadepsipeptides has also been reported. The valinomycin analogs obtained by cyclopolymerization are listed in Table 3 [925].

As one can see from Table 2, the valinomycin analogs can differ considerably from each other and from the parent antibiotic in both physicochemical and biological properties. One group of compounds (3—10, 27, 49, 54, 56, 58, 60, 61, 75—78) practically forms no complexes with potassium ions in ethanol and does not inhibit microbial growth. Of the remainder, compounds (26, 35, 36, 39, 41, 46, 48, 50, 67, 71, 72 and 82) form comparatively weak K^+ complexes (stability constants 10^2-10^4 $l \cdot mole^{-1}$), whereas others, compounds (15, 22, 28, 57, 64, 74, 79 and 80), show almost no difference in this property from valinomycin. Still other analogs (compounds 14, 18, 19, 21, 23—25, 37, 38, 40, 42—45, 47, 52) assume an intermediate position in respect of complex stability. Special mention should be made of compound (53) whose K^+ complex is much more stable than that of valinomycin or any of the other analogs. The reasons for such sharp differences are discussed in Part III.A.1.b. All K^+-complexing analogs retain the high K/Na-selectivity of valinomycin; certain of them display enhanced Na^+-complex stability. A distinctive feature of analogs (30) and (32) is their possession of ionogenic side chains. Hence the stability of their complexes should depend upon the pH of the medium. Moreover, the side chains can be utilized for covalently attaching the valinomycin ring to various carriers and for preparing spin- or fluorescence-labeled valinomycin derivatives.

The antimicrobial activity of the valinomycin analogs varies within wide limits, but almost none of them exceeds valinomycin in either the minimal required dosage or the antibiotic spectrum. The only exceptions are compounds (23), (45) and (53), which, contrary to the parent antibiotic, inhibit *Staphylococcus aureus* 209 P.

TABLE 2

STRUCTURE AND SOME PROPERTIES OF THE VALINOMYCIN CYCLODEPSIPEPTIDES

A = D-Val—L-Lac—L-Val—D-Hylv.

The six columns under "Minimal growth-inhibiting concentration (μg/ml)" are: S. aureus 209 P | S. aureus UV-3 | B. mycoides | E. coli | M. phlei | C. albicans

No.	Compound	m.p. (°C)	$[\alpha]_D$	c (%)	Solvent	S. aureus 209 P	S. aureus UV-3	B. mycoides	E. coli	M. phlei	C. albicans	Stability constants of K$^+$ complexes*	Free energy** of complex formation with K$^+$	References
1	cyclo[—(D-Val—L-Lac—L-Val—D-Hylv)$_3$—] ≡ cyclo(—A$_3$—) (valinomycin)	187	+32.8°	1	C_6H_6	>50	0.2—0.4	>50	>50	0.3	0.2—0.4	2 000 000	8.6	[289, 896, 922]
2	cyclo(—D-Val—L-Lac—^{15}N—L-Val—D-Hylv—)	190	+31.0	1	C_6H_6									[98]
		188	+31.8	1	C_6H_6									[433]
3	cyclo(—D-Val—D-Hylv—D-Val—D-Hylv—)	291	+190	0.75	$CHCl_3$	—	—	—	—	—	>100	—	—	[921, 922]
4	cyclo(—L-Val—L-Hylv—L-Val—L-Hylv—)	291	−191	0.7	$CHCl_3$	>100	>100	>100	>100	>100	>100	<50	<2.3	[921, 922, 926]
5	cyclo(—D-Val—L-Hylv—L-Val—D-Hylv—)	318 (decomp.)	meso-form			>100	>100	>100	>100	>100	>100	<50	<2.3	[925, 926]
6	cyclo(—D-Val—L-Lac—L-Val—L-Hylv—)	252	−51	1	$CHCl_3$	>100	>100	>100	>100	>100	>100	<50	<2.3	[922, 925, 926]
7	cyclo(—D-Val—L-Lac—L-Val—L-Lac—)	246	−73	1	$CHCl_3$	>100	>100	>100	>100	>100	>100	<50	<2.3	[922, 925, 926]
8	cyclo[—(L-Val—L-Hylv)$_4$—]	236	−130	0.75	$CHCl_3$	>100	>100	>100	>100	>100	>100	<50	<2.3	[712, 921, 922]
9	cyclo[—(D-Val—D-Hylv)$_4$—]	236	+131	0.7	$CHCl_3$	>100	>100	>100	>100	>100	>100	<50	<2.3	[712, 921, 922]
10	cyclo(—A$_2$—) ("octa-valinomycin")	246	+143	0.5	$CHCl_3$	>100	>100	>100	>100	>100	>100	<50	<2.3	[904, 921, 922]
		219	−6	1	C_6H_6									[872]
11	cyclo[—L-Val—L-Lac—L-Val—D-Hylv)$_2$—]	232	−37	1	$CHCl_3$	>100	>100	>100	>100	>100	>100	—	—	[922, 925]
12	cyclo[—(D-Val—L-Hylv—L-Val—D-Hylv)$_2$—]	241	0	2	$CHCl_3$	>100	>100	>100	>100	>100	>100	—	—	[922, 925]
13	cyclo(—D-Val—L-Lac—L-Val—L-Lac—L-Val—D-Hylv—D-Val—D-Hylv—)	222	+2	1	$CHCl_3$	>50	>50	>50	>50	>50	>50	—	—	[896, 920, 925]
14	cyclo(—L-Val—L-Lac—L-Val—D-Hylv—A$_2$—)	219	+24	1	C_6H_6	>50	2	>50	>50	1.5	3—5	11 000	5.5	[712, 922, 925]
15	cyclo(—D-Ala—L-Lac—L-Val—D-Hylv—A$_2$—)	173	+36	1	C_6H_6	>50	0.1—0.2	>50	>50	0.4	0.2—0.3	2 000 000	8.6	[712, 922, 925, 926]
16	cyclo(—D-Leu—L-Lac—L-Val—D-Hylv—A$_2$—)	167	+40	1	C_6H_6	>50	0.2	>50	>50	0.3	0.2—0.4	—	—	[922, 925]
17	cyclo(—D-aIle—L-Lac—L-Val—D-Hylv—A$_2$—)	164	+25	1	C_6H_6	>50	0.2	>50	>50	0.3	0.3—0.5	—	—	[922, 925]
18	cyclo(—D-Hylv—L-Lac—L-Val—D-Hylv—A$_2$—)	Amorph.	−4.5	0.2	$CHCl_3$	>25	>25	>25	>25	>25	>25	42 000	6.3	[1038]
19	cyclo(—D-Val—D-Lac—L-Val—D-Hylv—A$_2$—)	155	+44	1	C_6H_6	>50	>50	>50	>50	>50	>50	75 000	6.7	[922, 925, 926]
20	cyclo(—D-Val—Glyc—L-Val—D-Hylv—A$_2$—)	159	+31	1	C_6H_6	>50	0.2	>50	>50	0.3	4—6	—	—	[267]
21	cyclo(—D-Val—L-Hylv—L-Val—D-Hylv—A$_2$—)	210	+25	1	C_6H_6	>50	>50	>50	>50	>50	>50	400 000	7.7	[712, 922, 925, 926]
22	cyclo(—D-Val—Gly—L-Val—D-Hylv—A$_2$—)	101	+43	1	C_6H_6	>50	1—2	>50	>50	2	4.5	1 200 000	8.3	[267]
23	cyclo(—D-Val—L-Ala—L-Val—D-Hylv—A$_2$—)	192	+40	1	C_2H_5OH	4.5	0.1	>50	>50	0.2—0.4	4.5	300 000	7.5	[267, 712, 904, 922, 926]
24	cyclo(—D-Val—L-Val—L-Val—D-Hylv—A$_2$—)	203	+26	1	C_2H_5OH	>50	0.5—0.7	>50	>50	0.4	10—12	100 000	6.9	[267, 926]
25	cyclo(—D-Val—L-MeAla—L-Val—D-Hylv—A$_2$—)	204	+6	0.1	C_2H_5OH	>10	—	10	—	4	0.8	550 000	7.9	[1038]

No.	Compound	m.p.	[α]	c	Solvent										Refs.
26	cyclo(−D-Val−L-Lac−D-Val−D-HyIv−A₂−)	187	+29	0.6	C₆H₆	>50	1−2	>50	>50	>50	1−2	0.2−0.4	4400	5.0	[712, 922, 925, 926]
27	cyclo(−D-Val−L-Lac−L-MeVal−D-HyIv−A₂−)	143	−12	1	CHCl₃	>50	>50	>50	>50	>50	>50	>50	<50	<2.3	[267, 712, 904, 926]
28	cyclo(−D-Val−L-Lac−L-Ala−D-HyIv−A₂−)	152	+38	0.1	C₆H₆	>50	0.3−0.5	>50	>50	0.2	0.3−0.5	0.2−0.5	3000 000	8.9	[267, 712, 926]
29	cyclo(−D-Val−L-Lac−D-aIle−D-HyIv−A₂−)	176	+33	1	C₆H₆	>50	0.6	>50	>50	0.3	0.6	0.8	—	—	[922, 925]
30	cyclo(−D-Val−L-Lac−L-Glu−D-HyIv−A₂−)	Amorph.	+15	0.1	C₆H₆	>25	—	>25	>25	6−9	—	>25	—	—	—
31	cyclo(−D-Val−L-Lac−L-Glu(OBzlNO₂)−D-HyIv−A₂−)	Amorph.	+20	0.1	C₆H₆	>40	—	>40	>40	>40	—	>40	—	—	—
32	cyclo(−D-Val−L-Lac−L-Lys−D-HyIv−A₂−)	Amorph.	+7.5	0.1	C₆H₆	25−37	—	>40	>40	25−37	—	>40	—	—	—
33	cyclo(−D-Val−L-Lac−L-Lys(Z)−D-HyIv−A₂−)	Amorph.	+32	0.09	C₆H₆	>25	—	>25	>25	>25	—	>25	—	—	—
34	cyclo(−D-Val−L-Lac−L-Lys(Phth)−D-HyIv−A₂−)	Amorph.	—	—	—	>10	—	>10	>10	0.3	—	0.2−0.4	—	—	—
35	cyclo(−D-Val−L-Lac−L-HyIv−D-HyIv−A₂−)	132	+17	1	CHCl₃	>50	>50	>50	>50	>50	>50	>50	2500	4.7	[267, 712, 904, 926]
36	cyclo(−D-Val−L-Lac−L-HyIv−L-HyIv−A₂−)	180	+2	1	C₆H₆	>50	>50	>50	>50	>50	>50	>50	100−150	2.7	[922, 925, 926]
37	cyclo(−D-Val−L-Lac−L-Val−Glyc−A₂−)	173	+24	2	C₆H₆	>50	0.5	>50	>50	1.5−2	0.5	4.5−6	100 000	6.9	[267]
38	cyclo(−D-Val−L-Lac−L-Val−D-Lac−A₂−)	189	+30	0.2	CHCl₃	>25	0.2	>25	>25	0.3	0.2	0.2	300 000	7.5	[712, 1038]
39	cyclo(−D-Val−L-Lac−L-Val−D-Val−A₂−)	173	+16	1	CHCl₃	>50	1	>50	>50	1.5	1	3	5200	5.1	[267, 712, 926]
40	cyclo(−D-Val−L-Lac−L-Val−D-MeVal−A₂−)	169	+31	0.1	C₂H₅OH	>10	—	>10	>10	>10	—	2	50 000	6.4	[1038]
41	cyclo(−Gly-Glyc−L-Val−D-HyIv−A₂−)	Amorph.	+16	1	C₂H₅OH	>50	6−9	>50	>50	9−12	6−9	9−12	6200	5.3	[922, 925, 926]
42	cyclo(−D-Ala−L-Lac−L-Ala−D-HyIv−A₂−)	Amorph.	+34	1	C₆H₆	>50	2	>50	>50	1.5	2	4	100 000	6.9	[922, 925, 926]
43	cyclo(−D-Leu−L-Lac−L-Leu−D-HyIv−A₂−)	165	+45	1	C₆H₆	>50	2	>50	>50	1−2	2	3−4	130 000	7.0	[267, 712, 926]
44	cyclo[−D-Ala−L-Lac−L-Val−D-HyIv)₂−A−]	Amorph.	+3.6	0.1	C₂H₅OH	>50	0.5−0.7	>50	>50	0.5−1	0.5−0.7	1−1.5	88 000	6.8	[267, 712, 926]
45	cyclo[−(D-Val−L-Ala−L-Val−D-HyIv)₂−A−]	203	+5	1	CHCl₃	>50	2	12	2−4	2	2	12−18	220 000	7.3	[267, 712, 926]
46	cyclo[−(D-Val−L-MeAla−L-Val−D-HyIv)₂−A−]	Amorph.	−32	0.1	C₂H₅OH	>20	—	>20	>20	15	—	20	~100	~2.7	[1038]
47	cyclo[−(D-Val−L-Lac−L-Ala−D-HyIv)₂−A−]	Amorph.	+2	0.1	C₆H₆	>50	0.5	>50	>50	1	0.5	1.5−2	200 000	7.3	[267, 712, 926]
48	cyclo[−(D-Val−L-Lac−L-Val−D-MeVal)₂−A−]	204	+38	0.1	C₂H₅OH	>10	—	>10	>10	>10	—	>10	3100	4.8	[1038]
49	cyclo[−(L-Val−L-Lac−L-Val−D-HyIv)₃−]	Amorph.	−36	1	C₂H₅OH	>50	>50	>50	>50	>50	>50	>50	<50	<2.3	[712, 922, 925, 926]
50	cyclo[−(D-Ala−L-Lac−L-Val−D-HyIv)₃−]	Amorph.	+14	1	C₆H₆	>50	>50	>50	>50	>50	>50	>50	160	3.0	[267, 712, 926]
51	cyclo[−(D-aIle−L-Lac−L-Val−D-HyIv)₃−]	173	+28	1	C₆H₆	>50	>50	>50	>50	>50	>50	>50	—	—	[922, 925]
52	cyclo[−(D-Val−L-HyIv−L-Val−D-HyIv)₃−] ≡ cyclo(−B₃−) (meso-HyIv-valinomycin)	284	0	2	CHCl₃	>50	>50	>50	>50	>50	>50	>50	370 000	7.6	[712, 848, 922, 925, 926]
53	cyclo[−(D-Val−L-MeAla−L-Val−D-HyIv)₃−]	199	−17	0.1	C₂H₅OH	>20	0.4−0.8	>20	>20	0.4	0.4−0.8	>20	>10 000 000†	>9.6†	[1038]
54	cyclo[−(D-Val−L-Lac−D-Val−D-HyIv)₃−]	Amorph.	+60	0.5	C₆H₆	>50	>50	>50	>50	>50	>50	>50	<50	<2.3	[712, 922, 925, 926]
55	cyclo[−(D-Val−L-Lac−L-Ala−D-HyIv)₃−]	Amorph.	+19	0.1	C₆H₆	>50	1−1.5	>50	>50	4.5	1−1.5	4.5	20 000	5.9	[267, 712, 926]
56	cyclo[−(D-Val−L-Lac−L-Val−L-HyIv)₃−]	223	−35	1	CHCl₃	>50	>50	>50	>50	>50	>50	>50	<50	2.3	[712, 922, 925, 926]
57	cyclo[−(D-Val−L-Lac−L-Val−D-Lac)₃−] (meso-Lac-valinomycin)	210	0	2	CHCl₃	>50	>50	>50	>50	>50	>50	>50	2300 000	8.7	[712, 848, 922, 926]
58	cyclo[−(D-Val−L-Lac−L-Val−L-Lac)₃−]	222	−23	1	CHCl₃	>50	>50	>50	>50	>50	>50	>50	<50	<2.3	[922, 925]

TABLE 2—continued

No. Compound	m.p. (°C)	[α]$_D$	c (%)	Solvent	Minimal growth-inhibiting concentration (μg/ml)						Stability constants of K$^+$ complexes*	Free energy** of complex formation with K$^+$	References
					S. aureus 209 P	S. aureus UV-3	B. myco-ides	E. coli	M. phlei	C. albicans			
59 cyclo[—(D-Val—L-Lac—L-Val—D-MeVal)$_3$—]††	209	—	—	—	—	—	—	—	—	—	—	—	[1038]
60 cyclo[—(D-MeVal—L-Lac—L-MeVal—D-HyIv)$_3$—]	268	−35	0.25	CHCl$_3$	>25	>25	>25	>25	>25	>25	<50	<2.3	[712, 1038]
61 cyclo[—(D-Val—Glyc—L-Val—Glyc)$_3$—] (meso-Glyc-valinomycin)	198	—	—	—	>25	>25	>25	>25	>25	>25	<50	<2.3	[848]
62 cyclo[—(D-Ala—L-HyIv—L-Ala—D-HyIv)$_3$—] (meso-Ala-valinomycin)	Amorph.	—	—	—	>25	>25	>25	>25	>25	>25	220 000	7.3	[712, 848, 926]
63 cyclo[—(L-Val—L-Lac—D-Val—L-HyIv)$_3$—]††	Amorph.	−51	0.2	CHCl$_3$	—	—	—	—	—	—	—	—	[1038]
64 cyclo[—L-Val—L-HyIv—L-Val—D-HyIv—B$_2$—]	222	−15	1	CHCl$_3$	>25	>25	>25	>25	>25	>25	1 000 000	8.2	[848, 926]
65 cyclo[—L-Val—L-HyIv—L-Val—D-HyIv$_2$—B—]	112	−5	1	CHCl$_3$	>25	>25	>25	>25	>25	>25	11 000	5.5	[848, 926]
66 cyclo[—L-Val—L-HyIv—D-Val—D-HyIv—B$_2$—]	207	−8	2	CHCl$_3$	>25	>25	>25	>25	>25	>25	10 000	5.5	[848, 926]
67 cyclo[—D-Val—L-HyIv—L-Val—L-HyIv—B$_2$—]	268	−53	0.1	CHCl$_3$	>25	>25	>25	>25	>25	>25	1500	4.3	[848, 926]
68 cyclo[—(D-Val—L-HyIv—L-Val—L-HyIv)$_2$—B—]	Amorph.	−32	0.2	C$_6$H$_6$	>25	>25	>25	>25	>25	>25	<50	<2.3	[848, 926]
69 cyclo[—D-Val—D-HyIv—L-Val—L-HyIv—B$_2$—]	236	+21	0.1	CHCl$_3$	>25	>25	>25	>25	>25	>25	<50	<2.3	[848, 926]
70 cyclo[—L-Val—L-HyIv—L-Val—D-HyIv—D-Val—L-HyIv—D-Val—D-HyIv—B—]	291	+12	1	CHCl$_3$	>25	>25	>25	>25	>25	>25	10 000	5.5	[848, 926]
71 cyclo[—D-Val—L-HyIv—L-Val—L-HyIv—D-Val—D-HyIv—L-Val—D-HyIv—B—]	241	−1.6	0.1	CHCl$_3$	>25	>25	>25	>25	>25	>25	180	3.5	[848, 926]
72 cyclo[—(L-Val—L-HyIv—L-Val—D-HyIv)$_3$—]	Amorph.	−28.3	0.2	CHCl$_3$	>25	>25	>25	>50	>25	>25	150	3.0	[1038]
73 cyclo[—(D-Val—L-Lac)$_3$—(L-Val—D-HyIv)$_3$—]	—	—	—	—	—	—	—	—	—	—	—	—	[289]
74 cyclo[—(L-Val—D-Lac—D-Val—L-HyIv)$_3$—] (enantio-valinomycin)	188	−30.4	0.2	C$_6$H$_6$	>50	0.8	>50	>50	0.3	0.8	2 000 000	8.6	[433]

No.	Compound				Solvent							Ref.
75	cyclo[-(D-HyIv-L-Val-L-Lac-D-Val-)₃-] ≡ cyclo[-(L-Val-L-Lac-D-Val-D-HyIv)₃-] (retro-valinomycin)	248	+13.0	0.2	C₂H₅OH	>25	>25	>25	>25	<50	<2.3	[433]
76	cyclo[-(L-HyIv-D-Val-D-Lac-L-Val-)₃-] ≡ cyclo[-(D-Val-D-Lac-L-Val-L-HyIv)₃-] (retro-enantio-valinomycin)	246	-12.5	0.2	C₂H₅OH	>25	>25	>25	>25	<50	<2.3	[433]
77	cyclo[-(D-HyIv-L-Ala-L-HyIv-D-Val)₃-] ≡ cyclo[-(D-Val-D-HyIv-L-Ala-L-HyIv)₃-] ("false" valinomycin)	279	+6.8	0.2	C₂H₅OH	>50	>50	>50	>50	<50	<2.3	[433]
78	cyclo[-(L-HyIv-D-Ala-D-HyIv-L-Val)₃-] ≡ cyclo[-(L-Val-L-HyIv-D-Ala-D-HyIv)₃-] ("false" enantio-valinomycin)	281	-6.2	0.2	C₂H₅OH	>50	>50	>50	>50	<50	<2.3	[433]
79	cyclo[-(D-Val-L-HyIv-L-Ala-D-HyIv)₃-] ("false" retro-valinomycin)	234	-7.6	0.2	C₆H₆	>50	>50	0.2-0.4	0.2-0.4	2900000	8.9	[433]
80	cyclo[-(L-Val-D-HyIv-D-Ala-L-HyIv)₃-] ("false" retro-enantio-valinomycin)	234	+7.2	0.2	C₆H₆	>50	>50	0.2-0.4	0.2-0.4	2900000	8.9	[433]
81	cyclo[-(D-Val-L-Pro-L-Val-D-Pro)₃-]	>330	-2.2	0.2	CF₃COOH	—	—	—	—	—	2.3-2.7	[288,1017, 712, 904, 922, 925, 926]
82	cyclo[-(A₄-)] ("hexadeca-valinomycin")	236	-23	1	CHCl₃	>50	>50	>50	>50	50-100§	2.3-2.7	
83	cyclo[-(D-Val-L-Lac-D-Val-D-HyIv)₄-]	238	+36	0.6	C₆H₆	>50	>50	>50	>50	—	—	[925]
84	cyclo[-(D-Val-L-Lac-D-Val-D-HyIv)₆-]	261	+37	2	CHCl₃	>50	>50	>50	>50	—	—	[925]
85	cyclo[-(D-Val-L-Lac-L-Val-L-Lac-L-Val-D-HyIv-D-Val-D-HyIv)₂-]	261	-23	2	CHCl₃	>50	>50	>50	>50	—	—	[925]

* K, l·mole⁻¹, C₂H₅OH, 25°C. Stability constants for complexes of compounds (1–85) with Na+ are less than 50 l·mole⁻¹ (except: analog 23, K_{Na^+} = 100 l·mole⁻¹; $-\Delta F_{Na^+}$ = 2.7 kcal·mole⁻¹; analog 45, K_{Na^+} = 600 l·mole⁻¹; $-\Delta F_{Na^+}$ = 3.8 kcal·mole⁻¹; analog 53, K_{Na^+} = 200 l·mole⁻¹; $-\Delta F_{Na^+}$ = 3.2 kcal·mole⁻¹; and analogs 73 and 81 whose ability to form complexes was not studied).

** $-\Delta F = RTlnK$, kcal·mole⁻¹.

† K_{MeOH} = 7000 000 l·mole⁻¹; $-\Delta F_{MeOH}$ = 9.4 kcal·mole⁻¹; cf. with the valinomycin data: K_{MeOH} = 27 000 l·mole⁻¹, $-\Delta F_{MeOH}$ = 6.1 kcal·mole⁻¹.

†† Antimicrobial activity and stability constants were not measured due to low solubility.

§ K_{Cs^+} = 500 l·mole⁻¹; $-\Delta F_{Cs^+}$ = 3.7 kcal·mole⁻¹.

TABLE 3

FORMATION OF CYCLOPOLYMER HOMOLOGS IN THE PREPARATION OF VALINOMYCIN ANALOGS

$$
\begin{array}{cccc}
R & R & R & R \\
| & | & | & | \\
\end{array}
$$
H—HNCHCO—OCHCO—NHCHCO—OCHCO—OH →

$$
cyclo \left[\begin{array}{cccc} R & R & R & R \\ | & | & | & | \\ \end{array} \text{—(HNCHCO—OCHCO—NHCHCO—OCHCO)}_n \text{—} \right]
$$

Starting tetradepsipeptide	Cyclization product	n	Yield (%)
H—D-Val—L-Lac—L-Val—D-HyIv—OH	(10)	2	20
	(1)	3	5
H—L-Val—L-Lac—L-Val—D-HyIv—OH	(11)	2	40
H—D-Val—L-Lac—D-Val—D-HyIv—OH	(54)	3	4
	(83)	4	2.5
H—D-Val—L-HyIv—L-Val—D-HyIv—OH	(5)	1	7
	(12)	2	24
	(52)	3	14
H—D-Val—L-Lac—L-Val—L-HyIv—OH	(6)	1	20
	(56)	3	2.5
H—D-Val—D-HyIv—D-Val—D-HyIv—OH	(3)	1	11
	(9)	2	5

I.A.3. Enniatins

The enniatins A (86) and B (87) (Fig. 2) were first isolated in 1947 by Plattner and coworkers from the cultural fluid of several *Fusarium* species (*F. avenaceum, F. scripi, F. oxysporum* Schlecht). *F. oxysporum* Schlecht was at first identified as *Fusarium orthaceras* var. *enniatum*, from whence the name "enniatins" [759—763]. Besides enniatins A and B, which each contain three D-α-hydroxyisovaleryl residues together with three residues of N-methylisoleucine (enniatin A) or N-methyl-L-valine (enniatin B), *Fusarium* also produces a number of closely related substances. Among these are the "mixed" cyclodepsipeptides, enniatins A₁ (102) and B₁ (103) (see Table 5), detected by mass spectrometry in biosynthetic samples of the enniatins [496], and the N-methyl-L-leucine-containing compound or mixture (not isolated in the pure state) named enniatin C [762]. Later this name was given to the synthetic cyclodepsipeptide (88) [637, 710]. English workers have described another series of antibiotics (lateritiins I and II, sambucinin, avenacein and fructigenin) which possess similar properties to the enniatins and are apparently closely related mixtures with differing contents of A, B and C [160—162, 520].

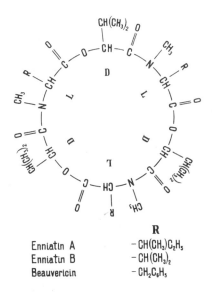

	R
Enniatin A	$-CH(CH_3)C_2H_5$
Enniatin B	$-CH(CH_3)_2$
Beauvericin	$-CH_2C_6H_5$

Fig. 2. The enniatin antibiotics.

The antibiotic preparation baccatin A produced by *Gibberella baccata* [336] is also a mixture of enniatins A and B; and the closely related fusafungin is produced by a species of *Fusarium* [344]. Finally, mention should be made of beauvericin (Fig. 2), a metabolite of the fungus *Beauveria bassiana* which differs from enniatins A, B and C in possessing an aromatic amino acid residue, N-methyl-L-phenylalanine [349].

The enniatins are colorless substances, practically insoluble in water, but readily soluble in most organic solvents. In contrast to enniatin B, enniatin A readily dissolves in petroleum ether, a property utilized for their separation [759, 760, 763].

Enniatins A and B are potent antibiotics [162, 280, 520, 913, 1004, 1005]. Enniatin B for instance, is effective against *Mycobacterium para-tuberculosis* and *Mycobacterium phlei* at doses of 3—5 µg/ml and enniatin A at doses of 1—1.25 µg/ml. Enniatin B inhibits the growth of a virulent species of *M. tuberculosis* in concentrations as low as 0.1 µg/ml. The enniatins are also active against *Staphylococcus aureus*, *Bacillus subtilis*, *Streptococcus pyogenes* and a number of other Gram-positive organisms. Enniatin B stops the development of experimental tuberculosis in mice at daily doses of 100 mg/kg [1005], and there are reports of its possessing antiprotozoic activity [456]. On the other hand, enniatin is ineffective against bacteriophages [345]. It has been reported that sambucinin, closely related to enniatins A and B, possesses strong hypotensive properties [1007]. In its antimicrobial action beauvericin is very similar to enniatin A [709] (see Table 5); its toxicity towards marine shrimps (*Artemia salina*) and wigglers has also been reported [349].

Chemical study of the enniatins has in many respects a very similar history to that of valinomycin (see Part I.A.1). They were first formulated as cyclotetradepsipeptides (94) and (91) [759, 760, 762] the synthesis of which yielded biologically inactive compounds differing considerably in physicochemical properties (m.p., $[\alpha]_D$, acid stability) from naturally occurring samples [908, 914]. Subsequent study ending in total synthesis showed that enniatins A and B consist of three, rather than two, didepsipeptide units (L-MeIle—D-HyIv and L-MeVal—D-HyIv); i.e. that they correspond to the formulae (86) and (87) [764, 765, 803, 804, 910, 916—918].

Enniatin B is the first naturally occurring metal-binding depsipeptide to have been prepared by total synthesis. The method used was similar to those discussed above for the valinomycin cyclodepsipeptides (Scheme 2). Of

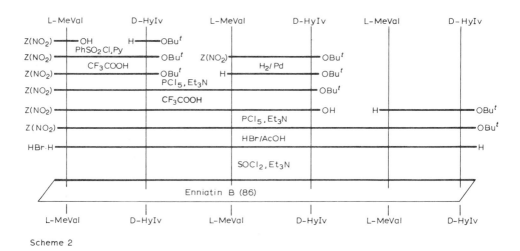

Scheme 2

particular advantage was the possibility of using the acid chloride method for the condensation stage, because acylation of *N*-methylamino acids by the usual peptide chemistry methods gives much lower yields than with ordinary amino acids. The presence of *N*-methyl groups also hinders the benzyloxy-carbonyl-*N*-methylamino acid acylation of hydroxy acid esters by the mixed anhydride method in the presence of benzenesulfonyl chloride. Thus, the reaction

$$\text{Z—L-MeVal—OH} + \text{H—D-HyIv—OBu}^t \xrightarrow[\text{Py}]{\text{PhSO}_2\text{Cl}} \text{Z—L-MeVal—D-HyIv—OBu}^t$$

TABLE 4

STABILITY CONSTANTS OF THE 1:1 ENNIATIN B COMPLEXES WITH DIFFERENT CATIONS*

Solvent	Li^+	Na^+	K^+	Rb^+	Cs^+	NH_4^+	Tl^+	Mg^{2+}	Ca^{2+}	Sr^{2+}	Ba^{2+}	Mn^{2+}
Methanol [326]	19	260 240[1083]	830 840[1083]	550	220	83	520	16	900	450	850	4
Ethanol [15, 714, 897, 903, 904]		1300	3700	4000	2200							
96% Aqueous ethanol	<50	340	2100	3000	740		330	13	120	250	2600	

* The values for the bivalent cation complexes are too low owing to neglect of (anion metal)$^+$ pair formation.

affords yields of only 30—50%, whereas the same reaction with Z—L-Val—OH is almost quantitative. Therefore, p-nitrobenzyloxycarbonyl was used as the protective group in the synthesis of enniatin B (as shown in Scheme 2) and this raised the first stage yield to 70—80%. By a similar scheme, ^{2}H-labeled enniatin B was prepared [906] as an aid in NMR signal assignment (see Part III.A.1.c). For ester bonding, use was made of the carbonyldiimidazole method which, although somewhat more complicated than using benzenesulfonyl chloride, gave excellent results with N-methyl-amide derivatives.

N,N'-Carbonyldiimidazole was also used as the condensing agent in the synthesis of beauvericin [709, 815], the synthesis confirming the proposed structure [349].

The enniatins form complexes in organic solvents with a wide range of cations including, besides the alkali metals (Li^+, Na^+, K^+, Rb^+, Cs^+) [15, 326, 700, 897, 903, 904, 1083] and ammonium [326, 700], also those of the alkaline earth metals (Mg^{2+}, Ca^{2+}, Sr^{2+}, Ba^{2+}) [326, 904] and the transition metals (Ag^+, Tl^+, Mn^{2+}, Zn^{2+}, Cd^{2+}) [326, 700].

In dilute solutions the enniatins formed 1:1 complexes with all the ions and in all the solvents investigated. However, in concentrated methanol or dimethylsulfoxide solutions of enniatin B containing $> 10^{-2}$ M K^+ or Cs^+ ions, significant amounts of 2:1 (macrocyclic compound):ion complexes were found.

The stability constants of equimolar enniatin B complexes with a number of ions in methanol and ethanol are presented in Table 4. From this it can be seen that, as with valinomycin, the stability of the complexes decreases with increasing polarity of the medium, particularly on passing over to aqueous solutions. In respect of the K^+-complex stability the enniatins lag far behind the majority of the other metal-binding macrocyclic compounds; they also display a low K/Na-selectivity. Crystalline salts of the 1:1 complexes have been obtained for enniatin A with KBr and for enniatin B with KI, KNCS and NaNCS [209, 903, 1083]. Isolation of the complexes in the pure state is facilitated by their lower solubility than the free antibiotics or the analogous valinomycin complexes in neutral organic solvents (acetone, ether, ethyl acetate).

I.A.4. Enniatin analogs

Enniatins have been synthesized which differ from the naturally occurring compounds in respect to ring size (compounds 91, 94, 96, 141, 143 and 145), and to the nature (compounds 104—106, 121—123, 125, 127, 129, 131 and 132) and configuration (compounds 107—120) of the N-methylamino and hydroxy acid residues, with the N-methyl amide groups replaced by amide (compounds 136—138) or ester groups (compound 139), or which differ in more than one of the above-mentioned features (compounds 90, 92, 93, 95, 97—101, 124, 126, 128, 130, 140, 142 and 144) (Table 5). Modification of the parent molecule can be local or it can concern the molecule as a whole, in which case it gives rise to so-called topochemical analogs (compounds 106—109).

The synthetic routes to the majority of the enniatin analogs are very similar in principle to that shown in Scheme 2 for enniatin B. In all cases the C-terminal carboxyl groups were protected by *tert*-butyl esters, and the N-terminal amino acid residues by benzyloxycarbonyl or *p*-nitrobenzyloxy-carbonyl groups. Ester bonds were formed by the benzenesulfonyl chloride method, and amide bonds by the acid chloride method. Exceptions are the peptide analogs of enniatin B (135—137) for which *N,N'*-dicyclohexyl-carbodiimide, phosphoryloxychloride or the mixed anhydride method was used. Compound (139) which has no amide bonds is the only one among those discussed in Part I.1 whose ring closure was effected by ester bonding. Compounds (127), (131) and (145) were prepared by treating the corresponding analogs (126), (130) and (144) with methyl iodide in the presence of silver oxide. The yields, however, (especially of compound 127) were low because under the *N*-methylating conditions the ester bonds of both reactants and products underwent rupture. A similar reaction was used for preparing enniatin B from its *N*-desmethyl analog (135). It is noteworthy that the cyclodepsipeptide (145) with doubled ring size was formed as a side product.

$$\overline{(L-Val-D-HyIv)_3} \xrightarrow[\text{Ag}_2\text{O}]{\text{CH}_3\text{I}} \overline{(L-MeVal-D-HyIv)_3} + \overline{(L-MeVal-D-HyIv)_6}$$

(135)   Enniatin B (87) , 30%   (145), 10%

As with the valinomycins, doubling and tripling reactions (Table 6) during the syntheses yielded several enniatin analogs.

In general the enniatin cyclodepsipeptides were isolated and purified by means of chromatography on alumina or silica gel followed by crystallization or reprecipitation. In separating the analogs (126, 127, 130 and 131) from non-complexing impurities, recourse was made to the poor solubility of their potassium thiocyanate complexes.

As one can see from Table 5, some of the analogs (104—109, 133 and 141) have antibiotic properties very similar to the parent compounds. At the same time there is a group of inactive compounds (90—101, 113, 115, 138, 142 and 144) and another group (compounds 110—112, 116—131, 134—137, 139, 140, 143, 144) that occupies an intermediate position with respect to antibiotic activity or complexing properties. The data obtained are

TABLE 5

STRUCTURE AND SOME PROPERTIES OF CYCLODEPSIPEPTIDES OF THE ENNIATIN GROUP

No.	Compound	m.p. (°C)	[α]_D	c (%)	Solvent	S. aureus 209 P	S. aureus UV-3	B. mycoides	E. coli	M. phlei	C. albicans	Stability Na+	constants K+	Free energy Na+	complexation K+	References
						Minimal growth-inhibiting concentration (μg/ml)						Stability constants of complexes*		Free energy of complexation**		
86	cyclo[—L-MeIle—D-HyIv)₃—] (enniatin A)	122; 124; 121	−92°; −94.5; −87	1; 0.7; 1	CHCl₃; CHCl₃; CHCl₃	9	1.5	4.5	>100	1.5–2.5	9	2900	9800	4.8	5.5	[762, 904, 913] [803] [918]
87	cyclo[—L-MeVal—D-HyIv)₃—] (enniatin B)	175; 174; 169	−108; −105.7; −110	0.6; 1.1; 0.8	CHCl₃; CHCl₃; CHCl₃	75–100	9	25–37	>100	9–12	37	1300	3700	4.3	4.9	[760, 904, 913] [765] [918]
88	cyclo[—L-MeLeu—D-HyIv)₃—] (enniatin C)	129	−24	0.1	CHCl₃	>50	>50	>50	>50	>50	>50	2500	5500	4.7	5.1	[637, 710, 903, 904]
89	cyclo[—L-MePhe—D-HyIv)₃—] (beauvericin)	148; 94	+69; +65.8	0.1; 1	CH₃OH; CH₃OH	2–4.5	2–4	—	>25	1–2	9–12	300	3100	3.4	4.8	[709] [349]
90	cyclo[—L-Val—D-HyIv)₂—]	320 (subl.)	−51	1.0	CHCl₃	>100	>100	>100	>100	>100	>100	<50	<50	<2.3	<2.3	[706]
91	cyclo[—L-MeVal—D-HyIv—L-MeIle—D-HyIv)₂—] ("tetra-enniatin B")	229	+4.8	0.9	CHCl₃	>100	>100	>100	>100	>100	>100	<50	<50	<2.3	<2.3	[712, 764, 765, 908, 913, 914, 915, 979]
92	cyclo[—L-MeVal—D-HyIv—L-MeIle—D-HyIv—)	207	+9	0.8	CHCl₃	>100	>100	>100	>100	>100	>100	<50	<50	<2.3	<2.3	[908, 913]
93	cyclo[—L-MeVal—D-HyIv—L-MeLeu—D-HyIv—)	165	+22	0.8	CHCl₃	>100	>100	>100	>100	>100	>100	<50	<50	<2.3	<2.3	[908, 913]
94	cyclo[—L-MeIle—D-HyIv)₂—] ("tetra-enniatin A'")	216	+13.5	0.8	CHCl₃	>100	>100	>100	>100	>100	>100	<50	<50	<2.3	<2.3	[705, 908, 913]
95	cyclo[—L-MeIle—D-HyIv—L-MeLeu—D-HyIv—)	165	+24	0.9	CHCl₃	100	>100	>100	>100	>100	>100	<50	<50	<2.3	<2.3	[908, 913]
96	cyclo[—L-MeLeu—D-HyIv)₂—] ("tetra-enniatin C'")	157	+35	1.0	CHCl₃	100	>100	>100	>100	>100	>100	<50	<50	<2.3	<2.3	[908, 913]
97	cyclo[—(MeVal—HyIv)₂—]; DDDD	158	+61	1.0	CHCl₃	>100	>100	>100	>100	>100	>100	<50	<50	<2.3	<2.3	[706, 707]
98	cyclo[—(MeVal—HyIv)₂—]; DLLD	230	0	2	CHCl₃	>100	>100	>100	>100	>100	>100	<50	<50	<2.3	<2.3	[707]
99	cyclo[—(MeVal—HyIv)₂—]; DDLL	230	0	2	CHCl₃	>100	>100	>100	>100	>100	>100	<50	<50	<2.3	<2.3	[707]
100	cyclo[—(MeVal—HyIv)₂—]; DDDL	129	+66	0.4	CHCl₃	>100	>100	>100	>100	>100	>100	<50	<50	<2.3	<2.3	[707]
101	cyclo[—(MeVal—HyIv)₂—]; LDDD	144	+218	0.7	CHCl₃	>100	>100	>100	>100	>100	>100	<50	<50	<2.3	<2.3	[707]
102	cyclo[—L-MeVal—D-HyIv—(L-MeIle—D-HyIv)₂—] (enniatin A₁)	117	−94	1	CHCl₃	12–8	2	9	100	1.5–2	9	<50	<50	<2.3	<2.3	[496, 913, 918]
103	cyclo[—L-MeLeu—D-HyIv—(L-MeVal—D-HyIv)₂—] (enniatin B₁)	—	—	—	—	—	—	—	—	—	—	—	—	—	—	[496]
104	cyclo[—L-MeLeu—D-HyIv—(L-MeVal—D-HyIv)₂—] (enniatin BC)	Amorph.	−65	0.1	CHCl₃	18	9	25	>50	9	37	2200	5500	4.6	5.1	[637, 712, 904]
105	cyclo[—L-MeIle—D-HyIv—(L-MeVal—D-HyIv)₂—] (enniatin CB)	146	−60	0.1	CHCl₃	9	4–6	18	>50	9	37	1700	5100	4.4	5.1	[637, 712, 904]
106	cyclo[—D-MeVal—L-HyMeV)₃—] ("false" enniatin A)	Amorph.	+84.5	0.1	CHCl₃	9–18	4.5–6	4.5–6	>50	4.5–6	4.5–9	100	1700	2.7	4.4	[904, 906]
107	cyclo[—D-MeIle—L-HyIv)₃—] (enantio-enniatin A)	122	+90	1	CHCl₃	9	1.5	4.5	>100	1.5–2.5	9	2900	9800	4.8	5.5	[904, 906]
108	cyclo[—D-MeVal—L-HyIv)₃—] (enantio-enniatin B)	175	+100	0.2	CHCl₃	75–100	9	25–37	>100	9–12	37	1300	3700	4.3	4.9	[904, 905]
109	cyclo[—D-MeLeu—L-HyIv)₃—] (enantio-enniatin C)	129	−24	0.1	CHCl₃	>50	>50	>50	>50	>50	>50	2500	5500	4.7	5.1	[904, 906]
110	cyclo[—(MeVal—HyIv)₃—]; LDLDLL	149	−79.1	0.2	C₂H₅OH	>50	18	>50	>50	18	>50	<50	<50	<2.3	<2.3	[906]
111	cyclo[—(MeVal—HyIv)₃—]; LDLDDD	197	−68.0	0.2	C₂H₅OH	>50	9–18	>50	>50	9–12	>50	100	500	2.7	3.7	[906]
112	cyclo[—(MeVal—HyIv)₃—]; LDLLDL	169	−94.4	0.2	C₂H₅OH	18	9–12	>50	>50	9–18	25–37	100	150	2.7	3.0	[906]

113	*cyclo*[–(MeVal–HyIv)$_3$–]; LLDDDL	202	+47.8	0.2	C_2H_5OH	> 50	> 50	> 50	> 50	> 50	> 50	< 50	< 50	< 2.3	< 2.3	[906]
114	*cyclo*[–(MeVal–HyIv)$_3$–]; LLLLDL	235	+3.3	0.2	C_2H_5OH	> 50	> 50	> 50	> 50	> 50	> 50	< 50	< 50	< 2.3	< 2.3	[906]
115	*cyclo*[–(MeVal–HyIv)$_3$–]; LLLLLD	167	−119.2	0.2	C_2H_5OH	> 50	> 50	> 50	> 50	> 50	> 50	< 50	< 50	< 2.3	< 2.3	[906]
116	*cyclo*[–(MeVal–HyIv)$_3$–]; LDLDDL	140	−10.0	0.2	C_2H_5OH	> 50	18	> 50	> 50	18	> 50	< 50	150	< 2.3	3.0	[906]
117	*cyclo*[–(MeVal–HyIv)$_3$–]; LLDDLD	186	−2.5	0.2	C_2H_5OH	> 50	> 50	> 50	> 50	> 50	> 50	< 50	100	< 2.3	2.7	[906]
118	*cyclo*[–(MeVal–HyIv)$_3$–]; LLLDDL	134	−24.6	0.2	C_2H_5OH	> 50	9–18	> 50	> 50	9–18	> 50	< 50	50	< 2.3	2.3	[906]
119	*cyclo*[–(MeVal–HyIv)$_3$–]; LLLLDD	169	−36.7	0.2	C_2H_5OH	> 50	9–18	> 50	> 50	9–12	> 50	< 50	< 50	< 2.3	< 2.3	[906]
120	*cyclo*[–(MeVal–HyIv)$_3$–]; DDDDDD	250	+138.0	0.06	$CHCl_3$	—	—	—	—	—	—	—	—	—	—	[906]
		185†	+73.7†	0.4	C_6H_6											[562]
121	*cyclo*[–L-MeVal–Glyc–(L-MeVal–D-HyIv)$_2$–]	Amorph.	−245	0.1	$CHCl_3$	> 100	> 100	> 100	> 100	> 100	> 100	700	300	3.9	3.4	[637, 700]
122	*cyclo*[–L-MeVal–D-HyIc–(L-MeVal–D-HyIv)$_2$–]	Amorph.	−70	0.1	$CHCl_3$	> 100	> 100	50–75	> 100	> 100	> 100	2200	4900	4.6	5.0	[637, 700]
123	*cyclo*[–L-MeAla–D-HyIv–(L-MeVal–D-HyIv)$_2$–]	Amorph.	−160	0.1	$CHCl_3$	> 100	75–100	> 100	> 100	> 100	> 100	2200	1600	4.6	4.4	[637, 700]
124	*cyclo*[–L-Glu–D-HyIv–(L-MeVal–D-HyIv)$_2$–]	Amorph.	−10.0	0.2	C_6H_6	> 40	—	> 40	> 40	> 40	> 40	—	—	—	—	—
125	*cyclo*[–L-MeGlu–D-HyIv–(L-MeVal–D-HyIv)$_2$–]	Amorph.	−72.0	0.2	C_6H_6	> 40	—	> 40	> 40	> 40	> 40	—	—	—	—	—
126	*cyclo*[–L-Glu(OBzlNO$_2$)–D-HyIv– (L-MeVal–D-HyIv)$_2$–]	Amorph.	−91.2	0.2	C_6H_6	> 40	—	> 40	> 40	> 40	> 40	—	—	—	—	—
127	*cyclo*[–L-MeGlu(OBzlNO$_2$)–D-HyIv– (L-MeVal–D-HyIv)$_2$–]	Amorph.	−69.2	0.2	C_6H_6	25–37	—	> 40	> 40	12–18	> 40	—	—	—	—	—
128	*cyclo*[–L-Lys–D-HyIv–(L-MeVal–D-HyIv)$_2$–]	Amorph.	−77.8	0.2	C_6H_6	> 40	—	> 40	> 40	> 40	> 40	—	—	—	—	—
129	*cyclo*[L-MeLys–D-HyIv–(L-MeVal–D-HyIv)$_2$–]	Amorph.	−52.5	0.2	C_6H_6	> 40	—	> 40	> 40	> 40	> 40	—	—	—	—	—
130	*cyclo*[–L-Lys(Phth)–D-HyIv– (L-MeVal–D-HyIv)$_2$–]	Amorph.	−8.2	0.2	C_6H_6	> 40	—	> 40	> 40	12–18	> 40	—	—	—	—	—
131	*cyclo*[–L-MeLys(Phth)–D-HyIv– (L-MeVal–D-HyIv)$_2$–]	Amorph.	−44.1	0.2	C_6H_6	7.5	—	> 40	> 40	7.5–12	12–18	—	—	—	—	--
132	*cyclo*[–(L-MeAla–D-HyIv)$_3$–]	Oil	−63.5	0.6	C_6H_6	—	—	—	—	—	—	—	—	—	—	[563]
133	*cyclo*[–L-Val–D-HyIv–(L-MeVal–D-HyIv)$_2$–]	129	−109	0.1	$CHCl_3$	> 100			> 100		> 100					[979]
	((mono-*N*-desmethyl)-enniatin B)	129	−110	0.1	$CHCl_3$	75	25	50–75	> 100	25–37	75	140	7800	2.9	5.3	[632, 700]
134	*cyclo*[–(L-Val–D-HyIv)$_2$–L-MeVal–D-HyIv–] ((di-*N*-desmethyl)-enniatin B)	196	−103.7	0.1	$CHCl_3$	Inactive	—	—	Inactive	—	Inactive	400	2800	3.6	4.7	[700, 979]
135	*cyclo*[–(L-Val–D-HyIv)$_3$–]	267	−22.3	0.15	$CHCl_3$	Inactive			Inactive		Inactive					[979]
	((tri-*N*-desmethyl)-enniatin B)	267	−22	1	$CHCl_3$	> 50	> 50	> 50	> 50	> 50	> 50	2500	2600	4.7	4.7	[712, 904, 906]
136	*cyclo*[–L-MeVal–D-Val–(L-MeVal–D-HyIv)$_2$–]	Oil	−100	1	C_6H_6	Weak activity	—	—	—	—	—	—	—	—	—	[563]
137	*cyclo*[–(L-MeVal–D-Val)$_2$–L-MeVal–D-HyIv–]	Amorph.	−88	0.4	C_6H_6	Weak activity	—	—	—	—	—	—	—	—	—	[563]
138	*cyclo*[–(L-MeVal–D-Val)$_3$–]	330	−200	1	C_6H_6–CH_2Cl (1:1)	Inactive	—	—	—	—	—	—	—	—	—	[563]
139	*cyclo*[–(L-HyIv–D-HyIv)$_3$–]	214	*meso*-form			> 50	> 50	> 50	> 50	> 50	> 50	170	1000	3.2	4.1	[712]
140	*cyclo*[–(L-Val–D-HyIv)$_4$–] ((tetra-*N*-desmethyl)-octa-enniatin B)	272	+28	1.3	$CHCl_3$	> 50	> 50	> 50	> 50	> 50	> 50	80	220	2.6	3.2	[700, 706, 765]
141	*cyclo*[–(L-MeVal–D-HyIv)$_4$–] (octa-enniatin B)	179	−106	1.3	$CHCl_3$	> 100	1.5–3	> 100	> 100	18–25	> 100	2600	4900	4.7	5.1	[712, 918, 922]
142	*cyclo*[–(D-MeVal–D-HyIv)$_4$–]	183	+124	0.8	$CHCl_3$	> 25	> 25	> 25	> 25	> 25	> 25	< 50	< 50	< 2.3	< 2.3	[706, 913]
143	*cyclo*[–(L-MeVal–D-HyIv)$_5$–] (deca-enniatin B)	176	−178	0.1	C_6H_6	> 50	> 50	50	50	50	50	50	430	2.3	3.6	[712, 906]
144	*cyclo*[–(L-Val–D-HyIv)$_6$–] ((hexa-*N*-desmethyl)-dodeca-enniatin B)	233	+41	0.9	$CHCl_3$	> 50	> 50	> 50	> 50	> 50	> 50	< 50	< 50	< 2.3	< 2.3	[706, 913]
145	*cyclo*[–(L-MeVal–D-HyIv)$_6$–] (dodeca-enniatin B)	128	+54.3	0.1	C_6H_6	> 50	> 50	> 50	> 50	> 50	> 50	< 50	100	< 2.3	2.7	[712, 906]

* K, l·mole^{-1}, C_2H_5OH, 25°C.
** $-\Delta F = RT\ln K$, kcal·mole^{-1}
† The sample prepared in ref. [562] and its enantiomer apparently have another structure [906].

TABLE 6

FORMATION OF CYCLOPOLYMER HOMOLOGS IN THE PREPARATION OF
THE ENNIATIN B ANALOGS

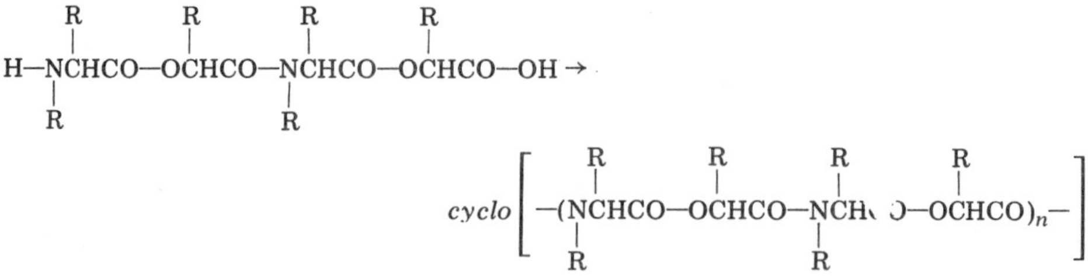

$$
\begin{array}{cccc}
R & R & R & R \\
| & | & | & | \\
\end{array}
$$
H—NCHCO—OCHCO—NCHCO—OCHCO—OH →
$$
\begin{array}{cc}
| & | \\
R & R
\end{array}
$$

$$
cyclo \left[-(\underset{|}{\overset{R}{N}}CHCO-OCHCO-\underset{|}{\overset{R}{N}}CH \; O-\underset{|}{\overset{R}{O}}CHCO)_n- \right]
$$

Starting tetradepsipeptide	Cyclization product	n	Yield (%)
H—L-Val—D-HyIv—L-Val—D-HyIv—OH	(90)	1	6
	(140)	2	19
	(143)	3	8
H—D-MeVal—D-HyIv—D-MeVal—D-HyIv—OH	(97)	1	8
	(142)	2	13

discussed from the standpoint of structure—activity relationships in Part
III.A.1.d. Compounds (124, 125, 128 and 129), with amino- and carboxyl-
carrying side chains, were prepared for studying the effect of charged
groupings on the stability of the complexes and the biological activity of the
membrane-affecting depsipeptides.

I.A.5. Monamycins

In 1959 the isolation of a crystalline antibiotic preparation, monamycin,
from a culture of *Streptomyces jamaicensis* was described, the antibiotic
being very active against Gram-positive bacteria (Table 7) [370]. Careful

TABLE 7

ANTIMICROBIAL ACTIVITY OF MONAMYCIN

Microorganism	Minimal growth-inhibiting concentration ($\mu g/ml$)
Staphylococcus aureus	0.05
Bacillus subtilis	0.07
Streptococcus pyogenes	0.10
Streptomyces lavendulae	0.12
Escherichia coli	>1
Pseudomonas fluorescens	>1

selection of the fermentation conditions sharply increased the productivity of the species (from 3—5 μg/ml to 160—170 μg/ml) [347], permitting sufficient amounts of the antibiotic to be obtained for its chemical investigation [59, 60, 372, 374]. Fractional crystallization separated monamycin into two major parts from each of which the individual

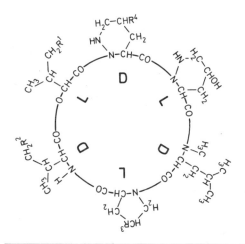

Compound		R^1	R^2	R^3	R^4
Nº	Monamycin				
146	A	H	H	Me	H
147	B_1	H	H	Me	H
148	B_2	H	Me	H	H
149	B_3	Me	H	H	H
150	C	Me	H	Me	H
151	D_1	Me	H	Me	H
152	D_2	H	Me	Me	H
153	E	Me	Me	Me	H
154	F	Me	Me	Me	H
155	G_1	H	H	Me	Cl
156	G_2	H	Me	H	Cl
157	G_3	Me	H	H	Cl
158	H_1	Me	H	Me	Cl
159	H_2	H	Me	Me	Cl
160	I	Me	Me	Me	Cl

Fig. 3. The monamycin antibiotics.

components were isolated by counter-current distribution (chlorine-free monamycins A—F and monamycins G_1—I containing covalently bound chlorine). The structures of the major components (D_1 and H_1) were elucidated by analysis of their partial- and total-hydrolysis products. The primary structures of the other monamycins were determined mainly from high-resolution mass-spectrometry data. As one can see from the formulae (146—160) (Fig. 3) monamycins A—I are 18-membered cyclohexa-depsipeptides containing five amino acid residues and one hydroxy acid

22

residue. Among the amino acids are unusual ones, such as D-isoleucine, *trans*-4-methyl-L-proline, (3R)-piperazinic acid, (3S,5S)-5-hydroxypiperazinic acid and (3R,5S)-5-chloropiperazinic acid (which in monamycins A (146), C (150) and E (153) is present in the form of the Δ^4- or Δ^5-dehydrochlorinated derivatives).

A noteworthy feature of the monamycin structure is the presence of usually protonated hydrazide groups enhancing the surface activity of these compounds [346]. As with the enniatin antibiotics the monamycins have alternating LDLDLD amino and hydroxy acid configurations, an apparently necessary condition for their manifestation of metal-complexing properties.

The interaction of monamycins with metal ions has not been investigated in detail, but mention has been made of their forming, in aqueous alcohol, sparingly soluble complexes with sodium, potassium, rubidium, cesium, silver and barium salts. No signs of complex formation with other bivalent cations or with lithium have been observed [346, 347]. The monamycin antibiotics differ somewhat among themselves in antimicrobial activity. Their high toxicity is a barrier to practical application in medicine [59].

I.A.6. Other depsipeptides

Several other cyclodepsipeptides (isariin (161) (Fig. 4) [1037, 1089], destruxins A—D (162—166) (Fig. 5) [499, 500, 519, 982, 998] etc.), whose structures should be conducive to the manifestation of complexing

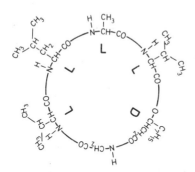

Fig. 4. Isariin.

properties, have been described in the literature, but no experimental evidence is available.

On the other hand, there are naturally occurring and synthetic cyclodepsipeptides such as the angolides (167) [493, 495, 571, 839], serratamolides (168) [20, 371, 375, 497, 898, 901, 1045] and sporidesmolides

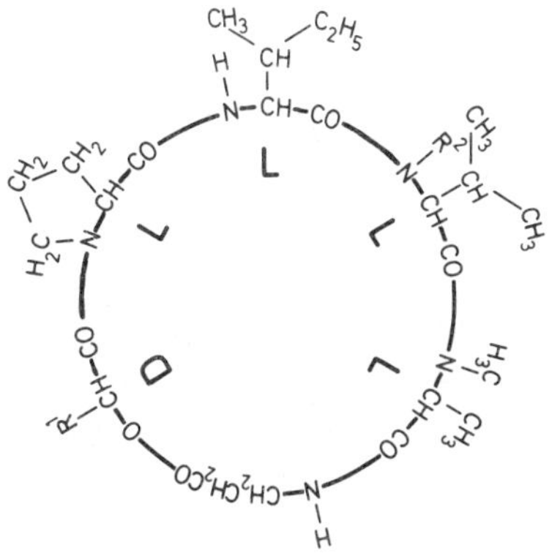

	R¹	R²
Destruxin A (162)	$C(CH_3)=CH_2$	CH_3
Destruxin B (163)	$CH(CH_3)C_2H_5$	CH_3
Desmethyldestruxin B(164)	$CH(CH_3)C_2H_5$	H
Destruxin C (165)	$CH_2CH(CH_3)CH_2OH$	CH_3
Destruxin D (166)	$CH_2CH(CH_3)COOH$	CH_3

Fig. 5. The destruxin depsipeptides.

(169—177) [58, 68, 494, 706, 708, 714, 837, 838, 840, 907, 909, 911, 912], with many structural features similar to the enniatins, that do not complex with alkali metal ions and (excepting serratamolide) do not inhibit bacterial growth [913].

Angolide (167): cyclo(—L-Ile—L-Val—D-aIle—L-HyIv—)
Serratamolide (168): cyclo(— L-Ser—D-HyDec—L-Ser—D-HyDec—)
Sporidesmolide I (169):
   cyclo(—D-Val—D-Leu—L-HyIv—L-Val—L-MeLeu—L-HyIv—)
Sporidesmolide II (170):
   cyclo(—D-aIle—D-Leu—L-HyIv—L-Val—L-MeLeu—L-HyIv—)
Sporidesmolide III (171):
   cyclo(—D-Val—D-Leu—L-HyIv—L-Val—L-Leu—L-HyIv—)
Sporidesmolide IV (172):
   cyclo(— D-Val—D-Leu—L-HyIc—L-Val—L-Leu—L-HyIv—)

(173): cyclo[—(L-Val—L-MeLeu—L-HyIv)₂—]
(174): cyclo[—(D-Val—L-MeLeu—L-HyIv)₂—]
(175): cyclo[—(L-Val—L-MeLeu—D-HyIv)₂—]
(176): cyclo[—(D-Val—L-MeIle—D-HyIv)₂—]
(177): cyclo[—(D-Val—D-Leu—L-HyIv)₂—]

I.B. Peptides

Some purely peptide compounds have also been found capable of inducing cation permeability in membranes. Among those of prime importance are the linear gramicidins (gramicidins A, B, C and D) and the macrocyclic peptide alamethicin. True, the mechanism by which these peptides interact with the metal cations is unknown and there is no experimental evidence of their forming complexes. More and more interest is being attracted by the antamanides — cyclic peptides which are known to complex sodium and potassium ions and which, moreover, clearly exhibit sodium selectivity. Although certain properties of antamanide and its analogs have impeded their use as tools for studying membrane systems, recent findings have shown that it is feasible to look further into this problem.

I.B.1. Gramicidins A, B, C and D

In 1939, Dubos [216, 217] showed that *Bacillus brevis* produces an antibiotic, which was first considered to be an individual substance and named tyrothricin (see reviews [407, 899]). In 1941, it was found that tyrothricin can be separated into two major crystalline fractions differing in physicochemical and antibiotic properties [408]. The fractions were named tyrocidin and gramicidin (gramicidin D, Dubos). The physical constants of the gramicidin fraction (m.p. 230—231°; $[\alpha]_D^{25} + 2.5°$ (c 1.5 in ethanol)) no longer change on subsequent repeated recrystallization and for a long time it was considered to be a single substance. In 1948, Craig and collaborators showed it to be a mixture which could be separated by counter-current distribution into three components, gramicidins A, B and C [168, 323]. In 1963, Ramachandran isolated still another component from gramicidin A, which he named gramicidin D [807]. Finally, in 1965, Gross and Witkop [332] established that each of the gramicidins A, B and C is a mixture of two closely related compounds and proposed the name "gramicidin D" for still another, particularly hydrophilic, minor component they had discovered.

Taking into account the absence of free amino or carboxyl groups in gramicidin, its structure was initially formulated as the cyclopeptide [406, 408]. The detection of ethanolamine hydrolysates [988] led Synge, in 1950, to the hypothesis that the gramicidin molecule should contain orthoamide (cyclol) groupings [989].

Only in 1964 did Sarges and Witkop find that there is a formyl grouping in the gramicidin molecule which blocks its N-terminal group and is readily removed under conditions of mild acid hydrolysis [850]. Consecutive

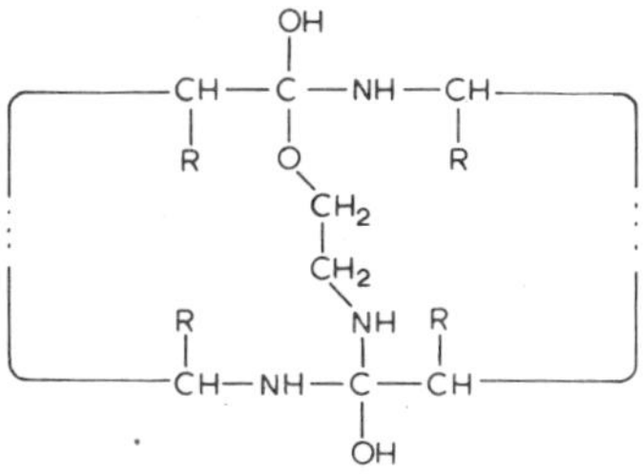

Edman elimination of the amino acid residues from the N-terminal desformylgramicidins A—C led to the rapid establishment of their structures [851, 852, 854, 855]:

$$
\begin{array}{ccccccccccccccc}
1 & 2 & 3 & 4 & 5 & 6 & 7 & 8 & 9 & 10 & 11 & 12 & 13 & 14 & 15
\end{array}
$$

HCO—ValGlyAlaLeuAlaValVal ValTrpLeuTrpLeuTrpLeuTrpNHCH$_2$CH$_2$OH

$$
\begin{array}{ccccccccccccccc}
L & & L & D & L & D & L & & L & D & L & D & L & D
\end{array}
$$

Valine-gramicidin A (178)

HCO—IleGlyAlaLeuAlaValValValTrpLeuTrpLeuTrpLeuTrpNHCH$_2$CH$_2$OH

Isoleucine-gramicidin A (179)

HCO—ValGlyAlaLeuAlaValValValTrpLeuPheLeuTrpLeuTrpNHCH$_2$CH$_2$OH
Valine-gramicidin B (180)

HCO—IleGlyAlaLeuAlaValValValTrpLeuPheLeuTrpLeuTrpNHCH$_2$CH$_2$OH
Isoleucine-gramicidin B (181)

HCO—ValGlyAlaLeuAlaValValValTrpLeuTyrLeuTrpLeuTrpNHCH$_2$CH$_2$OH
Valine-gramicidin C (182)

HCO—IleGlyAlaLeuAlaValValValTrpLeuTyrLeuTrpLeuTrpNHCH$_2$CH$_2$OH
Isoleucine-gramicidin C (183)

Thus gramicidins A, B and C have closely related structures, being the ethanolamides of formylpentadecapeptides (178—183) built up of hydrophobic amino acids and differing from each other only in the nature of one (the eleventh) amino acid residue. In turn, each of the gramicidins A, B and C are mixtures of the corresponding valine[1]- (80—95%) and isoleucine[1]- (5—20%) gramicidins. The structure of gramicidin D, distinguished by the presence of hydrophilic amino acids (Thr, Lys, Asp, Glu) and also of Met and Pro, has not yet been fully elucidated.

The major component of commercial gramicidin is gramicidin A (70—85%), gramicidin B being present in amounts of from 5 to 10%, gramicidin C from 7 to 20%, and gramicidin D less than 1%. In the subsequent discussion use of the term gramicidin will mean this mixture.

It is on this mixture before its separation into the individual components that most of the medical and biological work on the gramicidins has been done [266, 407, 412, 899]. Gramicidin is active against many Gram-positive organisms (Table 8); it does not lyse the bacteria, the action being bacteriostatic in nature. Gram-negative organisms are as a rule insensitive to gramicidin. The antibiotics are highly toxic, causing hemolysis in concentrations of 0.5—1 μg/ml. The toxicity is particularly manifested when administered intraveneously and much less on subcutaneous and peroral

TABLE 8

ANTIMICROBIAL ACTIVITY OF GRAMICIDIN

Microorganism	Minimal growth-inhibiting concentration (μg/ml)
Pneumococcus	0.01
Streptococcus hemolyticus	2
Staphylococcus aureus	1—10
Sarcina lutea	10
Bacillus subtilis	> 1000
Bacillus mycoides	> 1000
Escherichia coli	> 500
Eberthella typhi	> 500
Shigella dysenteriae	> 500
Pasteurella tularensis	250
Salmonella schottmuelleri	> 500

administration. Initially gramicidin was quite popular in medical practice, but it is now replaced by other, more potent antibiotics. Attempts to improve its pharmacological properties by treatment with various reagents (formaldehyde, succinic anhydride, etc.) have met with little success, although in certain cases the solubility was somewhat raised and the toxicity lowered (see [899] and references therein). There are a few reports of studies on the effect of gramicidin upon leucocytes, microphage, spermatozoids and on the embryos and tissue cultures of the frog (see [412]).

The biological activity of the individual components of gramicidin, while studied less fully, has shown (cf. Tables 8 and 9) that valine-gramicidin A is very similar to the overall mixture in antibiotic properties. Gross and Witkop showed that gramicidins A—D differ little in their effect on *Streptococcus faecalis*, but that gramicidin D is much more active (3—25-fold) against *Staphylococcus aureus* and *Streptococcus pyogenes* than the other antibiotics of this group [332].

TABLE 9

ANTIMICROBIAL ACTIVITY OF THE SYNTHETIC VALINE-GRAMICIDIN A

Microorganism	Minimal growth-inhibiting concentration ($\mu g/ml$)
Staphylococcus aureus 209 P (10^3 cells/ml)	0.002—0.05
Staphylococcus aureus 209 P (10^6 cells/ml)	4.7—9.4
Streptococcus lactis SC 1783	0.8—1.2
Streptococcus faecalis SC 1648	0.01
Salmonella schottmuelleri SC 3580	> 50
Proteus vulgaris SC 3855	> 50
Escherichia coli SC 2927	> 50

Gramicidins A and B are colorless, crystalline substances (m.p. after recrystallization from acetone 228 and 259°, respectively), readily soluble in the lower alcohols, acetic acid, and pyridine, moderately soluble in acetone and dioxane, and very sparingly soluble in water and petroleum ether. A striking feature of the gramicidins A, B and C (unique in linear, naturally occurring peptides) is the alternating configuration of their amino acid residues. Molecular weight measurements of gramicidin A (Table 10) have

TABLE 10

DATA ON ASSOCIATION OF GRAMICIDIN A UNDER DIFFERENT CONDITIONS [422]

Solvent	Temperature (°C)	Concentration		Observed molecular weight*	Association constant**
		mg/ml	mole/l		
Dioxane	40	9.5	$5.1 \cdot 10^{-2}$	3480	10^4
		14.9	$8.0 \cdot 10^{-2}$	3530	
	60	9.0	$4.8 \cdot 10^{-2}$	2990	10^3
		15.3	$8.2 \cdot 10^{-2}$	3090	
Methanol	30	5.0	$2.7 \cdot 10^{-2}$	1860	50
		30.0	$16.1 \cdot 10^{-2}$	2390	
	57	5.0	$2.7 \cdot 10^{-2}$	1880	1—5
		40.0	$21.5 \cdot 10^{-2}$	1900	
Ethanol	30	5.0	$2.7 \cdot 10^{-2}$	2340	$5 \cdot 10^2$
		30.0	$16.1 \cdot 10^{-2}$	3150	
n-Propanol	60	5.0	$2.7 \cdot 10^{-2}$	2530	$2 \cdot 10^2$
		26.6	$14.3 \cdot 10^{-2}$	2840	
Ethyl acetate	40	1.4	$0.75 \cdot 10^{-2}$	2080	10^3
		7.8	$4.2 \cdot 10^{-2}$	3010	

* Molecular weight of monomer is 1882.
** $l \cdot mole^{-1}$ calculated for dimer formation.

shown its molecules to be highly associated in many solvents. Sarges and Witkop considered that the tendency of the gramicidins to form unusually stable dimers, trimers, etc. is the most plausible explanation of their behavior in thin-layer silica-gel chromatography: with the system AcOH—CHCl₃ (2:1) pure gramicidins, as expected, give a single spot, but with the system methylethyl ketone—pyridine (7:3) each gives several spots [423, 852, 853]*.

The structures of valine-gramicidin A (178) and isoleucine-gramicidin A have been confirmed by total synthesis [853]. As was shown in Scheme 3, N-benzyloxycarbonyldesformylvaline-gramicidin A was prepared by the carbodiimide condensation of the C-terminal decapeptide and the N-terminal pentapeptide**, the latter being synthesized by stepwise chain build-up using the mixed anhydride method. Removal of the benzyloxycarbonyl group, N,O-diformylation and subsequent saponification of the O-formyl grouping led to a preparation identical with naturally occurring valine-gramicidin A in physical constants, chromatographic behavior and antimicrobial activity. The same procedure was used for synthesis of isoleucine-gramicidin A.

Several O-alkyl and O-acyl derivatives of gramicidin and gramicidin A have been described [406, 452, 699, 852] and by acylation of desformyl-gramicidin a series of N,O-diacyldesformylgramicidins A (184—186) [852] was obtained, while saponification of analog (184) led to N-acetyldesformyl-gramicidin A (187). Qualitative microbiological assays of compounds (184—187) (Table 11) showed that the presence of a formyl group and a free hydroxy group is not necessary for the manifestation of biological activity by the gramicidins.

Rambhav and Ramachandran have investigated the antimicrobial and hemolytic action of gramicidin analogs which they had prepared [810]. They found (Table 12) that the antimicrobial activity is much more sensitive than the hemolytic activity to structural changes in the tryptophan side chains; modification of the N-acyl group has only a weak effect on each type of activity.

Recently a series of shortened valine-gramicidin A analogs (compounds 178a—178k) has been synthesized. A study of their antimicrobial activities (Table 13) has shown that the 13-membered analogs (compounds 178a and 178b) are potent antimicrobial agents, although somewhat weaker than the naturally occurring antibiotic. Further shortening of the chain leads to practically complete inactivation. Synthesis of the membrane-active analog

---

* Veatch and Blout have shown that gramicidins form four distinct dimers (private communication).

** The same compound, although in much lower yield, was also obtained by condensation of the azide of the N-terminal benzyloxycarbonyl octapeptide and the C-terminal heptapeptide.

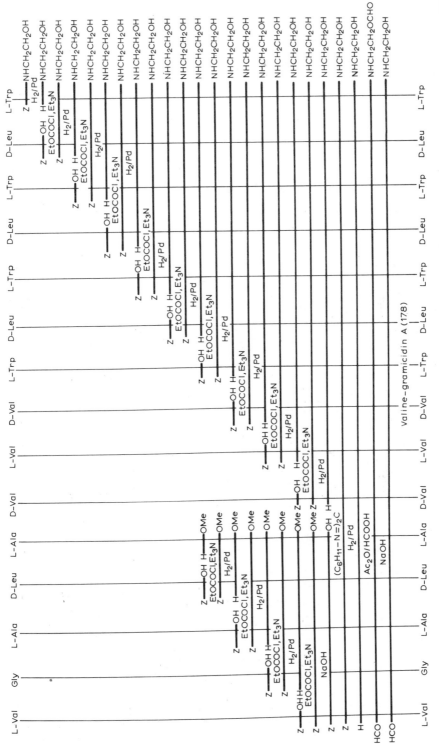

Scheme 3

TABLE 11

ANTIMICROBIAL ACTIVITY OF GRAMICIDIN A AND ITS DERIVATIVES

Compound	Growth inhibition by concentrations of 10 μg/ml									
	B. subtilis	S. faecalis	B. megatherium	B. cereus	M. flavus	S. aureus	S. fradiae	S. griseus	R. nigricans	S. epidermus
Gramicidin A (178)	+	+	+	+	+	±	+	+	−	+
N,O-diacetyldesformyl-gramicidin A (184)	−	+	+	+	+	+	+	+	−	+
N,O-di-(o-sulfobenzoyl)-desformylgramicidin A (185)	−	−	−	−	−	−	−	−	−	−
N,O-disuccinyldesformyl-gramicidin A (186)	−	+	+	+	+	+	±	+	+	+
N-acetyldesformyl-gramicidin A (187)	−	+	+	+	+	+	+	+	±	+

TABLE 12

BIOLOGIC ACTIVITY OF GRAMICIDIN ANALOGS WITH MODIFIED
TRYPTOPHAN RESIDUES AND N-ACYL GROUPS

Compound	Activity (% of gramicidin activity)	
	S. faecalis	Haemolysis
(Tetra-N^{Ind}-formyl)-gramicidin	0	66—88
[Tetra-2-C^{Ind}-(2-hydroxy-5-nitrobenzyl)]-gramicidin	3—5	42—100
N-acetyldesformylgramicidin	92	88
N-maleyldesformylgramicidin	69	51
N-(2,4,6-trinitrophenyl)-desformylgramicidin	60	83

of gramicidin A (HCO—(L-Ala—L-Ala—Gly)$_5$—OH) has also been reported
[1027].

```
      L     L  D  L  D  L  D  L  D  L  D  L  D  L
HCO—Val—Gly—Ala—Leu—Ala—Val—Val—Val—Trp—Leu—Trp—Leu—Trp—Leu—Trp—NHCH2CH2OH
```

Valine-gramicidin A (178)

HCO—Val—Gly—Ala————————Val—Val—Val—Trp—Leu—Trp—Leu—Trp—Leu—Trp—NHCH2CH2OH	(178a)
HCO—Val—Gly—Ala—Leu—Ala—Val————Trp—Leu—Trp—Leu—Trp—Leu—Trp—NHCH2CH2OH	(178b)
HCO—Val—Gly—Ala—Leu—Ala—Val—Val—Val————Trp—Leu—Trp—Leu—Trp—NHCH2CH2OH	(178c)
HCO—Val—Gly————————Val—Val—Trp—Leu—Trp—Leu—Trp—Leu—Trp—NHCH2CH2OH	(178d)
HCO—Val—Gly—Ala—Leu—Ala————Leu—Trp—Leu—Trp—Leu—Trp—NHCH2CH2OH	(178e)
HCO—Val—Gly—Ala—Leu—Ala--Val—Val—Val————Trp—Leu—Trp—NHCH2CH2OH	(178f)
HCO—Val—Gly————————Trp—Leu—Trp—Leu—Trp—Leu—Trp—NHCH2CH2OH	(178g)
HCO—Val—Gly—Ala—Leu—Ala————————Leu—Trp—Leu—Trp—NHCH2CH2OH	(178h)
HCO—Val—Gly—Ala—Leu—Ala—Val—Val—Val————————Trp—NHCH2CH2OH	(178i)
HCO--Val—Gly————————Trp—Leu—Trp—Leu—Trp—NHCH2CH2OH	(178j)
HCO—Val—Gly—Ala—Leu—Ala—Val————————Trp—NHCH2CH2OH	(178k)

There are as yet no concrete published data on the interaction of
gramicidins with metal ions, the conjectures regarding the formation of
complexes (see, for instance, [526]) being based only on the ability of
gramicidin to augment the alkali ion permeability of biological and artificial
membranes (see Parts V.C, VI.B and VI.C).

I.B.2. Antamanide

In the course of a chromatographic separation of the lipophilic components of the poisonous green mushroom (*Amanita phalloides*) Wieland and
coworkers found [1077] that the fraction containing one of the *A.
phalloides* poisons, phallin B, had no toxic effect. Moreover, after intra-
venous injection of this fraction into white mice they became insensitive to
lethal doses of another poison, phalloidine [1074]. This led to the surmise

TABLE 13

ANTIMICROBIAL ACTIVITY OF VALINE-GRAMICIDIN A AND ITS SHORTENED ANALOGS

Compound	Minimal growth-inhibiting concentration (μg/ml)							
	S. aureus 209 P	S. lutea	M. lysodeikticus	S. faecalis	B. subtilis	M. phlei	E. coli B	C. albicans
Val-gramicidin A	0.05–0.1	0.0007–0.0015	0.0007–0.0015	0.0007–0.0015	1–2	2	>10	>10
(178b)	1–2	0.3–0.5	0.1–0.2	0.2–0.4	2–4	2	>10	>10
(178c)	0.07–0.1	0.03	0.03	0.03–0.07	1	0.2–0.4	>10	>10
(178d)	>10	2	>10	>10	>10	>10	>10	>10
(178e—178j)	>25	>25	>25	>25	>25	>25	>25	>25

of the existence of a specific antitoxin among the *A. phalloides* metabolites. Later the antitoxin present in very slight amounts in the fungi was isolated in the individual state. It was shown that in 0.5 mg/kg doses the compound completely cancelled the effect of 0.5 mg/kg doses of amanitine, the basic toxic principle of the green mushroom. For this reason it was given the name of antamanide.

Antamanide has a protective effect not only on the whole body but also on the individual organs (liver, kidneys). Thus, the intake of phalloidine by

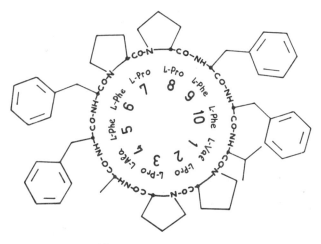

Fig. 6. Antamanide.

the perfused liver of rats is lowered to one tenth of the original value in the presence of antamanide, with a simultaneous sharp decrease in the toxic effects (release of potassium ions and enzymes from the cells) [1059, 1070].

Despite the simple amino acid constitution of antamanide (Ala, Val, Pro$_4$, Phe$_4$) elucidation of its primary structure (Fig. 6) was hampered by a number of serious obstacles [796]. For instance, because of its low solubility in aqueous solutions, enzymic degradation could not be employed, whereas partial acid hydrolysis, as was later ascertained, caused intra-molecular rearrangements (Fig. 7) yielding fragments with "false" sequences. Such rearrangements most likely proceed via the orthoamide (azacyclol) structure (b), resulting from transannular interaction of spatially close amide groups in antamanide (a). The cyclols transform into the acylated peptides (c) which after elimination of the linear derivatives (e), on hydrolysis give peptides with sequences non-existent in the original compound (a).

The formation of azacyclols (b) is accompanied by considerable racemiz-ation of the neighboring asymmetric centers because of reversible acylation

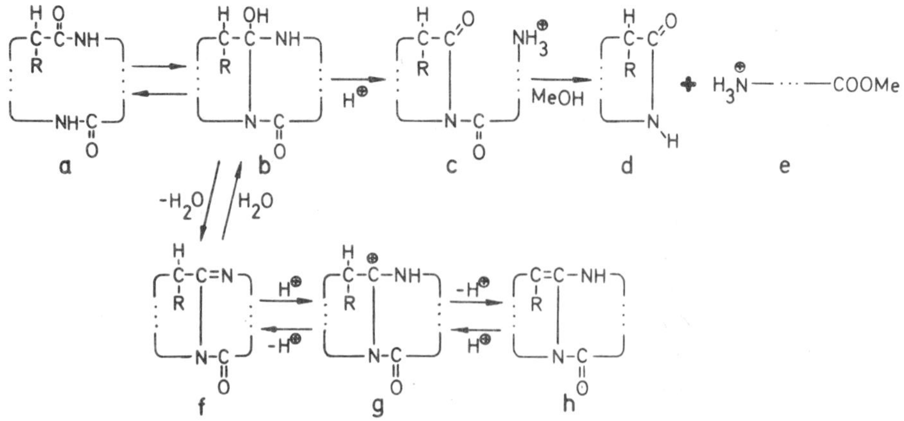

Fig. 7. Intramolecular rearrangements of antamanide under partial acid hydrolysis conditions.

wherein the resultant acylamidines (f) participate in the equilibrium (f) ⇄ (g) ⇄ (h). The presence of a large number of identical amino acid residues also presented obstacles in determining the amino acid sequence because of the numerous possible variants. As a result the initially proposed formula for antamanide, cyclo(—Phe$_4$—Ala—Val—Pro$_4$—), based on Edman degradation of the partial hydrolysis products turned out to be erroneous [796]. Careful mass-spectrometric analysis of both antamanide and its degradation products separated by gas—liquid chromatography led to a proposed structure (188) for this compound [796, 1066, 1073], which was soon confirmed by total synthesis carried out in a number of laboratories.

For the cyclization reaction, Wieland and coworkers [1066, 1067, 1073] selected two of the ten possible decapeptides: Phe—Pro—Pro—Phe—Phe—Val—Pro—Pro—Ala—Phe and Phe—Phe—Val—Pro—Pro—Ala—Phe—Phe—Pro—Pro, so that ring closure occurred at the Phe^5—Phe^6 or Pro^8—Phe^9 bonds (Schemes 4 and 5). However, it is to be noted that activated peptide derivatives with C-terminal phenylalanine residues were introduced at various stages of this synthesis, which could have led to racemization. The first of the above decapeptides was also obtained by Merrifield's method with the exceptionally high yield of 82% [66] (cf. [1064]). Soon after, other variants for ring closure of the decapeptides were described [254, 1072].

Koenig and Geiger [502] subjected the decapeptide Phe—Phe—Pro—Pro—Phe—Phe—Val—Pro—Pro—Ala to cyclization prompted by the consideration that there should be somewhat less steric hindrance to the formation of the Ala^4—Phe^5 bond than to the Phe^5—Phe^6 and Pro^8—Phe^9 bonds. Indeed, after cyclization using the dicyclohexylcarbodiimide method in the presence

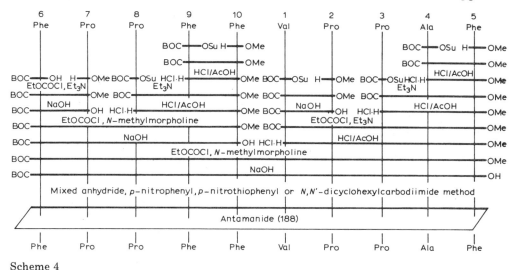

Scheme 4

of *N*-hydroxysuccinimide, antamanide was obtained in yields of 36.5% (the synthesis according to Schemes 4 and 5 gave 7—30% yields).

In order to exclude racemization, Ovchinnikov and coworkers chose for synthesis of antamanide the method shown in Scheme 6 [711].

An important property of this compound that was responsible for the considerable interest it attracted on the part of chemists and biologists is its clearly expressed preference for sodium ions (Table 14). The crystalline salt (antamanide·Na)$^+$·ClO$_4^-$ has been isolated in an analytically pure form

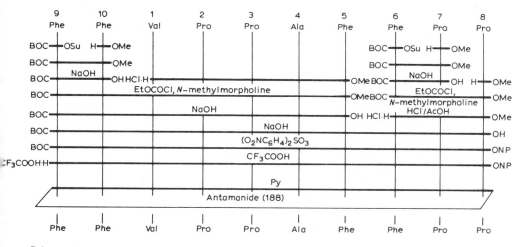

Scheme 5

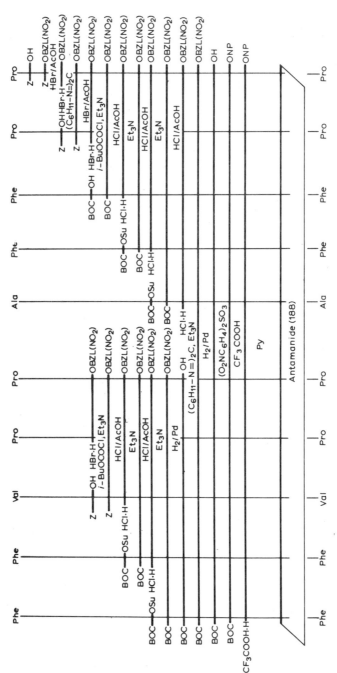

Scheme 6

[1068]. Antamanide also forms very stable complexes with Ca^{2+}; the Rb^+ and Tl^+ complexes are much less stable [1068, 1069]. According to the results of extraction experiments, antamanide also complexes, although very weakly, with Li^+, Cs^+, NH_4^+, Sr^{2+} and Ba^{2+}; no signs of complexing with Mg^{2+} have been detected.

TABLE 14

STABILITY CONSTANTS OF THE ANTAMANIDE COMPLEXES WITH DIFFERENT CATIONS

Solvent	Reference	Li^+	Na^+	K^+	Tl^+	Ca^{2+}
C_2H_5OH	[15, 639, 1069]	—	2800	220	—	—
$C_2H_5OH-H_2O$ (96:4)	[1068]	—	2000	180	—	—
	[1069]		1300			
$C_2H_5OH-H_2O$ (30:70)	[1069]	—	0	—	—	—
CH_3OH	[1069]	< 10	500	10	190	30
CH_3CN	[1069]	—	30 000	290	—	100 000
CH_3CN-H_2O (96:4)	[1069]	1300	2600	20	—	—
CH_3CN-H_2O (92:8)	[1069]	—	1200	280	—	—

The valuable property of antamanide as a Na^+-specific complexone has made it a promising tool for ion-permeability studies of membranes. However, in contrast to the depsipeptidic valinomycin and enniatin antibiotics, antamanide manifests little surface activity and displays no tendency to enter the phospholipid monolayer. This could explain why it has no influence on the ion permeability of phospholipid bilayers and does not inhibit microbial growth. However, ways are being developed for specifically modifying antamanide so as to increase the membrane activity while retaining its Na^+ specificity [701, 702, 712].

I.B.3. Antamanide analogs

Soon after the synthesis of antamanide, several dozens of its analogs were prepared (Table 15) which differed from the parent compound in one (189—196 and 198—200) or in several (201—209, 211—215 and 217—219) amino acid residues, in ring size (194, 197, 210, 216 and 220), in inversed configuration of all asymmetric centers (*enantio*-antamanide 221) or in reversed direction of acylation (*retro*-antamanide, 202). The analogs (204, 206, 209, 212, 218 and 219), consisting of two identical pentapeptide fragments, differ from antamanide also in having a twofold symmetry axis.

TABLE 15
STRUCTURE AND PROPERTIES OF ANTAMANIDE AND ITS ANALOGS

No. Compound

No.	1	2	3	4	5	6	7	8
188	cyclo(-L-Val-	L-Pro-	L-Pro-	L-Ala-	L-Phe———	L-Phe———	L-Pro-	L-Pro-
				(antamanide)				
189	Gly							
189a	Ala							
189b	Abu							
189c	Leu							
189d	Ile							
190				des				
191				D-Ala				
192				Gly				
192a				Abu				
192b				Phe				
193				Lac				
193a				D-Lac				
194					des			
194a					Tyr			
194b					Tyr(OCH_3)			
194c					Tyr(OCH_2COONH_4)			
194d					Tyr$(OC_3H_7SO_3Na)$			
195						Tyr		
195a						Tyr(OMe)		
195b						Tyr(OBzl)		
195c						Tyr$(OC_{12}H_{25})$		
195d						Tyr(OCH_2COONH_4)		
195e						Tyr$(OC_3H_7SO_3Na)$		
195f						Tyr(OSO_3Na)		
195g						Tyr(O-[β]-D-glucopyranosyl)		
195h						3',5'-Br_2-Tyr		
196							Gly	

10
e-L-Phe-)

m.p. (°C)	$[\alpha]_D$	c (%)	Solvent	Antitoxic dose (mg/kg) active against 5 mg/kg phalloidine (white mice)	Stability constants of the complexes (C_2H_5OH, 25°C) Na$^+$	K$^+$	References
172	−180	0.2	CH$_3$OH	0.5	2020—2800	185—270	[15, 639, 1066, 1067, 1069, 1073]
174	−148	1.0	CH$_3$OH				[502]
174	−168	0.5	C$_2$H$_5$OH				[711]
182	—	—	—	10	180	—	[1068, 1071]
178	—	—	—	15	150	—	[1068, 1071]
165	—	—	—	2.5	1000—2000	—	[1062, 1071]
169	—	—	—	0.5	1000	—	[1068, 1071]
167	—	—	—	0.5	2300	—	[1068, 1071]
—	—	—	—	$>$ 10	—	—	[1066]
—	—	—	—	2.5—5	—	—	[1066]
—	—	—	—	3	—	—	[1062]
158	—	—	—	0.5	—	—	[1062]
—	—	—	—	$>$ 20	—	—	[1062]
168—171	−124.6	0.21	C$_2$H$_5$OH	—	50—100	$\sim$ 50	—
143—146	−125	0.34	C$_2$H$_5$OH	—	$<$ 50	$<$ 50	—
—	—	—	—	5	—	—	[1066]
176—180	—	—	—	0.5—1	—	—	[1063, 1076]
—	—	—	—	0.5	—	—	[1076]
—	—	—	—	1	—	—	[1076]
—	—	—	—	10	—	—	[1076]
175—178	—	—	—	0.5—1	2000	—	[1060, 1063, 1068, 1076]
—	—	—	—	0.5	—	—	[1060, 1076]
160	—	—	—	$>$ 10	—	—	[1060]
—	—	—	—	$>$ 20	$<$ 10	—	[1068]
215—235	—	—	—	0.5	—	—	[1076]
228—240	—	—	—	0.5	—	—	[1076]
210—230	—	—	—	0.5	—	—	[1076]
180—210	—	—	—	1	1700	—	[1068, 1076]
180—187	—	—	—	2.5	3500	—	[1068, 1076]
—	—	—	—	$>$ 10	60	—	[1060, 1068]

TABLE 15—*continued*

No.	Compound

No.						
197						—des—
198						
199						
200						
201	—Gly—		—Gly—			
201a	—Ala—		—Gly—			
202	—Ala—		—Val—			
			(*retro*-antamanide)			
203	—Abu—		—Abu—			
204	—Phe—		—Phe—			
205	—Ile—			—Tyr—		
206	—Phe—					
207		—Ala—Pro—				
208		—Phe—Ala—				
209		—Phe—		—Val——		
210		—des—				
211			—Pro—		—Phe—	
212			—Val—			
213	—Gly—	—Gly—Tyr—				
214	—Gly—	—Gly—		—Tyr—		
215	—Ile—			—Tyr—		
216						—des—
217		—Cha—		—Cha—		—Cha—
		(perhydroantamanide)				
218		—Cha—		—Val—		
219		—D-Phe—		—Val—		
220		—Phe—		—des—		—des——des—
221	-D-Val-D-Pro-D-Pro-D-Ala-D-Phe—		—D-Phe—		—D-Pro-D-Pro—	
	(*enantio*-antamanide)					
222	-D-Ala-D-Pro-D-Pro-D-Val-D-Phe—		—D-Phe—		—D-Pro-D-Pro—	
	(all D *retro*-antamanide or *retro-enantio*-antamanide)					
222a	-D-Ala-D-Pro-D-Pro-D-Val-D-Phe—		—D-Tyr—		—D-Pro-D-Pro—	
222b	-D-Ala-D-Pro-D-Pro-D-Val-D-Phe—		—D-Tyr (OSO$_3$Na)—		—D-Pro-D-Pro—	

	m.p. (°C)	$[\alpha]_D$	c (%)	Solvent	Antitoxic dose (mg/kg) active against 5 mg/kg phalloidine (white mice)	Stability constants of the complexes (C$_2$H$_5$OH, 25°C)		References
						Na$^+$	K$^+$	
	—	—	—	—	> 10	—	—	[1060]
Tyr———	173—180	—	—	—	2	—	—	[1063]
———Tyr-	179—190	—	—	—	5	—	—	[1063]
———MePhe	158—159	−176.5	0.34	C$_2$H$_5$OH	—	11000	—	—
	180	—	—	—	>20	100	—	[1062, 1068]
	183	—	—	—	15	120	—	[1062, 1068]
	177	−176	0.08	C$_2$H$_5$OH	>20	400	100	[711, 1062]
	152	—	—	—	10—20	—	—	[1062]
	—	—	—	—	> 20	—	—	[1062]
	171—183	—	—	—.	1	—	—	[1063]
Ala———	179	−108	1	C$_2$H$_5$OH	10	6000	500	[639, 1060]
	—	—	—	—	> 10	—	—	[1066]
	—	—	—	—	> 10	—	—	[1066]
	167	—	—	—	1.5—2	—	—	[1065]
Tyr———	192—200	—	—	—	5—10	—	—	[1063]
	—	—	—	—	> 20	50	—	[1060, 1068]
Ala———	322	−384	1	C$_2$H$_5$OH	> 10	25 000	1000	[639, 712, 1060]
	187—210	—	—	—	> 20	—	—	[1063]
	197—203	—	—	—	> 20	—	—	[1063]
———Tyr-	189—197	—	—	—	5	—	—	[1063]
es——des-	—	—	—	—	> 10	—	—	[1066]
Cha———	166	−86	0.5	C$_2$H$_5$OH	> 10	2000	50	[639, 711, 1060, 1068]
Ala——Cha-	305	−153	1	C$_2$H$_5$OH	> 10	1700	100	[639, 1060]
Ala-D-Phe-	157—159	−68.8	0.3	C$_2$H$_5$OH	—	26 000	26 000	—
es——des-	150	—	—	—	> 20	—	—	[1065]
Phe-D-Phe-	173	+166	1	C$_2$H$_5$OH	10	2500	200	[1060]
Phe-D-Phe-	181—185	—	—	—	—	—	—	[1075]
Phe-D-Phe-	201—203	—	—	—	—	—	—	[1075]
Phe-D-Pro-	—	—	—	—	—	—	—	[1075]

One of these (compound 209) was prepared by doubling the linear pentapeptide [702]:

$$CF_3COOH \cdot HValProProPhePheONP \xrightarrow{Py}$$

$$cyclo(-ValProProPhePhe-) + cyclo[-(ValProProPhePhe)_2-]$$
$$\quad\quad 2\% \quad\quad\quad\quad\quad\quad (209); 3\%$$

As one can see from Table 15, although many of the resultant compounds have a protective action against phalloidine, none exceeds the original cyclopeptide in this respect. All antamanide analogs (excepting 219) whose complexing properties have been investigated have been found to retain the high Na/K-specificity of the parent compound: the enhanced stability of the Val^6,Ala^9-antamanide—Na^+ complexes show the feasibility of the search in this series for new, powerful sodium complexones. The hydrogenated analog (217), in which all phenyl groups are displaced by cyclohexyl groups, is more lipophilic than antamanide itself. Apparently this property is responsible for its membrane and antimicrobial activities [701].

A structure similar to that of the nonapeptide analogs of antamanide is possessed by the cyclic nonapeptide, $cyclo(-L-Val-L-Pro-L-Pro-L-Phe-L-Phe-L-Leu-L-Ile-L-Ile-L-Leu-)$ (223), isolated from linseeds [476, 797]. However, its physicochemical and biological properties have been little investigated.

I.B.4. Alamethicin

Alamethicin (initially antibiotic U 22324) is produced in very high yield (10 mg/ml) by the microorganism *Trichoderma viride* [635, 818]. As with other low molecular weight peptides, its synthesis is apparently carried out by enzyme systems without the participation of ribosomes [818]. Alamethicin has a relatively low antibiotic activity (see Table 16).

The structure of alamethicin (Fig. 8) has been established independently by two groups of workers [715, 736]. In view of its resistance to peptidases, degradation of the antibiotic was in both cases carried out by partial acid hydrolysis. Isolation of the resultant peptides by paper electrophoresis [736], chromatography on ion-exchange resins or paper chromatography [715], followed by their structural determination by the "Dansyl"-Edman method [736] or by mass spectrometry [715], gave the amino acid sequence of this compound (all amino acids are of the L configuration) (Fig. 8).

It thus follows that alamethicin (224) is a cyclic octadecapeptide with a high content of the unusual amino acid C-methylalanine (α-aminoisobutyric acid — Aib). The 53-membered ring of alamethicin is closed by an amide

TABLE 16

ANTIMICROBIAL ACTIVITY OF ALAMETHICIN

Bacteria	Minimal growth-inhibiting concentration (μg/ml)	Fungi	Minimal growth-inhibiting concentration (μg/ml)
Escherichia coli	> 1000	*Nocardia asteroides*	1000
Klebsiella pneumoniae	> 1000	*Blastomyces dermatitidis*	100
Proteus vulgaris	> 1000	*Coccidioides immitis*	1000
Pseudomonas aeruginosa	> 1000	*Geotrychum sp.*	> 1000
Salmonella paratyphi	> 1000	*Hormodendrum compactum*	1000
Salmonella pullorum	> 1000	*Cryptococcus neoformans*	> 1000
Salmonella typhosa	> 1000	*Histoplasma capsulatum*	1000
Staphylococcus aureus	62.5—125	*Sporotrichum schenckii*	> 1000
Streptococcus faecalis	31	*Monosporium apiospermum*	> 1000
Streptococcus hemolyticus	31	*Trichophyton rubrum*	> 1000
Streptococcus viridans	31	*Trichophyton interdigitale*	> 1000
		Candida albicans	> 1000
		Trichophyton violaceum	> 1000
		Trichophyton asteroides	> 1000
		Trichophyton mentagrophytes UC-4797	> 1000
		Trichophyton mentagrophytes UC-4860	1000

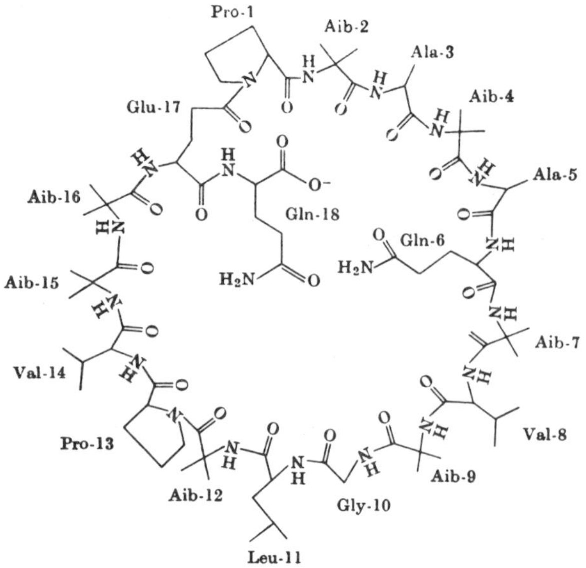

Fig. 8. Alamethicin.

bond formed by the γ-carboxyl of the Glu[17] residue and the amino group of Pro[1]. Alamethicin possesses the largest ring of all known homodetic cyclopeptides. Preparations of the naturally occurring antibiotic contain also its homolog, in which an Aib[5] residue takes the place of an Ala[5] residue [715] and apparently other homologs differing in the content of C^{α}-methyl-alanine [620, 736].

The (γ) Glu[17]—Pro[1] bond formed by the γ-carboxyl group of glutamic acid is more sensitive to acids and alkalis than the other amide groups of the antibiotic [715, 736].

In the course of the structural elucidation of alamethicin its methyl ester and a derivative with the carboxyl group substituted by a methyl group were prepared.

Two-phase ion-partition experiments showed that alamethicin facilitates Na^+, K^+, Rb^+ and Cs^+ transfer into the organic phase. It was therefore concluded that this substance apparently forms ion—dipole complexes with the ions [785, 791], although quite possibly the observed effect is merely due to the high solubility of the alamethicin salts in the organic phase or to micelle formation. No direct investigation of complex formation of alamethicin in solutions has been carried out, but it has been noted that neither its CD curves [621] nor its NMR spectra [127, 377] are affected by the presence of univalent ions in solution.

Sedimentation analysis of aqueous solutions of alamethicin disclosed strong self-association, the molecular weight for a 0.6% concentration being 27 000 whereas that of the monomer is 1691. The addition of urea (6M) diminishes the effective molecular weight to 6900 and no association at all is noted in alcoholic solution [620, 622]. Like many other peptide antibiotics, alamethicin possesses a high surface activity, readily forming monolayers on the air—water interface [127, 263, 377].

I.B.5. Cyclopeptide from mitochondrial membranes

The striking ability of valinomycin to be a "vehicle" for potassium ion transport through membranes was a very attractive reason for many workers to conjecture that valinomycin-like ion carriers might possibly be functioning in natural membranes. The fact that no such substances had been isolated they ascribed to its insignificant concentrations (a few molecules per cell). Despite this, in 1971, Blondin et al. [73] attempted to isolate such a carrier from mitochondrial membranes. After extraction of freeze-dried bovine heart mitochondria by pentane or ethyl acetate and repeated chromatography of the extract on alumina, silica gel and silicic acid columns, they succeeded in obtaining a ninhydrin-negative substance in insignificant yield $(2.10^{-8}—5.10^{-5}$ g/g mitochondrial protein) that induced approximately equal respiration-dependent sodium and potassium ion transport in mitochondria. The substance also caused passive swelling of the mitochondria in the presence of potassium and sodium.

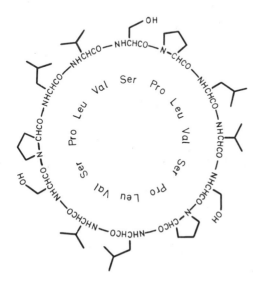

Fig. 9. Suggested structure of the ionophore isolated from mitochondrial membranes [73].

Because of the small amount isolated (< 2 mg) the structure of the substance was not fully elucidated. From data on its amino acid composition and the assumption that its molecular weight must be close to that of valinomycin, the authors indicated several possibilities for its structure, among them that shown in Fig. 9.

The results obtained by the American workers are undoubtedly of considerable interest. However, their poor reproducibility (yields from one experiment to another differed by more than three orders of magnitude) shows that the conclusions require further proof.

I.B.6. Synthetic cyclopeptides

Cyclic hexapeptides consisting of alanine and glycine residues do not form complexes with alkali metal ions [427]. However, the N-methylated cyclopeptides $cyclo[-(\text{L-MeAla}-\text{Sar})_3-]$ (225) [700, 904] and $cyclo[-(\text{Val}-\text{Sar})_3-]$ (225a), structurally resembling enniatin antibiotics, do complex Na^+ and K^+ in ethanol solution. $Cyclo[-(\text{L-Pro}-\text{Gly})_3-]$ (225b) also forms alkali metal ion complexes, the ion selectivity in dimethylsulfoxide solution being of the following order [191, 192]: $\text{Na}^+ > \text{K}^+ > \text{Rb}^+, \text{Li}^+ > \text{Cs}^+$. No information on the stoichiometry of the complexes is as yet available.

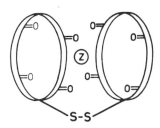

Fig. 10. The mode of formation of "sandwich complexes" from cysteine-containing cyclopeptides.

An interesting idea has been realized by Schwyzer *et al.* [876], which resulted in the discovery of a new type of alkali metal complexone. These authors suggested that the molecules of a small peptide consisting of 4—6 amino acid residues can in principle form 2:1 ("sandwich") complexes with alkali metal ions. The stability of such complexes should, of course, considerably increase when the two cyclopeptide molecules are joined by a covalent bond, one of the simplest ways of accomplishing this being formation of a S—S bond between two of the cysteine side chains (Fig. 10). In fact, $cyclo(-\text{Gly}-\text{Cys}-\text{Gly}-\text{Gly}-\text{Pro}-)$, synthesized according to Scheme 7 and manifesting no signs of complex formation in methanol solution, gives rise to an efficient complexone on oxidation to $\text{S}-\text{S}'-bis$-

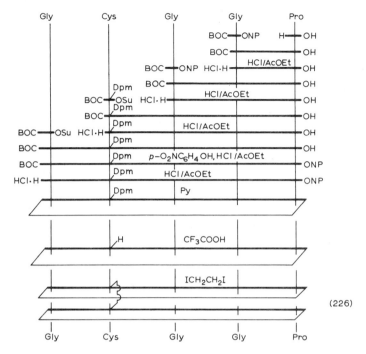

Scheme 7

[*cyclo*(—Gly—L-Ċys—Gly—Gly—Pro—)] (226). The compound, sparingly soluble in most neutral solvents, readily dissolves in water in the presence of potassium, sodium and lithium salts; the stability constant of the K^+ complex in H_2O is approximately 0.1 l·mole^{-1}. Preliminary experiments on bulk membranes (see Part V.2) showed the following order of ion selectivity: $K^+ > Na^+ > Li^+ > Ca^{2+}$. The action of compounds (225) and (226) on biological objects has not been investigated.

I.C. Depsides. The nonactins

Nonactin (227) and its homologs (228—236) produced by various *Actinomyces* sp. [13, 14, 164, 686, 695, 1043] form a group of macrotetrolide antibiotics of the general formula represented in Fig. 11 [52, 164, 210, 283, 480]*.

---

* To this group should also be referred the antibiotic preparations (peliomycin [864],verramycin [1043], SQ 15859 [55, 218], N-329A [686], fluorin [201], lustericin [928] and longisporin [629]) which, as far as their sparsely investigated properties show, are apparently closely related to the macrotetrolide antibiotics.

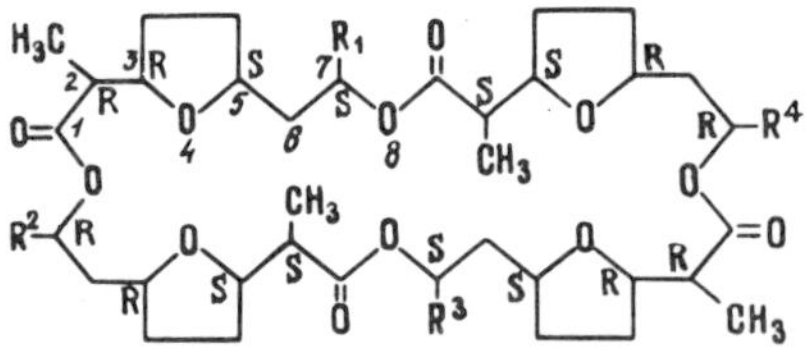

Fig. 11. Structure of macrotetrolides.

No.	Compound	R^1	R^2	R^3	R^4
227	Nonactin	CH_3	CH_3	CH_3	CH_3
228	Monactin	CH_3	CH_3	CH_3	C_2H_5
229	Dinactin	CH_3	C_2H_5	CH_3	C_2H_5
230	Trinactin	CH_3	C_2H_5	C_2H_5	C_2H_5
231	Tetranactin	C_2H_5	C_2H_5	C_2H_5	C_2H_5
232	Substance G	CH_3	C_2H_5	CH_3	$CH(CH_3)_2$
233	Substance D	CH_3	C_2H_5	$CH(CH_3)_2$	C_2H_5
234	Substance C	CH_3	C_2H_5	$CH(CH_3)_2$	$CH(CH_3)_2$
				or	
		CH_3	$CH(CH_3)_2$	$CH(CH_3)_2$	C_2H_5

The macrotetrolides are colorless, largely crystalline, low-melting substances (Table 17) readily soluble in the usual organic solvents, but poorly soluble in water. The parent compound of this group, nonactin, was discovered in 1955 [164] and somewhat later (1962) its homologs monactin (228), dinactin (229) and trinactin (230) were isolated from the mother liquor after crystallization of nonactin [52, 210]. Following this four new macrotetrolides — compounds B (235), C (234), D (233) and G (232) — were isolated by counter-current distribution [283, 481] and,

TABLE 17

PHYSICOCHEMICAL CHARACTERISTICS OF MACROTETROLIDES [13, 480]

No.	Macrotetrolide	m.p. (°C)	Empirical formula	$[\alpha]_D$ (CHCl$_3$)	R_f*	Mass number of molecular ion in mass spectrum
227	Nonactin	148	$C_{40}H_{64}O_{12}$	0	0.62	736
228	Monactin	64	$C_{41}H_{66}O_{12}$	+2	0.48	750
229	Dinactin	67	$C_{42}H_{68}O_{12}$	+2.5	0.32	764
230	Trinactin	68	$C_{43}H_{70}O_{12}$	+1.5	0.15	778
231	Tetranactin	106	$C_{44}H_{72}O_{12}$	0	0.31	792
232	Substance G	oil	$C_{43}H_{70}O_{12}$	—	0.29	778
233	Substance D	65	$C_{44}H_{72}O_{12}$	—	0.11	792
234	Substance C	71	$C_{45}H_{74}O_{12}$	—	0.13	806
235	Substance B	58	$C_{46}H_{76}O_{12}$	—	0.10	820

* Thin-layer chromatography on kiselgel G (Merck) in chloroform—ethyl acetate (1:2).

finally, in 1970, still another antibiotic of this group, tetranactin (231), was found among the metabolites of actinomycetes [13, 14, 695].

A characteristic structural feature of the macrotetrolides is the presence of a 32-membered ring built up of four hydroxy acid residues and containing four ether and four ester bonds. Swiss workers [52, 210, 282—284] were able to show that naturally occurring macrotetrolides possess, as constitutent parts, (+) and (—) nonactinic, homononactinic and *bis*-homononactinic acid residues in the 2*S*, 3*S*, 6*R*, 8*R* and 2*R*, 3*R*, 6*S*, 8*S* configurations, respectively.

R = CH₃ nonactinic acid
R = C₂H₅ homononactinic acid
R = CH(CH₃)₂ bis-homononactinic acid

Up to now, for only two antibiotics of this group, nonactin and dinactin, have the configurations been completely elucidated. X-ray analysis of the K⁺ complex showed the hydroxy acid residues in nonactin to be alternatingly (+) and (—) [208, 488]; the same alternating sequence was shown by mass spectrometry for dinactin [481]. Similar stereochemical features of the depside chain have been very plausibly assumed on biogenetic grounds for the other nonactin homologs (228, 230—236) [480]. However, the sequence of nonactinic acid and *bis*-homononactinic acid residues in compound (234) still remains to be ascertained, and for compound B (235) the configurations of its homononactinic and *bis*-homononactinic acid residues are also unknown.

TABLE 18

BIOLOGICAL ACTIVITY OF MACROTETROLIDES [636]

No.	Macrotetrolide	Minimal growth-inhibiting concentration (μg/ml)		
		Staphylococcus aureus	*Mycobacterium bovis*	L cells
227	Nonactin	0.95	0.96	0.060
228	Monactin	0.08	0.12	0.010
229	Dinactin	0.05	0.04	0.004
230	Trinactin	0.04	0.04	0.005

TABLE 19

ANTIMICROBIAL ACTIVITY OF TETRANACTIN

Bacteria	Minimal growth-inhibiting concentration (μg/ml)	Fungi	Minimal growth-inhibiting concentration (μg/ml)
Arthrobacter simplex	0.9	Candida albicans	> 24.3
Brevibacterium ammoniagenes	0.1	Candida krusei	8.1
Microbacterium flavum	0.9	Cryptococcus neoformans	> 24.3
Micrococcus flavus	0.9	Hansenula anomala	24.3
Serratia marcescens	> 24.3	Saccharomyces cerevisiae	> 24.3
Staphylococcus aureus	< 0.1	Cercospora oryzae	> 24.3
Sarcina lutea	0.3	Gibberella saubinetii	8.1
Bacillus megaterium	0.9	Gibberella fujikurai	> 24.3
Bacillus cereus	0.9	Botrytis cinerea	24.3
Bacillus subtilis	0.9	Botrytis tulipae	> 24.3
Bacillus roseus	0.9	Sclerotinia arachidis	> 24.3
Bacillus circulans	0.1	Fusarium lini	> 24.3
Bacillus firmus	0.9	Cochliobolus miyabeanus	0.9
Shigella sonnei	< 0.1	Macrosporium bataticola	8.1
Aerobacter aerogenes	> 24.3	Alternaria kikuchiana	> 24.3
Agrobacter radiobacter	> 24.3	Corynespora vignicola	> 24.3
Alcaligenes faecalis	> 24.3	Glomerella lagenarium	> 24.3
Escherichia coli	> 24.3	Cladosporium fluvum	> 24.3
Salmonella typhosa	> 24.3	Colletotrichum atromentarium	> 24.3
Klebsiella pneumoniae	> 24.3	Rosedlinis necatrix	> 24.3
Proteus vulgaris	> 24.3	Piricularia oryzae	2.7
Pseudomonas aeruginosa	> 24.3	Pellicularia sasakii	8.1
Pseudomonas ovalis	> 24.3	Gloeosporium kaki	> 24.3
Xanthomonas oryzae	> 24.3	Helminthosporium sesanum	> 24.3
Xanthomonas citri	> 24.3	Rhizoctonia solani	0.9
Xanthomonas faecalis	> 24.3	Phoma citricarpa	> 24.3
Erwinia aroidea	> 24.3	Pestalotia diospyri	> 24.3
Corynebacterium michiganensis	> 24.3	Mycospherella arachidicola	> 24.3
Mycobacterium tuberculosis	8.1	Aspergillus oryzae	> 24.3
Mycobacterium 607	> 24.3	Aspergillus fumigatus	> 24.3
		Penicillium citrinum	> 24.3
		Aspergillus niger	> 24.3

In the early biological studies of the macrotetrolides no antibiotic activity was observed, whence the name "nonactin" for one of them [164]. However, it was subsequently found that the apparent non-activity was the result of their low water solubility [55] and that they actually possess broad antimicrobial spectra, inhibiting Gram-positive and acid-fast bacteria, fungi and cancer cells in low concentrations [14, 501, 636, 686, 1043]. It can be seen from Table 18 that nonactin is less active than its homologs. Lengthening the nonactin side chain by a CH_2 unit (transition to monactin) increases the antimicrobial activity 8- to 11-fold and the cytotoxic activity 6-fold. The increase is not so large on passing further to dinactin and trinactin. Table 19 summarizes the biological data for one of the best known of these compounds, tetranactin.

TABLE 20

INSECTICIDE ACTIVITY OF MACROTETROLIDES AGAINST *TETRANYCHUS KANZAWA* IN FIELD CONDITIONS

Insecticide	Conc. (µg/ml)	Number of mites after spraying			
		0 days	2 days	6 days	10 days
Macrotetrolide mixture	200	156	0	2	11
	133	815	0	4	5
Celtan	200	86	0	3	6
	200	54	0	3	7
Control		133	244	578	413

Recently, Japanese investigators have shown that di-, tri- and tetranactins manifest high activity against mites, exceeding in this respect some of the widely used insecticides and thus disclosing wide possibilities for their application in farming practice. For example, as can be seen from Table 20, Celtan (one of the most potent DDT analogs) is somewhat less active than the naturally occurring mixture of macrotetrolides against the mite *Tetranychus kanzawa*. Of particular importance is the fact that the macrotetrolides break down much more easily than many of the common insecticides and have the advantage of low peroral toxicity against warm-blooded animals (mice, rats, quail).

When *Streptomyces* Tü 10 (ETH 7796) was grown in a [^{14}C]glucose medium, a labeled preparation was obtained of specific activity 8750 counts per min per mg, containing 75% nonactin, 29% monactin and 2% dinactin; it was utilized in the study of the mode of macrotetrolide action [1082].

TABLE 21

STABILITY CONSTANTS OF MACROTETROLIDE COMPLEXES WITH ALKALINE METAL IONS AND BARIUM [666]

Compound	Solvent	$l \cdot mole^{-1}$ (measured at $30°C$)				
		Na^+	K^+	Rb^+	Cs^+	Ba^{2+}
Nonactin	Methanol	240	3800	3400	900	52
	Ethanol	2000	13 000	—	—	250
Monactin	Methanol	380	6800	3700	1400	19
	Ethanol	3400	19 000	—	—	270
Dinactin	Methanol	870	5000	5000	1900	150
	Ethanol	4300	—	—	—	—
Trinactin	Methanol	4400	—	5900	2500	—
	Ethanol	4100	—	—	—	—

With alkali metal ions macrotetrolides form equimolar complexes which readily lend themselves to study by physical methods (IR and NMR spectroscopy, gas—liquid osmometry) [676, 756, 792, 793]. As with other metal-binding macrocyclic compounds, increasing water content of the solution is accompanied by decreasing complex stability. At the same time, as one can see from Table 22, the K^+ complex of nonactin is more stable in water-containing solutions than the corresponding sodium or cesium complexes. In their K^+-complex stability and K/Na-selectivity the nactins assume an intermediate position between valinomycin and the enniatins (cf. Tables 2, 5 and 21). Macrotetrolides complex with divalent cations (Ca^{2+}, Ba^{2+} and Cu^{2+}) [676] as well as alkali cations. From studies in solutions

TABLE 22

STABILITY CONSTANTS OF NONACTIN COMPLEXES WITH Na^+, K^+ AND Cs^+ IN AQUEOUS ACETONE [793]

Solvent CH_3COCH_3: H_2O (mole fractions)	$l \cdot mole^{-1}$		
	Na^+	K^+	Cs^+
0.999:0.001	70 000	—	—
0.997:0.003	—	—	15 000
0.993:0.007	—	70 000	—
0.66:0.34	—	17 000	—
0.61:0.39	210	—	—
0.45:0.55	—	—	400

and on membranes the following ion-selectivity sequence has been established [233, 258, 666, 753, 866]: $NH_4^+ > K^+ > Rb^+ > Cs^+ > Ba^{2+}$.

The ability of the macrotetrolides to form lipophilic complexes with potassium picrate has served as a basis for their quantitative analysis in biological systems [984]. For the nactins the readily formed crystalline complexes with sodium, potassium, ammonium and potassium rhodanides and with $Ba(ClO_4)_2$ and $CuBr_2$ [676, 756], utilized in X-ray studies, have also been described [208, 488, 676].

I.D. Macrocyclic polyethers and polythioethers

Macrocyclic polyethers form a large group of synthetic compounds composed of 9- to 60-membered rings, containing 3—20 ether oxygen atoms. The synthesis of various representatives of this group has been carried out at

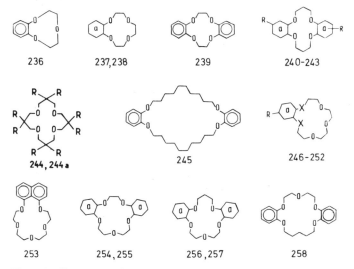

Fig. 12. Structure of cyclic polyethers (236—258).

No.	a	R	X	No.	a	R	X
237	Benzo	—	—	247a	Benzo	Me	O
238	Cyclohexylene	—	—	248	2,3-Naphtho	—	O
240	Benzo	H	—	249	Cyclohexylene	H	O
241	Benzo	Bu^t	—	250	Cyclohexylene	Bu^t	O
242	Cyclohexylene	H	—	251	2,3-Decalylene	—	O
243	Cyclohexylene	Bu^t	—	252	Benzo	Me	S
244	—	H	—	254	Benzo	—	—
244a	—	Me	—	255	Cyclohexylene	—	—
246	Benzo	H	O	256	Benzo	—	—
247	Benzo	Bu^t	O	257	Cyclohexylene	—	—

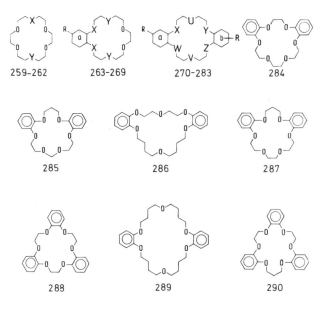

Fig. 13. Structure of cyclic polyethers (259–290).

No.	a	b	R	X	Y	Z	W	V	U
259	—	—	—	O	O	—	—	—	—
260	—	—	—	S	S	—	—	—	—
261	—	—	—	NH	O	—	—	—	—
262	—	—	—	NH	NH	—	—	—	—
263	Benzo	—	H	O	O	—	—	—	—
264	Benzo	—	Bu^t	O	O	—	—	—	—
264a	Benzo	—	Me	O	O	—	—	—	—
265	2,3-Naphtho	—	—	O	O	—	—	—	—
266	Cyclohexylene	—	H	O	O	—	—	—	—
267	Cyclohexylene	—	Bu^t	O	O	—	—	—	—
268	Benzo	—	Me	S	O	—	—	—	—
269	Benzo	—	H	O	S	—	—	—	—
270	Benzo	Benzo	H	O	O	O	O	O	O
271	Benzo	Benzo	Me	O	O	O	O	O	O
272	Benzo	Benzo	Bu^t	O	O	O	O	O	O
273	2,3-Naphtho	2,3-Naphtho	—	O	O	O	O	O	O
274	Cyclohexylene	Cyclohexylene	H	O	O	O	O	O	O
275	Cyclohexylene	Cyclohexylene	Bu^t	O	O	O	O	O	O
276	Benzo	Cyclohexylene	H	O	O	O	O	O	O
277	Benzo	Benzo	H	S	S	O	O	O	O
278	Benzo	Benzo	H	S	O	S	O	O	O
279	Benzo	Benzo	Me	S	S	S	S	O	O
280	Cyclohexylene	Cyclohexylene	H	S	S	S	S	O	O
281	Benzo	Benzo	H	O	O	O	O	O	NH
282	Benzo	Benzo	H	O	O	O	O	O	N—C$_8$H$_{16}$
283	Benzo	Benzo	H	O	O	O	O	NH	NH

different times by several teams of investigators (see, for instance, [1, 2, 81, 104, 180, 211, 564, 565, 566, 821, 973, 995, 1093]), but the most extensive work in both synthesis and study has been done by Pedersen [737—742]. Polyethers are acid- alkali-fast, colorless crystals or oils. The formulae of some of them are represented in Figs 12—14.

Since the full names of the cyclic polyethers are too cumbersome for everyday use, Pedersen introduced an abbreviated nomenclature, giving these compounds the generic name of *crowns* because of the peculiar shape of their molecular models (resembling a crown) and adding to this name the number of ring atoms, ether groupings and aromatic (or alicyclic) substituents. For instance, compound (270) has the abbreviated name of dibenzo-18-*crown*-6, whereas its full name is 2,3,11,12-dibenzo-1,4,7,10,13,16-hexa-oxacyclo-octadeca-2,11-diene. Usually the name refers to the most symmetrical arrangement of the hydrocarbon chains and oxygen atoms; when not, this is designated by the suffix *asym*. It can be readily seen that if the quotient of the first number divided by the second equals three, the cyclic polyethers, as a rule, are derivatives of the cyclic oligomer ethylene oxide.

Five basic methods for the synthesis of cyclic polyethers have been proposed as shown below, in which R, T, U and V are divalent groups.

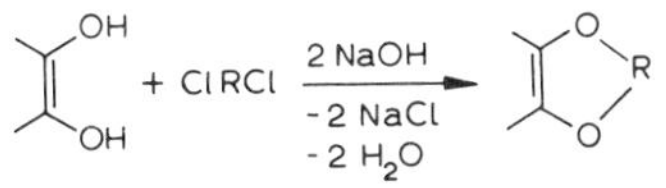

(compounds 236, 237, 239, 246 – 248, 253, 265 – 267)

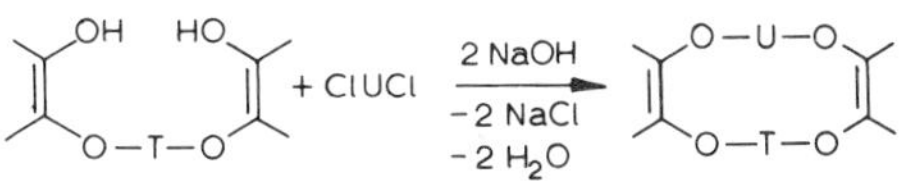

(compounds 239 – 241, 244, 254, 256, 258, 270, 272, 273, 284 – 288, 290, 291, 295, 297, 302, 305, 307)

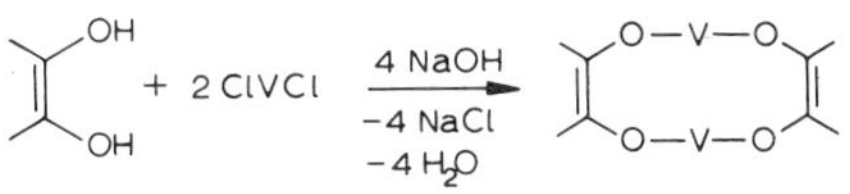

(compounds 270, 271, 289, 299, 301 – 303)

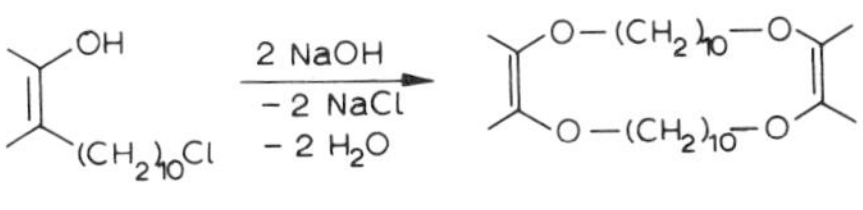

(compound 245)

56

(5) Hydrogenation of aryl-containing cyclic polyethers in the presence of a rhenium catalyst (Re_2O_3).
(compounds 238, 242, 243, 249, 251, 255, 257, 266, 267, 274–276, 296, 301, 304, 306, 308).

The second method is the most reliable for preparing cyclic polyethers with two or more aromatic rings, but the third is also frequently employed because of the greater availability of the starting materials. Saturated polyethers obtained by hydrogenation of the corresponding aromatic compounds have several asymmetric centers and may therefore be mixtures of numerous diastereoisomers. The actual number, however, has turned out to be less than the maximum because stereospecificity of the hydrogenation reaction leads to products with *cis* fusion of the polyether and alicyclic rings [182, 260]. The diastereoisomers of compound (274), so called isomer (A) and isomer (B) (the racemic and *meso* forms, respectively, Fig. 15) have been separated by chromatography on alumina [273, 442, 744].

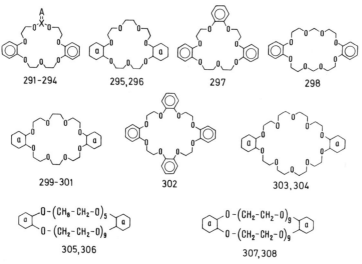

Fig. 14. Structure of cyclic polyethers (291–308).

No.	X	A	No.	a	No.	a
291	C	H_2	295	Benzo	303	Benzo
292	C	$(CH_2)_5$	296	Cyclohexylene	304	Cyclohexylene
293	C	O	299	Benzo	305	Benzo
294	S	O	300	Cyclohexylene	306	Cyclohexylene
			301	2,3-Naphtho	307	Benzo
					308	Cyclohexylene

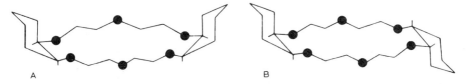

Fig. 15. A and B isomers of dicyclohexyl-18-*crown*-6 (274). A, A isomer, *cis-syn-cis*-conformation; B, B isomer, *cis-anti-cis*-conformation.

In order to assign the NMR signals of compound (271) its derivative with deuterated methylenes adjacent to the phenolic oxygens was synthesized [1093].

Dale, Christiansen and Green obtained cyclic polyethers not containing side chains by a modification of method 1, using ditosyl compounds instead of the chlorides [180, 181, 323].

$$HO-(CH_2CH_2O)_{n-m}-H + TosO-(CH_2CH_2O)_m-Tos \xrightarrow[-\ 2KOTos]{KOBu^t}$$

$$n = 5\ (309)$$
$$n = 6\ (259)$$
$$n = 7\ (310)$$
$$n = 8\ (311)$$

$$\overline{}-(CH_2CH_2O)_n\overline{}$$

The high yields of the polyethers obtained thereby are attributed by the authors to the "matrix effect" of the potassium ions present in the reaction mixture. The method has also been used for the synthesis of the aza derivative of 18-*crown*-6 (1-aza-4,7,10,13,16-pentaoxycyclo-octadecane, 261) [322].

$$\xrightarrow[45\%]{KOBu^t}$$

$$\xrightarrow[65\%]{HCl}$$

(261)

A number of cyclic polyether derivatives with modified aromatic rings were prepared for use in the synthesis of metal-complexing polymers. Thus, on nitrating the ether (270) a mixture of *cis* and *trans* dinitro derivatives was formed which, separated by recrystallization and reduced to the corresponding diamines (314) and (315), were polycondensed with isophthalyl and terephthalyl chlorides [257]. Radical or anionic polymerization of the vinyl derivatives of ethers (240) and (263) afforded poly(vinylbenzo-15-*crown*-5) (316) and poly(vinylbenzo-18-*crown*-6) (317), with a degree of polymerization $n > 300$ [504].

Bromination of the dibenzo-18-*crown*-6 affords the dibromo derivative (270a) together with ~10% of the mono and ~7% of the tribromo derivatives [892].

Starting with the appropriate vicinal mercaptophenols or dithiols, Pedersen, using methods similar to 1 and 3, has synthesized thia analogs (252, 268, 269, 277—280) of the polyesters containing 2—4 sulfur atoms in the molecule [742]*. Also Frensdorff has investigated the thia analog (260), but did not indicate the method of its preparation [272]. In the course of synthesis of macrobicyclic complexones (see Part I.E.) Lehn *et al.* prepared several cyclic polyethers (262, 318—320) each with two NH groups in the ring [204]; other such polyethers with amino functions (261, 262, 281—283) are mentioned in the paper [272] without giving the methods of synthesis. By means of the method of Lehn *et al.* synthesis has also been carried out of the amino ether (259a) [1032] and of several macrocyclic complexones with five ligand groups (ether, sulfide, amino and/or methyl amino) [745].

* Several macrocyclic polysulfides had been described earlier by other workers (see, for instance, [71, 72, 185, 554, 667, 822—824, 1018]), but their behavior towards alkali ions was not investigated.

Recently, the first optically active metal-complexing polyethers were synthesized starting with L-proline and D-ψ-ephedrine [1097].

Whereas the cation-complexing depsipeptides, peptides and depsides, have found application mainly in biological and pharmaceutical studies, interest in the cyclic polyethers, at least in the initial stages of study, was centered mostly on their use in organic chemistry and technology as effective, chemically stable and readily accessible complexones of a wide range of usefulness. Apparently this is the reason why the complexing properties of the cyclic polyethers have been investigated far more than their biological activity. Regarding the latter, it is only known that small doses of compound (274) cause irreversible damage to the cornea and that large doses (300 mg/kg perorally and 130 mg/kg subcutaneously) are lethal to experimental animals (rats) [738].

The ability of cyclic polyethers to form complexes has been qualitatively demonstrated on a large number of uni-, di- and trivalent metal ions, including members of group IA (Li^+, Na^+, K^+, Rb^+, Cs^+), Group IB (Ag^+), Group IIA (Mg^{2+}, Ca^{2+}, Sr^{2+}, Ba^{2+}), Group IIB (Zn^{2+}, Cd^{2+}, Hg^+, Hg^{2+}), Group IIIA (La^{3+}), Group IIIB (Tl^+), Group IVA (Ti^{3+}), Group IVB (Pb^{2+}), Group VA (V^{3+}), and Group VIII (Fe^{2+}, Fe^{3+}, Co^{2+}, Co^{3+}, Ni^{2+}) of Mendeleev's Periodic Table [738, 980]. Some of the thia analogs also complexed with gold ions [742]. On the other hand, no signs of cyclic polyether interaction with Cu^+, Be^{2+} or Mn^{2+} have been observed [732]. As well as with metal ions the cyclic polyethers effectively interact with H^+ [890—892], H_3O^+ [440], with NH_4^+ and with substituted ammonium salts containing $R—NH_3^+$, but not $R_2—NH_2^+$, $R_3—NH^+$ or $R_4—N^+$ groups (Table 23) [738].

TABLE 23

INTERACTION OF DIBENZO-18-*crown*-6 (270) WITH SUBSTITUTED AMMONIUM SALTS [738]

Cation	Ultraviolet absorption spectrum of ether (270)
$HONH_3^+$	spectral shift
$NH_2NH_3^+$	spectral shift
$(CH_3)_2CHCH_2NH_3^+$	spectral shift
$(CH_3)_2CHCH_2NH_3^+$	spectral shift
$(CH_3)_3CNH_3^+$	spectral shift
$C_6H_5NH_3^+$	spectral shift
$NH_2CH_2CH_2NH_3^+$	spectral shift
$(CH_3)_4N^+$	no changes
(morpholine ring) O⟨ ⟩NH_2^+	no changes
(morpholine ring) O⟨ ⟩$\overset{+}{N}H{-}CH_3$	no changes
(piperazine ring) N⟨ ⟩$\overset{+}{N}H$	no changes
$HOOCCH_2NH_3^+$	spectral shift
$[NH_2C(=NH_2)NH_2]^+$	spectral shift
$[(CH_3)_2NC(=NH_2)N(CH_3)_2]^+$	no changes

From the ultraviolet spectral data and extraction data it followed that the broadest and most readily complexing tendencies are manifested by 18-*crown*-6 derivatives (for instance, compounds 270 and 274) [738—742], and this has been fully confirmed in a quantitative study of the complexing properties of cyclic polyethers in solution [272, 441, 442, 744, 811]. A comparison of the data summarized in Tables 1, 2, 4, 5, 14, 21 and 24 shows that, with respect to stability of the alkali metal complexes in methanol and in water, the cyclic polyethers as a whole markedly exceed the metal-binding macrocyclic compounds discussed in the previous sections. Among the ethers are compounds selectively complexing Na^+ ions (compound 242) and K^+ ions (compounds 259, 266, 270, 274, 275, 295 and 303). However, it should be borne in mind that the complexing selectivity is highly solvent-dependent so that, for instance, compounds (271) and (274) which like compound (270) show a noticeable K/Na selectivity in methanol and water, display the following order of affinity in tetrahydrofuran: $Na^+ \geqslant K^+ > Cs^+ > Li^+$ (see Part III.C) [1093].

TABLE 24

DATA ON THE STABILITY OF 1:1 CYCLIC POLYETHER COMPLEXES WITH UNIVALENT AND BIVALENT IONS (25°C) [206, 272, 442]

No	Cyclic polyether (abbreviated name)	Li⁺ H₂O lgK	Li⁺ H₂O K (l·mole⁻¹)	Na⁺ H₂O lgK	Na⁺ H₂O K (l·mole⁻¹)	Na⁺ CH₃OH lgK	Na⁺ CH₃OH K (l·mole⁻¹)	K⁺ H₂O lgK	K⁺ H₂O K (l·mole⁻¹)	K⁺ CH₃OH lgK	K⁺ CH₃OH K (l·mole⁻¹)	Cs⁺ H₂O lgK	Cs⁺ H₂O K (l·mole⁻¹)	Cs⁺ CH₃OH lgK	Cs⁺ CH₃OH K (l·mole⁻¹)	NH₄⁺ H₂O lgK	NH₄⁺ H₂O K (l·mole⁻¹)	Ag⁺ H₂O lgK	Ag⁺ H₂O K (l·mole⁻¹)
242	Dicyclohexyl-14-crown-4	<1.0	<10	—	—	2.18	150	—	—	1.30	20	—	—	—	—	—	—	—	—
249	Cyclohexyl-15-crown-5	—	—	—	—	3.71	5100	0.6	4.0	3.58	3800	—	—	2.78	600	—	—	—	—
259	18-crown-6	—	—	<0.3	<2.0	4.32	20 900	2.06	115	6.10	1 260 000	0.8	6.3	4.62	41 700	1.1	12	1.6	40
259ᵃ		—	—	<0.3	<2.0	3.26ᵇ	1820ᵇ	—	—	4.38ᵇ	24 000ᵇ	—	—	—	—	—	—	—	—
260		—	—	—	—	—	—	—	—	1.65	45	—	—	—	—	—	—	4.34	22 000
261		—	—	—	—	—	—	—	—	3.90	7900	—	—	—	—	—	—	3.3	2100
262		—	—	—	—	—	—	—	—	2.04	110	—	—	—	—	—	—	7.8	63 100 000
266	Cyclohexyl-18-crown-6	<0.7	<5	0.8	6.3	4.09	12 300	1.90	80	5.89	776 000	0.8	6.3	4.30	20 000	1.1	12	1.8	63
270	Dibenzo-18-crown-6ᶜ·ᵈ	—	—	—	—	4.36	22 900	—	—	5.00	100 000	—	—	3.55	3550	—	—	—	—
274A	Dicyclohexyl-18-crown-6 (A isomer)ᵉ	0.6	4.0	1.7	50	4.08	12 000	2.18	150	6.01	1 020 000	1.25	18	4.61	40 700	1.4	25	2.3	200
274B	Dicyclohexyl-18-crown-6 (B isomer)ᶠ	—	—	1.4	25	3.68	4800	1.78	60	5.38	240 000	0.9	7.9	3.49	3100	0.8	6.3	1.7	50
275	Di(tert-butylcyclohexyl)-18-crown-6	<0.9	<8	1.42	26	—	—	2.08	120	—	—	0.9	7.9	—	—	1.28	19	—	—
281		—	—	—	—	—	—	—	—	3.20	1600	—	—	—	—	—	—	—	—
282		—	—	—	—	—	—	—	—	4.10	12 600	—	—	—	—	—	—	—	—
283		—	—	—	—	—	—	—	—	1.63	42	—	—	—	—	—	—	—	—
295	Dibenzo-21-crown-7	—	—	—	—	2.40	250	—	—	4.30	20 000	—	—	4.20	15 900	—	—	—	—
296	Dicyclohexyl-21-crown-7	—	—	—	—	—	—	—	—	—	—	—	—	—	—	—	—	—	—
299	Dibenzo-24-crown-8	—	—	—	—	—	—	—	—	3.49	3100	1.9	80	3.78	6030	—	—	—	—
300	Dicyclohexyl-24-crown-8	—	—	—	—	—	—	—	—	—	—	1.9	80	—	—	—	—	—	—
303	Dibenzo-30-crown-10ᵍ	—	—	—	—	2.00	100	—	—	4.60	29 000	—	—	—	—	—	—	—	—
307	Dibenzo-60-crown-20	—	—	—	—	—	—	—	—	3.90	7900	—	—	—	—	—	—	—	—

ᵃ For Ba²⁺ in 95% aqueous methanol lgK = 6.67 (K = 4700 000).

ᵇ Solutions in 95% aqueous methanol.

ᶜ For ethanolic solutions the following stability constants were obtained: Na⁺, 150 000; K⁺, 1 100 000; Rb⁺, 4300; Cs⁺, 5500 l·mole⁻¹

ᵈ For solutions in water—tetrahydrofuran: K⁺, —lgK = 1.87 (K = 74); Rb⁺, —lgK = 1.35 (K = 22) [811].

ᵉ For Sr²⁺ in aqueous solutions lgK = 3.24 (K = 1740); Ba²⁺, lgK = 3.57 (K = 3720).

ᶠ For Sr²⁺ in aqueous solutions lgK = 2.64 (K = 440); Ba²⁺, lgK = 3.27 (K = 1860).

ᵍ For solutions in water—tetrahydrofuran: K⁺, —lgK = 1.35 (K = 22); Rb⁺, —lgK = 1.56 (K = 35) [811].

Complexes with macrocyclic compound:cation ratios other than 1:1 (2:1, 3:2 or 1:2) [179, 181, 259, 740] are encountered more frequently among the cyclic polyethers than among other metal-binding macrocyclic compounds. A stronger tendency to form 2:1 complexes in solution is displayed by the larger cations (see Table 25).

TABLE 25

DATA ON THE STABILITY OF 2:1 COMPLEXES OF CROWN ETHERS WITH K^+ AND Cs^+ [272]

No.	Compound	Cation	Log stability constant $(CH_3OH, 25°C)$	Ratio of stability constants of 2:1 complexes to corresponding constants for 1:1 complexes
249	Cyclohexyl-15-*crown*-5	K^+	1.88	0.02
249	Cyclohexyl-15-*crown*-5	Cs^+	1.91	0.13
259	18-*crown*-6	Cs^+	1.30	0.0005
266	Cyclohexyl-18-*crown*-6	Cs^+	1.52	0.002
270	Dibenzo-18-*crown*-6	Cs^+	2.92	0.23
274A	Dicyclohexyl-18-*crown*-6 (A isomer)	Cs^+	0.59	0.0001
274B	Dicyclohexyl-18-*crown*-6 (B isomer)	Cs^+	0	0
295	Dibenzo-21-*crown*-7	Cs^+	1.9	0.005

A large number of cyclic polyether complex salts has been isolated in the crystalline state (Table 26). For salts that decompose during crystallization Pedersen has proposed a number of procedures permitting their preparation in the crystalline form. Such methods are (a) heating the polyether with the initial salt in the absence of solvent, (b) crystallization of the polyether complex in the presence of excess amounts of soluble metal salts, (c) evaporation under vacuum of a solution of equimolar amounts of polyether and metal salt, and (d) crystallization in a two-phase water—organic solvent system.

Attempts to obtain crystalline cyclic polyether complexes from the fluorides, phosphates, carbonates and in many cases the nitrates and acetates, have frequently proved unsuccessful, apparently owing to the high lattice energy of the corresponding metal salts. On the other hand, the crystalline cyclic polyether complex salt can sometimes be isolated even when spectral methods do not reveal complex formation in solution (for instance, compound (270) with $CdCl_2$ [738]). As one can see from Table 26, the melting point of the crystalline salts is sometimes markedly higher than that of the initial polyether, facilitating their separation. Usually in

TABLE 26

CRYSTALLINE COMPLEXES OF CYCLIC POLYETHERS WITH METAL SALTS [181, 738, 740]

No.	Cyclic polyether	m.p.(°C)	Salt	Complex salt m.p.(°C)	Polyether to salt ratio
240	Dibenzo-14-crown-4	152	LiNCS	300	1:1
246	Benzo-15-crown-5	79	NaI	152—156	1:1
246	Benzo-15-crown-5	79	NaNCS	162—165	1:1
246	Benzo-15-crown-5	79	NaNCS	~80 softening; 150—164	2:1
246	Benzo-15-crown-5	79	KNCS	176	2:1
246	Benzo-15-crown-5	79	NH_4NCS	131—132	2:1
246	Benzo-15-crown-5	79	$CsNO_3$	127	2:1
246	Benzo-15-crown-5	79	$AgNO_3$	134—135	1:1
246	Benzo-15-crown-5	79	$Ba(NCS)_2$	169—171	2:1
247	Butylbenzo-15-crown-5	44	LiNCS	169—174	1:1
247	Butylbenzo-15-crown-5	44	NaBr	77—84	1:1
247	Butylbenzo-15-crown-5	44	NaI	86—134	1:1
247	Butylbenzo-15-crown-5	44	NaNCS	163—167	1:1
247	Butylbenzo-15-crown-5	44	KNCS	—	1:1
249	Cyclohexyl-15-crown-5	Oil	NaNCS	94—110	1:1
250	Butylcyclohexyl-15-crown-5	Oil	KI	100—133	1:1
254	Dibenzo-15-crown-5	115	KNCS	143—144	2:1
259	18-Crown-6	39	$NaNCS \cdot H_2O$	142	1:1
259	18-Crown-6	39	$NaO_3SC_6H_4CH_3$	40 (decomp.)	3:2
259	18-Crown-6	39	KF	57 (hygroscopic)	1:1
259	18-Crown-6	39	$KCl \cdot H_2O$	98	1:1
259	18-Crown-6	39	KBr	196	1:1
259	18-Crown-6	39	KI	>260	1:1
259	18-Crown-6	39	KNCS	190	1:1
259	18-Crown-6	39	$KO_3SC_6H_4CH_3$	164	1:1
259	18-Crown-6	39	$RbNCS \cdot H_2O$	193	1:1

259	18-*Crown*-6	39	$RbO_3SC_6H_4CH_3$	153	1:1
259	18-*Crown*-6	39	$CsNCS \cdot H_2O$	200	1:1
259	18-*Crown*-6	39	$CsO_3SC_6H_4CH_3$	130	1:1
259	18-*Crown*-6	39	$Ca(O_3SC_6H_4CH_3)_2$	decomposition	2:1
259	18-*Crown*-6	39	$Sr(O_3SC_6H_4CH_3)_2$	decomposition	1:1
259	18-*Crown*-6	39	$Ba(O_3SC_6H_4CH_3)_2$	decomposition	1:1
263	Benzo-18-*crown*-6	44	KNCS	134.5–136.5	1:1
263	Benzo-18-*crown*-6	44	CsNCS	145–146	3:2
264	Butylbenzo-18-*crown*-6	37	KNCS	134–136	1:1
266	Cyclohexyl-18-*crown*-6	Oil	NH_4NCS	124–147	1:1
266	Cyclohexyl-18-*crown*-6	Oil	$Ba(NCS)_2$	282.5	1:1
267	Butylcyclohexyl-18-*crown*-6	Oil	KI	52–98	1:1
270	Dibenzo-18-*crown*-6	164	LiI_3	134–139	1:1
270	Dibenzo-18-*crown*-6	164	NaNCS	230–232	1:1
270	Dibenzo-18-*crown*-6	164	$NaNO_2$	154–157	1:1
270	Dibenzo-18-*crown*-6	164	KI	232–234	1:1
270	Dibenzo-18-*crown*-6	164	$KI_{1,5}$	152–153	1:1
270	Dibenzo-18-*crown*-6	164	KI_2	156–238 (decomp.)	1:1
270	Dibenzo-18-*crown*-6	164	KI_3	258–267 (decomp.)	1:1
270	Dibenzo-18-*crown*-6	164	KNCS	248–249	1:1
270	Dibenzo-18-*crown*-6	164	KNCS	164–233	2:1
270	Dibenzo-18-*crown*-6	164	$(CH_3)_3CCOOK$	—	1:1
270	Dibenzo-18-*crown*-6	164	NH_4NCS	187–189	1:1
270	Dibenzo-18-*crown*-6	164	CH_3NH_3NCS	128–155	1:1
270	Dibenzo-18-*crown*-6	164	RbNCS	184–185	1:1
270	Dibenzo-18-*crown*-6	164	RbNCS	175–176	2:1
270	Dibenzo-18-*crown*-6	164	CsNCS	146–147	2:1
270	Dibenzo-18-*crown*-6	164	CsNCS	145–146	3:2
270	Dibenzo-18-*crown*-6	164	CsI	115–116	2:1
270	Dibenzo-18-*crown*-6	164	$CaCl_2$	> 300	1:1
270	Dibenzo-18-*crown*-6	164	$Ba(OH)_2$	~180 (softening)	1:1
270	Dibenzo-18-*crown*-6	164	BaI_6	153–159 (decomp.)	1:1
270	Dibenzo-18-*crown*-6	164	$Ba(NCS)_2$	> 360	1:1
270	Dibenzo-18-*crown*-6	164	$CdCl_2$	> 360	1:1

TABLE 26—*continued*

No.	Cyclic polyether	m.p.($^\circ$C)	Salt	Complex salt m.p.($^\circ$C)	Polyether to salt ratio
270	Dibenzo-18-crown-6	164	HgCl$_2$	238–249	1:1
270	Dibenzo-18-crown-6	164	PbAc$_2$	167–198	1:1
272	Bisbutylbenzo-18-crown-6	137	KNCS	186–189	1:1
272	Bisbutylbenzo-18-crown-6	137	CsNCS	108–116	2:1
274	Dicyclohexyl-18-crown-6	36–56	KI	123–170	1:1
274	Dicyclohexyl-18-crown-6	36–56	KNCS	72–122	1:1
274	Dicyclohexyl-18-crown-6	36–56	KI$_3$	194–195	1:1
274	Dicyclohexyl-18-crown-6	36–56	RbI$_3$	193–195	1:1
274	Dicyclohexyl-18-crown-6	36–56	CsI$_3$	112–114	3:2
274	Dicyclohexyl-18-crown-6	36–56	NH$_4$NCS	107–110	1:1
295	Dibenzo-21-crown-7	107	CsNCS	~40 (softening)	2:1
299	Dibenzo-24-crown-8	114	KNCS	113–114	1:1
299	Dibenzo-24-crown-8	114	CsNCS	88–89	2:1
303	Dibenzo-30-crown-10	107	KNCS	176–177.5	1:1

such cases 1:1 macrocyclic compound—metal complexes precipitate out owing to their higher stability constants (Tables 24 and 25). Nevertheless, by special procedures it is sometimes possible to obtain crystalline preparations with the polyether:metal salt with 2:1 and 3:2 stoichiometry [740], which, as in solution, the majority of compounds most readily display with the bulky cesium ion (Table 26). An exception is 18-*crown*-6 (259) with which Na^+ and Ca^{2+} give non-equimolar complexes [181].

Cyclic polyethers retain their ability to complex metal ions on insertion into a polymer chain, but the complexing selectivity and the nature of the macrocyclic compound—cation interaction are somewhat modified. For instance, polymer films formed by polycondensation of the ethers (314) and (315) effectively extract potassium ions from aqueous solutions, but in contrast with the monomeric cyclic polyether, do not interact with Ca^{2+}, Ba^{2+} and Cs^+ ions [257]. Extraction experiments showed the polymeric cyclic polyethers (316) and (317) to be more efficient complexones than the corresponding monomers (246) and (263). Moreover, the polymer state is conducive to the formation of sandwich compounds (see Part III.B) [504].

Pedersen also showed [743] that cyclic polyethers cocrystallize with neutral organic molecules. Thus, on heating dibenzo-18-*crown*-6 (270) with a saturated methanolic solution of thiourea crystals of the composition $(270) \cdot H_2 N \overset{\overset{\displaystyle S}{\|}}{C} NH_2$ will precipitate after cooling. Other polyethers (246, 254, 256, 263, 272, 276 and 298) also form crystalline complexes with thiourea, and the ether (274) complexes as well with a number of derivatives of the latter, with thiobenzamide, 2-thiazolidinethione and with 1-phenylsemi-carbazide, the polyether:second component ratio varying from 3:1 to 1:6. Cyclic polyether complexes with Na^+ and K^+ can further add 1 to 6 molecules of thiourea, but with cesium and ammonium complexes the thiourea displaces the inorganic salt. The mechanism underlying these inter-actions is still unclear.

I.E. Macrobicyclic diaminopolyethers

Lehn and collaborators have prepared an interesting series of bicyclic polyethers (321—329g) with amino nitrogen atoms at the bridgehead [135, 203]. The synthesis was carried out as follows (Scheme 8). The appropriate diamines and oxadiacyl chlorides were subjected to condensation under high dilution conditions. Reduction of the resultant diamides by lithium alumo-hydride afforded the cyclic oxadiamines (262) and (318—320). The latter were again condensed with diacyl chlorides; the resulting bicyclic diamides were reduced by diborane and the products on refluxing with 6N

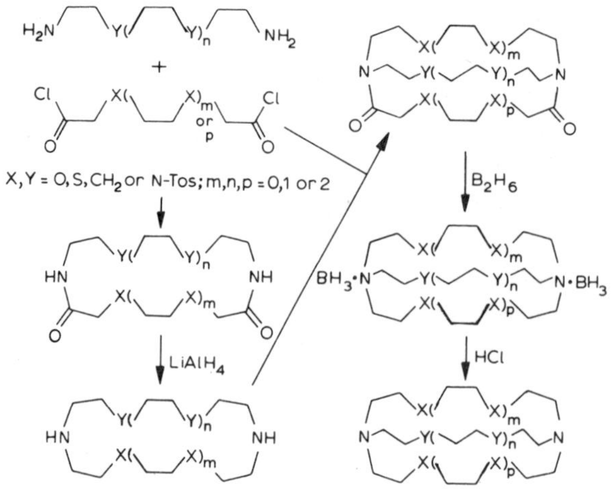

Scheme 8

(318) X=Y=O; m=n=0
(262) X=Y=O; m=n=1
(319) X=Y=O; m=1; n=2
(320) X=Y=O; m=n=2
(320a) X=O; Y=S; m=n=1
(320b) X=Y=S; m=n=1
(320c) X=O; Y=N–Tos; m=n=1

(321) X=Y=O; m=n=p=0
(322) X=Y=O; m=n=O; p=1
(323) X=Y=O; m=n=1; p=0
(324) X=Y=O; m=n=p=1
(324a) X=O; Y=CH$_2$; m=n=p=1
(325) X=Y=O; m=n=1; p=2
(326) X=Y=O; m=1; n=p=2
(327) X=Y=O; m=n=p=2
(328) X=O; Y=S; m=n=p=1
(328a) X=S; Y=O; m=n=p=1
(328b) X=Y=S; m=n=p=1
(329) X=O; Y=N–Tos; m=n=p=1
(329a) X=N-Tos; Y=O; m=n=p=1
(329b) X=O; Y=NH; m=n=p=1
(329c) X=NH; Y=O; m=n=p=1
(329d) X=O; Y=N—COOCH$_3$; m=n=p=1
(329e) X=N—COOCH$_3$; Y=O; m=n=p=1
(329f) X=O; Y=N—CH$_3$; m=n=p=1
(329g) X=N—CH$_3$; Y=O; m=n=p=1

hydrochloric acid yielded the bicyclic amines (321) and (324—327). The compounds (322) and (323) described in paper [536] were apparently synthesized in a similar way.

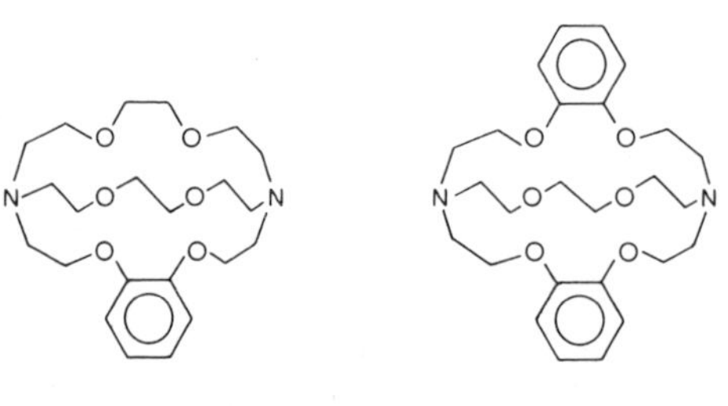

324b   324c

An analogous scheme but with the thia derivatives and the *N*-tosyl analog instead of the oxadiamines and the oxadiacyl chlorides was used for making oxathia macrobicyclic amines (328—328b) and the *N*-tosyl derivatives (329 and 329a) [205, 535]. Detosylation of the macrobicyclic compounds (329 and 329a) yielded the amines (329b and 329c) which on treatment with ethyl chlorocarbonate were converted into carbomethoxy derivatives (329d and 329e), and after reduction by lithium alumohydride to the *N*-methyl compounds (329f and 329g).

Condensation of the diamine (318) with diglycolyl dichloride yielded, besides the corresponding bicyclic diamide, a product of doubled molecular weight which, by conventional methods, was transformed into the tricyclic tetramine (330) (Scheme 9) [134].

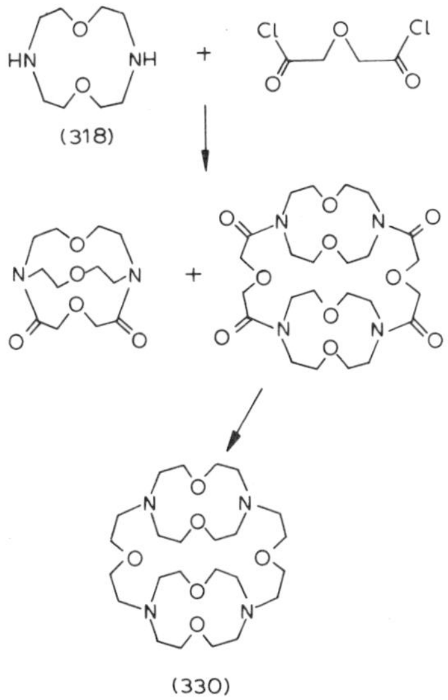

(318)

(330)

Scheme 9

The most important feature in the structure of the macrobicyclic polyaminoethers responsible for their being classified into a group separate from the cyclic polyethers, is the participation of the ligand heteroatoms in the covalent bridge structures, easily leading to a three-dimensional cavity in the center of the molecule. It is doubtless this property which is responsible for the exceptional stability of the complexes of this group of compounds and for the variety of atoms with which they complex. The French authors proposed the name cryptates for such complexes (from the Greek κρυπτοσ, hidden, secret, or the Latin *crypta*, depression, cave [204, 534]). The formation of cryptates of compounds (324—330) with lithium, sodium, potassium, rubidium, cesium, monovalent thallium, silver, ammonium, calcium, strontium, barium and divalent lead salts, has been qualitatively established in aqueous and chloroform solutions by NMR [204, 205]. Cryptates are readily soluble in organic solvents. For example, the complex,

[(324)·K$^+$] MnO$_4^-$, dissolves in even such a non-polar solvent as benzene [204]. Quantitative measurements, mainly in aqueous solutions (Tables 27 and 28), have shown that many cryptates are considerably more stable than the complexes of other known metal-binding macrocyclic compounds with alkali and alkali earth ions. In particular, the effectiveness of "cryptation" is well illustrated by the fact that such a classically insoluble compound as

TABLE 27

DATA ON THE STABILITY OF UNIVALENT ION CRYPTATES IN WATER [134, 206, 536]

Macrobicyclic ligand	lg K				
	Li$^+$	Na$^+$	K$^+$	Rb$^+$	Cs$^+$
(322)	4.30	2.55—2.8	< 1.0 2.7^a	< 2.0	< 2.0
(323)^b	2.50	5.40 > 9.0^a	3.9	2.55	< 2.0
(324)^c	< 2.0	3.7—3.9 > 9.0^a	5.40 9.45^d	4.35—4.8	< 2.0
(324a)	—	3.0^d	4.35^d	—	—
(324b)^d	—	4.0^d 7.4^d	4.9 9.05^d	3.4	—
(324c)^e	—	7.3^d	8.6^d	—	—
(325)	< 2.0	1.65 4.80^a	2.1—2.2	2.05	1.8—2.2
(326)	< 2.0	< 2.0 2.8^a	< 2.0	≤ 0.7	< 2.0
(327)	< 2.0	< 2.0	< 2.0	≤ 0.5	< 2.0
(328b)^e	—	1 4.5^a	— 5.8^a	— 6.2^a	1.7 > 6.0^a

^{a} Data for methanolic solutions.
^{b} For Ag$^+$, lg K = 10.6.
^{c} For Ag$^+$, lg K = 9.6; for Tl$^+$, lg K = 6.3.
^{d} Data for solutions in 95% aqueous methanol.
^{e} For Ag$^+$, lg K = 6 (for CH$_3$OH, lg K = 9.5).

barium sulfate can be solubilized in water in the presence of the aminoether (324) [204].

In conformity with the size of the intramolecular cavity, maximum stability of the cryptates on passing from (322) to (330) shifts from the Li$^+$ to the Cs$^+$ complex (Table 27). A similar tendency is observed with cations of the alkaline earth metals. Of all the cations investigated (Li$^+$, Na$^+$, K$^+$, Rb$^+$, Cs$^+$, Mg^{2+}, Ca^{2+}, Sr^{2+} and Ba^{2+}), compound (321) complexes only lithium ions. Interestingly, this compound forms an amazingly stable

TABLE 28

DATA ON THE STABILITY OF BIVALENT ION CRYPTATES IN WATER
[206, 536]

Macrobicyclic ligand[a]	lg K		
	Ca^{2+}	Sr^{2+}	Ba^{2+}
(322)	2.80	< 2.0	< 2.0
(323)	6.95	7.35	6.30
(324)[b]	4.40	8.0	9.50
(324a)	—	—	11.5[c]
			< 2.0[c]
(324b)	3.8	6.9	7.4
(324c)	—	—	11.05[c]
			8.5[c]
(325)	~ 2.0	3.40	6.00
(326)	< 2.0	~ 2.0	3.65
(327)	< 2.0	< 2.0	—

[a] For compounds (322—324) and (325—327), $lg K_{Mg^{2+}} < 2$.
[b] For Pb^{2+}, lg K = 12.
[c] Data for solutions in 95% aqueous methanol.

dication (321a) on protonation (Fig. 16). It is not affected by acids or alkalis and does not give up a proton even on standing for many days at room temperature in 5N KOH solution. Only on heating in this solution at 60°C for 80 hours is it converted into the monocation (321b) with a yield of 50%.

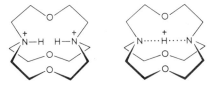

Fig. 16. Structure of di- and monocations (321a) and (321b).

According to NMR data the proton of the monocation (321b) can undergo exchange between the two N atoms. The mono- and dications may be regarded as kinds of protonic cryptates [135].

The cryptates (330)·Ag$^+$ and (330)·Tl$^+$ are capable of binding a second metal ion to form binuclear complexes; the second stability constant of the silver complex in methanol exceeds 10^6 l·mole^{-1} [134].

Cryptates readily form crystalline salts. Some of these (NaI·(324) [634], KI·(324) [634], RbNCS·(324)·H$_2$O [632], CsNCS·(324)·H$_2$O [634], Ba(NCS)$_2$·(324)·H$_2$O [633] and Ba(NCS)$_2$·(325)·H$_2$O [633]) have been successfully subjected to X-ray analysis. A 2:1 complex with cobaltous rhodanide has been described for the oxathia macrobicyclic compound (330), apparently of the structure $[(330) \cdot Co]^{2+} [Co(NCS)_4]^{2-}$ [205].

The biological properties of the macrobicyclic polyaminoethers are practically unknown. The only exception is compound (324) which has been found to be applicable for elimination of ^{85}Sr from the animal organism (experiments on rats) [672].

I.F. The nigericin antibiotics

The antibiotics considered in this section, nigericin, X-206, X-537 A, monensin, dianemycin, etc., form a group of related compounds interacting with alkali metal ions like the metal-binding macrocyclic compounds discussed heretofore, but differing considerably from them in structure, physicochemical properties and biological action. Since nigericin was the first of these compounds to be discovered (1950), it lent its name to the entire family. All the nigericin antibiotics contain tetrahydrofuran and tetrahydropyran rings and a carboxyl function. A characteristic feature is their ability to form with various cations water-insoluble salts of exceptionally high solubility in organic solvents. A number of physicochemical characteristics of the antibiotics and of their sodium salts are given in Table 29. The carboxyl group of the nigericin antibiotics dissociates under complexing conditions and its charge neutralizes that of the complexed ion. Hence, contrary to the other complexes, these are electrically neutral.

Owing to the complicated structure and large number of asymmetric centers, chemical study of the nigericin antibiotics has shed little light on their structure, not to mention their stereochemistry. Structural elucidation was achieved by X-ray analysis of the crystalline complexes, which also yielded valuable spatial data and mechanistic information on their complexing reactions. It was found that, while the compounds of this group are not formally macrocyclic compounds, a macrocyclic system results from the formation of stable intramolecular "head to tail" hydrogen bonds (see Part III.A.5). Because of this the term metal-binding macrocyclic compounds has also been extended to the nigericin antibiotics. In what follows, a brief description will be given of individual compounds of this group, separated into subgroups in conformity with their structural characteristics and with historical considerations.

I.F.1. Nigericin (331) and grisoryxin (332)

Nigericin (Fig. 17) was first isolated in 1950 from the metabolites of an unidentified streptomycete found in Nigerian soil [352, 353], the streptomycete later receiving the name of *Streptomyces* Nig. 1 [53].

TABLE 29

PHYSICOCHEMICAL CHARACTERISTICS OF THE NIGERICIN ANTIBIOTICS AND THEIR SODIUM SALTS

No.	Compound	Empirical formula	Molecular weight	m.p.(°C)	$[\alpha]_D$	c	Solvent
331	Nigericin	$C_{40}H_{68}O_{11}$	724	185 [516, 517, 932]	+35.2° +36.2	0.9 0.8	CHCl$_3$ [1054] CHCl$_3$ [517]
	Na salt of nigericin	$C_{40}H_{67}O_{11}Na$	744	255 [516] 254 [353]	+9.21 +7.8	1.03 1.05	CH$_3$OH [1054] CH$_3$OH [1054]
332	Grisoryxin	$C_{40}H_{68}O_{10}$	708	amorph. [420]	+16	4	CH$_3$COCH$_3$ [420]
333	X-206	$C_{45}H_{78}O_{13}$	826	133–145 [76] 128 [57] 190 [76] 187 [57]	+15.0	2	CH$_3$OH [57]
	Na salt of X-206	$C_{45}H_{77}O_{13}Na$	848				
334	X-537 A	$C_{34}H_{54}O_6$	558	114 [1054] 100–109 [57]	−7.2	1	CH$_3$OH [57]
	Na salt of X-537 A	$C_{34}H_{53}O_6Na$	570	171 [57, 1054]			
335	Monensin	$C_{36}H_{62}O_{11}$	670	105 [5, 351]	+47.7	1	CH$_3$OH [351]
	Na salt of monensin	$C_{36}H_{61}O_{11}Na$	692	269 [312]	+57.3	1	CH$_3$OH [351]
336	Monensin B	$C_{35}H_{60}O_{11}$	656	—	—	—	—
	Na salt of monensin B	$C_{35}H_{59}O_{11}Na$	678	228 [312]	—	—	—
337	Monensin C	$C_{37}H_{64}O_{11}$	684	—	—	—	—
	Na salt of monensin C	$C_{37}H_{63}O_{11}Na$	706	214 [312]	—	—	—
338	Monensin D	$C_{37}H_{64}O_{11}$	684	—	—	—	—
	Na salt of monensin D	$C_{37}H_{63}O_{11}Na$	706	252 [312]	—	—	—
339	Dianemycin	$C_{47}H_{78}O_{14}$	866	157 [312, 350]	+39.9	2	CH$_3$OH [350]
	Na salt of dianemycin	$C_{47}H_{77}O_{14}Na$	888	220 [312] 212 [350]	+37.1	1	CH$_3$OH [350]
	Antibiotic K 178	—	—	187 [576]	−7.8	2	CH$_3$OH
	Na salt of K 178	—	—	260 [576]			

Obtained in yields of only 0.05—0.1 mg/ml, nigericin displayed a very high potency against Gram-positive bacteria, mycobacteria and fungi. In higher concentrations it also inhibits the growth of certain Gram-negative

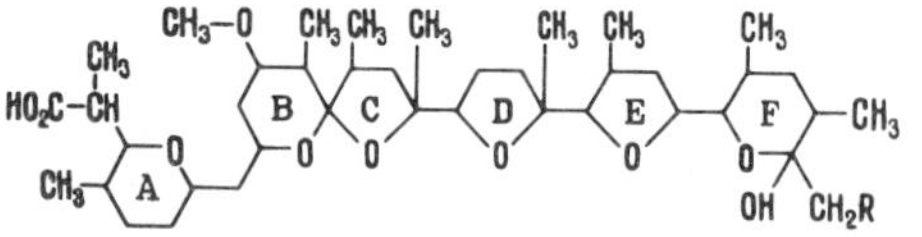

Fig. 17. Nigericin (331, R = OH), grisoryxin (332, R = H).

organisms (Table 30), but has no effect on bacteriophage even in concentrations as high as 500 μg/ml [345]. Like other allied antibiotics, nigericin is successfully used in the prevention and treatment of coccidiosis in chicken* [935].

TABLE 30

ANTIMICROBIAL ACTIVITY OF NIGERICIN [935]

Microorganism	Minimal growth-inhibiting concentration (μg/ml)
Staphylococcus aureus 209 P	0.50
Bacillus subtilis NRRL 558 R	0.25
Bacillus subtilis ATCC 6633	0.12
Micrococcus flavus	0.50
Bacillus mycoides	0.25
Sarcina lutea	0.50
Brucella bronchiseptica	64.0
Escherichia coli NRRL 4348	64.0
Pseudomonas aeruginosa	64.0
Proteus vulgaris	64.0
Mycobacterium 607 sp.	4.0
Mycobacterium smegmatis	2.0
Candida albicans, Emmons 3149	2.0
Trichophyton mentagrophytes	16.0

Although nigericin had been utilized in biochemical studies [244, 314, 315, 527, 623, 845, 846, 888] for a number of years after its isolation and preliminary chemical characterization, its structure remained unknown until 1968 when Steinrauf and collaborators carried out the X-ray analysis of its silver salt [970].

---

* According to Shunnard and Callender [935], in the U.S.A. the damage inflicted by this disease to the poultry industry amounts to 38 million dollars a year.

TABLE 31

ANTIMICROBIAL ACTIVITY OF POLYETHERIN A [932]

Bacteria	Minimal growth-inhibiting concentration (µg/ml)	Fungi	Minimal growth-inhibiting concentration (µg/ml)
Shigella dysenteriae	> 50	Piricularia oryzae	0.8
Salmonella typhosa	> 50	Colletotrichum sp.	3.2
Escherichia coli, Umezawa	> 50	Phytophora infestans	> 100
Pseudomonas aeruginosa	> 50	Corticium sasakii	> 100
Klebsiella pneumoniae	> 50	Cochliobolus miyabeanus	1.6
Bacillus subtilis, PCI-219	1.0	Fusarium oxysporum	> 100
Bacillus anthracis	0.5	Sclerotinia libertiana	6.3
Staphylococcus aureus, FDA 209 P	0.5	Alternaria kikuchiana	> 100
Sarcina lutea, PCI-1001	0.5	Trichophyton mentagrophytes	> 100
Diplococcus pneumoniae, Type I	0.5	Trichophyton rubrum	> 100
Streptococcus hemolyticus, Denken	1.0	Trichophyton purpureum	> 100
Corynebacterium diphteriae, Tront	1.0	Trichophyton ferrugineum	> 100
Mycobacterium 607	5.0	Epidermophyton floccosum	25
Mycobacterium phlei	2.0	Microsporium gypseum	> 100
Mycobacterium smegmatis	5.0	Aspergillus niger	> 100
Mycobacterium avium	2.0	Candida albicans	12.5
Mycobacterium tuberculosis, H$_{37}$Rv	5.0	Cryptococcus neoformans	> 100
		Saccharomyces cerevisiae	12.5

In the meantime a group of Japanese workers had isolated from a culture of *Streptomyces hygroscopicus* E-749 an antibiotic called polyetherin A, with an antibiotic spectrum similar to nigericin (Table 31), but, according to these authors, differing from the latter in physicochemical properties [932]. However, X-ray analysis of the polyetherin A silver salt [517, 930], performed soon after, disclosed no differences in the two antibiotics whose identicalness was confirmed also by direct comparison of the infrared spectra, optical rotatory dispersion curves and chromatographic behavior [971]. It was shown in the same paper that still another antibiotic preparation, isolated in 1951 and at first called X-464 ([57], see Part I.F.2), is also nigericin.

TABLE 32

STABILITY CONSTANTS OF NIGERICIN COMPLEXES WITH SODIUM AND POTASSIUM IONS [569]

Solvent	$l \cdot mole^{-1}$	
	Na^+	K^+
CH_3OH	25000	150000
$CH_3OCH_2CH_2OH-H_2O$	6600	30000
(4:1, v:v)		

Like the other antibiotics of this group, nigericin has no symmetry elements. Seven ether groups are distributed more or less evenly along the hydrocarbon chain and there are two hydroxyl groups and one carboxyl group at its ends. A distinctive feature of nigericin is the presence of a spiroketal grouping linking rings D and E. The ion-selectivity of nigericin is in the following order: $K^+ > Rb^+ > Na^+ > Cs^+ > Li^+$ [23, 258, 386, 448, 569, 785, 786]. Measurement of the stability constants of the complexes in organic solvents showed that the potassium complex of nigericin is more stable than that of all other naturally occurring metal-binding macrocyclic compounds, but that this antibiotic falls behind valinomycin in K/Na-selectivity (cf. Tables 1 and 32). The affinity of nigericin for potassium ions in aqueous solutions is also much higher (10—20-fold) than that of valinomycin [258].

Recently, the isolation from *Streptomyces griseus* of the antibiotic grisoryxin (332) was reported, and was found to be similar in chemical and biological properties to nigericin [279]. X-ray analysis of the silver [8] and thallium [9] salts of grisoryxin showed it to differ from nigericin only in the absence of a hydroxyl group (Fig. 17).

It was found by mass spectrometry [404] that the same structure can be ascribed to the antibiotic K-178, isolated by Hungarian workers in 1964 from several *Streptomyces albus* species [405, 576]. Its antimicrobial spectrum is very similar to that of nigericin. No published data are available on the complexing properties of grisoryxin (sometimes called deoxynigericin).

I.F.2. The antibiotics X-206 (333) and X-537 A (334)

In 1951, Berger *et al.* [57] isolated from unidentified *Streptomyces* species three antibiotics called X-206 (R_0 2-2879), X-464 and X-537 A. Their paper contained a brief physicochemical characterization (results of elementary and functional analyses, of the action of acids and alkalis and of

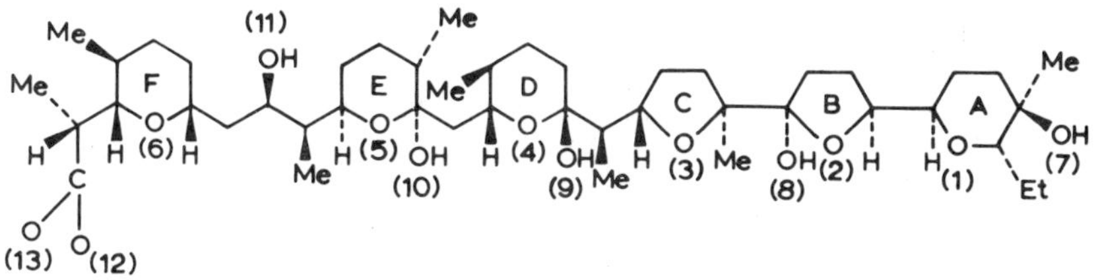

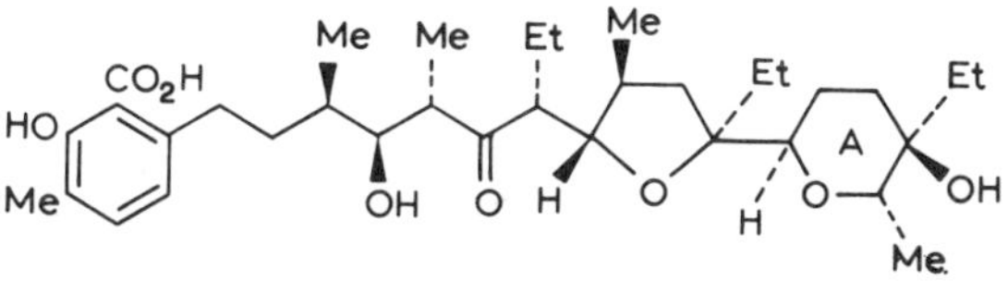

Fig. 18 (above). Antibiotic X-206 (333).
Fig. 19 (below). Antibiotic X-537 A (334).

oxidizing and reducing agents, solubility data, and the ultraviolet and infrared spectra), as well as the results of preliminary antimicrobial tests of these substances. It has already been mentioned in the previous section that X-464 proved to be identical with nigericin [971]; the structure of the other two antibiotics (Figs 18 and 19) was established only in 1970—1971 by the combined use of X-ray and chemical methods [76, 77, 462, 463, 577, 1054]. As one can see from a comparison of formulae (331) and (333), the most important differences between X-206 and nigericin are its higher molecular weight (Table 29), the possession of five (instead of two) hydroxyl groups and the lack of spiroketal grouping. Generally speaking, however, both antibiotics have quite similar structures. As for the antibiotic X-537 A (334), it has only one tetrahydrofuranic ring and one tetrahydropyranic ring, and therefore the lowest molecular weight of the antibiotics

(331—339); also a characteristic feature of this compound is the presence of an aromatic ring and a phenolic hydroxyl.

Only a few papers on the properties of the antibiotics X-206 and X-537 A have been published. Among them are the studies by Westley *et al.* [1055] concerning the biosynthesis of X-537 A, and of Lardy *et al.* on the effect of both substances on mitochondria [247, 314, 315, 526, 527]. X-206 is very close to nigericin in complexing properties and in the lipophilicity of the complexes, whereas X-537 A, while in general displaying very low selectivity, nevertheless preferably complexes cesium ions [786], in which respect it has no analogs among the naturally occurring metal-binding macrocyclic compounds.

I.F.3. *Monensin (335), monensin B (336), monensin C (337) and monensin D (338)*

Despite the only comparatively recent discovery (in 1967) [5] of monensin (335) and its homologs (336—338), it has already become the subject matter of a large number of papers and together with nigericin may be considered to be the best known member of the nigericin antibiotics. Monensin (initially "Antibiotic 3823 A" and then "monensinic acid" [5]) was isolated from a *Streptomyces cinnamonensis* culture (ATCC 15413) [351]. Proper choice of fermentation conditions allowed the yield of the composite antibiotic, in which monensin was the principal component, to be raised from 0.3—0.6 to ~5 mg/ml [963]. Highly active against a number of Gram-positive bacteria, mycobacteria and fungi (Table 33), monensin also displays cytotoxic activity against HeLa cells and murine cells NCTC 1742 in tissue culture [351]. The minor components of the mixture monensins B—D (factors A 3823B, A 3823C and A 3823D, 336—338), possess activity against *Bacillus subtilis* comparable with monensin*; to be more exact, the first has one half the activity of the principal component, whereas the second exceeds it by twofold [312]. Extensive tests have shown that monensin is very effective in the treatment of chicken coccidiosis caused by *Eimeria tenella, E. necatrix, E. brunetti, E. maxima, E. acerulina* and *E. mivati*, or by more than one of the above-mentioned infections (therapeutic dose 110 g antibiotic per 1 ton of food stuff) [935]. In such doses monensin was found to be only weakly toxic perorally (LD_{50} = 43.8 ± 5.2 mg/kg, experiments on rats; LD_{50} = 284 ± 47 mg/kg, tests on chickens).

The structure of monensin (Fig. 20) was established by X-ray analysis of both the free antibiotic [568] and of its silver salt [5, 752]. From a comparison of formulae (331) and (335) it can be seen that monensin is

---

* For the sake of uniformity monensin is sometimes called monensin A.

TABLE 33

ANTIMICROBIAL ACTIVITY OF MONENSIN [351]

Microorganism	Minimal growth-inhibiting concentration (μg/ml)		
	After 24 h	After 48 h	After 72 h
Staphylococcus aureus 3055	< 0.78	< 0.78	
Bacillus subtilis ATCC 6633	1.56	1.56	
Mycobacterium avium ATCC 7992	—	< 0.78	
Streptococcus faecalis	3.13	12.5	
Lactobacillus casei ATCC 7469	0.78	< 0.78	
Leuconostoc citrovorum ATCC 8081	0.78	3.13	
Proteus vulgaris sp.	50	> 100	
Vibrio metschnikovii	50	50	
Alternaria solani			6.25
Botrytis cinerea			3.13
Helminthosporium sativum			50
Pullularia sp.			1.56
Penicillium expansum			12.5
Sclerotinia fructicola			3.13

structurally very close to nigericin, having practically the same A—D rings as the latter. Agtarap and Chamberlin thoroughly investigated various methods for the chemical degradation of monensin and showed that the spiroketal grouping of rings D and E is very susceptible to acids and responsible for the instability of the antibiotic at low pH values [4]. Mass spectrometry of the

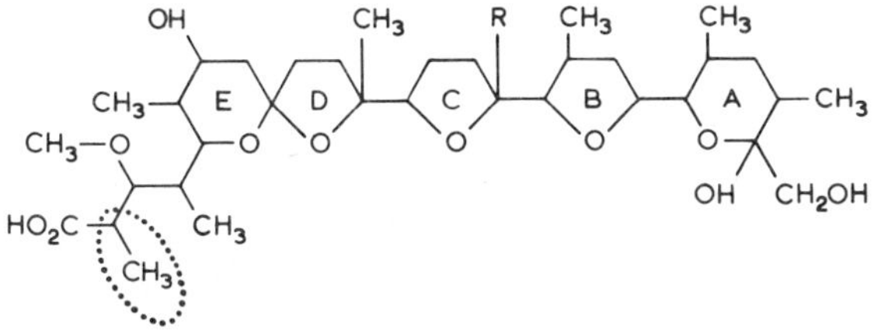

Fig. 20. Monensin (335, R = Et), monensin B (336, R = Me) and monensin C (337, R = Et); closed dotted line encircles the C_3H_7 group.

sodium salts of monensins A—D and the methyl esters of their diacetyl derivatives has led to the conclusion that compounds (335—337) differ by one CH_2 unit and that compounds (337) and (338) are isomers [4, 124]. Further exploration into the structures of monensins B and C (336 and 337) was carried out by NMR spectroscopy of their consecutive acid and oxidative degradation products. Owing to the difficulty of obtaining pure monensin D, its structure is still a matter of conjecture. Apparently it does not differ from monensin A in the carboxyl terminus region but has an extra methylene group between rings C and D [4].

Monensin is one of the few naturally occurring metal-binding macrocyclic compounds that preferentially complex sodium ions, displaying high Na/K-selectivity (Table 34) [23, 386, 569, 785, 786, 791].

TABLE 34

STABILITY CONSTANTS OF MONENSIN COMPLEXES WITH SODIUM AND POTASSIUM IONS [569]

Solvent	$l \cdot mole^{-1}$	
	Na^+	K^+
CH_3OH	700 000	95 000
$CH_3OCH_2CH_2OH–H_2O$	85 000	6600
(4:1, v:v)		

Besides the above-mentioned Na^+ and Ag^+ salts, monensin also forms crystalline complexes with K^+, Rb^+ and Tl^+; no crystalline complexes could be obtained with Li^+, Cs^+ and Cu^{2+} [752].

I.F.4. Dianemycin (339)

Reports of the use of dianemycin for biochemical studies had been creeping into the literature for a number of years [527, 845, 846, 888], but for a long time there had been no description of the method of its preparation, antibiotic spectra, physicochemical properties or structure. Only in 1969 did Hamill and coworkers [350] describe the conditions for isolating dianemycin from *Streptomyces hygroscopicus* species 3444 and the results of its biological testing. From a comparison of Tables 30, 31, 33 and 35, it can be seen that in its antimicrobial action dianemycin resembles other nigericin compounds. In addition, it has a certain antifungal and insecticidal activity [350].

The formula for dianemycin, elucidated by X-ray analysis of its Na^+, K^+ and Tl^+ salts [177, 969], is shown in Fig. 21. In general, dianemycin is very similar to nigericin and monensin in the region of the A, B, D and E rings, especially if one takes into account that the same asymmetric centers are of identical configuration. Distinctive structural features of dianemycin are mainly in ring C (larger ring with bulky side chain), in the carboxyl-containing moiety and in the presence of an α,β-unsaturated ketonic grouping instead of the nigericin ring F. It can easily be seen that the real or

TABLE 35

ANTIMICROBIAL ACTIVITY OF DIANEMYCIN [350]

Microorganism	Minimal growth-inhibiting concentration (μg/ml)
Bacteria:	
Staphylococcus aureus 3055	1.56
Bacillus subtilis	3.12
Mycobacterium avium	0.78
Streptococcus faecalis	1.56
Lactobacillus casei	0.78
Leuconostoc citrovorum	6.25
Vibrio coli Iowa No. 10	12.5
Mycoplasma gallisepticum	25
Mycoplasma, strain N	50
Saccharomyces pastorianus	50
Fungi:	
Trichophyton mentagrophytes	100
Pseudomonas solanacearum	50
Alternaria solani	12.5
Botrytis cinerea	6.25
Ceratostomella ulmi	50
Colletotrichum pisi	50
Helminthosporium sativum	12.5
Penicillium expansum	50
Spicaria divaricata	50

potential carbonyl functions (a ketonic at C_5, spiroketalic at C_{13} and C_{21} and hemi-ketalic at C_{29}) are regularly distributed along the dianemycin hydrocarbon chain at eight-atom intervals. The other oxygen atoms are also positioned at the same distances from one another (C_1, C_9, C_{17} and C_{25}). With such regularity it is only natural to suppose that the biosynthesis of

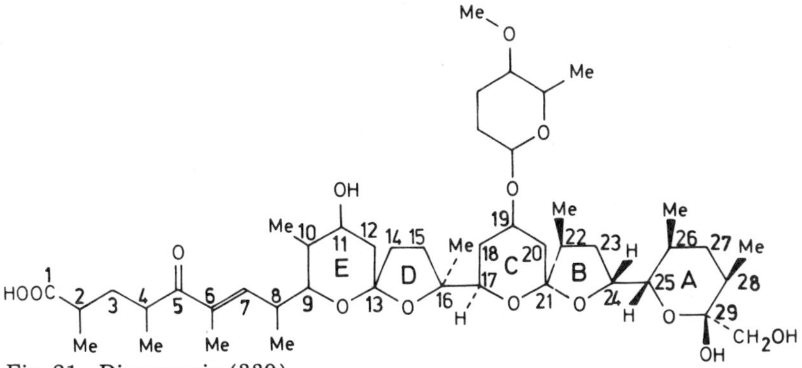

Fig. 21. Dianemycin (339).

dianemycin proceeds by the participation of four or eight carbonyl-containing units. From a biosynthetic viewpoint, dianemycin can thus be considered to be more primitive than nigericin, antibiotic X-206 or monensin, because it still retains a "non-utilized" carbonyl.

The sparse data on the complexing reaction of dianemycin show that, while there is some evidence of Na/K-selectivity, it is much less expressed than in the case of monensin [785, 786, 791].

I.F.5. Other nigericin-like compounds

There are several more naturally occurring compounds which in structure or biological properties resemble the above-discussed antibiotics, but for most of them no information is available regarding their complexing ability.

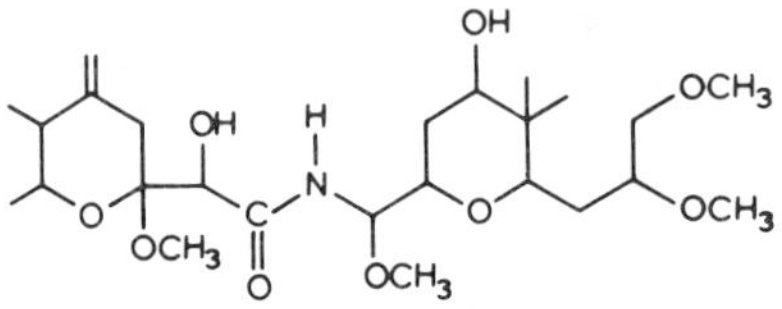

Fig. 22. Pederin.

Among such compounds may be included the insect poison pederin [278, 605] (Fig. 22) and antibiotics of unknown structure — helixin [951], K 358 [340] and azalomycin [696]. Finally, to the group of nigericins should be referred the novel antibiotic A-204, similar to compounds (331—339) in its effect on mitochondria and displaying Rb$^+$ and Cs$^+$ specificity [1092].

Chapter II

METHODS FOR STUDYING THE COMPLEXING REACTION

Alkali metal ion complexing macrocyclic compounds possess a number of specific properties which put their imprint on the approach to such compounds by the traditional techniques of complexone chemistry. Consequently this chapter will give a brief summary of the methods presently used for studying the interaction of cations with the neutral macrocyclic complexones and with the nigericin antibiotics. Special attention is paid to the quantitative aspects of the complexing reaction in homogeneous solutions and in two-phase systems. However, methods for determining the spatial structures of the complexes or the free molecules will not be discussed as they are dealt with in detail in Chapter III. It should be mentioned that for membrane-active complexones, information on ion selectivity or other characteristics of complex formation may be obtained from studies of the properties of artificial membranes modified by these compounds and of their effects on various biological objects (see Chapters V and VI).

II.A. Solutions

II.A.1. *Spectral methods*

The application of spectral methods in studying complexation is based primarily on the existence of differences between the conformational and stereoelectronic characteristics of the free molecules and their cation complexes. Hence, by observing the spectral changes one can usually obtain, besides information on the composition and stability of the complexes, also information on their structure and on the nature of their metal—ligand interactions. For instance, the infrared spectra of valinomycin and its synthetic analogs revealed a considerable bathochromic shift of the ester carbonyl bands on formation of a complex by these cyclodepsipeptides, which was taken as evidence that the carbonyl groups are interacting with the bound cation (see Fig. 23). It is noteworthy that similar carbonyl frequency shifts are displayed by acetone and other alkali ion solvating ketones [638, 1101, 1102]. Characteristic shifts of the carbonyl bands also occur in complexation by the nactins [756]. Sometimes, such spectral changes which are not attributable to the groups directly engaged in the cation bonding can serve as indicators of complex formation. For instance, in the formation of valinomycin complexes, all the amide NH groups become incorporated into a system of intramolecular hydrogen bonds. This leads to the disappearance in the course of the complexing reaction of the free NH absorption band in the 3400 cm^{-1} region (see Fig. 23; cf. Part III.A.1.a).

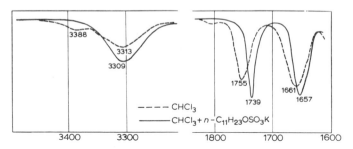

Fig. 23. Infrared spectra of valinomycin (— —) and of its K$^+$ complex (———) in chloroform [904].

Valuable information on the mode of the cation coordination and the nature of the ion—ligand interactions in the complexes can be obtained from their far infrared spectra. It is well known that alkali metal ions solvated by polar solvents display characteristic absorption in the 100—500 cm^{-1} region [221, 528, 606, 1091, 1098] (see Table 36). When the cations undergo complexation and the solvent molecules are displaced from their coordination spheres by the ligand groups of the macrocycle, a shift occurs of the corresponding bands in the spectrum and they become independent of the solvent species. This effect was first observed by Tsatsas et al. [1021] in the binding of sodium and potassium ions by the cyclic polyether (270) in pyridine and dimethylsulfoxide (see Table 36). Ivanov et al. [428] have

TABLE 36

CHARACTERISTIC FAR INFRARED ABSORPTION FREQUENCIES (in cm^{-1}) OF SOLVATED CATIONS IN SOLUTIONS OF DIBENZO-18-crown-6 (270) COMPLEXES AND OF THE ALKALI METAL SALTS [1021]

Solvent	Compound (270)	Na$^+$			K$^+$		
		NCS$^-$	BPh$_4^-$	PF$_6^-$	NCS$^-$	BPh$_4^-$	PF$_6^-$
Dimethylsulfoxide	—	205	203	198	150	152	147
	+	215	213	210	169	167	170
Pyridine	—	183	180	180	139	135	134
	+	217	212	214	170	165	167

recently found that a splitting can occur in the absorption bands of the cation ligand system in the spectra of the macrocyclic complexes, the number of resulting bands apparently depending only on the symmetry of the ligand spacing (see Table 37); these findings are discussed in greater detail in Part III.A.

TABLE 37
CHARACTERISTIC COMPLEXED CATION ABSORPTION FREQUENCIES (in cm^{-1}) IN THE FAR INFRARED SPECTRA OF THE CHLOROFORM SOLUTIONS OF MACROCYCLIC COMPLEX SALTS [428]

Complexone	Li$^+$	Na$^+$	K$^+$	Rb$^+$	Cs$^+$
Valinomycin			171, 117	148, 111	145, 104
Beauvericin	445, 397	165	145, 128		195, 140, 115, 95
Nonactin			150		142, 102
Perhydroantamanide		205			

In some cases it is convenient to track the complexing reaction by following the changes in the ultraviolet spectra. Thus, in the case of valinomycin, enniatin B and antamanide, absorption in the 200 nm region has been used as the parameter reflecting the degree of cation binding [325, 327, 1068]. The formation of complexes by cyclic polyethers containing aromatic nuclei is accompanied by the appearance of additional absorption bands in the ultraviolet spectra in the 275 nm region [738, 739, 741] (see Fig. 24).

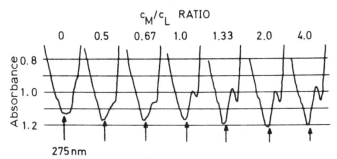

Fig. 24. Complexation-induced change in the ultraviolet spectrum of $2.14 \cdot 10^{-4}$ M dibenzo-18-*crown*-6 in methanol. Dependence on ratio of salt (KNCS) to complexone concentrations (c_M/c_L) [738].

For the aromatic chromophore-containing antibiotics X-537 A and A 23187, the stability constants of the complexes with alkaline earth metal ions have been determined by a spectrofluorometric method based on the disappearance during the complexing reaction of the band at approximately 420 nm in the fluorescence spectrum [122].

For investigating complexation in solution, particularly extensive use is being made of nuclear magnetic resonance (NMR) spectroscopy. Chapter III includes numerous examples of how proton spin—spin coupling data have been highly instrumental in unraveling the spatial structure of metal-binding macrocyclic compounds. Interesting possibilities have been disclosed by the discovery of spin—spin coupling between the complexone protons and the bound $^{203,205}Tl^+$ ion ($J = 12$—14 Hz) [537]. One may expect that ^1H- and ^{13}C-NMR spectroscopy of thallium complexes will give further indications regarding the location of the ligand groupings.

Valuable information on the complexing-induced electron-density changes near the ligand atoms or groups can be obtained from analysis of the chemical shifts. Thus, in ^{13}C-NMR studies of valinomycin, beauvericin and [Val^6,Ala^9]-antamanide complexes, it was shown that changes in the chemical shifts of the carbonyl carbon can serve as indicators of carbonyl groups participating in the cation binding. Moreover, the magnitude of the shifts depends on the cation species, an observation of considerable interest from the standpoint of the nature of the metal—ligand interactions [114] (see also [12, 324, 794]). Cation-dependent signal positions are also observed in the ^1H-NMR spectra of cryptates. When potassium ions are complexed by the bicyclic polyether (328), the largest frequency change is displayed by the proton signal of the methylene groups adjacent to the N atoms; when silver complexes are formed, it is the CH_2S lines which are most affected [205]. Naturally, complexing has a bearing on the spectral characteristics not only of the complexone, but also of the complexing cation. Thus, Haynes et al. [384] have found the chemical shift of the ^{23}Na signal in the spectra of the macrocyclic Na^+ complexes to be linearly dependent on the free energy of complexation. The calibration curve obtained by these authors can be used for evaluating the sodium complex stability constants.

Line-width and position data from the ^1H- and ^{23}Na-NMR spectra are widely used for studying cation exchange kinetics of the macrocyclic complexes [205, 238, 383, 402, 792, 793, 893, 952, 1093]. The conventional approach is to study the temperature dependence of the NMR spectra of solutions which contain the complexone molecules in concentrations exceeding those required for binding all the complexable cations present. Then, with increasing temperature, the exchange rate will approach nearer and nearer the internal chemical shifts of the respective nuclei in the complex and the free molecule as these signals first broaden and then coalesce [537] (see Fig. 25). Analogous effects are observed in ^{23}Na-NMR spectra when free cations are present in Na^+ complex-containing solutions. A typical example is afforded by the broadening of the free ^{23}Na signal

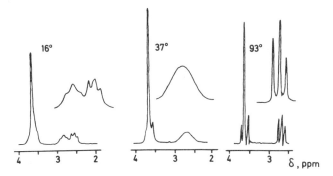

Fig. 25. Temperature dependence of the ^{1}H-NMR spectra (60 MHz) of a ^{2}H$_2$O solution of the bicyclic polyether (324) and KF (2:1) [537].

observed when the cyclic polyether (274) is added to a NaBF$_4$ solution in nitromethane [238]. Haynes [381] made use of "initial" line broadening for kinetic studies of relatively rapid exchange in valinomycin and nactin complexes in a CH$_3$OH—CHCl$_3$ (4:1) mixture. The theory predicted that the line half-width ($\Delta\nu_{1/2}$) for low degrees of complexation should obey Eqn 1.

$$\Delta\nu_{1/2} = \frac{1}{\pi} (k_D + k_E \cdot c_L) \cdot \frac{c_{ML}}{c_L} \tag{1}$$

where k_D is the rate constant of complex dissociation and k_E the rate constant of the bimolecular exchange reaction M$^+$L + L* $\rightleftharpoons$ M$^+$L* + L.

The data obtained by this author are given in Table 38. It can be seen from Fig. 26 that under the experimental conditions $\Delta\nu_{1/2}$ is a linear function of c_{ML}/c_L, i.e. there is only an insignificant contribution by the bimolecular exchange reaction.

The concrete manner of applying NMR spectroscopy to the evaluation of stability constants by titration with metal salts is determined by the magnitude of the exchange rate. For slow exchange (nonactin + KNCS,

TABLE 38

DISSOCIATION RATE CONSTANTS OF THE
VALINOMYCIN· AND NACTIN · K$^+$ COMPLEXES
IN CH$_3$OH—C^{2}HCl$_3$ (4:1, v:v) MIXTURE [381]

K$^+$ complex	k_D (sec^{-1})
Valinomycin	21 ± 0.5
Nonactin	32.3 ± 0.6
Monactin	22.9 ± 1.6
Dinactin	21 ± 6
Trinactin	18 ± 5

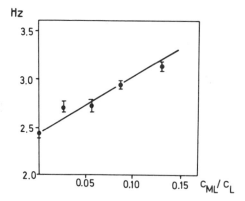

Fig. 26. Effect of degree of K^+ complexation on the lactyl methyl line-width in the ^{1}H-NMR spectra (60 MHz) of a 22.7 mM solution of valinomycin in a KNCS-containing CH_3OH—C^2HCl_3 (4:1) mixture.

$CHCl_3$ [383]) the spectra display patterns due to both the free molecules and the complexes, and one can assess the degree of complexing directly from the intensity ratios. In the case of rapid exchange (nonactin + KNCS, acetone [792]), when the corresponding signals coalesce into a single narrow line, the characteristic parameter becomes the chemical shift, linearly dependent on the degree of complexation.

With nuclei possessing quadrupole moments (for instance, [23]Na) further information on the heterogeneity of the local electrical field can be obtained from the resonance line-width. When the correlation time, τ_c, is sufficiently small ($\tau_c \ll (2\pi\nu_0)^{-1}$, where ν_0 is the radio frequency) the line half-widths are described by Eqn 2 [236].

$$\Delta\nu_{1/2} \sim \left(\frac{e^2 \cdot q \cdot Q}{\hbar}\right)^2 \cdot \tau_c \tag{2}$$

where e is the electron charge, Q the nuclear quadrupole moment, q the electric field gradient near the nucleus, and $\hbar$ Planck's constant.

Since the complexes are of little flexibility (see Chapter III), the correlation time of the bound [23]Na ion will depend mainly on the rotational diffusion rate of the complex ($\tau_R = 4\pi\eta r_e^3/3kT$) [236]. Consequently, Eqn 2 can be expressed in the form of Eqn 3. To make the formula amenable for quantitative calculations, the radii of the complexes can be taken from X-ray or conductimetric data (see Part II.A.2).

$$\Delta\nu_{1/2} \sim \left(\frac{e^2 \cdot q \cdot Q}{\hbar}\right)^2 \cdot r_e^3 \cdot \frac{\eta}{T} \tag{3}$$

where r_e is the effective radius of the complex, η the solution viscosity, T the temperature (°K).

The heterogeneity of the local field in the Na$^+$ complexes depends to a large extent on the symmetry of disposition of the ligand atoms. Thus, for example, the line-widths corresponding to the complex of cyclic polyether (272) are much greater in the case of sodium ions solvated by tetrahydrofuran or heptaglyme molecules [333] (see Fig. 27; cf. [893]). The

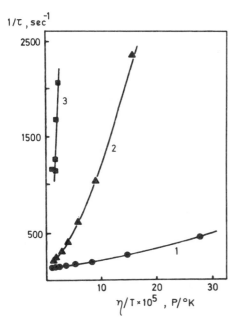

Fig. 27. Effect of complexation on the line-width in the ^{23}Na-NMR spectra (15.1 MHz) of a 0.2 M NaBPh$_4$ solution in tetrahydrofuran [333]. 1, control; 2, in the presence of 0.2 M CH$_3$O(CH$_2$CH$_2$O)$_6$CH$_3$; 3, in the presence of 0.2 M cyclic polyether (272).

broadening is explained by the ligand O atoms being in a planar arrangement in the complex (see Part III.A.3), whereas they are in a nearly spherically symmetric disposition in the solvate sheath. Similar results were obtained with respect to the ^{23}Na line-widths of the Na$^+$-complex spectra in methanol solutions [384]. Under these conditions ($\tau_R = (1-4)\cdot10^{-10}$ sec, $(2\pi\nu_0)^{-1} = 10^{-8}$ sec), the quantity $q\cdot Q$ of the nigericin complex is approximately two times that of the monensin, valinomycin, monactin and enniatin complexes. The difference is apparently due to direct participation of the ionized carboxyl group in the binding of the cation in the nigericin complex (see Part III.A.5). On these grounds one could expect that ^{23}Na line broadening could be used for detecting ion pair formation between anions and positively charged Na$^+$ complexes of neutral macrocyclic

90

compounds in weakly polar solvents. That this is so can be seen from the findings of Sam and Simmons [847], i.e. that the line-width of the quadrupolar ^{55}Mn nuclei in the NMR spectra of a benzene solution of the cyclic polyether complex salt $[(274) \cdot K^+]$ MnO_4^- is much greater than for the aqueous solution of $KMnO_4$. This broadening is evidence of a certain loss in symmetry of the anion environment, most likely due to ion pair formation. It is quite evident that this effect is of similar nature to the one just discussed.

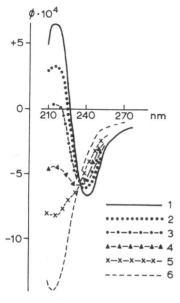

Fig. 28. Effect of complexation on the ORD curves of enniatin B in KNCS-containing ethanol solutions [903]. The curves correspond to the following salt to antibiotic concentration ratios (antibiotic concentration $1.5 \cdot 10^{-4}$ M): 1, 0; 2, 0.5; 3, 1; 4, 2; 5, 5; 6, 20.

Sometimes optical rotatory dispersion (ORD) and circular dichroism (CD) can prove convenient for the quantitative evaluation of complex formation. These spectropolarimetric methods have been successfully used for determining the stability constants of the enniatins and related cyclohexadepsipeptides [700, 704, 897, 903, 904] (see Fig. 28), valinomycin [325, 327], antamanide [1068, 1069], and other cyclic peptides [876, 904]. An interesting effect has been discovered by Pressman [788], who found that in binding of K^+ by the antibiotic X-537A the CD spectra display a considerable increase in intensity of the 245 and 295 nm bands originating from the asymmetric enclosure of the aromatic chromophore.

Wudl [1096] has recently proposed L-1-(2'-hydroxyphenoxyethyl)-2-

hydroxymethylpyrrolidine as reagent in the spectropolarimetric determination of sodium in the presence of lithium and potassium. Quite possibly, although this has not yet been proved, the changes observed in the ORD curves of the solutions in CH_2Cl_2 are due to the formation of a sodium salt such that the heteroatoms of this compound enter into the coordination sphere of the cation and the phenolate O atom is intramolecularly hydrogen-bonded to the alcoholic hydroxyl. The changes in the ORD curves are much weaker in the case of Li^+ ions and are absent in the case of the K^+ and $(CH_3)_4N^+$ ions.

Naturally, selection of the most effective spectral method for observing the complexing reaction depends primarily on the structure and conformational characteristics of the complexone. At the same time, these methods have the merit of permitting detection of complexes independent of whether or not they are ion paired with the anions, thus facilitating study of the complexing reaction in weakly polar solvents. The choice of the anion is determined by its spectral characteristics and the solubility of its salt with the respective metal. Frequently use is made of such "fat soluble" anions as BF_4^- [333], Ph_4B^- [238] and dodecylsulfate [904]. With the latter anion equilibration is sometimes sluggish, apparently because of the micellar nature of alkali metal alkylsulfates in non-polar solvents.

Since in spectroscopy in practice there is, as a rule, little variation in either complexone concentration or the volume of the absorption cell, the experiments are mostly performed with varying concentrations of salts of the complexable cation. In the general case, when the complexing reaction is followed by recording the change in the spectral parameters of the complexone, the stability constants of the M^+L type complexes (L, neutral ligand molecule) are calculated from Eqn 4.

$$K = \frac{c_{ML}}{c_M \cdot c_L} = \frac{\alpha}{(1-\alpha)(c_M^{tot} - \alpha \cdot c_L^{tot})} \tag{4}$$

where $\alpha = \dfrac{\beta_L - \beta_{obs}}{\beta_L - \beta_{ML}}$, is the degree of complexation; β_L, β_{ML} and β_{obs} are

the characteristic spectral parameters (chemical shift, molecular rotation, etc.) determined for the free complexone (or cation), the complex and equilibrium mixture, respectively; and c_i is the concentration of the ith component.

This equation can be transformed into Eqn 5.

$$\alpha = \frac{1}{2} \{(1 + \varphi + \vartheta) + [(1 + \varphi + \vartheta)^2 - 4\varphi]^{1/2}\} \tag{5}$$

where $\varphi = c_M / c_L$, and $\vartheta = 1/K \cdot c_L$.

With the aid of Eqn 5, Prestegard and Chan [792] calculated the family of curves depicting the α–φ dependence for different ϑ values; by means of such curves the stability constants can be obtained graphically (see Fig. 29).

It is rather difficult to determine the stability constants in non-polar solvents owing to the necessity of correcting for ion pair formation. Because of this one can often find data only on the degree of complexation under certain standard conditions.

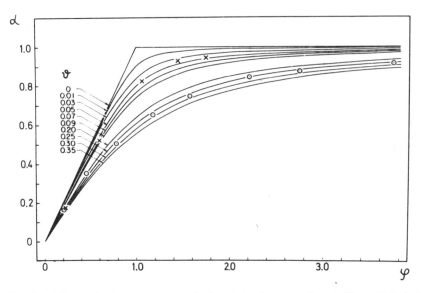

Fig. 29. Theoretical α–φ curves calculated by Prestegard and Chan [792]. The points designate experimental values obtained in a study of K^+ complexing by nonactin in dry (x, $7 \cdot 10^{-3}$ mole fraction of H_2O) and in moist ($\odot$, 0.34 mole fraction of ^2H_2O) acetone.

Sometimes changes in the anion spectra may also be used as indicators of complex formation. Thus, Smid *et al.* [402, 952, 995, 996, 1093] have developed an original method for following cation complexing by cyclic polyethers in aprotic solvents based on the spectral differences between the fluorenyl carbanions (Fl^-) in the contact ion pairs M^+Fl^- and in contact and "separated" pairs of the type $(M^+L)Fl^-$. As is well known, when charge-delocalized anions couple with cations to form ion pairs, their electron shells become less polarized as the cation radius increases. If, for example, the fluorenyl salts of the alkali metals are in the form of contact ion pairs as, say, in tetrahydrofuran solutions, the passing from lithium to cesium is accompanied by a bathochromic shift of the Fl^- anion bands from 349 to 364 nm [400] (for analogous data on the picrate, see p. 243). The bathochromic shift is particularly large when the contact ion pairs are transformed into solvent separated pairs in which the anion is displaced from direct contact with the cation by the solvating molecules (we shall designate these pairs by $M^+\|Fl^-$). In such pairs the anions can be so screened from the polarizing effect of the cation as to resemble spectroscopically the free anions (in the case of free Fl^- anions, $\lambda_{max} = 372$ nm [400]). The fraction of contact ion pairs diminishes with increase in the solvent's solvating capacity; for example, whereas in tetrahydrofuran solutions Na^+Fl^- is in the

TABLE 39

STABILITY CONSTANTS OF THE IONIC PAIRS AND SANDWICH COMPLEXES FORMED BY CYCLIC POLYETHERS AND ALKALI ION FLUORENYLATES [995]

$$M^+Fl^- + L \underset{}{\overset{K_{cp}}{\rightleftarrows}} (M^+L)Fl^-$$

$$M^+Fl^- + L \underset{}{\overset{K_{sp}}{\rightleftarrows}} (M^+L) \| Fl^-$$

$$(M^+L)Fl^- + L \underset{}{\overset{K_s}{\rightleftarrows}} (M^+L_2) \| Fl^-$$

Com-plexone	Solvent	Na^+Fl^-			K^+Fl^-		
		$K_{cp}\cdot 10^{-3}$	K_s	$K_{sp}\cdot 10^{-3}$	$K_{cp}\cdot 10^{-3}$	K_s	$K_{sp}\cdot 10^{-3}$
247	Tetrahydrofuran	9.2	3.5	16.5	~5	1840	~1
	Tetrahydropyran	>20	2.8	>10			
264a	Tetrahydrofuran			$>2\cdot10^4$	>10		>9
	Tetrahydropyran				>10		>5.5

form of contact ion pairs, in trimethylene oxide solutions it is mostly in the form of solvent separated pairs [1093].

Smid and coworkers have shown that the spectral characteristics of the ion pairs $(M^+L)Fl^-$ formed by cyclic polyether complexes depend on the ring size and the bound cation radius. If the metal ion in the complex does not interact directly with the anion, then even in low polar solvents the complex salts resemble in their spectral characteristics solvent-separated ion pairs $M^+\|Fl^-$. Otherwise the $(M^+L)Fl^-$ ion pairs are of the contact type, differing from the M^+L contact pairs only by a small bathochromic shift, evidence of slight augmentation of the cation radius (in the complex salt [(4'-methyl-benzo-15-*crown*-5)-Na^+]Fl^-, $\lambda_{max} = 359$ nm [995]). Taking advantage of these spectral differences Smid *et al.* were able to determine the stability constants of contact and "separated" ion pairs for a number of cyclic polyether complexes (see Table 39). This method also permits study of the interaction between the bound cations and other ligands capable of displacing the Fl^- anions from their contact ion pairs with the complex. The solvent molecules can act as such ligands as well as other anions or cyclic polyether molecules (for more details see p. 225).

If the metal fluorenylates are present in solution in the form of solvent-separated ion pairs, complexation has practically no effect on the ultraviolet spectra. The existence of ion pairs can, however, be detected by the diamagnetic shift of the CH_2 signals of the cyclic polyether in the NMR spectra. This shift is due to the π currents of the fluorenyl ion [952, 1093].

Besides Fl^-, other charge-delocalized anions can apparently be utilized as complexation indicators. Thus, Hyman [416, 417] showed that in chloroform solutions the cation radius has an extraordinarily strong effect on the spectral characteristics of the ethyl ester of the tetrabromophenolphthaleinate anion. When the cations are the bulky complexes of neutral macrocyclic compounds, the spectra contain a characteristic band in the region of 600 nm (for the K^+ complexes of valinomycin and nonactin, λ_{max} = 612 nm; for that of cyclic polyether (274), λ_{max} = 604 nm). Spraying with chloroform solutions of this dye makes it possible to detect spots of the macrocyclic complexes on thin-layer silica-gel chromatography (sky-blue coloring). This method allows, for instance, valinomycin to be detected in a concentration of $3 \cdot 10^{-11}$ M. Interestingly, such a method is also applicable to nigericin despite the fact that the salts of the antibiotic have no effect on the spectrum of this dye in chloroform solution.

In some cases changes in the anion spectrum can be brought about by dissociation of the M^+X^- contact ion pairs on formation of the complex. Very likely it is such an effect that has been observed by Smith and Hanson [955] when the cyclic polyether (270) is added to a solution of the Na derivative of fluorenone oxime in an acetonitrile—*tert*-butanol mixture (2:1).

Complex—anion interaction also forms the basis for a method of studying cation binding by hydrophobic complexones in aqueous solutions proposed by Feinstein and Felsenfeld [258]. These authors found that, in the presence of valinomycin, nactins and nigericin, an increase in the complexable cation concentration leads to an increase in the fluorescence quantum yield of the 8-anilino-1-naphthalenesulfonic acid anions (ANS). The method permits the determination of stability constants ranging from 0.1 to 10 l·mole^{-1}. It is noteworthy that the ANS fluorescence spectra in the presence of complexes of the above-mentioned antibiotics approach those characteristic of ANS in non-polar media. This suggests that the interaction of ANS anions occurs not only by means of electrostatic forces but also by hydrophobic forces and possibly causes micelle formation (cf. [382] and p. 301). Support for such a concept is to be found also in the similar behavior of the positively charged valinomycin and nactin complexes and the neutral nigericin complexes. Interestingly, complexes of monensin, structurally very close to those of nigericin, do not affect the ANS fluorescence. It is as yet uncertain as to whether this is due to differences in the hydrophobicity of the complexes, in the coordination number of the bound cations or in the degree of participation of the carboxylate groups in the cation binding (cf. Part III.A.5). Complexes of the type 18-*crown*-6 cyclic polyethers also have no effect on the ANS fluorescence in aqueous solutions whereas alamethicin, which has a tendency to associate under these conditions [622], increases the ANS fluorescence even in the absence of alkali metal salts (cf. [61]).

II.A.2 Conductimetry

When the ligand molecules lack ionogenic groups the application of conductimetry to the study of the complexing reaction rests on the decrease in mobility of the bulky complex relative to the free cations.

In polar media, where the metal salts are considerably dissociated, complexing lowers the electroconductivity. This effect has been utilized for detecting the formation and determining the stability constants, limiting mobilities and Stokes' radii of complexes of the cyclodepsipeptides [3, 15, 700, 897, 903, 904], cyclic polyethers [15, 273, 741, 893] and antamanide [15, 1069] (see Table 40). A typical curve depicting specific solution electroconductivity changes ($\Delta\kappa$) as a function of the complexone concentrations, c_L^{tot}, for c_M^{tot} = constant is represented in Fig. 30A. The stability constants of complexes such as M^+L in dilute solutions are calculated by means of Eqn 6 [15, 903, 904].

$$K = \frac{\frac{\Delta\kappa}{\delta} \cdot 10^3}{\left(c_M^{tot} - \frac{\Delta\kappa}{\delta} \cdot 10^3\right)\left(c_L^{tot} - \frac{\Delta\kappa}{\delta} \cdot 10^3\right)} \tag{6}$$

where

$$\delta = \frac{(\Delta\kappa)c_L \to \infty}{c_M^{tot}} \cdot 10^3 = (\Lambda_0^M - \Lambda_0^{ML})(1 - B \cdot \sqrt{c_M^{tot}}),$$

B is the Onsager coefficient, and Λ_0^M and Λ_0^{ML} are the limiting free cation and complex mobilities. The δ values follow from $\Delta\kappa_{lim}$, corresponding to the height of the plateau on the experimental $\Delta\kappa - c_L^{tot}$ curves (see Fig. 30A). When the plateau is not reached due to low stability of the complexes or poor solubility of the complexone, δ is determined from the values of $\Delta\kappa$ obtained for varying c_L^{tot} values by equating the right-hand parts of Eqn 6.

On passing to low polar media, the conductimetric determination of the stability constants becomes much more complicated owing to incomplete dissociation of the ion pairs M^+X^- and $(M^+L)X^-$. A computer method has now been devised for calculating the dissociation constants of the M^+X^- ion

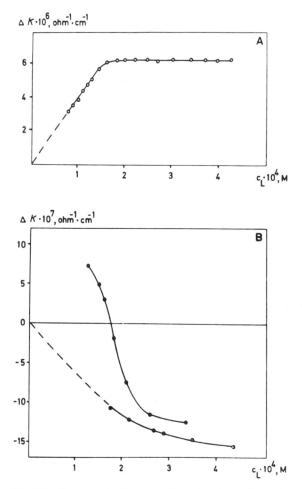

Fig. 30. Changes in the specific electroconductivity of potassium salt solutions with increase in valinomycin concentration [3]. A, $1.55 \cdot 10^{-4}$ M $KClO_4$ in acetonitrile; B, $1.64 \cdot 10^{-4}$ M KNCS (●), and $0.82 \cdot 10^{-4}$ M KBr (◑) in isopropanol.

pairs and the stability constants of the complexes, giving best agreement between the theoretical curves and the experimental data [3].

Since the complex is larger than the free ion, dissociation of the $(M^+L)X^-$ ion pairs can be sometimes markedly greater than that of the M^+X^- pairs. When this is so, complexation may cause an increase rather than a decrease

TABLE 40

MOBILITIES AND STOKES RADII OF COMPLEXES AS DETERMINED CON-
DUCTOMETRICALLY IN ETHANOLIC SOLUTIONS AT 25°C [15]

Compound	Stokes radius (Å)				Limiting mobility ($cm^2 \cdot ohm^{-1} \cdot mole^{-1}$)			
	Na^+	K^+	Rb^+	Cs^+	Na^+	K^+	Rb^+	Cs^+
Valinomycin		5.68	5.38	5.25		13.15	13.90	14.22
Enniatin B	4.80	5.15	5.15	5.15	16.00	14.50	14.50	14.50
Dibenzo-18-crown-6 (270)	4.36	4.05	4.40	4.11	17.00	18.40	16.90	18.10
Antamanide	5.95	5.50			12.55	13.15		

in electroconductivity, an effect encountered, for example, on adding the cyclic polyether (274) to a 10^{-3} M solution of KCl in a chloroform—methanol (9:1) mixture [273], and valinomycin to a 10^{-4} M solution of KNO_3 in acetone, or of KBr in isopropanol (see Fig. 30B).

Now and then the change in electroconductivity can be of a non-monotonic nature. Thus, when the c_{MX}/c_L ratio increases on addition of valinomycin to $1.6 \cdot 10^{-4}$ M KNCS in isopropanol, the electroconductivity first diminishes and then increases, reaching a limiting value higher than the electroconductivity of the initial solution (see Fig. 30B).

II.A.3. Relaxation methods

Relaxation methods which are widely used for studying fast reaction kinetics are now also being applied to the complexing of macrocyclic compounds. One of the most convenient of such methods for this purpose is the temperature jump technique (see [142] and references therein), where after abrupt increase in the solution temperature by several degrees one follows the establishment of reaction equilibrium. In the case of cation binding by macrocyclic compounds this can be done by directly recording the changes in the ultraviolet spectra of the complexone [326, 327], or by utilizing a special indicator complexone such as murexide [202] (see Fig. 31). In the latter case the limiting measurable relaxation times are determined by the forming and breaking rates of the indicator—metal complexes (for the Na^+-murexide complex these are $1.4 \cdot 10^{10}$ $mole^{-1} \cdot l \cdot sec^{-1}$ and $5.6 \cdot 10^6 sec^{-1}$, respectively [202]). Also highly promising is the acoustic method [225], based on measurement of the ultrasonic absorption by solutions containing an equilibrium mixture of the free and complexed molecules. This method, permitting measurement of relaxation times up to

10^{-2} microseconds, has been successfully used in studying conformational changes and complex formation in valinomycin and the enniatins. A kinetic analysis by Grell *et al.* [326, 327] has shown that the formation and

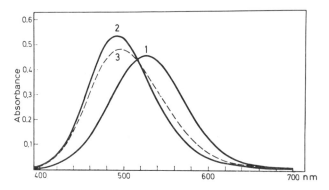

Fig. 31. Murexide as indicator of cation binding by complexones [202]. 1, spectrum of a $4 \cdot 10^{-5}$ M solution of murexide in methanol; 2, the same in the presence of $1.4 \cdot 10^{-3}$ M sodium ions; 3, the same in the presence of $1.4 \cdot 10^{-3}$ M sodium ions and $3.2 \cdot 10^{-3}$ M nonactin.

breakdown of valinomycin and enniatin complexes proceed through some kind of intermediate complex and are described by the scheme:

$$M^+ + L \underset{k_{-1}}{\overset{k_{+1}}{\rightleftarrows}} M^+ \cdots L \underset{k_{-2}}{\overset{k_{+2}}{\rightleftarrows}} M^+L$$

The rate constants of the direct (k_{12}) and reverse (k_{21}) overall reaction can be presented by Eqns 7 and 8.

$$k_{12} = \frac{k_{+1} \cdot k_{+2}}{k_{-1} + k_{+2}} \tag{7}$$

$$k_{21} = \frac{k_{-1} \cdot k_{-2}}{k_{-1} + k_{+2}} \tag{8}$$

There is every reason to assume that intermediate complex formation is not the rate-limiting stage of the overall complexing reaction, since it precedes a relatively slow change in conformation of the cyclodepsipeptides. Eqns 7 and 8 can therefore be simplified to Eqns 9 and 10.

$$k_{12} = \frac{k_{+1} \cdot k_{+2}}{k_{-1}} \tag{9}$$

$$k_{21} = k_{-2} \tag{10}$$

For small deviations from the equilibrium the largest relaxation time is given by Eqn 11.

$$1/\tau = k_{-2} + \frac{k_{+1} \cdot k_{+2} \cdot (c_M + c_L)}{k_{-1} + k_{+1} \cdot (c_M + c_L)} \tag{11}$$

from which it follows that for $(c_M + c_L) \to 0$, $1/\tau$ approaches k_{-2}, for large values of $(c_M + c_L)$, i.e. for $(c_M + c_L) \gg k_{-2}/k_{+1}$, it approaches $(k_{-2} + k_{+2})$ and for a certain range of $(c_M + c_L)$ values it depends linearly upon $(c_M + c_L)$ (see Figs 32 and 33). One can easily see that analysis of the

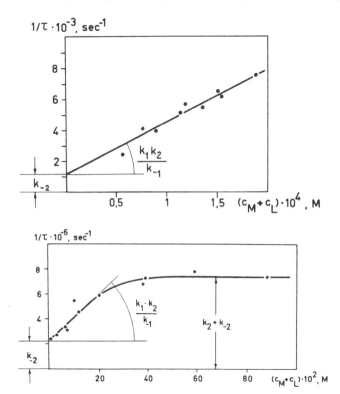

Fig. 32. (above). Kinetics of the formation of the K^+ complex of valinomycin in methanol containing 0.1 M tetrabutylammonium perchlorate as determined by the temperature jump method at 25°C ($\Delta T = 3.5$°C) using spectrophotometry at 224 nm [326, 327].

Fig. 33. (below). Kinetics of the formation of the Na^+ complex of valinomycin in methanol at 25°C by means of the acoustic method (1 MHz) [326, 327].

$1/\tau$—$(c_M + c_L)$ dependence permits the absolute or relative rates to be determined not only of the elementary reaction stages, but also of the overall complex forming and breaking reactions and hence gives the values of the stability constants (see Table 41).

TABLE 41

KINETIC PARAMETERS OF VALINOMYCIN COMPLEXING REACTIONS IN METHANOL AT $25°C$ [326, 327]

Cation	k_{12} (mole$^{-1}\cdot$sec^{-1})	k_{21} (sec^{-1})	$K = \dfrac{k_{12}}{k_{21}}$ (l$\cdot$mole^{-1})	k_1 (mole$^{-1}\cdot$sec^{-1})	k_{-1} (sec^{-1})	k_{+2} (sec^{-1})	k_{-2} (sec^{-1})
NH$_4^+$			47	$1\cdot10^9$	$1.5\cdot10^8$	$2\cdot10^6$	$2.5\cdot10$
Na$^+$			4.7	$7\cdot10^7$	$2\cdot10^7$	$4\cdot10^6$	$2\cdot10^6$
K$^+$	$3.5\cdot10^7$	$1.3\cdot10^3$	$3\cdot10^4$				
Rb$^+$	$5.5\cdot10^7$	$7.5\cdot10^2$	$6.5\cdot10^4$				
Cs$^+$	$2\cdot10^7$	$2.2\cdot10^3$	$8\cdot10^3$				

II.A.4. Other methods

Cation binding by macrocyclic complexones has been studied osmometrically [756, 1068] and potentiometrically by means of glass [248, 272, 1068] or "valinomycin" [248] cation-sensitive electrodes. Use can apparently also be made of the spectrophotometric titration of macrocyclic complexones by employing as indicators murexide or other compounds which change color upon binding cations. In case of the nigericin antibiotics, various pH-metric techniques can be applied [569]. The pK_a of these compounds is lowered when complexable cations are present, the magnitude of the shift reflecting that of the complex stability constants. Pressman [788] has shown that in water—ethanol (1:9) solutions the addition of 10 mM potassium thiocyanate diminishes the pK_a of nigericin from 8.45 to 5.75 (for the antibiotics X-206 and X-537 A under analogous conditions, the pK_a shifts are from 8.30 → 5.60 and 5.80 → 4.35, respectively).

Shifts in pH are also observed in the formation of cryptates, because only such molecules of the diazabicyclic polyethers can bind cations whose N atoms are deprotonated [204]. Evidently this should be a general property of all macrocyclic complexones containing primary, secondary or tertiary amino ligands.

Microcalorimetry is highly promising as a method for studying the thermodynamic parameters of complexation [21, 22, 274, 441, 442, 567, 668]. Its value lies, in particular, in the possibility of determining the enthalpy of the process without changing the temperature, so that one does not have to account for the temperature-dependent conformational equilibrium of the free complexone molecules. Modern microcalorimeters are highly sensitive, able to record the heat evolved from the formation of 0.2 micromoles of the valinomycin $\cdot$ K$^+$ complex in ethanol solution. Simon *et al.* [274] have developed a special automated system for transducing a

TABLE 42

THERMODYNAMIC PARAMETERS OF COMPLEXATION DETERMINED MICROCALORIMETRICALLY AT 25°C

Compound	Solvent	Cation	ΔH (kcal · mole^{-1})	ΔS (kcal · mole^{-1} · degree^{-1})	Reference
Valinomycin	methanol	K$^+$	-4.54		[274]
	ethanol	K$^+$	-8.9	-2.16	[668]
Nonactin	methanol	K$^+$	-11	-0.017	[274]
		Na$^+$	-3.4		
Monactin	methanol	Na$^+$	-5.45		[274]
Nigericin	methanol	K$^+$	-0.98	$+22.3$	[567]
		Na$^+$	$+1.65$	$+23.5$	
Monensin	methanol	K$^+$	-3.73	$+8.36$	[567]
		Na$^+$	-3.84	$+14.6$	
274 A	water	K$^+$(10°C)	-4.14	-4.8	[441]
		K$^+$	-3.88	-3.8	
		K$^+$(40°C)	-3.58	-2.7	
		Rb$^+$	-3.33	-4.2	
		Cs$^+$	-2.41	-3.7	
		NH$_4^+$	-2.16	-1.2	
		Sr^{2+}	-3.68	$+2.5$	
		Ba^{2+}	-4.92	-0.2	
274 B	water	K$^+$	-5.07	-9.6	[441]
		Rb$^+$	-3.97	-9.3	
		NH$_4^+$	-3.41	-7.8	
		Ag$^+$	-2.09	$+0.3$	
		Sr^{2+}	-3.16	$+1.5$	
		Ba^{2+}	-6.20	-5.8	
274 (A + B)	dimethylsulfoxide	Na$^+$(I$^-$)	-0.48 ± 0.45		[22]
		K$^+$(I$^-$)	-7.40 ± 0.40		
		K$^+$(BPh$_4^-$)	-7.45 ± 0.60		
	acetone	Na$^+$(I$^-$)	-6.43 ± 0.50		
		Na$^+$(BPh$_4^-$)	-6.00 ± 0.36		
		K$^+$(I$^-$)	-9.70 ± 0.33		
		K$^+$(BPh$_4^-$)	-9.30 ± 0.36		
		Cs$^+$(BPh$_4^-$)	-8.40 ± 0.33		
		NH$_4^+$(BPh$_4^-$)	-10.00 ± 0.5		
	tetrahydrofuran	Na$^+$(BPh$_4^-$)	-5.00 ± 0.55		

thermocouple signal initiated by the complexing reactions. The stability constants of the M^+L complexes are calculated by means of Eqn 12 (cf. Eqn 6).

$$K = \frac{\Delta q / \Delta H_0}{(c_M^{tot} - \Delta q / \Delta H_0)(c_L^{tot} - \Delta q / \Delta H_0)} \tag{12}$$

where Δq is the heat evolved at $c_M^{tot} \approx c_L^{tot}$, and ΔH_0 is the heat evolved at $c_{ML} = c_L^{tot}$ ($c_M \gg c_L^{tot}$).

Since the complexing reaction is started by mixing solutions of the salt and the complexone, in calculating values of q and ΔH_0, corrections are made for the heat of dilution of the salt solution. The data obtained by the calorimetric method are summarized in Table 42. In some cases the heat production is so small that this method becomes unsuitable for determining stability constants. Such is the case, for example, in the complexing of Na^+, Mg^{2+}, Ca^{2+} and Ag^+ ions in aqueous solutions by compound (274) [441].

Low-angle X-ray scattering can give certain information on the conformational transitions taking place during complexation [514].

II.B. Two-phase systems

High lipophilicity is the most striking feature of the cation complexes of neutral macrocyclic compounds such as valinomycin and of the acid nigericin antibiotics. As a result the salts of the latter are easily soluble in diethyl ether and even saturated hydrocarbons, and the former augment the

TABLE 43

SOLUBILITY OF SALTS IN APROTIC SOLVENTS IN THE PRESENCE OF CYCLIC POLYETHER (274) AT $26°C$ [738]

Salt	Solvent	Solubility (M)
$C_6H_5K^*$	benzene	0.2
KOH	toluene	0.32
KOH	dimethylformamide	0.3
CsOH	benzene	0.065
$(CH_3)_3COK$	benzene	0.12
KCN	nitromethane	0.19
NaI	benzene	0.1
K_2PdCl_4	o-dichlorobenzene	0.18
K_2PtCl_4	o-dichlorobenzene	0.047
$KMnO_4$	benzene	Intensive colouring**

 * Decomposes in several hours probably due to ether bond rupture.
 ** Decomposes slowly with precipitation of MnO_2.

solibility of various salts and hydroxides in the different weakly polar solvents (see Tables 43 and 44). In a two-phase system both acid and neutral complexones transfer complexable cations from the aqueous to the organic layer. This process has been investigated for various cyclic polyethers [232, 273, 738, 739, 785], for the non-ionic polyether detergents [91], for valinomycin [235, 337, 415, 784, 790, 791, 889, 976, 1011, 1014], the nactins [232, 233, 785, 790, 791, 991], antamanide [1068] and for nigericin and related compounds [784, 785, 790, 791]*.

TABLE 44

INCREASE IN SALT SOLUBILITIES IN THE PRESENCE OF EXCESS CYCLIC POLYETHER (270) [738]

Salt		Solubility (M)		Solubility ratio
Cation	Anion	Without complexone	With complexone	
Sodium	oxalate	$3.6 \cdot 10^{-4}$	$1.1 \cdot 10^{-3}$	3.1
	carbonate	$2.8 \cdot 10^{-2}$	$1.4 \cdot 10^{-1}$	5
	chloride	$2.1 \cdot 10^{-1}$	$5.2 \cdot 10^{-1}$	2.5
Potassium	iodate	$4 \cdot 10^{-5}$	$1.5 \cdot 10^{-4}$	38
	nitrate	$3.1 \cdot 10^{-2}$	$2.8 \cdot 10^{-1}$	9
	chloride	$5.7 \cdot 10^{-2}$	$2.3 \cdot 10^{-1}$	4

A theoretical treatment of the extraction process has been given by Eisenman and coworkers [232, 233, 991]. When an organic phase containing a neutral complexone (L) comes into contact with an aqueous phase in which the anion and complexable cation activities are a_X and a_M, respectively, the equilibrium $M_w + L_o + X_w \rightleftharpoons (M^+L)_o + X_o$, where the subscripts "w" and "o" refer to the aqueous and organic phases, respectively, is established, characterized by a "bulk extraction" constant of Eqn 13.

$$K_{w/o} = \frac{c^o_{ML} \cdot c^o_X}{c^o_L \cdot a_M \cdot a_X} = K_o \cdot k_X \cdot k_M = K_w \cdot \frac{k_{ML} \cdot k_X}{k_L} = \bar{K}_{w/o} \cdot k_X \qquad (13)$$

where K_w and K_o are the stability constants of the complex in the water (w) and organic (o) phases; k_i is the partition coefficient for ith component; and

$\bar{K}_{w/o} = \dfrac{c^o_{ML}}{c^o_L \cdot a_M}$ is the membrane solubilization constant.

---

* Saha et al. [843] (see also [1050]) were not able to detect the extraction of cations in the presence of gramicidins.

The different names for $K_{w/o}$ and $\bar{K}_{w/o}$ emphasizes the fact that due to electroneutrality requirements for each phase, cation transfer must be stoichiometrically coupled with anion transfer. Only when the linear dimensions of one of the phases is commensurate with the thickness of the electrical double layer, as, for instance, in the case of bimolecular phospholipid membranes (see Part V.C), is the electroneutrality requirement for this phase annuled and cation transfer no longer becomes obligatorily coupled to anion transfer. Usually, the c_{ML}^o and c_L^o values are very small

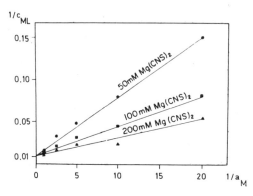

Fig. 34. Effect of thiocyanate anion concentration on rubidium label extraction by a 10 mM solution of cyclic polyether (270) in glycine—tricine buffer (pH 7.0) [785]. The curves in the figure correspond to $1/\bar{K}_{w/o}$ values in M: ●, 0.68; ■, 0.33; ▲, 0.2.

under the experimental conditions so that the constant $K_{w/o}$ in Eqn 13 can be regarded as differing only very insignificantly from the thermodynamic value. Studying neutral complexone induced cation transfer into organic solvents, Pressman et al. (see, for instance, [791]) treated the experimental results by calculation of $1/\bar{K}_{w/o}$ values. Naturally their "constants" so found depended upon the a_X values (see Fig. 34).

From Eqn 13 it can be seen that extraction of the cations into the organic phase is determined by the partition coefficients not only of their complexes but also of the X^- anions. Effective transfer has been shown to occur only in the presence of lipophilic anions such as laurate [790], thiocyanate [337, 784, 785, 791], 2,4-dinitrophenolate [232, 233, 991], picrate [232, 233, 235, 504, 739, 889, 976, 991] and 2,4,6-trinitro-m-cresolate [337, 1011, 1014]. The last three anions are colored and can therefore be utilized for spectrophotometric determination of c_X^o. In this respect, particularly convenient are the picrate and the trinitrocresolate ions because of their larger extinction coefficients and higher lipophilicities (in the system CH_2Cl_2—water, $k_{picrate}/k_{DNP} = 60$—70 [233]; cf. [1011, 1014]). In weakly polar media the complexes may also be colorimetrically determined with the aid of the ethyl ester of tetrabromophenolphthalein [416] (cf. [258]). It is also possible to use the fluorescent 8-anilino-1-naphthalene sulfonate (ANS) anions [1011]. For direct determination of c_{ML}^o in extraction experiments, and especially in studying the competitive displacement of cations from complexes, it is convenient to use the radioactive isotopes ^{86}Rb, ^{137}Cs, [42]K and [22]Na [784, 785, 790, 791]. The application of these labels coupled with the use of a γ-ray spectrometer and amplitude analyzer, considerably simplifies and standardizes determinations of cation extraction selectivity, because the different cation species in a sample can thereby be determined simultaneously [889].

Extraction experiments are usually performed at $c_M \gg c_L$ and $a_X \gg c_L$; the numerical values of the constants can be calculated by means of Eqns 14 and 15.

$$c_{ML}^o = c_X^o = \sqrt{K_{w/o}} \cdot \sqrt{a_M \cdot a_X \cdot [(c_L^o)^{tot} - c_X^o]} \quad [233] \tag{14}$$

$$\frac{1}{c_{ML}^o} = \frac{1}{c_X^o} = \frac{1}{\bar{K}_{w/o} \cdot (c_L^o)^{tot}} \cdot \frac{1}{a_M} + \frac{1}{(c_L^o)^{tot}} \quad [785, 791] \tag{15}$$

These equations are valid if (a) of the anions present in the aqueous solution only X^- has a non-negligible partition coefficient, (b) the partition coefficient of a complexone is large, and (c) the formation of type $(M^+L)X^-$ ion pairs in the organic phase and of complexes in the aqueous phase can be neglected. When the organic phase is of high polarity (for instance, n-butanol—toluene mixtures containing more than 50% butanol [791]) correction must be made for extraction of the M^+X^- salt.

The conditions enumerated above are obeyed when complexes of bulky, hydrophobic complexones like valinomycin and the nactins are extracted by relatively polar solvents such as CH_2Cl_2, $CHCl_3$ or a n-butanol—toluene (3:7) mixture (see Figs 34 and 35A). Nevertheless, the effect of "foreign" cations and anions in the aqueous solution upon the extractivity of the complex salts is a problem which still has not received an unequivocal solution. Eisenman et al. [233] state that extraction of picrates of the monactin complexes by methylene chloride is independent of the Cl^- and OH^- concentration in aqueous solution. However, under similar conditions, the picture of the valinomycin potassium complex unquestionably displays extractivity dependence upon the ion strength of the aqueous solution. Moreover, analysis of the data of Eisenman and his coworkers shows that such a dependence is also observed with monactin [889, 976].

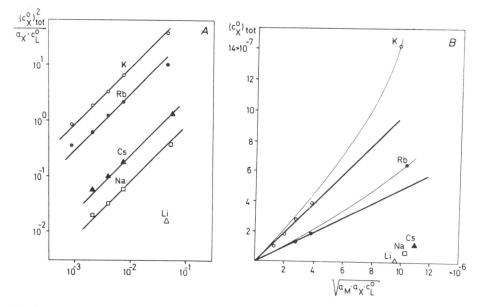

Fig. 35. Extraction of the salts by organic solvent solutions of monactin [233]. A: extraction of dinitrophenolates by CH_2Cl_2 solution obeys Eqn 14 (the slope of the curves in the double logarithmic scale is unity). B: deviation of experimental data on extraction of picrates by a (36:64, v:v) mixture of CH_2Cl_2 and n-hexane from those calculated according to Eqn 14. The theoretical straight lines correspond to $K_{w/o}$ values found from the initial portions of the experimental curves.

As the polarity of the organic phase decreases the complexes and anions in it become increasingly associated into ion pairs. The total anion or complex concentration in this phase is then described by Eqn 16 [233].

$$(c_X^o)^{tot} = (c_{ML}^o)^{tot} = c_{ML}^o + c_{MLX}^o = (K_{w/o} \cdot \xi)^{1/2} + K_{w/o} \cdot K' \cdot \xi \qquad (16)$$

where

$$K' = \frac{c_{MLX}^o}{c_{ML}^o \cdot c_X^o};$$

and

$$\xi = a_M \cdot a_X \cdot c_L^o \, [(c_{ML}^o)^{tot} \ll c_L^o].$$

A comparison of Eqns 14 and 16 shows that with an increasing degree of ion pairing the linear $(c_X^o)^2_{tot} - \xi$ dependence will approach more and more a parabola. Such is the case, for instance, in the extraction of monactin complex picrates by a CH_2Cl_2—n-hexane (36:64) mixture (see Fig. 35B), the stability constants of the ion pairs of K^+ and Rb^+ complexes as estimated by Eisenman et al. being $5.0 \cdot 10^5$ and $6.9 \cdot 10^5$ l·mole^{-1}, respectively.

Gunn and Tosteson [337] have shown that the nature of the anion affects the temperature dependence of the extraction. When a decane solution of valinomycin was equilibrated with an aqueous trinitrocresolate-containing KCl solution the potassium concentration in the organic phase fell to about 1/20 of the original value on increasing the temperature from 0 to 37°C; with thiocyanate, on the other hand, the degree of extraction was independent of the temperature. Temperature independence of extraction was observed also for trinitrocresolate with chloroform as organic phase.

It is important to note that extraction of the salts of macrocyclic complexes can no longer be described by Eqns 14 or 16 with solvents of such weak polarity as saturated hydrocarbons [542, 889]. The reasons for this are still unknown (for further discussion see p. 274).

In contrast to the neutral complexones, nigericin and related antibiotics do not require a lipophilic anion for transfer of the complexable cations from aqueous solution to the organic phase [790, 791]. This is due to the salt-like nature of their complexes. The behavior of these substances in the two-phase system can be represented as an equilibrium of the type $M_w^+ + HL_o \rightleftharpoons (M^+L^-)_o + H_w^+$.

The equilibrium can be described by Eqn 17 providing $a_M \gg c_{HL}$, $k_{HL} \gg 1$, $k_{ML} \gg 1$, and the values of c_L^w and c_L^o are very small.

$$\frac{1}{c_{ML}^o} = \frac{a_H(1 + k_{HL}) + K_D}{K_{w/o} \cdot k_{HL} \cdot (c_{HL}^o)^{tot}} \cdot \frac{1}{a_M} + \left(1 + \frac{1}{k_{ML}}\right) \cdot \frac{1}{(c_{HL}^o)^{tot}}$$

$$\approx \frac{a_H \cdot k_{HL} + K_D}{K_{w/o} \cdot k_{HL} \cdot (c_{HL}^o)^{tot}} \cdot \frac{1}{a_M} + \frac{1}{(c_{HL}^o)^{tot}} \qquad (17)$$

where

$$K_{w/o} = \bar{K}_{w/o} = \frac{a_H \cdot c_{ML}^o}{a_M \cdot c_{HL}^o}, \quad \text{and} \quad K_D = \frac{a_H \cdot c_L^w}{c_{HL}^w}.$$

It follows from this equation that metal ion transfer into the organic phase is a function of the pH (cf. Eqn 15 for neutral complexones). In order to avoid correction for the pH effect in determining $K_{w/o}$, Pressman et al. [791] have studied cation extraction in the presence of nigericin antibiotics at pH 10 (adjusted with $[(CH_3)_4N^+]OH^-$) at which pH practically all complexone molecules are ionized.

Extraction data may be used for evaluating K_w when its low values bar direct measurement in aqueous solutions. If in the absence of lipophilic anions complexable cation salts are added to the aqueous phase, the partition coefficient of the neutral complexone should diminish to the value k'_L determined by the expression

$$k'_L = \frac{c^o_L}{c^w_L + c^w_{ML}} = \frac{k_L}{1 + K_w \cdot a_M}$$

whence

$$K_w = \frac{k_L - k'_L}{k_L} \cdot \frac{1}{a_M} \tag{18}$$

In the presence of lipophilic anion

$$k'_L = \frac{c^o_L + c^o_{ML}}{c^w_L + c^w_{ML}} = k_L \frac{1 + \bar{K}_{w/o} \cdot a_M}{1 + K_w \cdot a_M}$$

and

$$K_w = \frac{k_L}{k'_L \cdot a_M} (1 + \bar{K}_{w/o} \cdot a_M) - \frac{1}{a_M} \tag{19}$$

Utilizing Pressman's data [791] for the labeled valinomycin partition in cyclohexane/water and toluene—butanol (7:3)/water systems, one can obtain with the aid of Eqns 18 and 19 values of $K_w = 1$—30 $l \cdot mole^{-1}$ for the Rb^+ complex, which are in quite satisfactory agreement with Feinstein and Felsenfeld's data (2 $l \cdot mole^{-1}$) obtained in another way (see p. 5).

The peculiarities of the monamycin antibiotics have allowed Hall [346] to apply what is in a sense a mirror image of the above approach for studying their interaction with alkali metal ions. Complexation in aqueous solution changes the amphiphilicity of the monamycin molecules and causes gelation of the complex salts, which in contrast with the free antibiotics are insoluble in ether. Hence, the amount of monamycin passing into the ether extract from the gel can be used for evaluating the degree of complex formation.

II.C. Mass spectrometry of complexes

A property specific to the salts of the nigericin antibiotics is their ability to sublime and to undergo fragmentation on impact of electrons without ejection of the bound cation [124, 404]. Comparison of the spectra of the various metal salts or use of high-resolution spectrometers provides the means for detecting cation-containing fragments.

The fragmentation paths of the salt and the free acid molecules differ considerably, greatly increasing the possibilities of the mass spectrometric method for structure determination in this class of compounds. The presence of the metal ion turned out to be a factor strongly stabilizing the segment of the molecule containing the ligand ether groupings. Indeed, whereas the characteristic processes in fragmentation of monensin and nigericin are inter-ring scission and step-by-step dehydration, the most prominent peaks in the spectra of the salts of these antibiotics are due to those metal-containing fragments which arise from bond ruptures in the carboxylate-carrying side chain.

The mechanism whereby the bound cation affects the fragmentation route deserves further study. Obviously the salt-like cation complex and the free acid differ in location of the charge in the molecular ion radical. In the case of the complex, the electron impact apparently transforms the anionic carboxylate group into a radical. The charge of the resultant molecular ion and its main daughter fragments is localized on the bound metal ion. Besides these main fragments which retain all ether ligands there are minor metal-containing fragments that have retained only a small part of the molecule. In such fragments the charge is apparently localized in the C or O atoms [124].

An M + 23 ion peak has recently been observed in the mass spectrum of antamanide recrystallized from NaCl containing ethanol solution [1069]. This peak corresponds to the sodium complex; however, it is unknown whether the peak-forming ion has been released on the electron impact or is due to dissociation of the $(M^+L)X^-$ ion pair.

Chapter III

THE SPATIAL STRUCTURE AND COMPLEXONE PROPERTIES OF MACROCYCLIC COMPOUNDS

Ordinarily the chemistry of organic complexones concerns the various acid and chelate groupings that are responsible for the binding of metal ions by these molecules. The efficiency and selectivity of the complexing is thus, as a rule, dependent on the properties of these "active centers" and hardly at all on the spatial structure of the rest of the molecule. A completely different picture is presented by the macrocyclic complexones discussed in this book. Already in the early stages of study of these compounds it became clear that their complexing properties depend not only on the number and nature of the groups directly interacting with the cations, but also to a large extent on the spatial structure of the entire molecule. This chapter will accordingly be divided into two principal parts. The first will be devoted to a discussion of the conformational states of the metal-binding macrocyclic compounds and their complexes. This discussion will then serve as the basis for the second part which will be devoted to the more general aspects of the complexing reaction, such as the nature of the ion binding, the causes for the selectivity, the mechanism of the complexing reaction, etc.

III.A. The conformational states of the metal-binding macrocyclic compounds and of their metal complexes

There are two principal approaches to the spatial structure of metal-binding macrocyclic compounds and of their complexes. The first involves X-ray analysis and gives the most exact information on the three-dimensional structure of molecules in a crystal; but this method is unable to cope with such questions as conformational equilibrium, the dynamics of the complexing reaction and their relation to biological function. These can be handled by another approach which was first used in studying the depsipeptide antibiotics. It is based on the composite use of a number of physicochemical methods (nuclear magnetic resonance, optical rotatory dispersion, circular dichroism, infrared and ultraviolet spectroscopy and dipole moment measurements) in combination with theoretical conformational analysis. Although this approach as yet lacks the accuracy and lucidity of the X-ray method, it more than compensates for this shortcoming by permitting the effect of such important factors as the solvent species, temperature, etc., on the conformational characteristics to be determined and extensive comparative studies to be performed on the complexing of the metal-binding macrocyclic compounds with different cations. Naturally the most complete and reliable information is obtained when the physico-chemical studies in solution are carried out with a background of the X-ray analysis, as has been demonstrated particularly clearly with the valinomycin and enniatin compounds.

III.A.1. Depsipeptides and peptides

Before dealing with the spatial structure of the complexones we shall briefly summarize the present conventions for describing conformational states of peptide systems and the methods being used for their study.

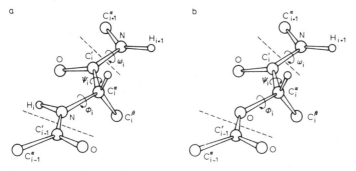

Fig. 36. Rotation angles in peptides (a) and depsipeptides (b).

In peptides built of α-amino acid residues there are three types of bonds, $N-C^\alpha$, $C^\alpha-C'$ and $C'-N$, rotation about which causes changes in the backbone conformation (Fig. 36a). According to the nomenclature proposed in 1966 [222] and used in the majority of papers on peptide complexones, the torsional angles about these bonds, determining the orientations of the substituents on the N, C^α and C' atoms, are denoted by ϕ, ψ and ω, respectively.

The origin ($\phi = \psi = \omega = 0°$) was taken for a completely stretched chain with a planar, *trans* amide group and the $C^\alpha-C'$ and $C^\alpha-N$ bonds *cis* with respect to the N—H and C'—O bonds (Fig. 36a). The angles are measured clockwise when one looks from the N- to the C-terminals along the corresponding bond of the backbone. A similar nomenclature was also used for the conformational states of depsipeptides, differing from the usual, homodetic peptides by the presence of ester groups (Fig. 36b). In 1970 a new system was proposed [484], with the origin ($\phi = \psi = \omega = 0°$) taken to be for the conformation which differs from that represented in Fig. 36a by a 180° turn about all three bonds. The angles are now measured either

TABLE 45

RELATION BETWEEN THE θ AND ϕ PARAMETERS FOR L- AND D-AMINO ACID RESIDUES

	θ (degrees):	ϕ (degrees) 0	30	60	90	120	150	180
Nomenclature of 1966 [222]	L	240	210, 270	180, 300	150, 330	0, 120, 360	30, 90	60
	D	120	90, 150	60, 180	30, 210	0, 240, 360	270, 330	300
Nomenclature of 1970 [484]	L	60	30, 90	0, 120	−30, 150	−180, −60, 180	−150, −90	−120
	D	−60	−30, −90	−120, 0	−150, 30	−180, 60, 180	90, 150	120

clockwise (positive values from $0°$ to $180°$) or counter clockwise (negative values from $0°$ to $-180°$). The new torsional angles are connected with the old ones by the following relation $(\phi, \psi, \omega)_{new} = (\phi, \psi, \omega)_{old} - 180°$ (see Table 45). In what follows, use will be made of the 1970 nomenclature.

Since the valency angles and the bond lengths are similar for the different peptides and the amide and ester bonds are as a rule in the energetically preferable *trans* configuration ($\omega = 180°$), the conformation of the peptide backbone is determined mainly by the ϕ and ψ parameters. The side chain (R) carbon atoms are designated by the Greek letters $\alpha, \beta, \gamma \ldots$ in alphabetic order; the bonds $C^{\alpha}-C^{\beta}$, $C^{\beta}-C^{\gamma}$, $C^{\gamma}-C^{\delta} \ldots$ by the numerals 1, 2, 3 $\ldots$, and the torsion angles about these bonds, by $\chi_1, \chi_2, \chi_3 \ldots$

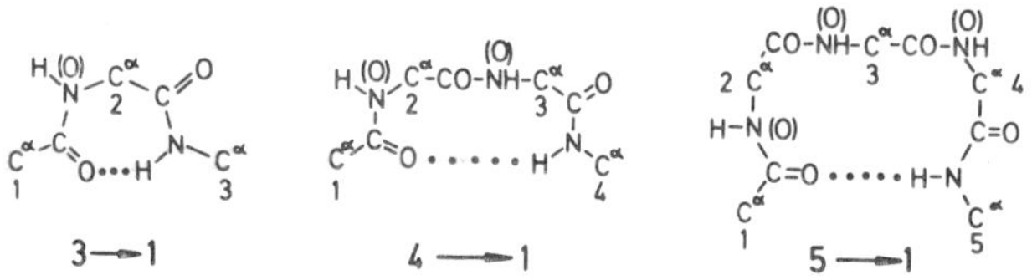

Fig. 37. Peptide (depsipeptide) fragments containing intramolecular hydrogen bonds.

Important structural elements in the peptides and depsipeptides are intramolecular hydrogen bonds $NH\cdots CO$ between the amide NH groups and the amide or ester carbonyls. If the bond is between an NH of an amino acid residue of sequence number m and the carbonyl of an amino or hydroxy acid residue of sequence number n, it is designated by $m-n$ or $m \rightarrow n$. The resulting structures are illustrated in Fig. 37 by intramolecular hydrogen bonds of types $3 \rightarrow 1$, $4 \rightarrow 1$ and $5 \rightarrow 1$.

In the early days of peptide structure study a decisive part was played by X-ray analysis, the use of which led to the elucidation of the geometrical parameters of the amide group, of the $NH\cdots CO$ hydrogen bond and also of the C^{α} bond lengths and valency angles (see review [600]). Subsequently, as evidence for the conformational flexibility of peptides began to accumulate, increasing recourse was made to their physicochemical study in solution. Evidence of solvent-caused shifts in conformational equilibrium is most directly obtained from CD and ORD curves, whose shape is determined mainly by the mutual orientation of the amide and ester chromophores (see, for instance, [51, 427, 429, 575]). Recently, it has been reported that ultrasonic absorption can be used for determining the number of conformers participating in the conformational equilibrium of peptides [325, 326]. The infrared spectra of peptides in solvents which do not form hydrogen bonds with amide groups (for example, CCl_4 or $CHCl_3$) can yield valuable information about the presence of intramolecular hydrogen bonds therein [24, 25, 224, 437]. Thus, an NH stretching band (so-called amide A region) at 3420—3480 cm^{-1} is evidence of free NH, whereas bands in the 3300—3380 cm^{-1} region are due to hydrogen-bonded NH. Care must be taken in making assignments in the 3380—3420 cm^{-1} region, where absorption can be due to both free or weakly hydrogen-bonded NH groups. Of considerable importance is that a quantitative appraisal of the different types of NH absorbing in the amide A region can be obtained from the integral band intensities [437].

Metal—oxygen bonds display characteristic stretching vibrations in the 100—450 cm^{-1} region of the far infrared [221, 528, 606, 1091, 1098], showing the considerable potentialities held in far infrared studies for determining force constants of the complexes in question, the symmetry of their internal coordination sphere, etc.

A wealth of possibilities for studying peptide spatial structure is offered by the vigorously developing nuclear magnetic resonance spectroscopy. The number and intensity of the signals in the NMR spectrum provide information about the number of conformers in the equilibrium mixture, their relative amounts and symmetry. The equilibrium kinetic parameters are determined from the temperature dependence of the shape of the signal. Since the C^α protons are spatially close to the amide or ester groups situated on either side along the length of the chain, their chemical shifts are strongly dependent upon the orientation of the corresponding carbonyls, i.e. they are ϕ- and ψ-dependent. The signals most sensitive to the peptide conformations are those of the amide NH proton. Their participation in intramolecular hydrogen bonding results in their being screened from the medium, lowering the rate of their deuterium exchange in mobile-deuterium-containing solvents (^2H_2O, $C^2H_3O^2H$, $C^2H_3COO^2H$, etc.). This has led to the extensive use of two NMR techniques for studying intramolecular hydrogen bonding in peptide systems, namely:

1. Measurement of the temperature dependence of the NH chemical shifts (δ) in hydrogen-bonding solvents (for instance, dimethylsulfoxide or methanol) [693, 1024]. High $\Delta\delta/\Delta T$ values ($(6 \div 12)\cdot10^{-3}$ ppm/deg.) correspond to solvated NH groups, whereas low values ($(0 \div 2)\cdot10^{-3}$ ppm/deg.) are due to groups participating in intramolecular hydrogen bonds. Intermediate values of $(2 \div 6)\cdot10^{-3}$ ppm/deg. could indicate an equilibrium of several conformational forms differing in the number and/or position of the intramolecular hydrogen bonds [86, 121, 213, 430, 434, 435, 505, 559, 776, 777, 1033, 1049].

2. Measurement of the deuterium exchange rate of the NH group by following the diminution in intensity of the corresponding signal [121, 434, 435, 505, 703, 776, 777, 972]. Differences in half-exchange times ($\tau_{1/2}$) of one order of magnitude or more permit location of the free and hydrogen bonded NH groups in the molecule. With lesser differences additional information is needed, because deuterium exchange kinetics depend not only upon the presence of intramolecular hydrogen bonds, but also upon other factors (steric hindrances, electronic structure of the given amide group, etc.).

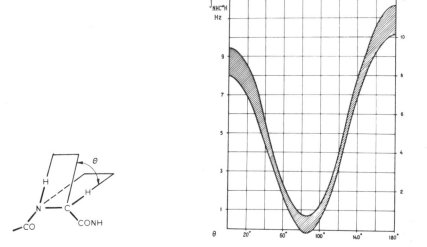

Fig. 38 (left). Dihedral angle θ between the H—N—C$^\alpha$ and N—C$^\alpha$—H planes.
Fig. 39 (right). Dependence of the vicinal spin—spin coupling constant $^3J_{NH-CH}$ on the dihedral angle θ.

The peptide NH protons are spin coupled to the C$^\alpha$H protons, the coupling constant depending upon the dihedral angle between the H—N—C and N—C$^\alpha$—H planes (Fig. 38). Several correlations have been proposed [116, 117, 808, 1001, 1049], but the one in best agreement with experimental results is that shown in Fig. 39 [113]. Hence, by determining the $^3J_{NH-CH}$ constant from the NMR spectrum it is possible to estimate the upper and lower limits of the corresponding θ values, and thus to obtain the conformational parameter ϕ for the given compound (the relation between the θ and ϕ angles for the L- and D-amino acid residues is shown in Table 45). Of course, the results are most unequivocal in the case of conformationally rigid systems, for which the region of possible ϕ values is obtained directly from the curve shown in Fig. 39. When there is an equilibrium of several forms a straight-forward interpretation can be made only for the values of $^3J_{NH-CH} > 9$ Hz and $^3J_{NH-CH} < 3$ Hz, corresponding to θ values of $\sim 180°$ and $\sim 90°$, respectively, whereas for analysis of the $^3J_{NH-CH}$ values from 3 to 9 Hz one must know the relative amounts of the conformers in the equilibrium mixture.

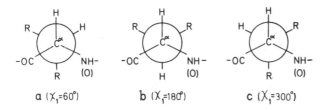

Fig. 40. C^α—C^β rotational isomers.

Similarly, from the corresponding stereochemical relationship, there follow the mutual orientations of the $C^\alpha H$—$C^\beta H$ protons, determining the conformations of the amino and hydroxy acid side chains. In the present case, of the three most probable C^α—C^β rotamers the *gauche* (a and c in Fig. 40) have $^3J_{C^\alpha H - C^\beta H}$ values of 3 Hz, while the third (*trans* rotamer b) has a $^3J_{C^\alpha H - C^\beta H}$ value of 13 Hz [113].

At present ^{13}C-NMR spectroscopy is beginning to be used for investigating complex formation, the main attention being focused on the chemical shifts of the carbons bound to the ligand oxygens. One can thereby determine the number of ligands in the complex and the relative strengths of their interaction with various ions [114, 325, 692, 794].

The methods described above give an insight into the conformational mobility of peptides and allow quantitative assessment of the principal conformers in the equilibrium mixture, characterization of the internal coordination sphere, identification of the NH groups forming intramolecular hydrogen bonds, and also determination of the NH—$C^\alpha H$ and $C^\alpha H$—$C^\beta H$ rotational states. While the information thus obtained sharply restricts the number of conformations possible in a given peptide system, it still does not directly give the spatial structure, i.e. the ϕ, ψ, ω and χ parameters. For this one requires a theoretical conformational analysis (see reviews [188, 809]) giving the most preferable forms from which there can be selected the one in best accord with the experimental data. The first steps in the conformational analysis are usually calculation of the so-called conformational maps, two-dimensional diagrams describing the potential energies of the amino and

$$X \quad R \qquad\qquad R$$
$$| \quad | \qquad\qquad\quad |$$

hydroxy acid fragments (—N—CH—CO— and O—CH—CO—) or their model compounds as functions of ϕ and ψ, the bond lengths, valency angles and the angle ω (0 or 180°) being taken as constant. The energy of individual forms of the more complicated peptide systems in a given conformational state is then considered to be approximated by the sum of the energies of their fragments (in the appropriate conformations) found from the corresponding conformational maps. The values obtained are further refined by minimizing the total energy of the system with respect to a number of variables (ordinarily with respect to all the ϕ, ψ, ω and χ angles and the C^{α} valency angles). It was in this way that the preferred forms of a number of fragments of the valinomycin and antamanide groups of metal-binding macrocyclic compounds (such as intramolecular hydrogen bond-stabilized 10-membered rings) have been determined. Undoubtedly, the most reliable calculations of the energies are for the entire peptide molecule rather than for its component parts. However, the considerable methodological difficulties and the excessive computer time requirements have greatly restricted the number of studies in this field, one of them being the analysis of the structure of enniatin B discussed in Part III.A.1.c.

After determining the conformational parameters of the energetically favorable forms, it is easy to calculate the dipole moments of their peptide skeleton by vector addition of moments of the individual amide and ester groups [223]. Since in most of the peptide complexones the side chains are hydrocarbon radicals contributing little to the total moment, a comparison of the dipole moment thus calculated with the experimental value serves as an important criterion of the correctness of the proposed conformation.

III.A.1.a. Valinomycin

The valinomycin molecule is in the form of a 36-membered ring with a wealth of conformational possibilities, and the complete elucidation of its spatial structure required considerable effort. At the same time it exemplified quite dramatically the potentialities inherent in the composite use of physicochemical methods, valinomycin being the first large peptide whose structure was established without resorting to X-ray analysis [430, 431, 607, 687, 693, 1024]. The findings served as the basis for studying the structure—function relation of this antibiotic, for which purpose a number of its membrane-active analogs with unique, predetermined properties were synthesized.

Spectroscopic data revealed that the valinomycin conformation is highly solvent-dependent. This could be seen, for example, by the sharp changes in the ORD curves on passing from heptane to alcohol and, further, to aqueous mixture (curves 1, 3 and 5 in Fig. 41). The composite use of ORD, CD, infrared, ultraviolet and NMR showed that in solution valinomycin exists as an equilibrium mixture of 3 major forms (A, B and C) (Fig. 42). In form A, predominant in non-polar solvents (CCl_4; $CHCl_3$; heptane—dioxane, 10:1) all NH groups are participating in hydrogen bonding with the amide carbonyls. Form B, with the valyl NH groups forming three intramolecular hydrogen bonds, is predominant in solvents of medium polarity (CH_3CN; C_2H_5OH; CCl_4—$(C^2H_3)_2SO$, 7:3), the L-valyl NH groups being solvated by

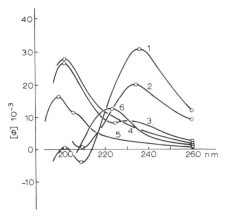

Fig. 41. ORD curves of valinomycin in different solvents (1, heptane; 2, heptane—dioxane, 6:1; 3, ethanol; 4, acetonitrile; 5, water—trifluoroethanol, 2:1) and of valinomycin·K^+ complex in ethanol (curve 6).

the solvent. In form C, predominant in polar solvents ($(CH_3)_2SO$; trifluoroethanol—water, 1:2), all the NH groups, particularly after heating, are hydrogen bonded to the solvent.

A number of additional forms in solution, intermediate between A, B and C and possessing five, four, two or one intramolecular hydrogen bond, have apparently been revealed in relaxation studies of valinomycin by ultrasonic absorption [326].

The structures of A and B were established by comparing the results of a theoretical analysis of the valinomycin molecule with its NMR data. In A, the cyclodepsipeptide chain forms a fused system of six 10-membered rings, each closed by a hydrogen bond resulting from the interaction of amide carbonyl with the neighboring amide NH in the "direction of acylation" (intramolecular hydrogen bond of the 4 → 1 type). Thus, in non-polar media the valinomycin molecule is in compact form, resembling a bracelet about 8 Å in diameter and 4 Å high. In principle, form A presents two ways for the folding of the depsipeptide chain (A_1 and A_2, Fig. 43), differing in chirality of the ring system and orientation of the side chains. In turn (as shown in Table 46) within each of forms A_1 and A_2 there can be four different conformers with respect to orientation of the ester carbonyls. As one can see

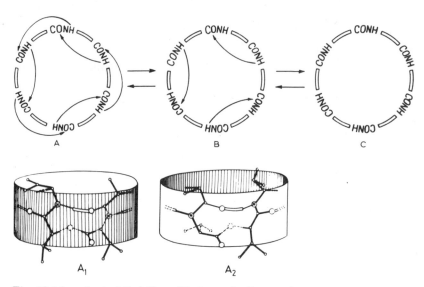

Fig. 42 (above). $A \rightleftharpoons B \rightleftharpoons C$ equilibrium of valinomycin.
Fig. 43 (below). Schematic drawing of the A_1 and A_2 forms of valinomycin.

from Fig. 43, when the molecule is so positioned that the L-lactic acid residues are in its upper part, acylation is clockwise in all the A_1 conformations and counterclockwise in all the A_2 conformations. Calculation of the optimal conformations of the two protected didepsipeptides,

TABLE 46

POSSIBLE CONFORMATIONS FOR VALINOMYCIN C_3-SYMMETRIC A FORM

Conformer	Orientation of the D-valine CO groups	Orientation of the L-valine CO groups	Relative energy (kcal·mole^{-1})
A_1	Out	Out	12.8
	In	Out	15.0
	Out	In	25.2
	In	In	27.6
A_2	Out	Out	>60
	In	Out	>40
	Out	In	>18
	In	In	0

Ac—D-Val—L-Lac—NHMe and Ac—L-Val—D-HyIv—NHMe, possessing 4 → 1 hydrogen bonds and simulating the hydrogen bond-stabilized 10-membered ring fragments of form A, permitted estimation of the relative conformational energies shown in Table 46. The most preferred structure from the standpoint of the fragment energies is that of A_2, with the ester groups directed into the molecule. Next to it, but energetically considerably less favorable are two A_1 conformations. The other conformations are of too high energy to be of any significant probability.

A similar conclusion has been reached by Mayers and Urry [607] after calculating the energies of the bracelet conformations of valinomycin. According to their findings the most advantageous conformation from the standpoint of non-bonded interactions is form A_2 ("all in"); form A_1 ("all out") is less preferable by 6.1 kcal/mole whereas the other conformations possess energies exceeding those of the above two forms by 400 kcal/mole. The sharper differentiation in the stabilities of the forms found by the American authors than that in Table 46 is apparently due to the simpler scheme they used for calculation, which did not include potential energy minimization.

In form A_2 the dihedral angle θ between the H—N—C^α and N—C^α—H planes in the NH—C^αH fragments can vary within the limits of 60 to 135°, which, according to the $^3J_{NH-CH}$ stereochemical dependence, corresponds to coupling constants from 0 to 6 Hz. The A_1 conformations can have a wider range of $^3J_{NH-CH}$ values (0 to 9.4 Hz, taking into account that the θ values vary from 0 to 60°), the most preferable conformations corresponding to $^3J_{NH-CH}$ = 7—9.4 Hz. The experimental $^3J_{NH-CH}$ values for form A are 6.6—8.8 Hz, showing unequivocal evidence in favor of the A_1 conformation.

It thus appears that structure A_2, which is the most favorable from the standpoint of nearest-neighbor interactions (A_2, all ester groups "in"; see Table 46), is in fact not realized in non-polar solvents. The reason for this is the appearance of additional destabilizing interactions, prominent among which are the electrostatic repulsions of the six oxygens of the inwardly directed carbonyls (see Part III.B). Moreover, a certain part is possibly played by steric interactions of the valylisopropyl groups which in all the A_2 conformers are oriented in the direction of the symmetry axis*. Of the two possible type A_1 conformations (all ester CO groups "out", and three ester CO groups "in" and three "out"), the second, which appeared preferable

---

* The existence of a threefold axis in form A is evidenced by the invariance of the NMR spectra recorded on cooling to low temperatures (cf. the low-temperature enniatin B spectrum; see Part III.A.1.c).

on the basis of a simple analysis of molecular models [431, 904], turned out to be the less probable one because of the destabilizing effect of the electrostatic interactions of the inside oriented carbonyl groups (similar to the case of A_2 "all in"). This corresponds approximately to the following parameters.

	D-Val	L-Lac	L-Val	D-HyIv
ϕ	−40	−100	25	100
ψ	−70	40	70	−30

The dipole moment calculated for this conformation (0—0.7 D) agrees with the experimental value of 3.5±0.1 D, if one takes into account that under the conditions of measurement (CCl_4, 25°C) significant amounts of the less symmetric forms (for instance form B with a dipole moment as high as 7—9 D) are present.

From an analysis of the spin—spin coupling constants of the $C^{\alpha}H$—$C^{\beta}H$ protons of valinomycin shown in Fig. 44, it follows in agreement with the theoretical calculations that in the A form the isopropyl side chains of the amino acid residues assume a *trans* conformation while those of the hydroxy acid residues are in *gauche* conformation.

Form B, stabilized by three intramolecular hydrogen bonds, is more flexible than form A. It is characterized by the following ϕ and ψ parameters.

	D-Val	L-Lac	L-Val	D-HyIv
ϕ	120	−60	50	100
ψ	85	120	65	−40

In the proposed conformation (Fig. 45) one may discern a hydrophobic "core" of the D-valyl and L-lactyl side chains, encased in the depsipeptide chain with its polar groupings. Under such circumstances the 10-membered hydrogen-bonded rings are on the molecular periphery, making form B resemble the "propeller" conformation of trisalicylides [212]. As in form A, the L-valyl and D-α-hydroxyisovaleryl isopropyl groups assume *trans* and *gauche* orientations, respectively, the most advantageous from the stand-point of nearest neighbor interactions. However, steric hindrance forces the spatially neared D-valyl side chains to assume the less preferable *gauche* orientation ($^3J_{C\alpha H - C\beta H}$ = 4.6 Hz).

On the basis of the NMR spectra of valinomycin in $(C^2H_3)_2SO$, Urry and Ohnishi [693] proposed a structure for this compound, outwardly resembling the B form but differing from it considerably in the orientations of the carbonyl groups and the side chains (Fig. 46). However, calculation

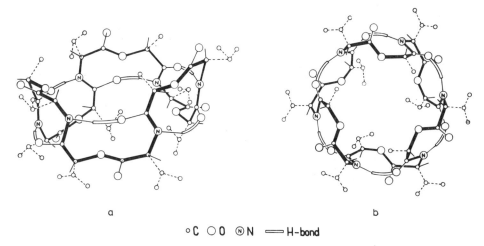

°C ○O ⓝN ══ H-bond

Fig. 44. Conformation of valinomycin in non-polar solvents. a, side view; b, view along the symmetry axis.

has shown it to possess high energy so that it is not likely to exist in significant amounts. Moreover, these workers assigned a *cis* conformation to the NH—C$^\alpha$H fragment of the L-valine residues in disagreement with the experimentally found *trans* orientation [430].

The C form of valinomycin devoid of intramolecular hydrogen bonds has apparently no fixed structure, being an equilibrium mixture of a large number of energetically similar conformers. Evidence of this can, for instance, be seen in the averaging out of the $^3J_{NH-CH}$ constants on passing over to polar solvents.

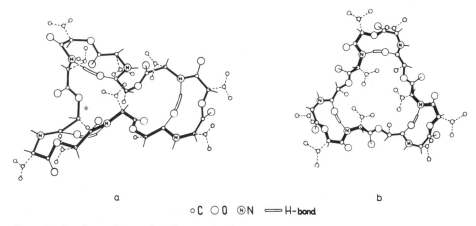

°C ○O ⓝN ══ H-bond

Fig. 45. Conformation of valinomycin in solvents of medium polarity. a, side view; b, view along the symmetry axis.

124

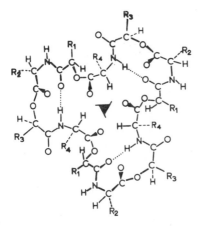

Fig. 46. Conformation of valinomycin in $(C^2 H_3)_2 SO$, according to Urry and Ohnishi [694].

The arguments presented above in support of the structure of forms A, B and C have been confirmed by the work of Patel and Tonelli [734]. For form C these authors presented the following ranges of possible ϕ and ψ values.

	D-Val	L-Lac	L-Val	D-HyIv
ϕ	140 — 150	−100 − −90	−150 − −140	80 − 90
ψ	−140 − −170	60 − 80	70 − 140	−80 − −60

Issuing from the principle of maximum hydrophobic interactions, Warner proposed a peculiar conformation for valinomycin with a sharp borderline between the hydrophobic and polar regions of the molecule [1044]. However, since he did not take into account the other non-bonding interactions, this purely hypothetical structure is at present considered to be of little likelihood.

Interesting results have been obtained from the X-ray analysis of crystalline valinomycin [214, 215]. The conformation found is represented in Fig. 47. It has the following torsional angles as determined from molecular models (accuracy of the measurements ±30°), the readings being given in the direction of acylation, beginning with D-Val [734].

	1	2	3	4	5	6	7	8	9	10	11	12
ϕ	70	−70	−80	110	70	−70	−70	70	80	−110	−70	70
ψ	−110	−20	60	−20	−110	−20	110	30	−60	35	110	30

The molecule has no C_3 axis, but does have a pseudosymmetry center; it has six intramolecular hydrogen bonds, of which four are of the $4 \rightarrow 1$ type formed by CO and NH groups. The other two H bonds, however, are of the $5 \rightarrow 1$ type, and are formed by ester groups closing 13-membered rings. In general the conformation of crystalline valinomycin can be regarded as a distorted A_2 "all in" conformation wherein the ester carbonyls have been somewhat shifted to the periphery of the bracelet by electrostatic

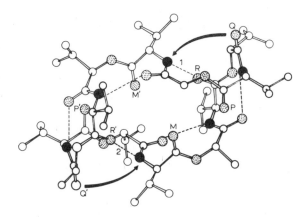

Fig. 47. Conformation of crystalline valinomycin. o, carbon atoms; ⊚, oxygen atoms; •, nitrogen atoms; – – –, hydrogen bonds. M and M', P and P', R and R', are ester carbonyl oxygens related by the pseudosymmetry center; Q and Q' are the amide oxygens not participating in hydrogen bonding. 1 and 2, hydrogen bonds of the $5 \rightarrow 1$ type. Arrows mark the direction of Q and Q' displacement after crystalline conformer rearrangement into the complexed form.

interaction. Such rotation has caused some changes in the intramolecular hydrogen bond system, the $5 \rightarrow 1$ bonded CO and NH groups being at considerable angles to each other and in different planes [734]. Thus, by a certain weakening of the H bonds, the valinomycin molecule has been able to retain the optimal conformations of the individual amino and hydroxy acid residues.

The following reasoning shows that this possibility does not materialize in solutions (considering only non-polar solvents where all six intramolecular hydrogen bonds occur): (1) In the crystalline state all amino and hydroxy acid residues are conformationally non-equivalent. However, the NMR spectra of valinomycin, even when the solution has been cooled to $-95°C$, disclosed no signs of peak separation [430]. (2) The infrared spectra showed only a single symmetric ester CO stretching band, in the region of relatively high frequencies (1757 cm^{-1} in CCl$_4$ [426, 431], 1755 cm^{-1} in CHCl$_3$; Fig.

23). This result is not in accord with the presence of two hydrogen-bonded and four free carbonyls. (3) Calculation of the $^3J_{NH-CH}$ constants on the basis of the above cited ϕ angles for the D- and L-valine residues, yields average values of 5.7 Hz, which is significantly less than the experimental 8.5 and 6.6 Hz in CCl_4 and 8.8 and 6.6 Hz in $CHCl_3$ [426]. (4) The calculated dipole moment for the crystalline conformation (6.8 D [734]) considerably exceeds the experimental value of 3.5 ± 0.1 D (p. 122).

There is no doubt that the A_1 "all out" conformation existing in heptane, CCl_4 or $CHCl_3$ is less advantageous from the standpoint of non-bonding interactions than the crystalline conformation (see Table 46) and that the differences in stability of two hydrogen bonds should not have a significant bearing on the energy relations of the two forms. Why then should the first form, which appears to be less favorable, be the more preferable in solution? This question has been partially answered above (p. 121) when the reasons for the instability of form A_2 "all in" in non-polar media were discussed. The point is that in form A_1 "all out" the ester carbonyl oxygens are at maximum distance apart, i.e. this conformation is more favorable from the standpoint of electrostatic interactions than is the crystalline conformation, where these atoms are nearer to each other. Hence, in non-polar media, where the contribution from the Coulomb interactions is felt the most, the first structure is found to be the more preferable; in the crystal, where the effective local dielectric constant is much higher, the Coulomb contribution is weakened and the more preferable form becomes the A_2 "all in" form. From this it follows that, in polar solvents, one might have expected the appearance of similar structures if the intramolecular hydrogen bonds had not been disrupted. It is also possible that significant amounts of the A_2 "all in" form are present in solutions of medium polarity (for instance, tetrahydrofuran or acetonitrile) which are weak donors and/or acceptors of protons in hydrogen bonding.

In concluding the discussion of free valinomycin it should be stressed that it is the existence of a conformational equilibrium shifting with change in the environment that determines the entire specificity of its behavior, above all, in the complexing of metal ions and in its interaction with membrane systems. This must, therefore, always be borne in mind when interpreting the results of physicochemical and biological studies of this antibiotic.

Of decisive importance in understanding the complexing behavior of valinomycin and its membrane-affecting properties is the structure of its alkali ion complexes. The ORD curves of its K^+ complex are little influenced by change of solvent. From this follows that contrary to the non-complexed antibiotic the mutual orientation of the chromophoric (amide and ester) groups of the complex is only weakly solvent-dependent. A comparison of the infrared spectra of a $CHCl_3$ solution of valinomycin and its K^+ complex (see Fig. 23) points to the existence in the latter of the bracelet system of hydrogen bonds and to interaction of the ester carbonyls with the cation (a shift in the stretching vibrations from 1755 to 1739 cm^{-1}). The chemical shifts of the NH protons in CH_3OH and CH_3CN—$CHCl_3$ (1:1) are practically temperature-independent, evidence of retention of the bracelet conformation in these solvents. The relatively low $^3J_{NH-CH}$ values (5.2—5.6 Hz) show the protons of the corresponding fragments to be *gauche* oriented.

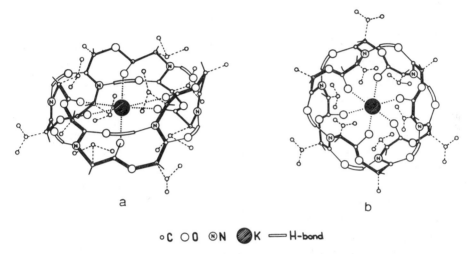

°C OO ⊗N ◍K ══H-bond

Fig. 48. Conformation of valinomycin·K^+ complex. a, side view; b, view along the symmetry axis.

Two of the eight conformations listed in Table 46 are in accord with these data: namely, the A_1 "all in" and the A_2 "all in", for these have *gauche* NH—CH protons and their inwardly pointing carbonyls provide for effective ion—dipole interaction with the centrally located cation. From a comparison of the energies of these two forms it can be seen that A_2 is much more preferable (by ~28 kcal/mole) than A_1, and therefore that the structure of the K^+ complex of valinomycin is the one represented in Fig. 48. It has approximately the following ϕ and ψ parameters.

	D-Val	L-Lac	L-Val	D-Hylv
ϕ	55	−70	−60	75
ψ	−110	−20	135	15

As in form A of free valinomycin the $C^\alpha H$—$C^\beta H$ protons of the valyl side chains in the K^+ complex are *trans* ($^3J_{C\alpha H-C\beta H}$ = 10.7—11.0 Hz), and their hydroxy acid counterparts are *gauche* ($^3J_{C\alpha H-C\beta H}$ = 3.6—3.8 Hz).

A similar structure has been proposed by Urry and Ohnishi [607, 694, 1024] based on the weak temperature dependence of the NH proton chemical shifts in CH_3OH solution and on the assumption of interaction of all the ester carbonyls with the central cation*. The choice between A_1 and A_2 conformations was made on the basis of theoretical analysis (see p. 121). It was shown, moreover, that when the ester carbonyls are oriented so as to give minimum non-bonded interactions the internal cavity diameter (~2.8 Å) is very near to the size of the potassium and rubidium ions [607].

As one might have expected the ester carbonyl signals in the ^{13}C-NMR spectra of valinomycin display a strong paramagnetic shift (3.1—5.5 ppm) [114] on complexation. The signals of the amide carbonyls (0.5—3.1 ppm) are also shifted down-field, although to a lesser extent, indicating their possible participation in the interaction with the central ion (the $K^+ \cdots O$ distance is approx. 2.7—3.0 Å for the ester carbonyls and approx. 4.0—4.5 Å for the amide carbonyls).

Thus, independent of the solvent species, when complexing K^+ valinomycin assumes a conformation wherein the ester carbonyls form a cavity ~2.8 Å in diameter. Whereas, as shown earlier, such a conformation is destabilized by electrostatic interactions, in the complex the presence of a central positive charge not only annuls the electrostatic repulsion but is itself an additional stabilizing factor.

In principle the most advantageous disposition of the six ligand oxygens is at the vertices of a regular octahedron with the metal ion in the center. However, the macrocyclic system of the K^+ complex of valinomycin shows a less symmetric coordination of the type of a triangular antiprism. This conclusion has been arrived at from a study of the far infrared spectra, which show two stretching frequencies for the $K^+ \cdots O$ bond instead of one as would have been expected for complexes with octahedral coordination of the ligands [428]. The same results have been obtained in studies of the valinomycin·Rb^+ and valinomycin·Cs^+ complexes.

X-ray analysis of the complex salt valinomycin·$KAuCl_4$ by Pinkerton, Steinrauf and Dawkins [751], in complete agreement with the above data, showed that the system of six intramolecular hydrogen bonds ($N \cdots C$ distance 2.8—3.0 Å) is also present in the crystalline state of the complex and that here, too, the cation, located in the center of the cavity, is interacting with the ester carbonyls ($O \cdots K^+$ distance 2.7—2.8 Å). It was also demonstrated for the first time that the cyclodepsipeptide chain of the complex is in the A_2 conformation.

---

* In this review Urry [1024] mistakenly ascribes to the NH—$C^\alpha H$ fragments of the K^+ complex of valinomycin the dihedral angle θ —30° which would mean that $\phi_1 \sim -30°$ and $\phi_3 \sim 30°$.

A characteristic feature of the complexed conformation is the effective shielding of the central cation by the ester groups, the hydrogen bond system and the pendant valyl isopropyl groups; the hydrogen bonds are in turn shielded from solvent attack by the hydroxy acid side chains outwardly projecting from the bracelet (Fig. 48). The lipophilic nature of the molecular surface of the complex cation explains its high solubility in non-polar organic solvents and is apparently essential for the functioning of valino- mycin on membranes (see Chapter IV).

Comparison of the conformations of free valinomycin and of its complex shows straightforwardly that complexing is accompanied by serious con- formational rearrangements in the depsipeptide chain. This appears with

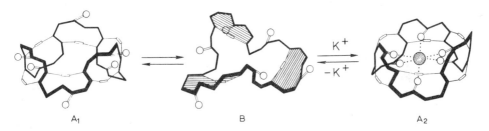

Fig. 49. Conformational equilibrium of valinomycin in the presence of potassium ions.

particular clarity in the opposite orientations of the ester carbonyls in form A of free valinomycin ("all out") and in the complex ("all in"). Mention should also be made of the opposite chiralities of the depsipeptide chain and of the hydrogen bond systems in the "non-polar" (A_1) conformation of free valinomycin and the complex (A_2). Interconversion of these two forms is impossible without rupturing at least three of the hydrogen bonds, from which follows that the intermediate stage of complex formation in non-polar media is apparently form B (Fig. 49). The long lifetimes of the K^+ complex and the slow rates of potassium migration between the valinomycin molecules [381], leading to considerable signal broadening in the NMR spectra under conditions of incomplete complexation [381, 383, 430] (see spectrum C in Fig. 50) (cf. Section II.A.1), are possibly due to the high energy barriers associated with these conformational reconstructions.

Interesting data have been obtained in studies of the valinomycin complexes with Rb^+, Cs^+ and Na^+ in CCl_4—CH_3CN and $CHCl_3$—CH_3CN mixtures. Despite their different stabilities, judging from the similarity of the $^3J_{NH-CH}$ constants, and from the low $\Delta\delta/\Delta T$ values, in non-polar solvents these complexes are all in the same conformation, similar to that of the K^+ complex. However, each complex has its own specificities, most clearly

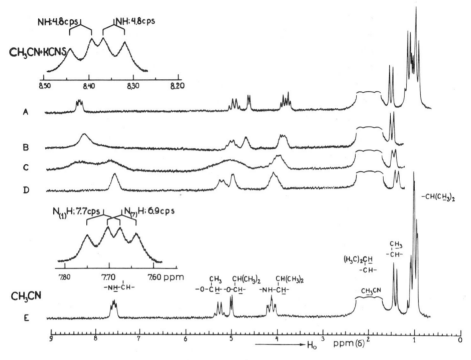

Fig. 50. NMR spectra of valinomycin, its K^+ complex (A) and equilibrium mixtures of the two forms in CH_3CN at different valinomycin to KNCS molar ratios. B, 1:0.73; C, 1:0.43; D, 1:0.14.

expressed in the infrared spectra (Fig. 51). For instance, $Rb^+ \cdot$valinomycin, very similar to $K^+ \cdot$valinomycin in all its characteristic features, displays a small (approx. 10 cm^{-1}) shift to the shorter wavelength region and a decrease in intensity of the amide A infrared band, indicating an increase in the $O \cdot \cdot N$ distance and a decrease in the hydrogen bonding energy. The tendency for the hydrogen bonding system to weaken with increase in cation size is particularly manifested by the Cs^+ complex ($\Delta\nu_{NH}$ is 24 cm^{-1}). The size of this cation exceeds the "normal" size of the internal cavity of valinomycin (~ 2.8 Å). It is apparently the weakening of the hydrogen bonds which explains the high field shift of the NH signal in the cesium complex relative to the potassium complex which was observed by Haines et al. in C^2HCl_3 solutions [383].

An interesting feature of the infrared spectra of the Na^+ complex taken in a CCl_4-CH_3CN (2:1) mixture is the asymmetry of the ester carbonyl stretching vibration, which indicates non-equivalence of the carbonyls in this structure. This can most likely be explained by the small size of the sodium cation, as a result of which it is shifted from the center to the periphery of the internal cavity so that the ion interacts differently with the different ester dipoles.

Without doubt the K^+, Rb^+ and Cs^+·valinomycin complexes also assume the A_2 "all in" conformation in the more polar solvents (ethanol, methanol). However, the CD curves of the Na^+ complex of valinomycin in methanol differ significantly from those of the K^+ complex, and the shift of the ^{13}C signals of the ester carbonyls upon formation of the Na^+ complex is much less (1.1 and 1.7 ppm) than in the case of the K^+ complex (4.3 and 4.6 ppm) [327]. On this basis, the inference has been made that on interaction with

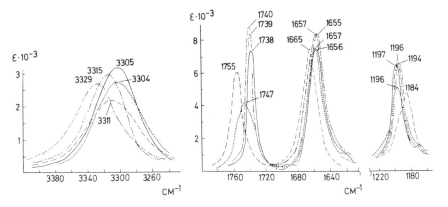

Fig. 51. Infrared spectra of valinomycin (- - - -) and its Na^+ (· · · ·), K^+ (———), Rb^+ (—·—·—) and Cs^+ (—··—···) complexes in a CCl_4—CH_3CN (2:1) mixture.

Na^+ in methanol, valinomycin assumes a new conformation, differing from that of the other complexes. One of the possible variants of such a structure is an equilibrium mixture of two open forms containing not more than three intramolecular hydrogen bonds. In one of these it is the D-valyl carbonyls which interact with Na^+, in the other, it is the L-valyl carbonyls which participate in the interaction. The sodium ion is partly solvated. This hypothesis, of course, requires further confirmation. In particular, of especial interest would be the investigation of the intramolecular hydrogen bonding system of the sodium complex in methanol by the conventional methods of NMR-^1H spectroscopy.

In general, despite the unfinished state of the studies aimed at determining the structure of Na^+·valinomycin, the data described here make quite clear the reason for the exceptional K/Na-complexing selectivity of valinomycin and the diminished stability of the Cs^+ complex, as compared with the K^+ complex (see Part III.C).

III.A.1.b. Valinomycin analogs

It turned out that the spatial structure of valinomycin could serve as the basis for explaining the complexing behavior of a large number of its analogs for which the Na^+ and K^+ stability constants were carefully determined. For further discussion it is convenient to divide these analogs into four types: 1. Analogs differing from valinomycin in ring size (compounds 10 and 82). 2. Analogs in which the methyl or isopropyl side chains have been substituted by related alkyl radicals (compounds 15—17, 21, 28, 29, 38, 42—44, 47,

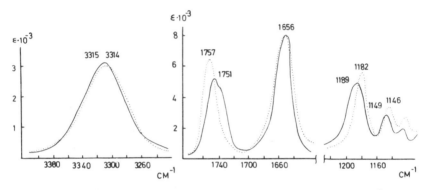

Fig. 52. Infrared spectra of "hexadeca-valinomycin" (· · · ·) and its K^+ complex (———) in a CCl_4—CH_3CN (2:1) mixture.

50—52, 54, 55, 57 and 62); actually, here also belong analog (79) ("false" *retro*-valinomycin) and its antipode (80), as well as the protected Glu^3 and Lys^3 derivatives of valinomycin (compounds 31, 33 and 34). 3. Analogs with different configurations of one or more amino and/or hydroxy acid residues (compounds 14, 19, 26, 36, 49, 54, 56 and 63); to these should also be added analogs with modified side chains (compounds 58 and 64—73) and several topochemical derivatives of valinomycin (75—78). 4. Analogs with *N*-methylamide or ester groups instead of the valinomycin amide and/or ester groups (compounds 18, 23—25, 27, 35, 39, 40, 45, 46, 48, 53, 59 and 60), and also the peptide analog of valinomycin obtained by replacing the L-lactic acid residues by L-proline residues and the D-hydroxyisovaleryl residues by D-proline residues (81).

Compound (82) built up of four instead of the valinomycinic three tetradepsipeptide fragments ("hexadeca-valinomycin") complexes potassium ions only very weakly, cesium ions somewhat better (Table 2). According to the temperature dependence of the NH proton chemical shifts, the Cs^+ complex possesses the bracelet system of hydrogen bonds and the depsipeptide chain is folded similarly to that of the valinomycin complexes (as follows from the close $^3J_{NH-CH}$ constants: 5.2—5.5 Hz for valinomycin and 4.8—6.1 Hz for compound (82)). The internal cavity of compound (82), however, (4—5 Å in diameter) is too large to provide for effective interaction of all the ester groups with the alkali metal ions. The situation closely resembles that for the interaction of valinomycin with sodium in non-polar

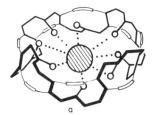 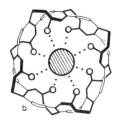

a   b

Fig. 53. Schematic drawing of "hexadeca-valinomycin" complex with Cs$^+$. a, side view; b, view along the symmetry axis.

media; the cation "rolling about" in the cavity being at any given moment in close contact with only part of the carbonyl groups, which thus explains the heterogeneous character of the C=O and C—O stretching bands in the infrared spectra of the "hexadeca-valinomycin" potassium complex (Fig. 52). It should be noted that eight carbonyls, instead of the valinomycinic six carbonyls, are participating in the ion—dipole interaction with Cs$^+$, forming a tetragonal antiprism with eight oxygen atoms at its vertices (Fig. 53).

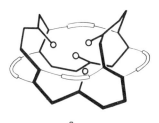 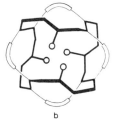

a   b

Fig. 54. Schematic drawing of the bracelet conformation of "octa-valinomycin". a, side view; b, view along the symmetry axis.

The formation of a bracelet conformation with inside-oriented carbonyl groups in "octa-valinomycin" (10) (Fig. 54) would have caused steric strain, making understandable the inability of analog (10) to complex Na$^+$ or K$^+$ ions efficiently.

The second group of analogs (for instance, compounds 15, 21, 28, 38, 52, 57, 79 and 80) often display no essential change in the complex stability since the position of the local energy minima on the conformational maps of the hydroxy and amino acid residues are only weakly dependent upon the structure of the side chains [773]. When there is a certain decrease in the free energy of complexation (up to say 3.0 kcal/mole in compound (50)) it is apparently due to increased flexibility of the initial cyclodepsipeptides rather than to steric hindrance in the complexes. In other words a decrease in free energy is here more probable on the left-hand side than an increase on

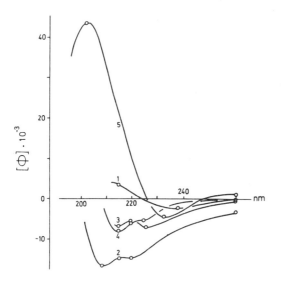

Fig. 55. ORD curves of the "false" *retro*-valinomycin in different solvents (1, heptane—dioxane (5:1); 2, ethanol; 3, acetonitrile; 4, water—trifluoroethanol (2:1)) and of its K^+ complex in ethanol (curve 5).

the right-hand side of the system $K^+ + L \rightleftarrows (K \cdot L)^+$, where L is the metal-binding macrocyclic compound (cf. p. 139).

Compound (79) is the most thoroughly investigated of the second group of analogs [3, 432, 521]. Structurally it is very similar to valinomycin, only interchange of the methyl groups of the L-lactic acid residues with the isopropyl groups of the valyl residues being required for conversion of the former to the latter. As in valinomycin, the spatial structure of analog (79) is highly environment-sensitive and (judging, for example, from the ORD curves in Fig. 55) changes radically on passing from non-polar to polar solvents. In CCl_4 and $CHCl_3$ (Fig. 56) its infrared spectra are very similar to those for valinomycin (cf. Fig. 23) and bear evidence of preference for the bracelet conformation of types A_1 and A_2. The $^3J_{NH-CH}$ constants (6.9 and 8.2 Hz) indicate that the NH—C^αH fragment is intermediate between *cis* and *gauche*, i.e. that A_1 is the preferable form. Like valinomycin, the most favorable conformation is with "all CO's out" (see Table 46). The relatively high dipole moment of analog (79) (6.0 D) is apparently due to a greater difference in the orientation of the alanyl and valyl carbonyls with respect to the symmetry axis than of the L- and D-valyl carbonyls of valinomycin. The K^+ complex of compound (79) has practically the same ϕ and ψ parameters as those of valinomycin·K^+ for it retains the hydrogen bonding system and its ester groups react equally with the cation (symmetric band at 1741 cm^{-1} in Fig. 56) and the $^3J_{NH-CH}$ constants (4.7—5.2 Hz) are close to those for the valinomycin complex (5.2—5.6 Hz) (see p. 127).

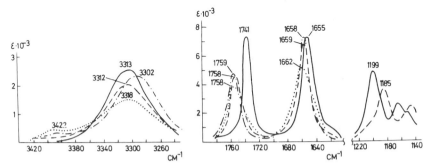

Fig. 56. Infrared spectra of "false" *retro*-valinomycin in CCl$_4$ (—·—·—), CHCl$_3$ (·····), CCl$_4$—CH$_3$CN (2:1) (— — —), and of its K^+ complex in CCl$_4$—CH$_3$CN (2:1) (———).

These similarities notwithstanding, a more detailed physicochemical study of valinomycin and compound (79) showed them to possess significant differences. It is these differences which, by using the latter compound as a probe, have shed considerable light on the mode of action of the valinomycin cyclodepsipeptides on membranes (see Part V.C).

Compound (79) displayed no type B forms when the solvent was gradually changed from C^{2}HCl$_3$ to (C^{2}H$_3$)$_2$SO (see Fig. 45), although the B form is predominant in valinomycin over a wide range of (C^{2}H$_3$)$_2$SO concentrations. Apparently lactic acid residues are needed for the compound to assume this conformation, which is destabilized in compound (79) by the steric hindrance of the inwardly projecting, bulky isopropyl groups of the α-hydroxyisovaleryl residues which have replaced the lactic acid residues. If it is assumed that the high surface activity of valinomycin is due to the structural peculiarities of form B (externally facing polar groups surrounding a hydrophobic core) [451], then it will be easy to understand the stability of analog (79) monolayers on the air/water interface and the decreased tendency of this compound to enter compressed lecithin monolayers [3].

136

It has been mentioned earlier that a distinguishing feature of the
K⁺·valinomycin molecular structure is its effective screening of the central
cation from solvent interaction, a particularly important part in the
screening effect being played by valyl isopropyl groups hanging over the
openings of the bracelet skeleton of the antibiotic. A completely different
situation holds for the K⁺ complex of compound (79). As can be seen in Fig.
57, instead of the pendant valyl isopropyls in the upper part of the complex,

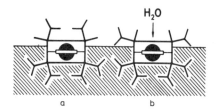

Fig. 57. Schematic drawing of the K⁺ complexes of valinomycin (a) and "false"
retro-valinomycin (b) on the membrane surface.

there are the less bulky alanyl methyls over the opening with their inferior
screening effect. As a result conditions become more favorable for
ion—dipole interaction between the solvent and the cation in the central
cavity. The energy of this interaction augments the stability of the analog
(79)·K⁺ complex in monolayers and in solution [3] and also tends to

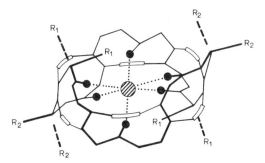

Fig. 58. Schematic representation of the change in configuration of the amino (R_1) and
hydroxy acid (R_2) residues in valinomycin·K⁺ complex (the dashed lines show the
position of the isopropyl groups after the change in configuration).

increase the difference between the free energies of complexation of analog
(79) and valinomycin with increasing solvent polarity. The asymmetric
solvation of the cation also enhances the optical activity (cf. Figs 41 and 55),
and is responsible for the appearance of a considerable effective dipole
moment in the monolayers and for the increase in surface activity of the K⁺
complex of compound (79) as compared with K⁺·valinomycin.

The configurational changes of the amino and hydroxy acid residues in the valinomycin complex characteristic of the third group analogs and schematically shown in Fig. 58 (to a first approximation energetically equivalent to the appearance of new fragments with conformational maps symmetric to the original ones with respect to the points $\phi = 0°$, $\psi = 0°$) are always accompanied by increased non-bonding interactions (Fig. 59). But,

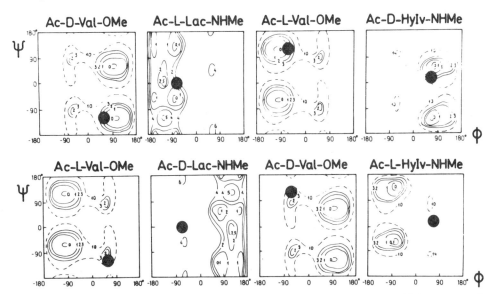

Fig. 59. Conformational maps of the enantiomeric amino and hydroxy acid fragments of valinomycin. Dark circles, coordinates of the respective fragments of the valinomycin·K^+ complex (upper row) and of its diastereomers (lower row).

whereas the energy increase of the valine fragments is rather small (1.9 kcal/mole), that of the hydroxy acid fragments is more significant (4 to 10 kcal/mole). This rationalizes the stepwise decrease in the stability constants with change in configuration of one (analogs 14 and 26*), two (analogs 65, 66 and 70) or more (analogs 49, 54 and 75—78) valyl residues and the greater sensitivity to configurational change of the α-hydroxyisovaleryl residues (analogs 36, 56, 67—69 and 71).

Highly interesting is the effect of substituting the amide and ester groups of valinomycin by related groupings (Figs 60 and 61). The similarity of the steric and electronic characteristics of the amide and ester groups can explain the fact that exchange of type L-Lac → L-Ala and D-HyIv → D-Val (compounds 23, 24, 39 and 45) yields complexing analogs; the hydrogen bonds are not affected and the carbonyls are free to participate in the

_____

* An exception is analog (64).

138

ion—dipole interaction. Interestingly, the Na$^+$ complexes of compounds (23 and 45) have a somewhat augmented stability (see footnotes to Table 2), possibly due to specific properties of the amide groups (for instance, their greater polarizability, or the enhanced point charge of the amide oxygen 0.415 compared with the ester oxygen 0.280 [775]).

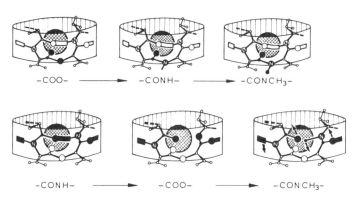

Fig. 60 (above). Schematic representation of the exchange of ester by amide or N-methylamide groups in the valinomycin·K$^+$ complex (the substituted atoms are designated by dark circles).
Fig. 61 (below). Schematic representation of the replacement of amide groups by ester or N-methylamide groups in the valinomycin·K$^+$ complex.

With substitutions of the type CONH → COO (Fig. 61) a hydrogen bond is lost, so that one could expect the energy of the complex to be raised by a value equivalent to the hydrogen bond energy, i.e. by about 4 kcal/mole. This is just what one observes with analog (32) (lowering of $-\Delta F_{\text{compl.}}$ from 8.6 to 4.7 kcal/mole). Modification of the antibiotic in this case (as well as in compound (18)) did not affect the other five hydrogen bonds or the octagonal system of carbonyl oxygens surrounding the cation. If, however, an amide group is substituted by a methylamide group, not only is there a hydrogen-bonded hydrogen lost, but steric hindrance to hydrogen bonding by the neighboring hydrogens arises. The resultant analog (27) manifests no signs of complex formation. Naturally, the exhaustively methylated analog is also not a complexone.

Thus, consideration of the synthesized analogs has led to an understanding of the conformational requirements for complexation in the valinomycin group of depsipeptides, opening the way for preparing new complexing analogs with multifarious properties. A dramatic example of the efficiency of such an approach are the results just obtained from a study of valinomycin analogs (compounds 25, 40, 46, 48, 53 and 59), wherein the ester groups are replaced by N-methylamide groups [1038]. The NH and CO groups required for the formation of the intramolecular hydrogen bonding system are retained and as before there are six carbonyls capable of interacting with the cation. Since this conversion is accomplished by the substitution L-Lac → L-MeAla and D-HyIv → D-MeVal, the new analogs contain the peptide fragments D-Val—L-MeAla and L-Val—D-MeVal instead of the depsipeptide fragments D-Val—L-Lac and L-Val—D-HyIv of valinomycin. Analysis of the corresponding conformational maps, Ac—D-Val—NMe$_2$ and Ac—L-MeAla—NHMe (instead of Ac—D-Val—OMe and Ac—L-Lac—NHMe of valinomycin) and of Ac—L-Val—NMe$_2$ and Ac—D-MeVal—NHMe (instead of Ac—L-Val—OMe and Ac—D—HyIv—NHMe) shows that the ϕ and ψ coordinates of the K$^+$ complexes of the valinomycin peptidic analogs are in allowed regions, i.e. the formation of the new complexes is not associated with any serious steric hindrances. In fact, all the new complexes (except possibly analog (59) which could not be investigated because of its poor solubility) form K$^+$ complexes and most of them inhibit microbial growth (see Table 2). Of particular interest is analog (53) whose stability constants have a "record" value which exceeds by about two orders of magnitude those of valinomycin or its other analogs. A predominant part in this effect is evidently played by transformations of the type Ac—Val—OMe → Ac—Val—NMe$_2$ leading to sharp conformational restriction of the corresponding cyclodepsipeptide (53) fragments [773] and augmentation of the free energy of the system represented by the left-hand side of the equation on p. 134. The same reasoning explains the stability higher than valinomycin of the analog (81)·K$^+$ complex, demonstrated indirectly by the results of extraction experiments.

The studies of valinomycin and its analogs have revealed striking analogies between these compounds and enzymes. In fact, the active center of an enzyme is a combination of functional groups having a definite order in space and time and interacting with the substrate molecule. The disposition of the functional groups is determined by the protein structure which plays the part of a framework or matrix. There is a somewhat similar situation in valinomycin where the ligand groups attached to the depsipeptide backbone form a specific center for binding the cation. As is often the case with enzymes, the substrate (the role of which is played here by the cation) is considerably shielded from the environment. The interaction of the cation with the valinomycin molecule is distinguished by high specificity and is accompanied by directed conformational rearrangements, a kind of "induced fit" in the Koshland sense. It is noteworthy that the analogy can be further extended to the kinetics of the interaction. According to Grell *et al.* [326], in polar media valinomycin rapidly forms an intermediate Michaelis-like complex with the cation, which then transforms into the "normal" complex, the rate of formation and breakdown of the complexes being dependent upon the cation species. Such a far reaching analogy suggests that Nature apparently might have found a way to "low molecular enzymes" containing active and/or binding centers attached to a framework of "natural" monomer residues such as amino and hydroxy acids. Why then are all the known enzymes more or less high molecular proteins? Was it that the matrix pathway of protein biosynthesis gave more freedom of evolution or is the enzyme molecule something more than merely a bedding for the active center? Perhaps in one of the approaches to this question, viz. model enzyme study, the data obtained for valinomycin and other macrocyclic complexones may prove quite significant.

III.A.1.c. Enniatins

The enniatins A, B and C display very high optical activity. The variability of their CD and ORD curves bears evidence of considerable flexibility of their depsipeptide backbone (see Fig. 62) [326, 704, 712, 897,

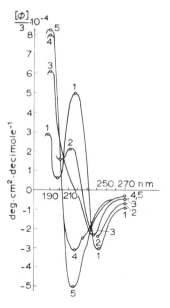

Fig. 62. ORD curves of enniatin B in different solvents (1, heptane; 2, 96% ethanol; 3, acetonitrile; 4, water—trifluoroethanol, 2:1) and of its K^+ complex in 96% ethanol (5).

904]. The presence therein of isosbestic points which, for instance, for the ORD curves of enniatin B are at approx. 234 nm, indicate with a high degree of probability that only two basic forms (a non-polar N and a polar P) are taking part in the conformational equilibrium. This N $\rightleftarrows$ P equilibrium is apparently also characteristic of beauvericin [709]; but its phenyl chromophores which possess several absorption bands in the 185—230 nm region render difficult comparison of its ORD and CD curves with those of enniatins A, B and C. The N → P transition, on adding trifluoroethanol to dioxane, occurs most easily for enniatin C and least easily for enniatin B (Fig. 63). As one can see from comparison of the corresponding ORD (Fig. 62) and CD curves, the enniatins assume the P type conformation not only in polar solvents but, when interacting with alkali ions, in all solvents [326, 383, 704, 712, 897, 903, 904].

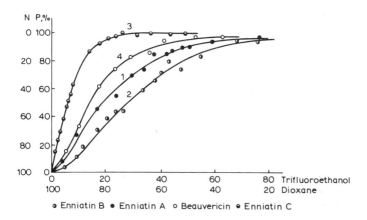

o Enniatin B • Enniatin A o Beauvericin • Enniatin C

Fig. 63. Solvent dependence of the relative amounts of the "non-polar"-N and "polar"-P conformers in enniatins A (1), B (2), C (3) and in beauvericin (4).

The conformation of enniatin B in the equimolar complex was first proposed by Müller and Rudin [669], who assumed the presence of an internal cavity with six inwardly pointing carbonyl groups*. This was confirmed by the X-ray analysis of the crystalline (enniatin B)·KI complex [209], which further showed that in the complexed state (and consequently also in polar solvents) enniatin B has a threefold symmetry axis and *trans-N*-methylamide and ester bonds; the rotational states about the single bonds are characterized by approximately the following parameters.

	L-MeVal	D-HyIv
ϕ	−60	60
ψ	120	−120

---

* The conformation of enniatin B in solution simultaneously proposed by Warner, in which polar and non-polar groups were situated on different sides of the average plane of the ring [1044], has not received experimental confirmation.

142

In general, the structure resembles a charged disc with lipophilic boundaries (Fig. 64) where the K$^+\cdots$O distance is 2.6—2.8 Å.

Spectral studies showed that the enniatin complexes are also in this conformation in solution (Fig. 65) [704, 904]. The most unequivocal proof of this has been obtained from the NMR spectra of the (tri-N-desmethyl)-

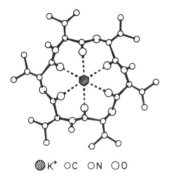

⬤K⁺ ○C ○N ○○

Fig. 64. Crystal structure of the (enniatin B)·KI complex salt.

enniatin B complexes with different alkali metal ions (135). Bearing in mind the conformational mobility of the enniatins, one should expect complexing of the larger ions to induce a corresponding increase in the diameter of the internal cavity. The conformational changes resulting from such reorientation of the carbonyl groups, outwardly resembling the opening of a flower

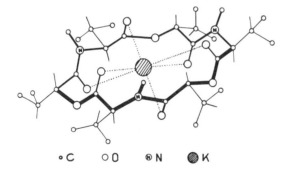

•C   ○○   ⊗N   ⬤K

Fig. 65. Conformation of the (enniatin B)·K$^+$ complex in solution.

bud, should cause in form P a synchronous rotation of the amide and ester planes; the angle θ should become closer to 180° and there should be an increase in the $^3J_{NH-CH}$ constant (Fig. 66). Actually the $^3J_{NH-CH}$ values in the series of Li$^+$, Na$^+$, K$^+$ and Cs$^+$ complexes of analog (135) do increase successively from 4.9 to 8.5 Hz, corresponding to an increase in θ from 130° to 150° and in the cavity diameter from approx. 2.0 to approx. 3.5 Å.

A similar conclusion has been arrived at from a study of the far infrared spectra of the Li^+, Na^+, K^+ and Cs^+ complexes of beauvericin in CH_3Cl, where two, one, two and four $M^+ \cdots O$ stretching bands, respectively, have been demonstrated [428]. Taking account of the selection rule for the active vibrations in the infrared, these data have made it possible to represent the

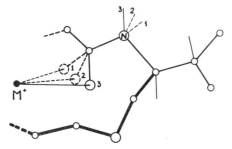

Fig. 66. Schematic presentation of the effect of complexed cation size on orientation of the amide groups in (tri-N-desmethyl)-enniatin B. 1, Li^+; 2, K^+; 3, Cs^+.

conformational transition accompanying the size increase in the complexing cations (Fig. 67, Table 47) in the following way. In the lithium complex, the ligand oxygen atoms form a flattened antiprism ($l_1 \approx l_2 > l_3$) which, by rotation of all the N-methylamide and ester groups pass over into a regular octahedron (Na^+, $l_1 \approx l_2 \approx l_3$) and further, in the potassium complex, again

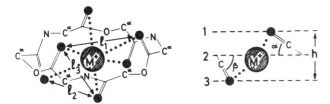

Fig. 67. Proposed model of the depsipeptide backbone of beauvericin complexed with alkaline metal ions. 1, plane of the ester carbonyl oxygens; 2, plane of the amide oxygens; 3, average plane of the C^α atoms; α, angle of inclination of ester carbonyls to plane 3; β, angle of inclination of N-methylamide carbonyls to plane 3; h, distance between planes 1 and 2.

into an antiprism (flattened or drawn out, $l_1 \approx l_2 \neq l_3$). In the cesium complex the necessary size of the cavity is attained mainly by rotation of the N-methylamide groups ($l_1 < l_2$ distorted antiprism). The geometric parameters of the beauvericin complexes in accord with this model are represented in Table 47; the ϕ and ψ values of the Cs^+ complex have been taken to be the same as for the theoretically calculated P conformation of the free cyclodepsipeptide (see Table 48).

TABLE 47

COMPUTED PARAMETERS OF BEAUVERICIN COMPLEXES WITH ALKALI METAL IONS

Size of the ion accommodated in the inner cavity (Å)	Amino acid residues*		Hydroxy acid residues*		h (Å)	l_1 (Å)	l_2 (Å)	l_3 (Å)	$\alpha°$	$\beta°$
	$\phi°$	$\psi°$	$\phi°$	$\psi°$						
0.68 (Li⁺)	−65 ± 5	180 ± 5	60 ± 5	−175 ± 5	1.9 ± 0.1	3.3 ± 0.2	3.3 ± 0.2	2.8 ± 0.1	35 ± 5	40 ± 5
0.98 (Na⁺)	−70 ± 10	165 ± 5	70 ± 10	−170 ± 10	2.5 ± 0.2	3.5 ± 0.2	3.5 ± 0.2	3.4 ± 0.2	50 ± 5	50 ± 5
1.33 (K⁺)	−90 ± 10	145 ± 10	90 ± 10	−150 ± 10	3.0 ± 0.2	4.3 ± 0.3	4.3 ± 0.3	3.9 ± 0.3	62 ± 5	66 ± 5
1.67 (Cs⁺)	−103	171	74	−136	2.8	3.5	5.0	3.6	46	74

* $\omega = 180° \pm 10$.

The NMR spectra also showed that in the L-*N*-methylvaline residues of the K^{+}·(enniatin B) complex the C$^{\alpha}$H—C$^{\beta}$H protons are preferably *trans* ($^{3}J_{C\alpha H - C\beta H}$ = 9.9 Hz), whereas in the D-α-hydroxyisovaleryl residues a significant part of the isopropyl side chains are in *gauche* conformation ($^{3}J_{C\alpha H - C\beta H}$ = 6.9 Hz).

In interpreting the above NMR and infrared data it was assumed that the enniatins form equimolar complexes. The possible existence of a certain amount of the 2:1 complex can apparently have no effect on the results obtained. There are at present no data available on the properties of such complexes, although on the basis of general considerations it is natural to

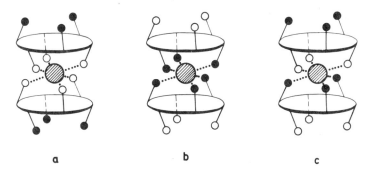

a   b   c

Fig. 68. Tentative structure of the 2:1 enniatin complexes. ⊘, cation; ●, ester carbonyl oxygens; ○, amide oxygens.

assume that two molecules of the depsipeptide in the P conformation form in such complexes a sandwich structure containing six ligand oxygens. As a result, one may imagine an "amide" type of coordination (Fig. 68a), an "ester" type of conformation (Fig. 68b) and a "mixed" coordination (Fig. 68c).

Contrary to form P, form N has no symmetry elements. This follows from a temperature study of the NMR spectra of enniatin B in CS$_{2}$ or CS$_{2}$—C^{2}H$_{3}$C$_{6}$^{2}H$_{5}$ (2:1), in which the temperature was gradually lowered to approx. —120°C (Fig. 69) [426, 704, 904]. In both solvents the signals first broaden (over the range —40 — —90°C) following which the spectrum assumes a pattern that, judging from the signals in the *N*-methyl region (2.3—3.4 ppm) and in the C$^{\alpha}$H region (4.2—5.6 ppm), corresponds to the presence of three non-equivalent L-MeVal—D-HyIv fragments. Hence, in form N, each of these fragments must assume with equal probability three different conformations. Rapid equilibration between these three conformers at room temperature causes an averaging of the chemical shifts and concomitant simplification of the spectra.

Theoretical analysis of the cyclodepsipeptide analog of enniatin B, carrying methyl instead of isopropyl side chains, not only yielded the ϕ and ψ parameters of forms N and P, but shed light on the nature of the interactions responsible for the N $\rightleftarrows$ P interconversions [775]. The calculation, which included potential energy minimization with respect to all the ϕ and ψ angles, showed that minimum energy was possessed by several structures (Table 48), of which one has a symmetry axis and therefore

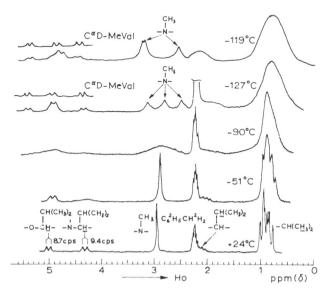

Fig. 69. NMR spectra of enniatin B in CS_2 at $-119°C$ (upper spectrum) and in $CS_2-C^2H_3C_6{}^2H_5$ (2:1) at different temperatures (lower spectra).

corresponds to form P, whereas the others (N_1-N_5) could in principle be assumed to form N. However, the data in Table 48 do not permit a straightforward choice among the N_1-N_5 conformations of the most preferable form because their energy differences do not exceed the error of the calculation, the latter being carried out without accounting for the presence of isopropyl groups (an exception might perhaps be form N_5 because of its exceedingly high energy in all solvents whatever the polarity). The choice was made from a comparison of the calculated $C^\alpha H$ proton chemical shifts in the N_1-N_5 conformations with those of their counterparts in the low-temperature enniatin B solutions. The signals in the 4.2—5.6 ppm region were assigned by taking the spectrum of a specially synthesized enniatin B analog deuterium-labeled in the $C^\alpha-H$ atoms of the N-methyl-valine residues. As the calculated and experimental spectra were in

TABLE 48

OPTIMAL CONFORMATIONS OF THE $cyclo[-(\text{L-MeAla}-\text{D-Lac})_3-]$

ϕ and ψ parameters*	Conformations					
	P	N_1	N_2	N_3	N_4	N_5
ϕ_1	$-103°$	$-131°$	$-67°$	$-133°$	$66°$	$-60°$
ψ_1	171	111	-166	-140	152	24
ϕ_2	74	118	-55	89	102	88
ψ_2	-136	55	-72	-109	-91	130
ϕ_3	-103	44	-96	-128	-149	59
ψ_3	-171	97	126	92	-60	79
ϕ_4	74	152	126	154	-73	128
ψ_4	-136	-118	-142	32	-132	22
ϕ_5	-103	-120	-113	73	-95	63
ψ_5	171	77	84	-55	113	86
ϕ_6	74	138	124	-58	121	139
ψ_6	-136	-101	-69	-92	42	-119
Total energy $\epsilon = 1$	3.9	0	7.2	4.4	10.9	34.6
(kcal · mole^{-1}) $\epsilon = 4$	0	2.8	5.5	8.0	12.0	31.0
$\epsilon = 10$	0	4.8	5.9	9.5	13.0	32.5
Dipole moment (D)	7.15	2.55	8.85	3.80	4.20	7.2

* Residues are listed in a clockwise order beginning with L-MeAla.

agreement only for form N_3 (Fig. 70) this structure was assigned to enniatin B in non-polar media [426].

Conformational analysis showed that from the standpoint of non-bonded interactions form P is preferable to form N_3, but that it is destabilized by electrostatic interaction of the spatially neared carbonyl groups (form P has the larger dipole moment, $7.15\ D$). It thus follows that form P is particularly advantageous in media of high dielectric constant. Hence in the enniatins the driving force for the $N \rightarrow P$ transition is solvation of the polar groups. We can

○ C   ⓃN   ○ O

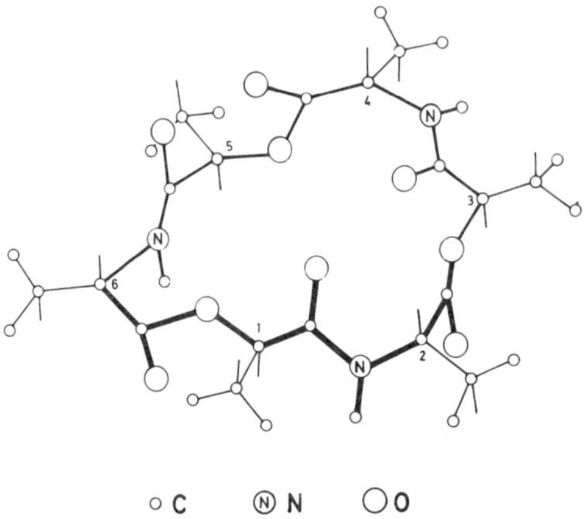

Fig. 70. Conformation of enniatin B in non-polar solvents.

now understand why the "polar" P form closely resembles the complex conformation; for, interacting with the methylamide and ester groups, the solvent and the cation fulfil essentially the same function, viz. neutralization of the mutual dipole—dipole repulsion of these groups. The dipole moment for form N_3 is calculated to be $3.80\ D$, in good agreement with the experimental value of $3.35\ D$ (in CCl_4).

The above considerations rationalize the different complexing efficiencies and selectivities of the enniatins and valinomycin. The main factors responsible for the higher stability of the valinomycin potassium complex are apparently the following:

1. In valinomycin more ligands are interacting with the cation than in the enniatins: six amide carbonyls additionally participate in the interaction, although at a much greater distance from the cation, as well as the six ester ligands common to both types of depsipeptides.

2. The conformational transition accompanying complexation results in a larger gain in non-bonding interaction energy for valinomycin than for the enniatins (see, for instance, the data in Tables 46 and 48).

3. The enniatin carbonyls form a larger angle, α, with the $K^+\cdots O$ direction $\left(K^+\cdots\overset{\alpha}{\underset{}{\cdots}}O_{\diagdown C}\right)$ than do the valinomycin ester carbonyls (cf. Figs 48 and 65) which might result in a lowered electrostatic interaction with the cation.

The comparatively low complexing selectivity of the enniatins is due to their high structural flexibility. For instance, when enniatin B complexes Na^+ and K^+, the higher desolvation energy of the former ion (by 17 kcal/mole, see p. 222) may be partly compensated for by stronger electrostatic interaction due to reorientation of the carbonyl groups so that both oxygen—cation distances are shortened and the α angles are diminished (Fig. 66).

The stronger interaction with the smaller cations is clearly manifested in the ^{13}C-NMR spectra of beauvericin and its Na^+ and K^+ complexes [114]. Formation of the Na^+ complex causes a markedly higher downfield shift in both the ester and N-methyl amide ^{13}CO resonances (1.7 and 2.3 ppm) than formation of the K^+ complex (1.4 and 2.0 ppm). As a result of all the above, the free energies of complexing and therefore the stability constants of $Na^{+\cdot}$ and $K^{+\cdot}$(enniatin B) complexes, are close in value, i.e. the K/Na selectivity is low. Since such conformational transformations are precluded by the rigid valinomycin structure, one can readily understand the reason for its exceptionally high K/Na selectivity (see Part III.C).

III.A.1.d. Enniatin B analogs

Although modification of the amino acid side chains (substitution of L-N-methylvaline by L-N-methylleucine, L-N-methylisoleucine or L-N-methylphenylalanine) affects the ease of the $N \rightarrow P$ transition, it has practically no effect on the stability sequence of any of the complexes in this group of compounds (86—89), the most stable always being complexes with K^+ and Rb^+, and the least stable with Li^+ and Na^+ (Table 49). The complexation non-sensitivity of the enniatin antibiotics to the structure of the amino acid side chains can be accounted for by the weak dependence upon the latter of the amino acid conformational maps [773].

The conformational maps also explain the similarity of the complexing properties of the N-desmethyl analogs of enniatin B to those of the parent compound, despite the marked differences in the maps between Ac—L-MeVal—OMe and Ac—L-Val—OMe, and between Ac—D-HyIv—NMe$_2$ and Ac—D-HyIv—NHMe (Fig. 71). In both cases there are deep potential wells in the region characteristic of the enniatin complexes ($\phi_1 \sim -80°$, $\psi_1 \sim 145°$,

TABLE 49

COMPLEXATION OF ENNIATIN CYCLODEPSIPEPTIDES WITH ALKALI METAL IONS

Compound	Stability constants of the complexes (K, $l \cdot mole^{-1}$; C_2H_5OH; 25°C)					Free energy of complex formation ($-\Delta F = 1.37 \lg K$, $kcal \cdot mole^{-1}$)				
	Li	Na	K	Rb	Cs	Li	Na	K	Rb	Cs
Enniatin A (86)	100	2900	9800	2900	2700	2.7	4.0	5.5	4.8	4.7
Enniatin B (87)	<50	1300	3700	4000	2200	<2.3	4.3	4.9	4.9	4.6
Enniatin C (88)	<50	2500	5500	7500	4100	<2.3	4.7	5.1	5.3	5.0
Beauvericin (89)	<50	300	3100	3500	3500	<2.3	3.4	4.8	4.9	4.9
"Tetra-enniatin B" (91)	<50	<50	<50	<50	<50	<2.3	<2.3	<2.3	<2.3	<2.3
Enniatin BC (104)	—	2200	5500	—	—	—	4.6	5.1	—	—
Enniatin CB (105)	—	1700	5100	—	—	—	4.4	5.1	—	—
(110)	—	<50	<50	—	—	—	<2.3	<2.3	—	—
(111)	—	100	500	—	—	—	2.7	3.7	—	—
(122)	—	2200	4900	—	—	—	4.6	5.0	—	—
(Tri-N-desmethyl)-enniatin B (135)	<50	2500	2600	1000	700	<2.3	4.7	4.7	4.1	3.9
(139)	<50	170	1000	1000	700	<2.3	3.2	4.1	4.1	3.9
(Tetra-N-desmethyl)-octa-enniatin B (140)	<50	80	220	350	5500	<2.3	2.6	3.2	3.5	5.1
Octa-enniatin B (141)	<50	2600	4900	800	200	<2.3	4.7	5.1	4.0	3.2
Deca-enniatin B (142)	<50	<50	430	150	250	<2.3	<2.3	3.6	3.0	3.3
(Hexa-N-desmethyl)-dodeca-enniatin B (143)	<50	<50	<50	<50	<50	<2.3	<2.3	<2.3	<2.3	<2.3
Dodeca-enniatin B (144)	<50	<50	100	<50	<50	<2.3	<2.3	2.7	<2.3	<2.3

$\phi_2 \sim 65°$, $\psi_2 \sim -130°$, which are the mean values between the experimental crystalline complex parameters and those theoretically calculated for the P form and given in Table 48); i.e. from the standpoint of complex stability, the L-valine analogs of enniatin B should not differ essentially from the parent compound.

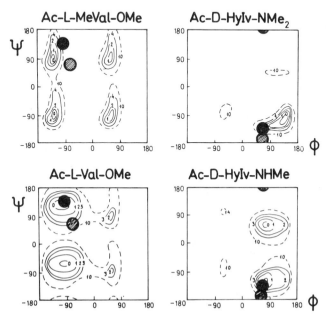

Fig. 71. Conformational maps of the amino and hydroxy acid fragments of enniatin B (above) and (tri-N-desmethyl)-enniatin B (below). The solid black circles represent the ϕ and ψ parameters of the (enniatin B)·K$^+$ complex; the hatched circles represent the φ and ψ parameters of the "(tetra-N-desmethyl)-octa-enniatin B"·Cs$^+$ complex.

A comparison of analogs (110) and (111) with enniatin B shows that change in the D-α-hydroxyisovaleryl configuration destabilizes the complex structure more than the same change in the L-N-methylvaline residue. This also stems directly from the nature of the enniatin hydroxy and amino acid conformational maps (Fig. 72). Indeed, with the proviso of retention of the overall complex structure and interaction with the cation of all six carbonyls, inversion of the D-hydroxyisovaleryl configuration means the appearance of a residue with a new conformational map (symmetrical to the former one with respect to ϕ 0°, ψ 0°), in accordance with which realization of the complex with $\phi \sim 65°$ and $\psi \sim -130°$ would be sterically prohibited. The maps for the L and D enantiomers of the amino acid fragments are very similar, which explains the lower sensitivity of complexation to configurational inversion of the latter.

The replacement of *N*-methylamide groups by ester groups (analog 139) diminishes to a certain extent stability of the complexes, possibly due to a somewhat stronger interaction of the cations with the amide than with the ester groups.

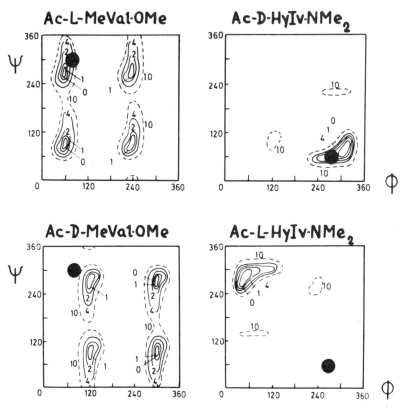

Fig. 72. Conformational maps of the amino and hydroxy acid fragments of enniatin B (above) and its diastereomers (below). The dark circles refer to the ϕ and ψ parameters of the corresponding K^+ complexes.

Enniatin analogs of lesser ring size (compounds 90—101) are incapable of complexing alkali metal ions because, according to both X-ray data [503] and theoretical analysis [774], these are rigid compounds with outwardly oriented carbonyls of *cis N*-methyl and the *trans* ester bonds (Fig. 73).

As could be expected, with increasing ring size and, therefore, internal cavity of the enniatin B analogs, the maximum value of the stability constants first shifts to cesium (compound 140), and then the ring becomes

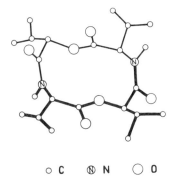

o C   Ⓝ N   ◯ O

Fig. 73. Conformation of "tetra-enniatin B".

too big for effective ion—dipole interaction in any of the optimal conformations with concomitant abrupt fall in complexing capacity (compound 144).

Somewhat unexpected is the behavior of the cyclopolymer homologs of enniatin B. Thus, the Na^+ and K^+ complexes of the analog expanded by one didepsipeptide fragment ("octa-enniatin B", 141) have a higher stability, whereas the stability of the Rb^+ and Cs^+ complexes diminishes. Interestingly, maximum stability of the potassium complexes is retained in still larger rings (analogs 143 and 145). Further study showed that the peculiar

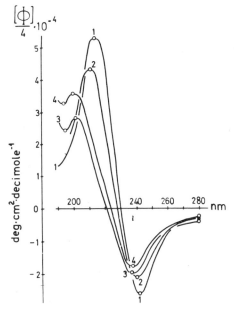

Fig. 74. ORD curves of "octa-enniatin B" in different solvents. 1, heptane; 2, 96% ethanol; 3, trifluoroethanol; 4, water—trifluoroethanol (2:1).

behavior of these homologs is due to their specific conformational properties. Analysis of the molecule $cyclo[-(\text{L-MeAla}-\text{D-Lac})_4-]$, simulating "octa-enniatin B", showed that it has several optimal conformations (N_1-N_6) lacking symmetry elements and differing only insignificantly in energy (Table 50). Moreover, variation of the effective dielectric constants

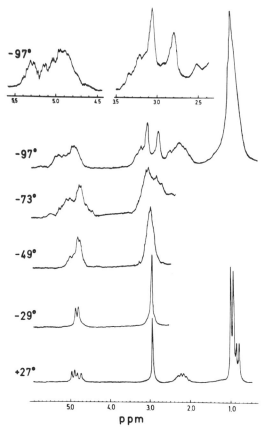

Fig. 75. NMR spectra of "octa-enniatin B" in CS_2 at different temperatures.

within limits most closely approaching the actually encountered local environments ($\epsilon = 4-10$) has only a weak influence on the energies of these forms. It is therefore clear why there is little change in the ORD curves of "octa-enniatin B" on passing from heptane to the more polar solvents (Fig. 74). When this analog in CS_2 solution is cooled, its NMR spectrum acquires a complicated pattern (Fig. 75) and examination of the N-methyl proton resonances (more than 8 signals in the 2.4—3.5 ppm region) shows that under these conditions at least three energetically similar forms are in

TABLE 50

OPTIMAL CONFORMATIONS OF *cyclo*[—(L-MeAla—D-Lac)$_4$—] [768]

ϕ and ψ parameters*	Conformation					
	N_1	N_2	N_3	N_4	N_5	N_6
ϕ_1	79°	77°	85°	73°	−120°	169°
ψ_1	−35	51	−142	108	114	−99
ϕ_2	119	70	83	−57	85	−61
ψ_2	−87	54	49	−53	48	−56
ϕ_3	−111	69	60	−99	60	−151
ψ_3	153	174	78	146	−83	−74
ϕ_4	109	89	158	99	64	56
ψ_4	−159	−166	−173	−167	−126	48
ϕ_5	−110	−109	−114	−127	−95	65
ψ_5	−176	169	138	−167	−131	−20
ϕ_6	93	97	147	93	−55	173
ψ_6	−163	53	−157	−164	−64	−68
ϕ_7	−119	58	−103	−87	−125	−102
ψ_7	111	44	101	120	−153	160
ϕ_8	89	83	138	73	115	133
ψ_8	−158	−118	−42	−155	−150	−58
Total energy $\epsilon = 1$	1.2	8.6	0	12.7	5.1	1.6
(kcal · mole^{-1}) $\epsilon = 4$	0	0.2	3.3	4.0	3.0	4.6
$\epsilon = 10$	0.8	0	4.9	3.3	3.6	6.3
Dipole moment (D)	6.8	8.9	5.5	9.1	8.4	7.7

* Residues are listed in a clockwise order beginning with L-MeAla.

equilibrium, thus confirming the results of theoretical analysis. On the other hand, it is unfavorable for "octa-enniatin B" to assume a conformation with a fourfold symmetry axis, similar to form P of the naturally occurring enniatins, and it is to this which is due the low stability of the large ion (Rb$^+$ and Cs$^+$) complexes. Such important differences between the complexing properties of "octa-enniatin B" and its N-desmethyl analog (140) are the result of the specificities of their amino and hydroxy acid conformational maps Ac—L-MeVal—OMe, Ac—L-Val—OMe and Ac—D-HyIv—NMe$_2$, Ac—D-HyIv—NHMe, respectively. It can be seen from Fig. 71 that the region of allowed ϕ and ψ values corresponding to conformation P (potential wells in the upper left-hand quadrants for the amino acid residues and in the lower right-hand quadrants for the hydroxy acid residues) are much broader in the case of the N-desmethyl derivatives. Consequently the ϕ and ψ parameters of the symmetric conformation of analog (140) with a large internal cavity turn out to be within the allowed limits, being the reason for the high stability of the Cs$^+$ complex. From the above it also follows that the highly stable complexes of "octa-enniatin B" with Na$^+$ and K$^+$ should have a peculiar unsymmetric structure wherein only a part of the carbonyls are effectively interacting with the cation. Its exact parameters, as also the parameters of the K$^+$ complexes of "deca-" (143) and "dodeca-enniatin B" (145), have not yet been established.

In concluding this discussion on the enniatins and their analogs, it can be said, in general, that the conformational mobility of the compounds of this group lays a significant impress on their complexing properties, leading to a mild (as compared to valinomycin) dependence of the complex stability constants on the structure of the macrocyclic compound and the nature of the cation, i.e. to a relatively low structural and cationic specificity of complexation.

III.A.1.e. Other cyclodepsipeptides

Hassall and Thomas, in their review devoted to the conformations of cyclic and "cylindrical" peptides [376], have represented a probable structure for monamycin D$_1$ (151) (Fig. 76) proposed on the basis of the infrared and NMR spectral data. The only peptide NH group forms a $4 \rightarrow 1$ type intramolecular hydrogen bond. There are no other available data on the structure of the monamycin antibiotics. The above-noted structural analogy between the monamycins and enniatins (p. 22) shows that their complexes are apparently constructed similarly.

For serratamolide (168) a symmetric structure has been proposed which, like the enniatins, has alternatingly oriented amide and ester groups and, moreover, two intramolecular hydrogen bonds between the serine OH and

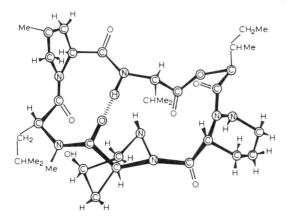

Fig. 76. Proposed structure for monamycin D_1.

CO groups (Fig. 77) [373]. The inability of serratamolide to complex sodium or potassium ions is explained by the small size of its internal cavity ($r = 0.6-0.8$ Å).

Cyclodepsipeptides of the sporidesmolide group (169—177) according to CD, infrared and NMR data, adopt a highly stable "pleated sheet" structure (like the cyclohexapeptides on p. 176) which, devoid of an internal cavity, precludes the possibility of stable complexes being formed.

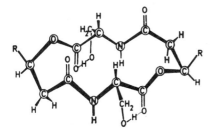

Fig. 77. The spatial structure of serratamolide in non-polar solvents.

III.A.1.f. Gramicidins A, B and C

For gramicidin—alkali metal complexes, Lardy *et al.*, on the basis of general considerations [526], proposed the pseudocyclic conformation shown in Fig. 78. This hypothetic structure has many features in common with that of the enniatin complexes: in both groups of compounds six carbonyls are interacting with the central cation, and the P type conformation (see p. 142) on the 1—6 segment, to which is due the complexing capacity of gramicidin, can be stabilized by the D-amino acid and glycine residues located over every other residue. The C-terminal segment of the chain with its tryptophan residues may be responsible for the formation of associates, stabilized by hydrophobic interactions or by hydrogen bonds.

A fundamentally different type of structure, so-called $\pi_{(L,D)}$ helices, was suggested for these complexes by Urry and collaborators [1025, 1026], which stemmed from the principle of maximum hydrogen bonding and the formation of a stable lipophilic dimer in the course of the functioning of

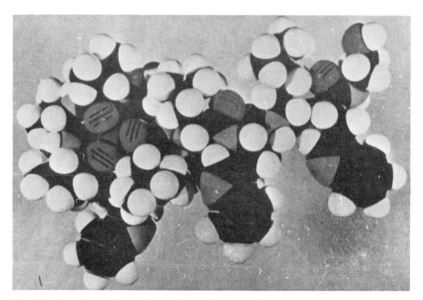

Fig. 78. Pseudocyclic conformation of gramicidin A.

these complexes on the membrane. The dimer is 30—50 Å long, i.e. its length is equal to the thickness of a bilayer membrane. As one may guess from the name, the peptide backbone adopts a helical conformation, but contrary to other helical structures, such as the classical α, γ or π helices [809], where the amide carbonyls are all oriented parallelly along the helical axis, in the $\pi_{(L,D)}$ helices of gramicidins A—C, the carbonyls are antiparallel all along the axis, the CO groups of the L-amino acid residues being oriented towards the C-terminus whereas their L-amino acid and glycine counterparts are oriented towards the N-terminus. Such an unusual build-up of the helix is due to alternation of the amino acid configurations in the gramicidin chain, for only then can sterically equivalent situations arise for the L and D residues, in which no steric hindrances are incurred upon incorporation of D residues into the conventional helical structures. This equivalence, which has been reflected in the name $\pi_{(L,D)}$, is manifested also in the orientation of the side chains, which are all at an angle of approx. 90° to the helical axis, giving rise to the high lipophilicity of the molecule. The principal characteristics of the theoretically possible $\pi_{(L,D)}$ helices are listed in Table 51. Two of them are represented in Fig. 79, from which it can be seen that the alternating orientations of the carbonyl groups determine the number of residues in a turn of the particular structure. This number appears in the designation of the helix type (left column, Table 51). The conformational parameters of the $\pi_{(L,D)}$ helices (ϕ $-120 - -145°$, ψ $85-150°$, for the L-residues; and ϕ $90-144°$, ψ $-50 - -125°$ for the D-residues) are in one of

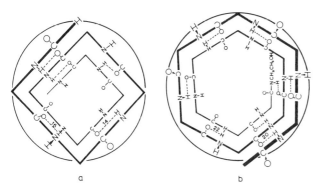

Fig. 79. Schematic drawings of the $\pi^4_{(L,D)}$ (a) and $\pi^6_{(L,D)}$ (b) helices of gramicidin A as viewed along the helical axis.

the most favorable regions of the conformational maps, from the standpoint of both depth and width of the potential wells. The intramolecular hydrogen bonding system in the proposed models comprise considerably varying (13—34-membered) rings, but all without exception are of the type of parallel β-chains (region a in Fig. 80).

The antiparallel orientations of the CO and NH groups of the neighboring amide fragments create conditions for effective "head to head" or "tail to tail" association (Fig. 81) owing to the cooperative formation of several intermolecular hydrogen bonds (four in the $\pi^4_{(L,D)}$, six in the $\pi^6_{(L,D)}$, eight in the $\pi^8_{(L,D)}$ and ten in the $\pi^{10}_{(L,D)}$ helices). The intermolecular hydrogen bonding system is thus of the antiparallel β-structural type (region b in Fig. 80). A strong argument in favor of the occurrence of "head to head" dimerization of gramicidins during their functioning on membranes is the high activity of the synthetic gramicidin A analog, in which two molecules of desformylgramicidin A are linked by a malonyl residue (see Part V.C). Analysis of molecular models shows that the insertion of a $-CO-CH_2-CO-$ group in place of two formyl groupings causes no steric hindrances in the

TABLE 51

PARAMETERS OF THE $\pi_{(L,D)}$ HELICES

Helix	No. residues per turn	Hydrogen bond type	Size of H-bonded rings (No. atoms)	Torsional angles (degrees)				Dimer length (Å)	Cavity diameter (Å)
				L residues		D residues			
				ϕ	ψ	ϕ	ψ		
$\pi^4_{(L,D)}$	4.4	$6 \to 1, 1 \to 4$	16, 14	−125	85	155	−115	35—39	1.4
$\pi^6_{(L,D)}$	6.3	$8 \to 1, 1 \to 6$	22, 20	−120	105	135	−125	25—30	4
$\pi^8_{(L,D)}$	8.4	$10 \to 1, 1 \to 8$	28, 26	−145	135	105	−90	18—24	6
$\pi^{10}_{(L,D)}$	10.4	$12 \to 1, 1 \to 10$	34, 32	−135	150	90	−100	14—21	8

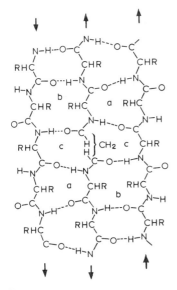

Fig. 80. The intra- and intermolecular hydrogen-bonding system of dimeric $\pi_{(L,D)}$ gramicidin A helices.

dimer's hydrogen bond system (shown in Fig. 80 in the extended form) because the resultant 11-membered hydrogen-bonded rings (segment c) are structurally very similar to the neighboring 10-, 12- and 14-membered rings of the parallel and antiparallel β-structures.

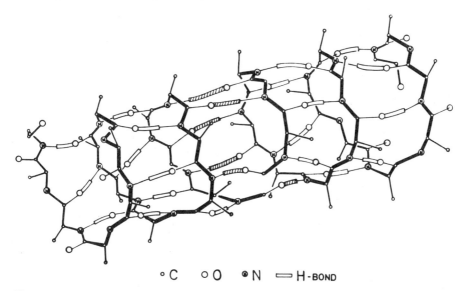

° C   ○ O   ◉ N   ⸦⸧ H-BOND

Fig. 81. Schematic drawing of a "head to head" dimer of gramicidin A in the $\pi_{(L,D)}^6$ helical conformation. Side chains are not fully shown for the sake of clarity.

All the $\pi_{(L,D)}$ helices have an internal cavity which, in the opinion of Urry and collaborators, plays the part of a channel through which the metal ion can pass in the functioning of gramicidins A—C on membranes. Cations in this cavity interact with the amide carbonyls which turn slightly towards the helical axis. The complexes thus formed resemble those of enniatin B if one takes into account the alternating "up-down" direction of the carbonyl ligands along the chain. It can readily be seen that the $\pi_{(L,D)}^4$ helix would coordinate the cation with four ligands, the $\pi_{(L,D)}^6$ with six, the $\pi_{(L,D)}^8$ with eight and the $\pi_{(L,D)}^{10}$ with ten. In this respect the $\pi_{(L,D)}^6$ helix resembles the structure proposed by Lardy $et\ al.$ [526]. It can be seen from Table 51 that

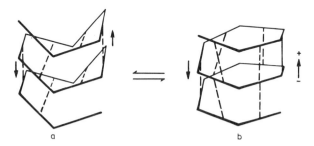

Fig. 82. Schematic representation of anti-β_2^6 helix $\rightleftarrows$ $\pi_{(L,D)}^6$ helix equilibrium of
(a)   (b)
gramicidin A; intramolecular hydrogen bonds are shown by dashed lines.

the cavity size of the $\pi_{(L,D)}^4$ helix is too small to accommodate even Na$^+$ ions without bulging of the corresponding region of the helix, whereas the cavities of $\pi_{(L,D)}^8$ and $\pi_{(L,D)}^{10}$ are too large to interact efficiently even with cesium cations, although the latter would be interacting with eight and ten carbonyls, respectively.

Based on the cavity size and the correspondence of the channel length with the actual thickness of the bilayer, Urry and his coworkers concluded that the $\pi_{(L,D)}^6$ helix was the preferable one for the gramicidins.

Elaborating this hypothesis further, Urry proposed that the $\pi_{(L,D)}^6$ helix can be in equilibrium with another structure which he called the $anti$-β_2^6 helix*, wherein some of the 8 → 1 and 1 → 6 H bonds are changed to 3 → 1 bonds (Fig. 82) [1027]. In the $anti$-β_2^6 helix there are no internal cavities, so that it cannot accommodate metal ions. This helix has a lesser dipole moment than the $\pi_{(L,D)}$ helix, whence according to Urry it follows that the equilibrium between these two forms can be shifted in one or the other direction under the influence of an external electrical field. It should also be

* In the paper [1027] Urry proposes a general way for designating helices containing β-structural elements. However, owing to the rather cumbersome and ambiguous nature of this nomenclature we shall not go into details of the meaning of the prefix "$anti$-β_2^6".

noted that the *anti-β_2^6* helix cannot give stable "head to head" or "tail to tail" dimers; the supposed formation of hydrogen bonds between the side surfaces of the helix is also of little probability because of steric interaction of the side chains.

In order to test their hypothesis Urry *et al.* carried out an ultraviolet, NMR and CD study of gramicidin A and its desformyl, malonyl and

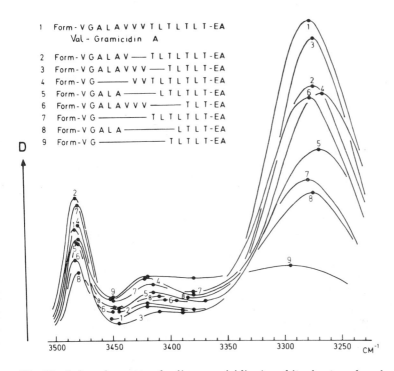

Fig. 83. Infrared spectra of valine-gramicidin A and its shortened analogs in CHCl₃.

hydrogenated derivatives [290, 1030]. A conformational transition was observed to occur on passing from dioxane or trimethyl phosphate to trifluoroethanol or dimethylsulfoxide, providing evidence of the existence of the antibiotic in at least two equilibrium forms. Leaving open the question of the gramicidin A structure in the two less-polar solvents, the American workers assigned it a left-handed $\pi_{(L,D)}$ helix in polar media.

They based this on a small ($\sim 15\%$) hypochromic shift in the ultraviolet absorption of perhydrogramicidin A in trifluoroethanol as compared with trimethyl phosphate solution and a slower deuterium exchange rate of the amide NH protons than for *cis*-3,6-dimethyl-2,5-diketopiperazine, (L-Ala)₂. They also cited as an argument in favor of the $\pi_{(L,D)}$ helical structure the

NMR spectral data for gramicidin A and desformylgramicidin A; namely, (a) higher $^3J_{NH-CH}$ constants (7.0—8.5 Hz) for four of the fifteen NH protons than in the case of a random coil (6.1 Hz) or α-helix conformation (2.4 Hz for a right-handed helix and 6.0 Hz for a left-handed helix), and (b) spread of the valyl and leucyl C-methyl chemical shifts, bearing evidence of interaction of some of the methyls with the tryptophan aromatic side chains.

In our opinion, however, the spectral data given by the authors are in full accord with a non-intramolecular hydrogen bond-containing extended

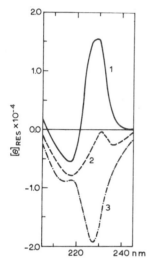

Fig. 84. CD curves of gramicidin A in dioxane in the 205—245 nm region. 1, 4.86 mg/ml; 2, 1.16 mg/ml; 3, 0.80 mg/ml.

conformation of gramicidin A, $\pi_{(L,D)}$ helices being most likely to occur in non-polar media. In fact, data on ultrasonic absorption by gramicidin A revealed no conformational transitions in di-n-butyl ether, pointing to conformational rigidity of the antibiotic in this solvent; in ethanol, however, an equilibrium mixture of several forms is observed [325]. Further support can be obtained from the dilute (10^{-4} M) chloroform infrared spectrum of gramicidin A (spectrum 1 in Fig. 83). It shows a strong NH stretching band at 3280 cm^{-1} and a weak free NH band at 3420 cm^{-1} (besides the indole NH vibration at 3480 cm^{-1}). Hence, under these conditions almost all the amide NH groups are participating in a system of strong hydrogen bonds (similar to that in the α-helix of poly-γ-benzyl-L-glutamate, $\nu_{NH}^{CHCl_3}$ 3290 cm^{-1} [138]). It is noteworthy that the relative free NH content increases somewhat with successive shortening of the gramicidin chain (spectra 2—8 in Fig. 83); at the stage of the heptapeptide (178 f) the intensity of the shorter wavelength band decreases abruptly, evidence of the breakdown of the secondary structure (spectrum 9 in Fig. 83).

Isbell *et al.* [422] have shown that the CD curve of the gramicidin A dimer differs sharply from that of the monomer (Fig. 84); in fact, the two curves are roughly antipodal (that of the dimer resembles the right-handed helix of poly-L-tryptophan [166]). This shows that on association, gramicidin A undergoes a conformational transition. The amide I regions of the infrared spectra of the monomer and dimer also display distinct differences. Such spectral characteristics of gramicidin A are in complete accord with its occurrence as an equilibrium mixture of the type *anti-β_2^6* helical monomer $\rightleftharpoons \pi_{(L,D)}^6$ helical dimer.

We have seen that there is much in favor of Urry's hypothesis. However, despite all its attractiveness there is still much that remains unresolved in the problem of the spatial structure of the gramicidins.

III.A.1.g. Antamanide

The spatial structures of antamanide (188), of some of its analogs (202, 212, 217) and of their Na$^+$ complexes have been investigated by the methods and techniques described in the introduction to this chapter. At first, on the basis of CD and NMR data, it was suggested that in all the solvents studied (dioxane, chloroform, methanol, etc.) antamanide assumes a unitype conformation characterized by absence of intramolecular hydrogen bonds and orientation of all the carbonyls to one side of the average plane of the ring [790].

	Val^1	Phe^6	Pro^2, Pro^7	Pro^3, Pro^8	Ala^4, Phe^9	Phe^5, Phe^{10}
ϕ	−90	−90	−78	−58	−90	−90
ψ	90	120	130	−55	−60	150
ω	180	180	180	180	180	180

Further study has shown, however, that the dipole moment (16.6 *D*) calculated for such a structure is in disaccord with the experimental value (5.2—5.8 *D* in chloroform); moreover, localization of the amide A bands in the 3350—3300 cm^{-1} region of the infrared spectrum of antamanide (Fig. 87) left no doubt as to its possessing intramolecular hydrogen bonds in this solvent [27, 407]. A more detailed analysis of the CD curves (Fig. 85), ORD curves (Fig. 86) and NMR spectra led to the conclusion that in solution antamanide is in the form of a complicated equilibrium mixture of conformers. In non-polar solvents (heptane—dioxane, 5:2; CHCl$_3$) predominant is form A with pseudo twofold symmetry and an intramolecular hydrogen bond system in which all six NH groups are taking part. Gradual addition of a hydroxyl-containing solvent (CH$_3$OH, C$_2$H$_5$OH, CF$_3$CH$_2$OH,

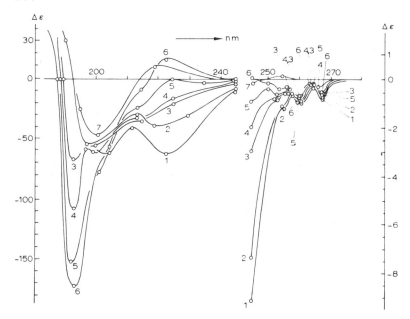

Fig. 85. CD curves of antamanide in different solvents (1, heptane—dioxane, 1:1; 2, acetonitrile; 3, heptane—ethanol, 4:1; 4, trifluoroethanol; 5, ethanol; 6, water—ethanol, 9:1) and of the antamanide·Na⁺ complex in ethanol (curve 7).

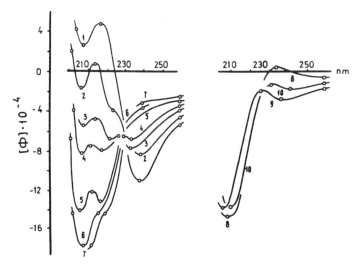

Fig. 86. ORD curves of antamanide and its Na⁺ complex. 1, heptane—dioxane, 5:2; 2, acetonitrile; 3, heptane—ethanol, 9:1; 4, trifluoroethanol; 5, 96% ethanol; 6, water—trifluoroethanol, 5:2; 7, water—ethanol, 3:1; 8, $1.4 \cdot 10^{-3}$ M sodium dodecyl sulfate in acetonitrile (4-fold excess of the salt); 9, $3.0 \cdot 10^{-3}$ M sodium dodecyl sulfate in trifluoroethanol (10-fold excess of the salt); 10, $2.0 \cdot 10^{-2}$ M sodium chloride in 96% ethanol (40-fold excess of the salt).

H_2O) leads to rupture first of the Ala[4] and Phe[9] hydrogen bonds and then of the remaining intramolecular hydrogen bonds. The NMR spectra in dimethyl sulfoxide-containing solutions of [Val[6], Ala[9]]-antamanide (213), an analog stereochemically very close to the naturally occurring cyclo-decapeptide, revealed large amounts (up to 60%) of forms which differ in the configurational set of tertiary amide bonds from the A, B and C forms [435].

Taking account of the values for the $^3J_{NH-CH}$ constants determined from the NMR spectra of antamanide and its Val[6],Ala[9]-analog and of reported data on proline-containing peptides [72, 228, 301, 472, 504] and

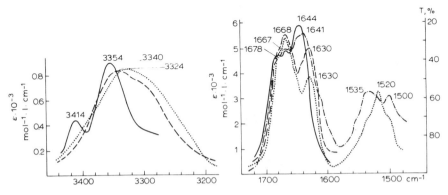

Fig. 87. Infrared spectra of antamanide and its Na$^+$ complex. · · · · · , CHCl$_3$; - - - - -, CCl$_4$—CH$_3$CN (2:1); ——, $1.5 \cdot 10^{-2}$ M NaNCS in CCl$_4$—CH$_3$CN (2:1) (5-fold excess of salt); — · — · —, crystalline complex, KBr disc. T, transmission.

intramolecular hydrogen bond-stabilized peptide fragments [72, 807], the following structure of form A with *trans*-tertiary amide bonds [27, 269, 581] (Fig. 88) has been proposed.

	Val[1], Phe[6]	Pro[2], Pro[7]	Pro[3], Pro[8]	Ala[4], Phe[9]	Phe[5], Phe[10]
ϕ	−80	−60	−55	−100	60
ψ	165	−40	−40	10	−70
ω	180	180	180	180	180

Elucidation of the conformational parameters of forms B and C, on the other hand, requires detailed analysis of the decapeptide molecules in the same way as has been done for the enniatins [75, 859].

Recently, Patel has compared the chemical shifts of the $C^\alpha H$ and ^{13}C atoms of the proline residues of antamanide with the corresponding data for proline peptides with known configuration of the tertiary amide bonds [1112]. On this basis the conclusion was drawn that in the solvents

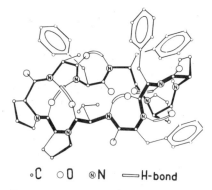

°C   ○O   ⊗N   ⇒H-bond

Fig. 88. Conformation proposed for antamanide in non-polar solvents.

investigated (acetonitrile, acetic acid, dioxane, dimethylformamide) antamanide contained two *cis* amide bonds. Below are the possible conformational parameters for such structural types [1112, 1114].

Structures with six intramolecular hydrogen bonds

		Val[1], Phe[6]	Pro[2], Pro[7]	Pro[3], Pro[8]	Ala[4], Phe[9]	Phe[5], Phe[10]
1,6-*cis*	ϕ	−150 − −120	−80 − −60	−80 − −60	−150	−150 − −90
	ψ	120 − 150	130 − 150	90	−60	−60
	ω	0	180	180	180	180
2,7-*cis*	ϕ	−120	−60	−60	−120 − −90	−120
	ψ	120	150	−30	90 − 120	120
	ω	180	0	180	180	180

Structures not containing intramolecular hydrogen bonds

		Val[1], Phe[6]	Pro[2], Pro[7]	Pro[3], Pro[8]	Ala[4], Phe[9]	Phe[5], Phe[10]
1,6-*cis*	ϕ	−150	−60	−60	−120 − −90	150
	ψ	120	150	−60	120	150 − 180
	ω	0	180	180	180	180
2,7-*cis*	ϕ	−150 − −120	−80 − −60	−60	−120	−90
	ψ	90 − 150	120	120 − 150	90 − 150	90 − 150
	ω	180	0	180	180	180

Since the Na⁺ complex of antamanide is of the 2,7-*cis* type conformation (see below), and the complexing reaction apparently has no effect on the configuration of the amide bonds, the most probable conformation of free antamanide for most solvents is very likely of such a type.

Recently a preliminary crystallographic study of antamanide crystals [558] has been reported, so that one can expect that soon there will appear an X-ray analysis of one of its forms.

Contrary to free antamanide, the CD and ORD curves of its Na⁺ complex are only weakly solvent-dependent (Figs 86 and 87), indicating stability of the amide chromophoric system [434]. The NMR spectral patterns of the Na^+ complex (Fig. 89), particularly in the NH region show that, despite the structural non-symmetry of antamanide, both the complex and form A possess a pseudo twofold axis. The $^3J_{NH-CH}$ constants indicate the corresponding protons in the Val^1 and Phe^5 residues to be *trans*, in the Ala^4 and Phe^9 residues to be *gauche*, and in the Phe^5 and Phe^{10} residues to be *cis* or a distorted *trans*.

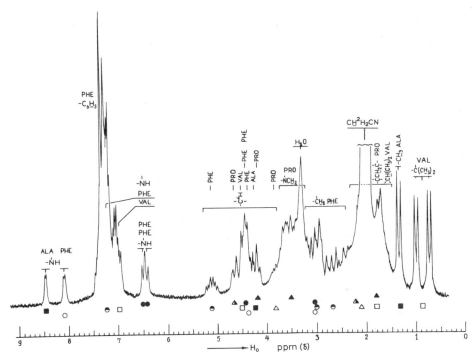

Fig. 89. NMR spectrum at 100-MHz of the Na^+·antamanide complex in C^2H_3CN. The squares and circles indicate signals of the spin coupled protons.

The similarity of the complexing properties of antamanide and of the other metal-binding macrocyclic compounds, together with the considerable NaNCS-induced changes in its CO stretching frequencies (Fig. 87), has led to the conclusion that its complex has an internal cavity with inwardly pointing 4—8 CO groups. From analysis of the amide A region of the infrared spectra and the $\Delta\delta/\Delta T$ and $\tau_{1/2}$ values for the NH signals in the PMR spectra the conclusion was drawn that the NH^1, NH^6, NH^5 and NH^{10} groups are participating in intramolecular hydrogen bonding; but it must be said that the data concerning the last two groups are not so definite [434].

All the above results are in agreement with the following conformational parameters of the Na$^+$ complex.

	Val^1, Phe^6	Pro^2, Pro^7	Pro^3, Pro^8	Ala^4, Phe^9	Phe^5, Phe^{10}
ϕ	−100	−60	−60	150	70
ψ	−160	−50	90	−30	−70
ω	180	180	180	180	180

In the proposed model all the amide bonds are *trans*, the Ala^4 and Phe^9 carbonyls form 3 → 1 intramolecular hydrogen bonds with NH^6 and NH^7 and the Pro^2 and Pro^7 carbonyls, 4 → 1 intramolecular hydrogen bonds with NH^5 and NH^{10}. The principal ligands are the Val^1 and Phe^6 carbonyls, the Na$^+\cdots$O distance being approximately 2.6 Å, the Pro^3, Phe^5, Pro^8 and Phe^{10} carbonyls with Na$^+\cdots$O distances 3.4—4.0 Å having a secondary significance. It has also been reported that the solvent (water) molecules may participate in the formation of the coordination sphere in the Na$^+\cdot$antamanide complex.

Further study of the Na$^+$ complexes of antamanide and [Val^6, Ala^9]-antamanide by means of ^{13}C-NMR spectroscopy [114, 1113] confirming the presence of 2—4 carbonyl ligands, indicated at the same time that there are only two rather than all four *trans* amide bonds. On these grounds, two more possible models have been suggested [1113, 1114].

		Val^1, Phe^6	Pro^2, Pro^7	Pro^3, Pro^8	Ala^4, Phe^9	Phe^5, Phe^{10}
1,6-*cis*	ϕ	−150 − −120	−60	−60	−60	−90
	ψ	120 − 150	150 − 180	−60	−60	−60 − −30
	ω	0	180	180	180	180
2,7-*cis*	ϕ	−120	−60	−60	−60	−120 − −90
	ψ	120 − 150	150	150	−60 − −30	30 − 90
	ω	180	0	180	180	180

A considerable step forward was made when an X-ray analysis was carried out of the Li$^+$ complex of antamanide and the closely related Na$^+$ complex of [Phe^4, Val^6]-antamanide [1115]. The structures elucidated are represented in Figs 90—92. The following rotational angles have been obtained for the Li$^+$ complex.

Residue	1	2	3	4	5	6	7	8	9	10
ϕ	−115	−65	−83	−67	−84	−123	−74	−69	−78	−88
ψ	138	139	147	−14	−6	139	144	144	−15	7
ω	178	−3	−173	176	−178	−171	−3	−176	172	173

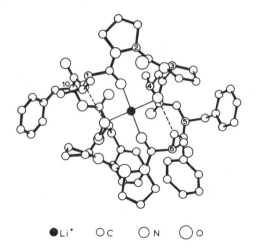

Li⁺ ○ C ○ N ○ O

Fig. 90. Conformation of the antama-nide·Li⁺ complex as viewed along the pseudo C_2 axis. The ligand CH_3CN molecule is not shown.

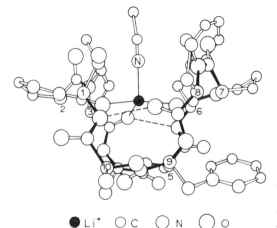

Li⁺ ○ C ○ N ○ O

Fig. 91. Conformation of the antama-nide·Li⁺ complex (side view). Side chains of the Val, Ala and Phe residues are omitted for clarity.

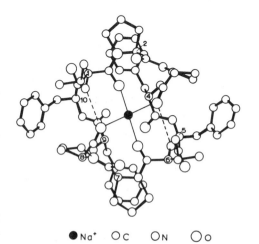

Na⁺ ○ C ○ N ○ O

Fig. 92. Conformation of [Phe⁴, Val⁶]-antamanide·Na⁺ complex as viewed along the C_2 axis.

In this way several of the structural elements were confirmed that had earlier been proposed for the Na^+ complex in solution, viz. the approximate C_2 symmetry, participation of the Val^1 and Phe^6 NH groups in intramolecular hydrogen bonds and CO groups in the ion—dipole interaction; also, the presence of two *cis* amide bonds, the rotational states of the NH—CH fragments as mentioned on p. 170, and the presence in the molecule of the ligand solvent molecules. In overall structure, however, the above models differed significantly from the conformation of the crystalline complex. As can be seen from Figs 90 and 92, the alkali metal ion coordinates the CO^7, CO^3, CO^6 and CO^8 oxygens located approximately in the corners of a square (average $Li^+ \cdots O$ distance 2.11 Å). Moreover, Li^+ interacts with acetonitrile molecules (see Fig. 91) and Na^+ complexed by [Phe^4, Val^6]-antamanide — with ethanol. The CO^2, CO^4, CO^7 and CO^9 carbonyls, oriented in the direction of the axis, can also weakly participate in the ion—dipole interaction. It is noteworthy that the ligand CO^3 and CO^8 carbonyls are simultaneously proton acceptors, taking part in $4 \rightarrow 1$ type intramolecular hydrogen bonds with NH^1 and NH^6.

The molecular surface of the antamanide complex is less lipophilic than that of valinomycin, the enniatins and the nactins (four outwardly oriented NH groups), which is apparently one of the reasons for its inability to function as an ionophore [701]. The molecular cavity of the antamanide complexes is smaller than that of valinomycin. This circumstance, besides the specific electronic structure of the amide ligands [234, 511], explains the high Na^+/K^+ selectivity of antamanide. It will be shown in the next section that the elucidated structure serves as a basis for interpretation of the complexing properties of a large series of synthetic antamanide analogs.

III.A.1.h. Antamanide analogs

The compounds listed in Table 15, for which the stability constants of the Na^+ complexes have been measured, differ from the naturally occurring cyclopeptide in the substitution of one or more amino acid residues by those of hydrophobic L-amino acids or glycine. For the example of the valinomycin cyclodepsipeptides it was shown that this type of substitution affects the metal-ion complexing capacity of the macrocyclic compound much less than does configurational inversion of the individual residues, change in ring size, replacement of the amide groups by ester groups, etc. No wonder, therefore, that all the antamanide analogs investigated have retained the ability to form complexes. At the same time there is a definitely expressed tendency for the stability of the complexes to fall when the sterically hindered Val^1 and Pro^7 residues are substituted by alanine or glycine residues with the highest conformational freedom (compounds 189,

189a, 196, 201 and 201a). As in the case of the valinomycins this effect is apparently due to a decrease in free energy of the left-hand side of the equation on p. 134.

The hydrogenated analog of antamanide (217) in which all phenyl radicals are replaced by cyclohexyl radicals has a quite similar Na^+ and K^+ affinity to the initial cyclodecapeptide. At the same time its higher lipophilicity gives rise to a certain degree of membrane activity [701]. The substitution of CONH groups, not donating protons for hydrogen bonding, by COO— and $CONCH_3$ groups renders analogs (193 and 200) capable of forming complexes. The exceptionally high complexing capacity of diastereomer (219) requires further study for its explanation.

According to Wieland and collaborators [1068], complexing ability is a necessary condition for the appearance of antitoxic activity in the antamanide analogs. It is very likely, therefore, that the phalloidine-counteracting analogs (190—192a, 194—194d, 195a, 195d—195f, 198, 199, 203, 205, 209, 210 and 215) in Table 15 are Na^+ complexones. Of particular interest in this series are compounds (191, 194 and 210), for *a priori* one should hardly have expected them to form complexes and manifest biological activity. However, it can readily be seen that if the substitution L-Ala⁴ → D-Ala⁴ (analog 191) be made, the parameters ϕ —67°, ψ —14° for the D-alanyl residue of the Li^+ complex would be in the vicinity

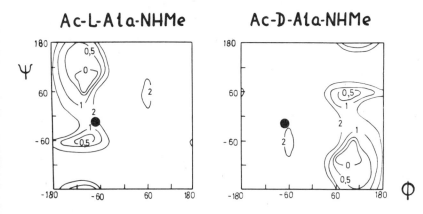

Fig. 93. Conformational maps of acetyl-L- and -D-alanine methylamides. Dark circles show the conformational coordinates of amino acids 4 and 9 in the Na^+ complexes of antamanide and its analogues.

of the allowed conformational region, bearing in mind that the conformational map of the D-alanine residues is symmetric to that of the L-alanine residue with respect to ϕ 0°, ψ 0° (Fig. 93). There is therefore nothing extraordinary in the fact that the resulting diastereomer of antamanide preserves its complexing ability.

As for the des-Phe[5]-antamanide analogs (194 and 210) examination of molecular models has shown that shortening of the peptide chain by a Phe[5] residue has little effect upon the general shape of the Na$^+$·antamanide complex, causing but slight distortion in the symmetry of ligand oxygens arrangement as well as replacing the $4 \rightarrow 1$ by $3 \rightarrow 1$ intramolecular hydrogen bonds.

It was mentioned in Chapter I that the primary structure of analog (194) is similar to the naturally occurring nonapeptide (223) in the presence of Pro—Pro fragments, the L-configuration and hydrophobicity of the amino acid residues, etc. Hence, one may also expect manifestation of complexing properties by the latter compound but this still awaits experimental confirmation. The flexibility of compound (223), as follows from its CD curves [675], NMR spectra [86] and conformational calculations [1008], does not contradict this proposal.

III.A.1.i. Alamethicin

The CD curves of alamethicin, similar in shape to those of α-helical polypeptides, are indicative of its possessing an ordered structure in ethanol, methanol, dioxane and acetonitrile [621]. The curves lose intensity in aqueous solutions (Fig. 94, curve 2) but the addition of detergents restores the original value (curve 4). On this basis it was suggested that in lipophilic

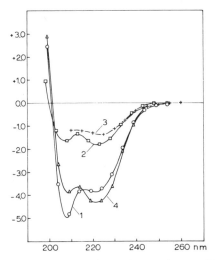

Fig. 94. CD curves of alamethicin in different solvents. 1, ethanol; 2, water—ethanol, 9:1; 3, 6.0 M guanidine hydrochloride; 4, 1% aqueous sodium dodecyl sulfate.

media alamethicin assumes a compact conformation with 40—60% of the amino acid residues localized in the $Aib^2—Aib^{12}$ segment participating in the formation of right-handed helical fragments. In aqueous media this conformation is partly disrupted. Alamethicin assumes a similar spatial structure in phospholipid membranes and also on interacting with detergent micelles in aqueous solutions. The smallest cross section of this structure has a surface area of about 250 Å^2, which is in accord with results obtained in the study of alamethicin monolayers [127].

Although the proposed model is purely hypothetical, it is nevertheless in accord with the recently calculated conformational map of N-acetyl-α-methylalanine methylamide (Fig. 95), from which it follows that the α-helical conformations (R, ϕ —60°, ψ —60° and L, ϕ 60°, ψ 60°) are the most advantageous for this specific amino acid [767].

III.A.1.j. Synthetic cyclopeptides

The lack of complexing ability of simple cyclohexapeptides built up of alanyl and glycyl residues [427] is due to their tendency to assume relatively planar "pleated sheet" structures, stabilized by two transannular H bonds and lacking an inner cavity (Fig. 96) [436, 437, 438, 776, 777].

It was concluded from the ^{1}H- and ^{13}C-NMR spectra of $cyclo$[—(L-Pro—Gly)$_3$—] (225b) that in CH_2Cl_2 this cyclopeptide assumes a symmetric conformation (S) with three cis amide bonds [191, 192]; its approximate conformational parameters are presented in Table 52. On passing to $(C^2H_3)_2SO$ one or two Gly—Pro bonds assume the $trans$ configuration. The

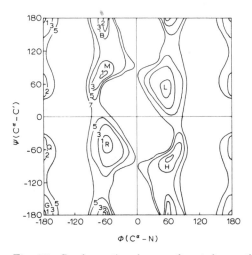

Fig. 95. Conformational map of acetyl-α-methylalanine methylamide.

symmetric conformation is again adopted in complexes of (225b) with alkali metal ions, but in this case all the amide bonds are in the *trans* configuration. Deber and coworkers consider three related conformations of this type, S_G, S_P and S* (Table 52). In conformation S_G^* maximum nearing is between the

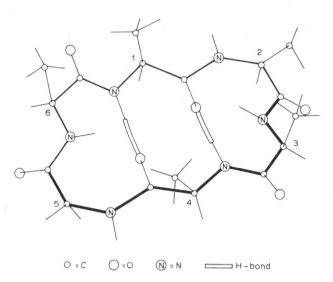

O = C   O = O   (N) = N   ▭ H – bond

Fig. 96. Preferred conformation of *cyclo*(—Gly—L-Ala—L-Ala—L-Ala—L-Ala—L-Ala—).

glycine O atoms to which are ascribed the role of ligands in the complex; in S_P^* this role is assumed by the neared O atoms of the proline residues, whereas conformation S* is intermediate between S_G^* and S_P^*. On passing from the S* conformation to the complex, the ^{13}C signal of the proline carbonyl is shifted more than is the corresponding glycine signal. This led to the conclusion that the cation is bound more strongly by the proline carbonyl groups [191]. However, such a conclusion is hardly justified, because the effect of the concomitant *cis* → *trans* transition of the amide groups had not been taken into account. Moreover, it is surprising that these authors have not considered the possibility of the cation being incorporated in the internal cavity to form a complex resembling that of the enniatins. We believe such a structure to be quite possible, as follows for instance from the similarity of the S_G^*, S_P^* and S* conformational parameters to their enniatin counterparts (see Table 52).

The spatial structure of sandwich complexes obtained by Schwyzer and coworkers is a problem still in the initial stages of solution. The parameters

TABLE 52

CONFORMATIONAL PARAMETERS OF DIFFERENT FORMS OF $cyclo[-(\text{L-Pro—Gly})_3-]$ AND THE RANGES OF POSSIBLE ϕ, ψ, ω VALUES FOR THE COMPLEXES OF ENNIATIN CYCLODEPSIPEPTIDES

Angle	S		S_G^*		S_P^*		S^*		Enniatin complexes[a]	
									$\begin{array}{c}\text{H}_3\text{C}\ \ \text{R}\\ \mid\ \ \mid\\ \text{—N—CH—CO—}\end{array}$	$\begin{array}{c}\text{R}'\\ \mid\\ \text{—O—CH—CO—}\end{array}$
	Pro	Gly	Pro	Gly	Pro	Gly	Pro	Gly		
ϕ	−60	180	−60	120	−60	60	−60	90	−60 — −103	60 — 100
ψ	−55	180	100	150	160	−150	120	180	120 — 185	−136 — −180
ω	180	0	180	180	180	180	180	180	180 ± 10	180 ± 10

[a] As obtained from X-ray analysis (see p. 141) and theoretical calculations (Table 47).

Fig. 97. Conformation of the C—S—S—C fragment in the complexes of cyclopeptide (226).

of the S—S bond (right-handed helix, Fig. 97 and dihedral angle of 75—90°) were determined from the ORD and CD curves, while for its attached pentapeptide rings a conformation has been proposed with four or five carbonyls oriented towards the central cation (Fig. 98) [876].

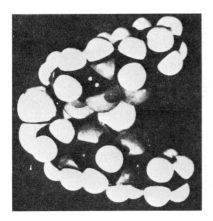 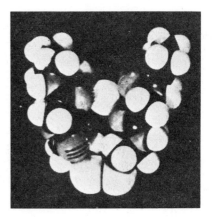

Fig. 98. Two possible conformations of the cyclopeptide (226).

III.A.2. Depsides. The nonactins

The conformation of the crystalline nonactin·KNCS complex established in 1967 by Dobler, Dunitz and Kilbourne [208, 488] served as a basis for elucidating several important features in the cation interaction of metal-binding macrocyclic compounds, subsequently found to be common to the other complexones. As one can see from Fig. 99, the non-hydrated central potassium ion is surrounded by four tetrahydrofuran and four carbonyl oxygens, forming an approximately cubic 8-coordination. Under these circumstances the nonactin molecule has a twofold symmetry axis and an almost exactly fourfold mirror axis. The $K^+ \cdots O$ distance is 2.81 and 2.88 Å for the ether oxygens, and 2.73 and 2.81 Å for the ester oxygens; the

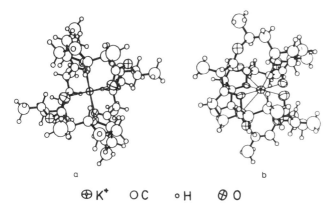

⊕ K⁺   O C   ∘ H   ⊘ O

Fig. 99. Conformation of the complex in crystalline nonactin·KNCS. a, view along the crystallographic axis a; b, view along the axis b.

C—O· · ·K⁺ angles are 107—115°, thus deviating considerably from 180°, in which respect they resemble the enniatin complexes (see p. 149). Fig. 100 and Table 53 show the substituent orientations along the chain of one of the (+)-nonactinic acid moieties. In accord with the molecular symmetry the second (+)-acid has the same dihedral angles, whereas the angles of the (—)-nonactinic acid moieties, while of similar value, are of the opposite sign. Fig. 100a shows that the ester groups have an almost planar *trans* configuration, and that the fragments with two tetrahedral carbon atoms (c, f and g) are all in a skew conformation. The tetrahydrofuran rings of the complex are in the envelope conformation. In general, the depside chain in

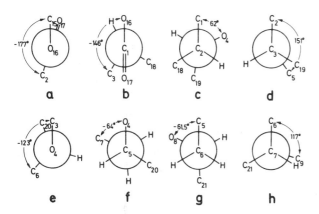

Fig. 100. Orientation of the substituents along the backbone of the nonactin·K⁺ complex.

TABLE 53

TORSIONAL ANGLES ALONG THE POLYETHER BACKBONE OF
THE (+) NONACTIN ACID RESIDUE IN THE FREE NONACTIN
[207] AND ITS K^+ COMPLEX [208]

Fragment	Torsional angle	
	Nonactin	K^+ complex
$O_{16}-C_1-C_2-C_3$	−65.7	−146.1
$C_1-C_2-C_3-O_4$	174.5	62.1
$C_2-C_3-O_4-C_5$	138.1	151.1
$C_3-O_4-C_5-C_6$	−153.8	−123.2
$O_4-C_5-C_6-C_7$	−71.6	−63.7
$C_5-C_6-C_7-O_8$	−67.0	−61.5
$C_6-C_7-O_8-C_9$	153.3	116.7
$C_7-O_8-C_9-C_{10}$	−177.1	178.2

the complexed antibiotic resembles the seam of a tennis ball, the methyl side
chains and tetrahydrofuran methylene groups forming a hydrophobic
exterior.

The nonactin sodium rhodanide complex is not isomorphic with the
potassium complex, but the close values of their cell parameters and of the
intensity distributions of their X-ray reflections bear evidence of structural
similarity between both compounds [488].

The crystalline non-complexed nonactin (Fig. 101) has the same symmetry as the K^+ complex (strictly C_2 and approximately S_4) but differs in
having a flatter structure. Their respective dimensions are 17 x 17 x 8.5 Å
and 17 x 17 x 12 Å; the carbonyl groups of nonactin are directed almost
normal to the average plane of the ring [207]. It can be seen from Table 53
that, on formation of a complex, only two rotational angles $-O_{16}-C_1-$
C_2-C_3 and $C_1-C_2-C_3-O_4$ of each nonactin acid moiety undergo
considerable change, the changes in the other angles not exceeding 37°. In
both the complex and free nonactin, the tetrahydrofuran rings adopt an
envelope conformation, whereas the ester bonds are in the *trans* configuration.

Indirect evidence for the interaction of the macrotetrolide's carbonyl and
ether groups with the cation are the significant changes in the stretching
vibrations of the C=O (1700—1750 cm^{-1}) (Fig. 102) and C—O groups [676,

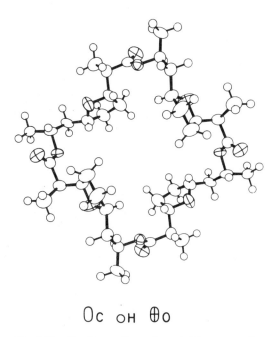

$\underset{\text{C}}{\text{O}} \quad \text{OH} \quad \underset{\text{O}}{\oplus}$

Fig. 101. Conformation of crystalline nonactin.

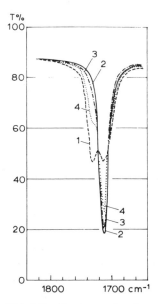

Fig. 102. Region of ester carbonyl stretching vibrations in the infrared spectra of nonactin and its K^+ complex in methanol. 1, $5.9\cdot10^{-3}$ M nonactin; 2, $5.9\cdot10^{-3}$ M nonactin, $2.9\cdot10^{-3}$ M KNCS; 3, $5.9\cdot10^{-2}$ M nonactin, $5.9\cdot10^{-3}$ M KNCS; 4, $5.9\cdot10^{-3}$ M nonactin, $2.9\cdot10^{-3}$ M KNCS.

756]. In conformity with the highly symmetric disposition of the ligand oxygen atoms in the K^+ complex of nonactin, its far infrared spectra reveal only one $K^+ \cdots O$ stretching band. Interestingly, in the cesium complex two bands are observed, evidence of the distortion of the ligand cubic symmetry with increase in size of the internal cavity [428].

A NMR study of nonactin and its Na^+, K^+ and Cs^+ complexes in dry and aqueous acetone [792, 793] also does not contradict the X-ray data, but does not allow for an independent conclusion as to the spatial structure. Thus, the spectral parameters of the nonactin complexes remain practically unchanged on adding water to the acetone solution, from which it may be inferred that in the complex the cation is non-hydrated.

TABLE 54

COMPLEXING INDUCED CHANGES IN THE CHEMICAL SHIFTS OF NONACTIN PROTONS

Cation	$\Delta\delta$ (Hz, as measured at 220 MHz)					
	H_7	H_3	H_5	H_{15}	H_{21}	H_2
Na^+	-113 ± 5	-62 ± 3	-89 ± 2	-23 ± 1	-8.8 ± 1	-4.4 ± 1
K^+	-115 ± 5	-103 ± 4	-56 ± 2	-20 ± 1	-1.9 ± 1	2.2 ± 1
Cs^+	-67 ± 13	-90 ± 18	-60 ± 12	-20 ± 4	1.8 ± 1	4.8 ± 1

The changes in the NMR spectra of nonactin observed on addition of salts and formation of complexes are difficult to interpret, being dependent on a variety of factors such as the effect of the electrical field of the cation, changes in the electronic state of the ligand groups, conformational changes in the macrotetrolide, etc. As one can see from Table 54, the largest $\Delta\delta$ values are experienced by the resonances of the C_7, C_5 and C_3 protons adjacent to the ligand groupings. However, such change has not been observed for the C_2H signals, apparently due to diminution of the carbonyl deshielding effect since the carbonyl group moves away from the corresponding proton when the complex is formed (the near transoidal disposition of the $C_1 = O_{17}$ and $C_2 - H$ groups in the complex can be seen in Fig. 100f). The C_7 protons, on the contrary, are *cis* oriented to the C_9 oxygens in the sodium and potassium complexes (Fig. 100) and their chemical shifts are, therefore, the most susceptible to complex formation (Table 54). The lower values of $\Delta\delta$ (H_7), found by Prestegard and Chan for the Cs^+ complex of nonactin, the authors ascribe to the increased size of the internal cavity as compared with the other complexes, with accompanying change in the

$C_6-C_7-O_8-C_9$ angle and increase in the distance between the C_9O carbonyl and the C_7H proton.

More definite conclusions could be drawn from a comparison of the vicinal proton spin—spin coupling constants of nonactin and its complex (Table 55). This refers especially to the $^3J_{H_2,H_3}$ constant of which the increase to 9.7 Hz is evidence of change in the H_2,H_3 orientation to *trans* in the complex (Fig. 100c). Interpretation of the other constants is hampered by their composite nature, although the different $^3J_{H_5,H_6}$ values observed in the series in question also bear evidence of the conformational mobility of the antibiotic molecule.

TABLE 55

COUPLING CONSTANTS FOR VICINAL PROTONS IN NONACTIN AND ITS COMPLEXES

Compound	$|{}^3J_{H_2,H_3}|$	$|{}^3J_{H_5,6} + {}^3J_{H_5,6'}|$	$|{}^3J_{H_7,6} + {}^3J_{H_7,6'}|$
Nonactin	7.6 ± 0.4	12.4 ± 0.4	13.0 ± 0.4
Na^+ complex	10.0 ± 0.4	10.9 ± 0.4	13.2 ± 0.4
K^+ complex	9.4 ± 0.4	11.0 ± 0.4	13.0 ± 0.4
Cs^+ complex	9.7 ± 0.8	9.3 ± 0.8	13.0 ± 0.4

In general, the conformations of the crystalline complexes of tetranactin with Na^+, K^+ and Rb^+ are close to the conformation of the crystalline K^+ complex of nonactin although some differences in the torsion angles along individual bonds and in the symmetry of the systems do exist [418]. The free, crystalline tetranactin has more important conformational differences from nonactin (cf. Figs 101 and 103) [418]. It is hard to say whether this is due to intramolecular interaction of the ethyl groups of tetranactin (for instance, Van der Waals interaction of the spatially neared oppositely lying ethyls) or to the peculiarities of the crystal packing. In any case a comparison of the structures of the free macrotetrolides (227) and (231) bear evidence of flexibility of the depside framework.

The conformational states of the other nonactin homologs (228—230) have not been investigated. Eisenman *et al.* [233] believe that the increased Na/K selectivity of their complexing reaction is due to change in the electron donating properties of the alkyl substituents on passing from methyl to ethyl. However, a more likely cause may perhaps be the decrease in flexibility of the macrotetrolide molecules with increasing bulk of the substituents.

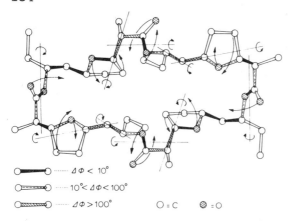

Fig. 103 (above). The conformation of crystalline tetranactin. Atomic shifts and bond rotations accompanying complexation are shown by arrows. Bond representations along the backbone correspond as shown in the lower left-hand corner to the complexing-induced changes in the torsional angles.

Fig. 104 (below). Possible orientations of atoms and groups along the C—C and C—O bonds of the polyester backbone.

III.A.3. Cyclic polyethers

The conformation of the crystalline tetraether (244) is shown in Fig. approximation to the problem of the character of the rotational isomerism along the C—C and C—O bonds of the backbone. Energetically favored are the skew conformations, *anti*, *gauche* (+), and *gauche* (−) (a, b, and c, respectively, in Fig. 104); the eclipsed conformation is sterically less preferable. The use of spectral methods in studies of the spatial structure of cyclic polyethers is more limited than, say, of the peptide ionophores. Consequently, most of the information obtained in this area is from X-ray analysis.

III.A.3.a. Tetraethers (244) and (244a)

The conformation of the crystalline tetraether (244) is shown in Fig. 105. As in the case of nonactin its symmetry is almost exactly S_4, i.e. it has a fourfold rotary-reflection axis [664]. Each monomer unit possesses in the crystal an *anti*, *gauche*, *gauche*, *anti* conformation; the exact values of the torsional angles for one quarter of the molecule are presented in Table 56.

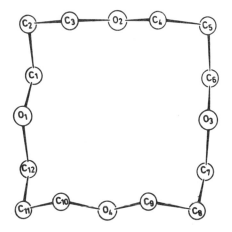

Fig. 105. Conformation of crystalline 1,5,9,13-tetraoxacyclohexadecane (244).

The infrared spectra of the tetraethers (244 and 244a) in the crystalline state and in solution are very similar and consist of narrow bands [81]. From this it follows that both compounds are conformationally homogeneous and assume a "square" structure independent of the state of aggregation, analogous to the structure shown in Fig. 105.

TABLE 56

TORSIONAL ANGLES FOR O_1-O_2 FRAGMENT IN CRYSTALLINE 1,5,9,13-TETRAOXACYCLOHEXADECANE (244)

Fragment	Angle (degrees)
$C_{12}-O_1-C_1-C_2$	171.0
$O_1-C_1-C_2-C_3$	−65.0
$C_1-C_2-C_3-O_2$	−65.4
$C_2-C_3-O_2-C_4$	174.6

Straightforward evidence that such conformations are realized in benzene is the practically zero value of their dipole moments (0.3 D for 244 and 0.0 D for 244a) [81].

In the complex (244a)·Li$^+$ the methylene and methyl protons are non-equivalent. On these grounds, a symmetric S_4 conformation has been proposed for the lithium complexes of (244) and (244a) with the O atoms neared and their unshaired electron pairs projecting into the molecular cavity (Fig. 106) [179]. It can also be seen in the figure that complexation is here accompanied by sign reversal of one of the *gauche* conformations in each

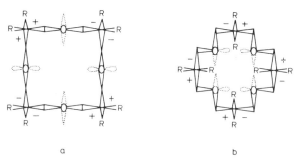

Fig. 106. Schematic drawing of tetraether conformations (244, R = H; 244a, R = CH$_3$) in non-complexed state (a) and on interaction with Li$^+$ ions (b). Dashed lines indicate lone electron pairs; + and − characterize the dihedral angles for fragments of *gauche* conformation. The remaining fragments are in the *anti* conformation.

monomer unit. Judging from the NMR spectra, the second polyether molecule adds to the 1:1 complex (to form a 1:2 complex) without conformational change. The nature of the O coordination remains unclear.

III.A.3.b. Benzo-15-crown-5 (246)

As one might have expected from general considerations, the oxygen atoms in the complex (246)·NaI are almost coplanar and form a pentagonal pyramid with the cation, the latter being displaced by 0.75 Å from the

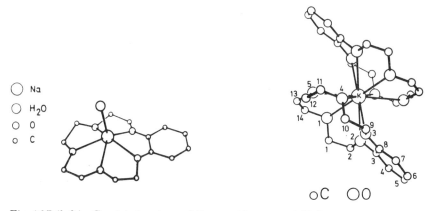

Fig. 107 (left). Crystal structure of (benzo-15-*crown*-5)·NaI.
Fig. 108 (right). Conformation of the complex in the crystalline salt (246)$_2$·KI.

oxygen plane (Fig. 107). The Na$^+$···O distance varies within limits of 2.35—2.43 Å, differing only slightly from the sum of the Na$^+$ and O crystallographic radii. The lattice cell of the complex salt also contains a water molecule interacting with the bound cation and the anion [105].

In the complex $(246)_2 \cdot K^+$ (Fig. 108) ten amide oxygen atoms form a somewhat distorted pentagonal antiprism [593]. The cation located in its center ($K^+ \cdots O$ distance 2.78—2.96 Å) does not interact with the I^- anion, statistically distributed in the holes of the crystal lattice. The conformations of the polyether (246) in the two complexes investigated differ somewhat, especially in the neighborhood of the C_1 and C_{10} atoms (Fig. 109).

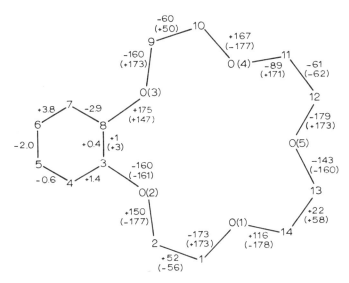

Fig. 109. Torsional angles of polyether (246) in the crystalline salts $(246)_2 \cdot KI$ and $(246) \cdot NaI \cdot H_2 O$ (in parentheses).

III.A.3.c. 18-Membered cyclic polyethers

The unsubstituted hexaether (259), when ground together with potassium bromide powder (it has been proposed that this results in the formation of a K^+ complex), resembles in the number, shape and position of the infrared bands in the 700—1500 cm^{-1} region, crystalline polyethylene glycol which has a helical structure and *anti,gauche,anti* conformations of the O—C—C—O fragments [181]. It has therefore been suggested that the complex has a symmetric S_6 conformation with six *anti,gauche,anti* fragments (Fig. 110). The structure has six unshared electron pairs (one pair from each of the oxygens) oriented towards the symmetry axis, which, according to Dale and Kristiansen [181], promotes the stability of the complex. Such a conformation has in fact been found in the crystalline complexes of cyclohexyl- and benzo-substituted polyethers (274a, 274b and 270) (see below), although the presence of aromatic substituents in the latter leads to a planar *cis*

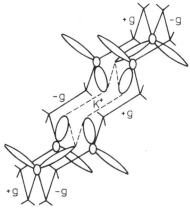

Fig. 110. Schematic drawing of the conformation of the crystalline (259)·K⁺ complex. Lone electron pairs are shown at oxygen atoms. Neighboring pairs are connected by dashed lines; the symbols +g and −g indicate the signs of the dihedral angles about the corresponding *gauche* bonds; the other fragments are in the *anti* conformation.

conformation of two of the O—C—C—O fragments. In the free polyether (259) interaction of the electron pairs causes conformational rearrangement to another, less symmetric form [181].

Correspondingly, the crystalline polyether (259) displays considerably more infrared bands than the K⁺ complex (cf. Figs 111a and 111b). In the

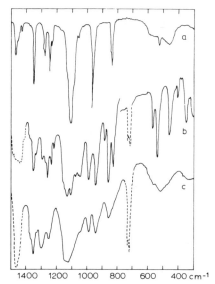

Fig. 111. Infrared spectra of 18-*crown*-6 (1,4,7,10,13,16-hexaoxacyclo-octadecane) (259). a, crystalline (259)·KBr complex; b, crystalline polyether (259) at −60°C; c, liquid polyether (259) at 50°C. Dashed curves indicate bands due to polyethylene film used to protect the KBr windows.

infrared spectrum of the liquid polyether the spectral bands are noticeably broadened (Fig. 111c), indicating that the specimen is conformationally non-homogeneous. The dipole moment of the polyether (259) in benzene is 2.65 D and does not permit selection of any preferable structure, being in accord, on the contrary, with the dipole moment calculated for a random set of conformations (3.18 D) [180]. A similar result has been obtained in a study of the dipole moments of the cyclopolymer homologs (309—311) constructed of five, seven and eight —O—CH_2—CH_2—O— units, respectively [180].

An entirely different behavior is exhibited by the aza analog of the polyether (259), 1,7,10,16-tetraoxa-4,13-diazacyclo-octadecane (262), which

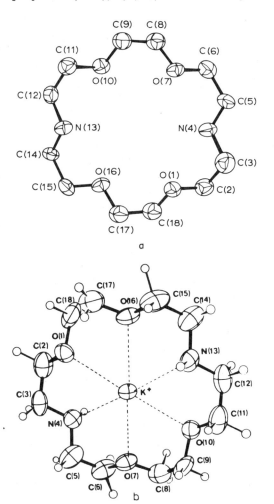

Fig. 112. Conformation of (a) the crystalline aminoether (262) and (b) its K^+ complex.

undergoes practically no conformational change on forming the potassium complex [388, 664]. As one can see from Fig. 112 and Table 57, in the free state and complexed with KNCS this compound assumes the (*anti,gauche, anti*)$_6$ conformation very similar to the conformation of the (259)·K$^+$ complex described above (Fig. 110). In the molecule of the free crystalline aminoether the distance from the nitrogen atoms to the symmetry axis is 2.92 Å, from O_1 and O_{10} 2.83 Å, and from O_7 and O_{16} 2.80 Å (i.e. they are close to the sums of the ionic radii of potassium and the corresponding atom). In addition, the cation is interacting, although much more weakly,

TABLE 57

TORSIONAL ANGLES FOR THE O_1–O_{10} FRAGMENT OF CRYSTALLINE POLYETHER (262) AND ITS K$^+$ COMPLEX

Fragment	Torsional angle (degrees)	
	Free polyether	K$^+$ complex
C_{18}–O_1–C_2–C_3	177	178
O_1–C_2–C_3–N_4	−64	−67
C_2–C_3–N_4–C_5	176	178
C_3–N_4–C_5–C_6	−179	179
N_4–C_5–C_6–O_7	67	66
C_5–C_6–O_7–C_8	−176	−178
C_6–O_7–C_8–C_9	−176	−175
O_7–C_8–C_9–O_{10}	−73	−72
C_8–C_9–O_{10}–C_{11}	−178	−176

with the thiocyanate anion, statistically distributed among the lattice holes (K$^+$···anion distance 3.33 Å). The reasons for the stability of the symmetric conformation of the aminoether (262) are not entirely understood. Apparently they must be sought for in the specific interactions of the unshared nitrogen and oxygen electron pairs. It is noteworthy that such similarity between the conformational parameters of the free macrocyclic compound and its complex has as yet no analogy in the chemistry of the complexes discussed in this book.

The free polyether (270), co-crystallizing with a mixture of the sodium and rubidium complexes, has a centrosymmetric conformation in which three non-equivalent pairs of oxygen atoms are situated at 2.3, 3.2 and 3.3 Å from the center of symmetry so that effective simultaneous interaction with one cation is impossible [87, 88]. On the contrary, in the Rb$^+$ and Na$^+$ complexes the oxygen atoms form regular hexagons with sides of 2.75 Å and are at almost equal distances from the Rb$^+$ and Na$^+$ ions located 0.94 and

0.54 Å, respectively underneath the oxygen plane (the $Rb^+\cdots O$ distance of 2.86—2.94 Å and the $Na^+\cdots O$ distance of 2.73—2.89 Å markedly exceed the sum of the corresponding crystallographic radii). The benzene rings are here also situated somewhat out of the oxygen plane, assuming a position on the side opposite to that of the cation (Fig. 113). It is important to bear in mind that, whereas in the valinomycin, enniatin and macrotetrolide crystalline complexes the cations are effectively screened from the anions, in compound (270) the anion is spatially close to Rb^+, forming a contact ion pair with the latter ($Rb^+\cdots N$ distance, 2.9 Å); in the Na^+ complex the interaction between the cation and the anion is weaker ($Na^+\cdots N$ distance, 3.3 Å).

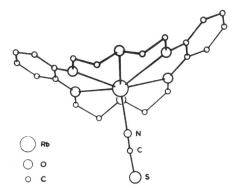

Fig. 113. Conformation of crystalline (dibenzo-18-*crown*-6)·RbNCS.

Bright and Truter have noted that if one takes into account the spatially close unshared electron pairs of the oxygen atoms in the complexes of (270) with RbNCS and NaNCS, the cation is found to be in an electron enriched region, since 10 out of the 12 electron pairs are situated closer to the cation than to the centers of the oxygen atoms [87].

Two forms of the crystalline polyether complex with NaBr are present in the elementary cell; in one (form B), the cation is interacting with two molecules of hydrate water, whereas in the other (form A), it is interacting with one water molecule and with one bromine anion (Fig. 114) [105, 106]. The A and B conformations of the cyclic polyether (270) practically coincide with its conformations in the RbNCS and NaNCS complexes. Such slight variability of the organic molecule leads to the conclusion that a similar structure should be present in the complexes of the polyether (270) in solution. However, the cation is in a considerably different position in forms A and B from that in the complex with NaNCS, being displaced from the oxygen plane in the opposite direction (i.e. in the same direction as the

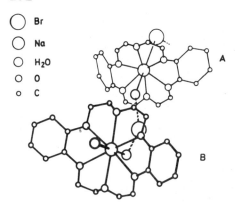

Fig. 114. Conformations A and B of crystalline (dibenzo-18-*crown*-6)·NaBr.

benzene rings). It is at much smaller distances from the plane (0.27 and 0.07 Å, respectively) than in compound (270)·NaNCS (0.54 Å). The $Na^+\cdots O$ distance, 2.56—2.89 Å in form A and 2.63—2.82 Å in form B, is on average also somewhat less than in the NaNCS complex. It is very likely that the observed positional differences of the cation in the complexes reflect the degree of its interaction with the anions and solvent (see Part III.B).

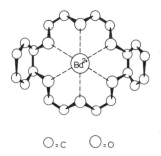

$\bigcirc = C \quad \bigcirc = O$

Fig. 115. Conformation of crystalline (*cis-syn-cis*-dicyclohexyl-18-*crown*-6)·Ba(NCS)$_2$.

X-ray analysis of the complexes (274a)·Ba(NCS)$_2$ [182] and (274b)·NaBr·2H$_2$O [260] has elucidated the fusion of the cyclohexyl rings onto the polyether cycle, which up to then had had contradictory interpretations [272, 273; see also 182]. It has been shown that isomer A has a *cis,syn,cis* configuration and isomer B a *cis,anti,cis* configuration. Corresponding to this the isomer A complex (Fig. 115) has a C$_2$ axis passing through the barium cation and the complex of the B isomer (Fig. 116), a center of symmetry positioned at the sodium ion. In isomer A the cyclohexyl rings are above the mean plane of the O atoms at an angle of

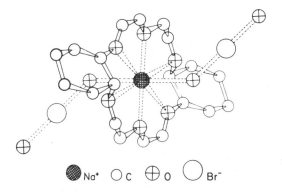

Fig. 116. Conformation of crystalline (*cis-anti-cis*-dicyclohexyl-18-*crown*-6)·NaBr·H$_2$O.

114°; evidently on this side also is a water molecule, whereas the NCS⁻ anions are interacting with the Ba^{2+} ions from under the O plane. The Ba^{2+}···O and Ba^{2+}···N distances are 2.77—2.89 Å and 2.89 Å, respectively. In the (274b)·NaBr·H$_2$O complex the polyether cycle also has a planar conformation and the cation is solvated by two symmetrically positioned water molecules.

III.A.3.d. 24-Membered cyclic polyethers

The infrared spectra of crystalline 24-*crown*-8 (311) and its K$^+$ complex are similar to the spectrum of the K$^+$ complex of 18-*crown*-6 (Fig. 111a) and of crystalline polyethylene glycol. Therefore, Dale and Kristiansen have proposed for it a centrosymmetric structure with an S$_4$ axis, each unit in the

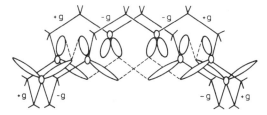

Fig. 117. Proposed conformations of crystalline 24-*crown*-8 and its K$^+$ complex.

structure having an *anti,gauche,anti* conformation (Fig. 117) [181]. They implied therein that the cyclic polyether is interacting with one potassium ion. However, Truter and coworkers [259] have shown that 24-membered octaethers are capable of forming dinuclear complexes in which two cations are residing in the cavity of the macromolecule. In the crystalline complex (dibenzo-24-*crown*-8)·(KNCS)$_2$ the ether oxygens and the potassium ions are approximately coplanar; each cation coordinates five oxygen atoms, two thiocyanate nitrogens and also apparently the aromatic rings of the neighboring molecules (Fig. 118). Two oxygens of the eight oxygen atoms and both nitrogen atoms are simultaneously interacting with both potassium ions. The $K^+\cdots O$, $K^+\cdots N$ and $K^+\cdots K^+$ distances are 2.74—3.01 Å, 2.84 and 2.89 Å, and 3.8 Å, respectively.

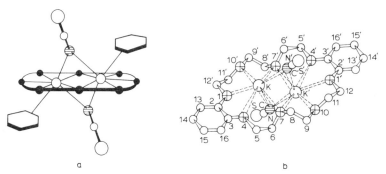

Fig. 118. Exact (a) and schematic (b) representation of crystalline (dibenzo-24-crown-8)·2KNCS.

III.A.3.e. Dibenzo-30-crown-10 (305) and its crystalline salt with KI [105, 108]

Further increase in the number of ligand oxygens and in the size of the polyether cycle on passing over to compound (305) brings about significant changes in the structure of the complex since the cation becomes effectively

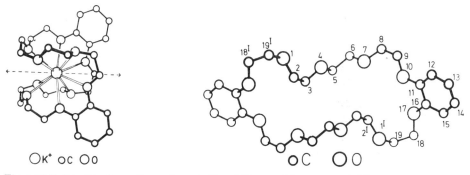

Fig. 119 (left). Conformation of crystalline (dibenzo-30-crown-10)·KI.
Fig. 120 (right). Conformation of crystalline dibenzo-30-crown-10.

shielded from the anion and solvent. The overall shape of the (305)·KI backbone, possessing a twofold symmetry axis, resembles the tennis ball seam conformation of the nonactin complexes (Fig. 119). The $K^+\cdots O$ distance is 2.85—2.93 Å.

TABLE 58

TORSIONAL ANGLES FOR $C_{19}^I-C_{19}$ FRAGMENTS IN CRYSTALLINE
DIBENZO-30-*crown*-10 (305) AND ITS K^+ COMPLEX

Fragment	Torsional angle (degrees)	
	Uncomplexed polyether	K^+ complex
$C_{19}^I-O_1-C_2-C_3$	−176	−175
$O_1-C_2-C_3-O_4$	62	−72
$C_2-C_3-O_4-C_5$	85	177
$C_3-O_4-C_5-C_6$	−169	−180
$O_4-C_5-C_6-O_7$	69	−70
$C_5-C_6-O_7-C_8$	172	−169
$C_6-O_7-C_8-C_9$	62	95
$O_7-C_8-C_9-O_{10}$	44	−67
$C_8-C_9-O_{10}-C_{11}$	178	148
$C_9-O_{10}-C_{11}-C_{16}$	147	−154
$O_{10}-C_{11}-C_{16}-O_{17}$	−4	−2
$C_{11}-C_{16}-O_{17}-C_{18}$	−174	−174
$C_{16}-O_{17}-C_{18}-C_{19}$	175	165
$O_{17}-C_{18}-C_{19}-O_1^I$	62	63
$C_{18}-C_{19}-O_1^I-C_2^I$	79	177

The free polyether (305) has a quite flexible structure and the crystalline
conformation of this compound (see Fig. 120 and parameters in Table 58)
is to a considerable extent determined by the peculiarities of the molecular
packing. A comparison of the torsional angles recorded in Table 58 shows
that considerable conformational rearrangement accompanies complexation
(the *anti* conformations of the C_3-C_4 and $C_{19}-O_1^I$ bonds become *gauche*
and four *gauche* conformations change sign).

III.A.4. Macrobicyclic polyethers

Bicyclic polyethers (322—330) are capable of *exo—endo* isomerism, as is
schematically shown in Fig. 121. This can be observed by variable-
temperature NMR experiments [203].

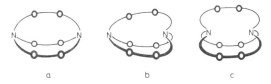

a   b   c

Fig. 121. Schematic drawing of the *exo—exo* (a), *exo—endo* (b) and *endo—endo* (c)
conformers of polyether (324).

Only the *endo—endo* isomer is present in the crystalline complexes of the polyether (324) with alkali metals [632, 634] and with barium [633] and also in the complex (325)·Ba(NCS)$_2$ [633]. X-ray analysis has shown that the complexes (324)·NaI, (324)·KI and (324)·CsNCS have structures based on the same principle (Fig. 122); the cation is in an internal cavity formed by six oxygen and two nitrogen atoms and is weakly interacting with the anion or solvent (the M$^+$—anion and M$^+$—OH$_2$ distances exceed 3.5 Å). The ligand oxygens assume an intermediate position between the vertices of a trigonal prism and a trigonal antiprism. As the cation radius decreases the

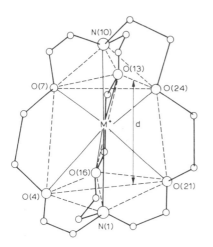
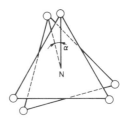

Fig. 122 (left). Structures of the crystalline complexes of polyether (324) with alkaline metal ions.

Fig. 123. (right). Disposition of the oxygen atoms in the crystalline complexes of polyether (324) with alkaline metal ions (as projected along the N—N axis).

complexone molecule undergoes gradual conformational adjustments, such that the internal cavity will best match the new cation. The conformational change involves a twist about the axis defined by both nitrogens. This results in an increase in the angle α shown in Fig. 123 and a shortening of the distance d (Fig. 122) between the planes of the oxygen atoms. Comparison of the N$^+$··· O distances listed in Table 59 with the sums of the relevant ionic and covalent radii leads to the conclusion that an optimum conformation of the cyclic polyether (324) is realized in the K$^+$ and Rb$^+$ complexes, because on passing over to Na$^+$ and Cs$^+$ the dimensional changes of the internal cavity "lag behind" the change in cation radii. Evidence of this can also be seen in the comparatively low stability constants of the corresponding cryptates (Table 23).

TABLE 59

STRUCTURAL PARAMETERS OF MACROCYCLIC POLYETHER (324)
COMPLEXES WITH ALKALI METAL IONS

Complex	α (degrees)	d (Å)	$M^+ \cdots O$ distance (Å)	$r_M + r_O$ (Å)	$M^+ \cdots N^+$ distance (Å)	$r_M + r_N$ (Å)
(324)·NaI	45	2.10	2.57	2.37	2.77	2.47
(324)·KI	21	2.65	2.79	2.73	2.87	2.83
(324)·RbNCS	15	2.81	2.90	2.83	3.00	2.99
(324)·CsNCS	15	2.87	2.97	3.09	3.03	3.19

An X-ray analysis of the cryptates (324)·Ba(NCS)$_2$·H$_2$O (Fig. 124) and
(325)·Ba(NCS)$_2$·2H$_2$O (Fig. 125) showed that the macrobicyclic structure
of the ligand does not exclude the possibility of contact pair formation such
as M^{n+}—anion or M^{n+}—OH$_2$, as one might have surmised from the study of

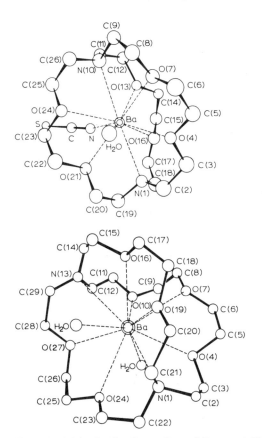

Fig. 124 (above). Conformation of the crystalline complex salt (324)·Ba(NCS)$_2$·H$_2$O.
Fig. 125 (below). Conformation of the crystalline complex salt (325)·Ba(NCS)$_2$·H$_2$O.

the alkali metal cryptates. In the first of the above-mentioned barium complexes the cation is interacting with the molecule of hydrate water ($Ba^{2+}\cdots OH_2$ distance, 2.86 Å) and with a rhodanide anion ($Ba^{2+}\cdots NCS$ distance, 2.88 Å) as well as with the eight ligand atoms of the macrocyclic compound. In the second complex, the central barium cation coordinates all nine of the surrounding polyether heteroatoms ($Ba^{2+}\cdots O$ distance, 2.80—3.09 Å; and $Ba^{2+}\cdots N$, 3.09—3.18 Å) and both water molecules ($Ba^{2+}\cdots OH_2$ distance, 2.81 and 2.88 Å). It is possible that these additional interactions are responsible for the exceptionally high complexing energies of the cryptates with the alkaline earth metals (Table 24; see also Part III.C).

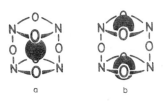

Fig. 126. Schematic drawing of the macrotricyclic ligand (330) complexes with (a) one, and (b) two metal ions.

In equimolecular complexes of the macrotricyclic polyether (330) the protons of the NCH_2 methylenes are magnetically equivalent, whereas the ring methylene protons are magnetically non-equivalent, as can be seen from the NMR spectra [134]. In accord with these findings is the model shown in Fig. 126a. Owing to the large size of the cavity (for example, it has a length of approx. 6 Å) it is natural to assume that the cation accommodated therein is not in a fixed position; an analogous situation has been discussed above for the example of the Na^+ complex of valinomycin (see Part III.A.1.a). The 1:2 complexes of polyether (330) resemble the dinuclear complex $(298)\cdot(KNCS)_2$, (discussed in Part III.A.3.d) in that part of its ligands are interacting only with one of the cations, whereas the two bridge ether oxygens are simultaneously interacting with both cations (Fig. 126).

III.A.5. The nigericin antibiotics

III.A.5.a. Nigericin (331) and grisoryxin (332)

X-ray analyses of the silver salt of nigericin have been independently performed by two groups of workers [517, 930, 970]. As one can see from Figs 127 and 128, the hydrocarbon chain of the antibiotic surrounds the cation such that most of the oxygen-containing functional groups are oriented towards the inner cavity, the exterior of the complex on the whole being of a lipophilic nature. The O_{43}, O_{44}, O_{45} and O_{49} ether atoms positioned at 2.47—2.66 Å from the cation, together with the fifth ligand O_{47} carboxylate oxygen (O· · ·Ag$^+$ distance, 2.26 Å), ensure high stability of the salt complex. The second oxygen atom of the carboxylate group (O_{46}) forms hydrogen bonds with two hydroxyls ($O_{50}H$ and $O_{51}H*$), thereby closing the pseudocyclic ring. The tetrahydrofuran rings D and E of the nigericin Ag salt are in the envelope conformation, whereas ring C is a half chair; all the tetrahydropyran rings are in the chair conformation.

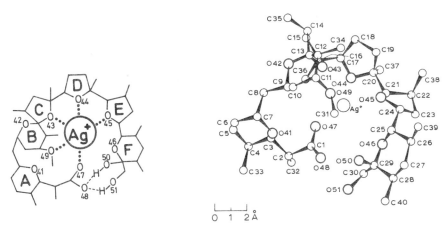

Fig. 127 (left). Metal—oxygen (· · · ·) and hydrogen (- - -) bonds in the crystalline nigericin silver salt.
Fig. 128 (right). Conformation of the crystalline silver salt of nigericin.

Grisoryxin differs from nigericin only by substitution of a methyl for the hydroxymethyl group of ring F. It is therefore to be expected that the spatial structure of the Ag$^+$ complex of these two antibiotics practically coincide, differing only in the number of hydroxyl groups hydrogen-bonding intramolecularly to a carboxyl function: in grisoryxin only one ($O_{50}H$— O_{48}, Fig. 129), in nigericin, two [8].

The thallium salt of grisoryxin is not isomorphous with the silver salt, but their conformations are very similar (cf. Figs 129 and 130) [9]. The Tl$^+$· · ·O distances are 2.6—3.0 Å, i.e. greater by 0.3 Å than the corresponding Ag$^+$· · ·O distances. The size of the internal cavity is increased by a ∼ 10° twist about the C_{16}—C_{17}, C_{20}—C_{21} and C_{24}—C_{25} bonds. The O_{48}

---

* Shiro and Koyama [930] do not mention the existence of O_{51}· · ·O_{48} intramolecular hydrogen bonding although the O_{51}· · ·O_{48} distance (2.61 Å) calculated from the atomic coordinates and cell parameters is almost the same as the O_{58}· · ·O_{48} distance (2.59 Å).

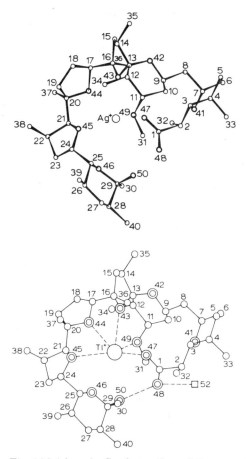

Fig. 129 (above). Conformation of the crystalline silver salt of grisoryxin.
Fig. 130 (below). Conformation of the crystalline thallium salt of grisoryxin.

carboxylate atom of the thallium complex apparently also forms a hydrogen bond with a water molecule (its atom O_{52} is marked with a square in Fig. 130).

The methods available at present are clearly insufficient for obtaining significant information of the solution structures of nigericin. The only exceptions are the infrared data on nigericin in chloroform (Fig. 131) [569]. The constancy of the carboxyl stretching vibrations (approx. 1717 cm^{-1}) with varying antibiotic concentrations leads to the conclusion that the carboxyl group participates in intramolecular hydrogen bonding rather than in intermolecular hydrogen bonding because intramolecular hydrogen bonds usually undergo rupture on dilution of the solution with concomitant shift of the CO band to higher frequencies [see, for instance, data on myristinic

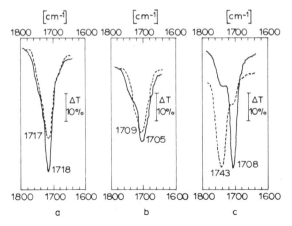

Fig. 131. Stretching vibration region in the infrared spectrum of (a) nigericin, (b) monensin and (c) myristinic acid in CHCl₃.

acid $(C_{14}H_{29}COOH)$, Fig. 131]. From the formula of nigericin it is seen that the only groups capable of hydrogen bonding the carboxyls are $O_{50}H$ and $O_{51}H$. However, one cannot exclude the possibility of the hydrogen bonding detected by infrared being due to formation of a stable hydrate, similar to the crystal hydrate of monensin (see Part III.A.5.d).

III.A.5.b. Antibiotic X-206 (333)

In the silver salt of the antibiotic X-206 (Figs 132 and 133) the cation is surrounded by two ether oxygens ($Me^+ \cdot \cdot \cdot O_1$ and O_2 distances, 2.5 and 2.8 Å), three hydroxyl oxygens and one carboxyl oxygen. The relatively long hydrocarbon chain of the antibiotic (see Table 61) is folded to a greater extent than that of the other complexones of this group; in this respect it

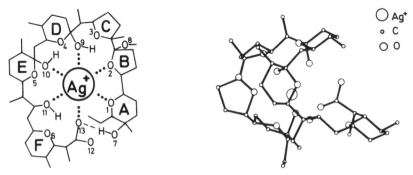

Fig. 132 (left). Metal—oxygen ($\cdots$) and hydrogen (- - -) bonds in the crystalline silver salt of antibiotic X-206.

Fig. 133 (right). Conformation of the crystalline silver salt of antibiotic X-206.

resembles the macrotetrolide and the polyether (303) complexes ("tennis ball seam" type conformation). Interacting with the cation, the O_{13} atom additionally participates in intramolecular hydrogen bonding ($O_{13} \cdots HO_{17}$ distance, 2.69 Å).

III.A.5.c. Antibiotic X-537 A (334)

The silver salt of antibiotic X-537 A crystallizes as the dimer with the monomer molecules in practically the same conformations (Figs 134 and 135) [577]. Each cation is interacting with all five heterocyclic oxygens of

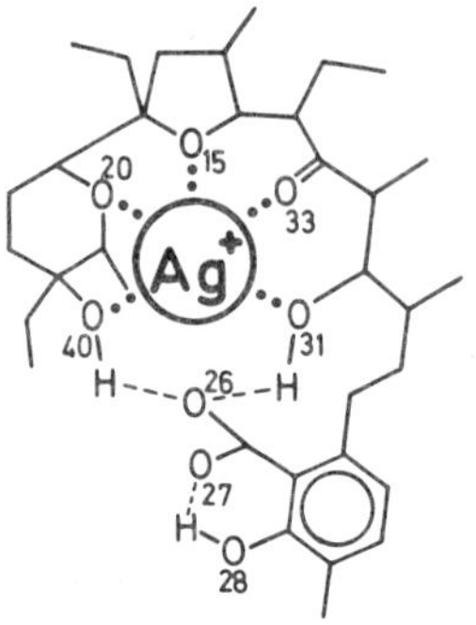

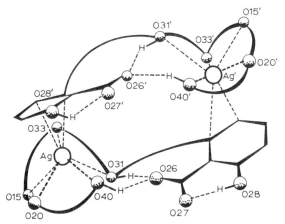

Fig. 134. Metal—oxygen ($\cdots$) and hydrogen (- - -) bonds in the crystalline silver salt of antibiotic X-537 A.

one of the antibiotic molecules (ether O_{15} and O_{20}, hydroxyl O_{31} and O_{41} and ketonic O_{33}); the $O \cdots Ag^+$ distance is less than 3.02 Å. The complexed structure is stabilized by an intramolecular hydrogen-bond system involving the carboxylate groups. The O_{27} atoms form hydrogen bonds with the

Fig. 135. Schematic drawing of the spatial structure of the dimeric crystalline complex $Ag_2^+ \cdot (X\text{-}537\ A)_2^-$.

phenolic hydroxyls (O_{27}···O_{28} distances, 2.39 and 2.41 Å) and the other carboxylate oxygens, with the $O_{31}H$ and $O_{40}H$ groups (O_{26}···O_{31}, 2.91 and 2.90 Å; O_{26}···O_{40}, 2.41 and 2.39 Å). Thus the oxygen ligands are in a 17-membered ring stabilized by two intramolecular hydrogen bonds. The aromatic rings are distanced from "their own" cations and at the same time are neared to the "other" cations, apparently π-complexing them (Ag^+···C_5, 2.41 and 2.46 Å; Ag^+···C_6, 2.61 and 2.79 Å).

In the crystalline Ba^{2+}·$(334)_2$ salt complex (Figs 136—138) the antibiotic molecules, while adopting very similar conformations, interact differently with the central barium atom (the more weakly interacting molecule has its atoms designated by primed numbers, Fig. 136) [462, 463].

The antibiotic conformation in the Ba^{2+} complex is stabilized by an intramolecular hydrogen-bond network as shown in Fig. 137: the O_{27} and $O_{27'}$ atoms form very stable bonds with the $O_{28}H$ and $O_{28'}H$ hydrogens (O_{27}···O_{28} and $O_{27'}$···$O_{28'}$, 2.49 and 2.41 Å) and the O_{26} and $O_{26'}$ atoms with the $O_{31}H$ and $O_{31'}H$ hydrogens (H bond lengths = 2.62 and 2.87 Å). The $O_{40}H$ and $O_{40'}H$ hydroxyls form intramolecular hydrogen bonds with O_{27} and $O_{27'}$ (O_{40}···O_{27} and $O_{40'}$···$O_{27'}$; 2.72 and 2.77 Å). Hence the intramolecular hydrogen-bond system in the Ba^{2+} and Ag^+ complexes is very similar, differing only in the interaction of the carboxylate O atoms with $O_{40}H$ (cf. Figs 134 and 136). All the tetrahydrofuran rings in the Ba^{2+} complex adopt the envelope conformation, and the tetrahydropyran rings the chair conformation.

The metal—oxygen bonding system of the first molecule (334) involves the O_{26} oxygen, as well as the ligand atoms of the Ag^+ complex (Fig. 137) (Ba^{2+}···O, 2.71—3.08 Å). In the second molecule (334) the barium ion effectively interacts with only two oxygens, $O_{26'}$ and $O_{40'}$ (Ba^{2+}···O, 2.64 and 2.84 Å). In addition, a hydrate water molecule is involved in the formation of the crystalline spatial structure of the complex, apparently forming H bonds with O_{20} and O_{33}; (O_{H_2O}···$O_{20'}$ and O_{H_2O}···$O_{33'}$, 2.80 and 3.20 Å) and interacting with the cation (Ba^{2+}···O_{H_2O}, 2.74 Å).

Bearing in mind the considerable differences in the Ba^{2+} interaction with the two antibiotic molecules, somewhat unexpected is the similarity of their conformational states, manifested in the same intramolecular hydrogen-bond systems and similarity of the rotation angles all along the backbone. One can see from Table 60 that, even in the region of the tetrahydrofuran rings where, judging from the degree of their participation in complex formation, one should have expected maximum differences, the $|\tau-\tau'|$ values do not exceed 24°.

The free antibiotic X-537 A, as the salt complexes discussed above, crystallizes as a dimer [919]. However, whereas the dimer complexes can be

204

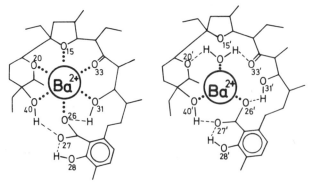

Fig. 136. Metal—oxygen ($\cdots$) and hydrogen (- - -) bonds in the crystalline complex $Ba^{2+} \cdot (X\text{-}537\ A)_2^-$.

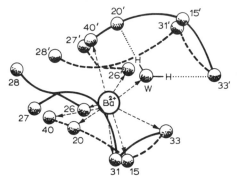

Fig. 137. Schematic presentation of the spatial structure of the crystalline complex $Ba^{2+} \cdot (X\text{-}537\ A)_2^-$.

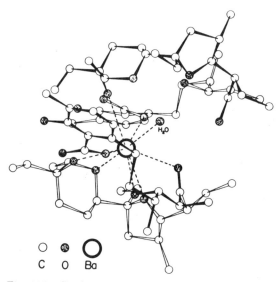

○   ⊛   ◯
C   O   Ba

Fig. 138. Conformation of the crystalline complex $Ba^{2+} \cdot (X\text{-}537\ A)_2^-$.

TABLE 60

TORSIONAL ANGLES ALONG THE BACKBONES OF TWO ANTIBIOTIC X-537 A MOLECULES IN THE $Ba^{2+}\cdot(X\text{-}537\ A)_2^-$ COMPLEX

Fragment	τ (degrees)	τ' (degrees)	$\lvert\tau-\tau'\rvert$ (degrees)
$C_1-C_6-C_7-C_8$	-70	-74	4
$C_6-C_7-C_8-C_9$	176	172	4
$C_7-C_8-C_9-C_{10}$	-60	-67	7
$C_8-C_9-C_{10}-C_{11}$	180	-174	6
$C_9-C_{10}-C_{11}-C_{12}$	179	177	2
$C_{10}-C_{11}-C_{12}-C_{13}$	-148	-133	15
$C_{11}-C_{12}-C_{13}-C_{14}$	92	78	14
$C_{12}-C_{13}-C_{14}-O_{15}$	56	80	24
$C_{12}-C_{13}-C_{14}-C_{18}$	178	-167	15
$C_{12}-C_{13}-C_{14}-C_{17}$	-155	-158	3
$C_{14}-C_{18}-C_{17}-C_{16}$	34	40	6
$C_{18}-C_{17}-C_{16}-O_{15}$	-26	-22	4
$C_{17}-C_{16}-O_{15}-C_{14}$	6	-6	12
$C_{16}-O_{15}-C_{14}-C_{18}$	15	32	17
$O_{15}-C_{14}-C_{18}-C_{17}$	-31	-45	14
$C_{18}-C_{17}-C_{16}-C_{19}$	-148	-140	8
$C_{17}-C_{16}-C_{19}-O_{20}$	-176	-178	2
$C_{16}-C_{19}-O_{20}-C_{21}$	176	172	4
$C_{19}-O_{20}-C_{21}-C_{22}$	65	60	5
$O_{20}-C_{21}-C_{22}-C_{23}$	-57	-50	7
$C_{21}-C_{22}-C_{23}-C_{24}$	51	49	2
$C_{22}-C_{23}-C_{24}-C_{19}$	-53	-56	3
$C_{23}-C_{24}-C_{19}-O_{20}$	58	62	4
$C_{24}-C_{19}-O_{20}-C_{21}$	-65	-67	2

defined as the "head to tail" type, without the cation, dimerization occurs by "head to head" (cf. Figs 135, 137 and 139). The conformation of the free acid is close to that of the complex, but some differences in the hydrogen bonding system do exist. Moreover, the dimeric structure of the antibiotic is stabilized by hydrate water which forms additional hydrogen bonds to both molecules of the dimer.

III.A.5.d. Monensin (335)

The silver salt of monensin served as the first example on which, in 1967, the basic principles of spatial structure formation for the series of antibiotics now being considered [5, 752] were elucidated. X-ray analysis showed that in the cyclic complex $Ag^+\cdot(335)^-$ (and, therefore, in its isomorphous K^+ and Tl^+ complexes*) the cation is close to six oxygen atoms: two hydroxyl

---

* Pinkerton and Steinrauf [752] have prepared another crystalline modification of the monensin Ag salt, isomorphous to the crystals of the sodium salt. The crystalline Rb salt differs from both these forms.

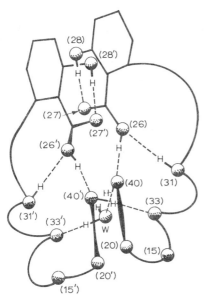

Fig. 139. Schematic drawing of the crystalline antibiotic X-537 A. W, oxygen atom of the hydrate water.

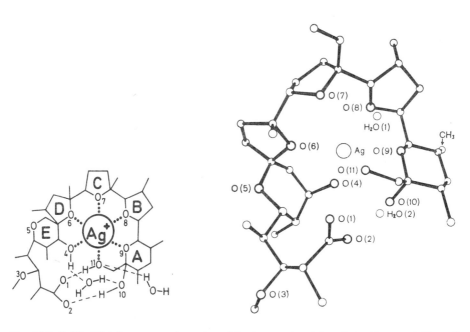

Fig. 140 (left). Metal—oxygen (· · · ·) and hydrogen (- - -) bonds in the crystalline silver salt of monensin.
Fig. 141 (right). Conformation of the crystalline silver salt of monensin.

oxygens (O_4 and O_{11}; $Ag^+\cdots O$, 2.43 and 2.45 Å) and four ether oxygens (O_6, O_7, O_8 and O_9; $Ag^+\cdots O$, 2.40, 2.69, 2.58 and 2.56 Å). The ionized carboxyl forms two hydrogen bonds with the ring A hydroxyl groups, giving rise to a 24-membered ring surrounding the cation (Figs 140 and 141) ($O_1\cdots O_{11}$ and $O_2\cdots O_{10}$, 2.51 and 2.65 Å). In the complexed molecule, the C and D tetrahydrofuran rings are flattened envelopes and the B ring is in a half chair conformation.

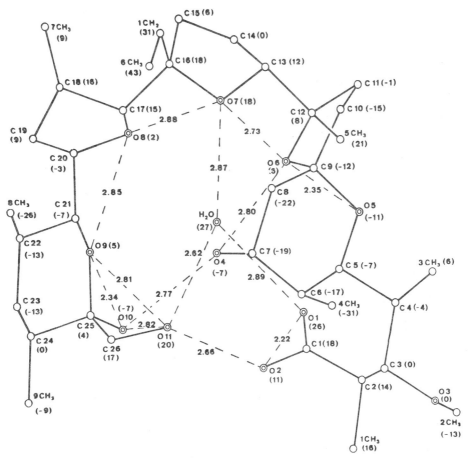

Fig. 142. Structure of crystalline monensin. Figures in parentheses are the distances (in 0.1 Å) of the atoms from the projection plane.

The hydrate water molecule present in the crystalline structure has apparently no marked influence on the antibiotic conformation since its hydrogen bonds are not very strong (bond lengths of 2.79—2.93 Å); the hydrated groups are fixed in their respective positions by other interactions.

In general, the conformation of non-complexed monensin (Fig. 142) resembles that of the Ag^+ complex, for the termini of the antibiotic are neared by intramolecular hydrogen bonds ($O_{11}\cdots O_2 = 2.62$ Å) and its rotation angles differ from those in the complex by not more than $17°$ [568]. However, comparison of the two structures shows that removal of the cation causes significant changes in the hydrogen bonding system (Fig. 142), an important part being played by the water molecule which now occupies the position of the cation. It is very natural to assign the nearly concentration-independent 1709—1705 cm^{-1} band in the infrared spectra of monensin [569] to the hydrated CO_1 carbonyl (see Fig. 131 in Part III.A.5.a).

III.A.5.e. Dianemycin (339)

The Na^+, K^+, Tl^+ and Rb^+ salts of dianemycin form isomorphic crystals; the structures of the first three having been investigated by Steinrauf et al. [177, 969]. Of the fourteen oxygen atoms of the antibiotic, six (the

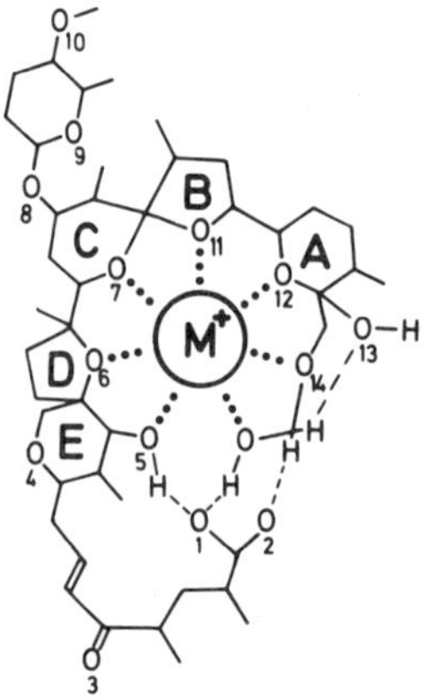

Fig. 143. Metal—oxygen (· · · ·) and hydrogen (- - -) bonds in the crystalline silver salt of dianemycin.

hydroxyl O_5 and O_{14} and the ether O_6, O_7, O_{11} and O_{12}, Fig. 144) in all instances surround the cation. A water molecule lodged between O_1 and O_{13} is also interacting with the cation. In the dianemycin complexes there is a rather complicated network of hydrogen bonds involving also the water molecule and the carboxyl group (Fig. 143). All 6-membered rings of dianemycin are in the chair conformation and the substituents of the pendant tetrahydropyran ring are oriented equatorially, the only axial substituent being the methyl group.

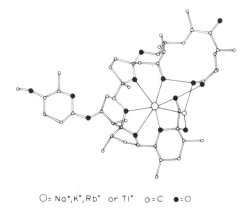

O = Na⁺,K⁺,Rb⁺ or Tl⁺   o = C   ●=O

Fig. 144. Conformation of the crystalline sodium, potassium, rubidium and thallium salts of dianemycin.

III.A.5.f. Comparative characteristics of the nigericin antibiotics

Examination of the data concerning the nigericin antibiotics leads to the conclusion that, despite the presence of a negatively charged carboxyl group and the irregular structure and numerous differences in the details of the covalent and spatial structures, all the compounds (331—339) have a number of important properties in common with the metal-binding macrocyclic compounds considered in the previous sections. Interacting with the metal ions, the nigericin antibiotics envelope them, as it were, so that the ions are surrounded by oxygen atoms of the functional groups instead of by the solvate sheath of the solvent. As a result the exterior of the complex is composed mainly of lipophilic groups which ensure its high solubility in organic solvents (and concomitant poor solubility in water). A characteristic feature of the nigericin antibiotics is that a compact arrangement of the molecule about the cation is stabilized by one or more H bonds between antipodally situated carboxyl and hydroxyl groups. The conformations obtained in the crystalline complexes must be very favorable from the viewpoint of non-bonded interactions, because very similar pseudocyclic conformations in the free antibiotic are observed in both the solution and crystalline states. The hindered internal rotation about single bonds of the antibiotics (331—339) due to the presence of spirano groupings and other segments with quaternary carbon atoms apparently facilitates realization of the complex conformation.

Comparing the structures of the nigericin complexes one cannot but note such features as the asymmetric disposition of the oxygen ligands, the variability of their number and the variety of the cation-interacting polar groups (carboxyl, ketone, hydroxyl* and ether functions), evidence of their mutual exchangeability. This leads to the notion that in general the factors (symmetry, number and nature of the ligands) are not crucial for the manifestation of complexing properties by the nigericin antibiotics. The mutual arrangement of the atoms in the hydrogen bond and metal—oxygen bond network, with participation of the carboxyl groups, differs in all the compounds investigated and thus also reveals no strict structural requirements (as one can see from Table 61, only the CO· · ·HO—A intramolecular hydrogen bond does not vary from compound to compound).

A more definite relationship can be discerned between the size of the hydrogen-bonded rings carrying the oxygen ligands and the complexing selectivity. For instance, the nigericin, monensin and dianemycin complexes are very similar in the regions of the A—E rings, but differ considerably at the site of closure of the pseudocyclic system, the largest number of atoms in the ring, and, consequently, the biggest internal cavity, being found in nigericin. Accordingly, of the alkali metal ions it prefers to complex potassium, whereas monensin and dianemycin preferentially complex sodium. The longer side chain of the latter antibiotic and the different ways of "coupling" rings A and E (Figs 140 and 143) makes the ring system of dianemycin more flexible than that of monensin and thus leads to the relatively low ionic specificity of the former.

The tendency on the part of antibiotic X-537 A to interact with cesium ions (see Part I.F.2), despite its small pseudocyclic system, can be explained by assuming the formation of "sandwich" complexes similar to the barium complexes (see data on the 2:1 type of cyclic polyether complexes, Table 25).

In concluding this part, it may be said that attempts to arrive at more fundamental inferences regarding the structure—function relation of the nigericins would be more on the side of conjecture than rigor. There are as yet no synthetic analogs of this class, whose systematic investigation could provide the basis for elucidating the structural prerequisites for complexing in this group of antibiotics.

III.B. The binding of alkali and alkaline earth metal ions by macrocyclic complexones. Theoretical aspects

Although varying widely both structurally and conformationally, the macrocyclic compounds binding alkali and alkaline earth ions have a common feature; viz. the coordination spheres of their bound cations are

---

* Of the compounds considered in this book, only the nigericins have hydroxyl ligands.

TABLE 61

SOME STRUCTURAL PARAMETERS OF NIGERICIN COMPLEXES

Antibiotic	Cation	Ligand groups						No. atoms in carbon chain	No. atoms in H-bonded cycles bearing ligand oxygens	Hydrogen (···) and metal–oxygen (ooo) bonds near the carboxyl group
		Total no.	Ether groups	Hydroxyls	Keto groups	Carboxyls	Water			
Nigericin	Ag^+	5	4	—	—	1	—	30	25, 26	
Grisoryxin	Ag^+	5	4	—	—	1	—	30	25	
X-206	Ag^+	6	2	3	—	1	—	35	29	
X-537 A	Ag^+	5	2	2	1	—	—	20	17	
X-537 A, X-537 A'	Ba^{2+}	9	2 / —	2 / 1	1 / —	1 / 1	— / 1	20	19 / 17	
Monensin	Ag^+ K^+ Tl^+	6	5	1	—	—	—	26	19, 23, 24	
Dianemycin	Na^+ K^+ Rb^+ Tl^+	7	4	2	—	—	1	30	20, 21	

filled by amide, ether, ester, etc. polar groups, just as in solvation of the cations by any organic molecules containing these groups. In view of this formation of the complexes can be likened to transfer of the cations from a weaker to a stronger solvating solvent; for instance, from alcohol to dimethylformamide in which, as is well known, the solvation energies of the cations are particularly high [277, 733]. Indeed, the complexes are structurally similar in many respects to the crystallosolvates which alkali metal (particularly sodium and lithium) salts form with a number of ketones, amides, amines, etc. [151, 294, 304, 391, 601, 611, 758, 929, 987, 1019,

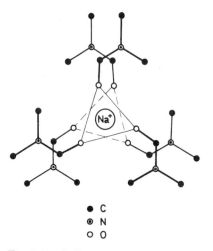

● C
⊙ N
○ O

Fig. 145. Sodium ion surrounding in the crystallosolvate [$HCON(CH_3)_2$]$_6$·NaI [294].

1103]*. A comparison of Figs 48 and 64 with Fig. 145 shows straight-forwardly that in both the macrocyclic complexes and crystallosolvates: (i.) the dipoles surrounding the cation are oriented towards it with their negative ends; (ii.) the cation—ligand atomic distances are usually close to the sum of the crystallographic radii (see also Table 62); and (iii.) the cation and ligand groups are surrounded by non-polar hydrocarbon radicals. Quite obviously, the difference between the solvate sheaths of the cations in solution, the crystallosolvates and the macrocyclic complexes is that in the complexes all ligand groups are fixed on a "skeleton". The macrocyclic molecule thus simply plays the part of a structurized solvate sheath.

From this standpoint one can readily understand the role played by entropy factors in the high stability of the cation complexes of such

---

* Various sugars and polyols are also known to complex salts of the alkali and alkaline earth metals (see [19, 103, 817] and references cited therein).

TABLE 62

INTERATOMIC DISTANCES IN CRYSTALLINE SALTS OF MACROCYCLIC COMPLEXES (Å)

Complex salt	$M^+ \cdots O$ distance	Deviation from the sum of crystallographic radii $(r_M + r_O)^a$	$M^+ \cdots N$ distance	Deviation from the sum of crystallographic radii $(r_M + r_N)^b$	$M^+ \cdots X^-$ distance	References
$(CH_3COCH_3)_6 \cdot NaI$	2.46	+0.08			6.78	[758]
$[HCON(CH_3)_2]_6 \cdot NaI$	2.40	+0.02			6.98	[294]
$[(p\text{-}NH_2C_6H_4)_2CH_2]_3 \cdot NaCl$			2.52	+0.04		[987]
Valinomycin $\cdot KAuCl_4$	2.7–2.8	−0.03—+0.07			>6	[751]
(Enniatin B) $\cdot KI$	2.6–2.8	−0.13—+0.07			>6	[209]
Nonactin $\cdot$ KNCS	2.73—2.88	0.00—+0.15			>6	[208, 488]
(246) $\cdot NaI \cdot H_2O$	2.35—2.43	−0.03—+0.05			2.29 (H_2O)	[105]
(246)$_2 \cdot KI$	2.78—2.96^c	+0.05—+0.23			e	[593]
(270) $\cdot NaBr \cdot 2H_2O$	2.54—2.89	+0.16—+0.51			2.82, 3.19 (H_2O)	[105, 106]
(270) $\cdot RbNCS \cdot H_2O$	2.86—2.94	−0.03—+0.05			2.94	[87, 88]
(303) $\cdot KI$	2.85—2.92	+0.12—+0.21			e	[105, 107]
(324) $\cdot NaI$	2.57^d	+0.20	2.77	+0.29	e	[634]
(324) $\cdot KI$	2.79^d	+0.06	2.87	+0.04	e	[634]
(324) $\cdot RbNCS$	2.90^d	+0.02	3.00	+0.01	e	[634]
(324) $\cdot CsI$	2.97^d	−0.06	3.03	−0.12	e	[634]
(324) $\cdot Ba(NCS)_2 \cdot H_2O$	2.74—2.88	−0.04—+0.10	2.94—3.00	+0.06——0.12	2.88, 2.86 (H_2O)	[633]

^{a} $r_O = 1.40$ Å.
^{b} $r_N = 1.50$ Å.
^{c} Deviation from the oxygen-atom plane 1.67 Å.
^{d} Mean value.
^{e} The distance greatly exceeds the sum of the crystallographic radii.

polyfunctional macrocyclic compounds. Let us compare a cyclic and a linear molecule with equal numbers of the same ligand groups and with similar complex conformations. These similarities notwithstanding, the former compound will have a higher free energy of complexation and, consequently, a higher complex stability constant because there is a larger set of conformational states for a linear molecule in solution than for a cyclic molecule. Hence, the formation of cation complexes by linear compounds is less favored from the standpoint of entropy changes of the system. As an illustration, it may be cited that the stability of the K^+ complex of compound (259) in methanol at 25°C is almost 10^4 times the stability of a similar complex of pentaglyme $(CH_3O(CH_2CH_2O)_5CH_3)$ whose molecule also possesses six ligand O atoms [272] (cf. [21, 1093]). In this connection it is to be emphasized that the complexone properties of the nigericin type antibiotics are intimately associated with the existence of intramolecular hydrogen bonds in their molecules promoting the stability of the folded (so-called pseudocyclic) conformation (see Part III.A.5)*. Moreover, stability of the cation complexes of linear compounds apparently increases if the chain has substituents conducive to its bending. Such instances are aromatic rings in the polyethers [741].

It was Mueller and Rudin who, in their pioneering paper [669], proposed that the complexes of metal-binding macrocyclic compounds structurally resemble molecular inclusion compounds formed by cyclodextrins [217]. The truth of this can be seen from the fact that in both cases the "guest" component is located within the intramolecular cavity of the "host" and is more or less effectively isolated from the environment. Nevertheless, this analogy is only superficial for in a certain sense it would be more correct to regard the cyclodextrins as metal-binding macrocyclic compounds turned inside out because of their hydrophobic molecular cavity and hydrophilic exterior. Consequently, they form hydrophilic complexes with non-polar molecules. Moreover, since the dipoles of the cyclodextrin polar groups are oriented with their positive ends towards the cavity [414], it is not surprising that the stability of anion complexes of the cyclodextrins is markedly higher than that of the complexes with cations or neutral molecules of similar structure [217, 218]. However, the difference is not very great because here the electrostatic contributions are small in comparison with the hydrophobic interactions.

With this reservation in mind, it is enticing to attempt further extension of the analogy between the cation complexes and molecular inclusion com-

---

* Concerning a possible nigericin-type synthetic complexone see p. 90.

pounds. As is well known, amylose macromolecules are in the form of a helix, each turn of which closely resembles a cyclodextrin molecule; "guest" anions and neutral molecules can be entrapped in the interior of such a helix as well as in the internal cavity of cyclodextrins [217, 553]. Probably helical chain polymers can also exist, such that the individual or adjacent turns of the helix can bind cations in the same way as the macrocyclic compounds discussed here. Urry [1026—1028] has recently pointed to the possibility of their being helical complexones of peptide nature. Realistic candidates for the role of such helical complexones are the ethylene glycol oligomers, in particular non-ionic detergents of the type $RO(CH_2CH_2O)_n H$ where R is an n-alkyl or p-n-alkylphenyl radical. In aqueous solutions these compounds associate with the potassium or barium salts of certain acids (tetraphenyl-borates, iodobismuthates, ferricyanides, etc.) to form precipitates [678, 679, 862, 868, 956, 1099]. On the other hand, the polyether detergents Triton X-100, Igepals CO430—CO990, Lubrol WX, and Tweens 20, 60 and 80, markedly promote the selective extraction of alkali metal picrates by organic solvents from aqueous solutions [91]. The partition coefficients of the picrates increase with increasing number of monomer units in the polyether chain, whereas the cation selectivity sequence remains unchanged (for Triton X-100, $K^+ > Rb^+ \approx Cs^+ > Na^+ > Li^+$). The underlying cause of these phenomena may be complexation such that the cations are enclosed within the polyether helix. In the cases, say, of potassium and barium ions, from the stoichiometry of the precipitates, each cation can be assumed to be bound by up to 6 and 12 monomer units, respectively [545, 546, 868, 1023].

It is to be stressed that the differences between the linear, macrocyclic and helical complexones are only quantitative and all of them may be regarded as the representatives of a single class of multidentate complexones binding alkali and alkaline earth metal ions with the formation of a monomolecular "solvate sheath".

From this standpoint the formation and dissociation of macrocyclic complexes can be compared with ligand substitution in the inner sphere of the cation solvate sheath. However, whereas individual solvent molecules therein are exchanging independently of one another, in the case of macrocyclic complexes the ion—ligand interaction is of a cooperative nature, because change in the number of cation-contacting ligand groups is associated with conformational transitions in considerable segments of the molecule. The degree of cooperativity depends upon the number of ligands present in the molecular segment undergoing the concerted conformational transition. As a limiting case, one may imagine the existence of macrocyclic complexones which, owing to thermal motion of independent chain units carrying isolated ligands, "enfold", as it were, the cations and gradually displace the solvent molecules from their solvate sheaths. At the other extreme, if rotation about all the bonds is concerted, the complexone molecule will function as a "trap", springing on contact with the cation and tearing off its solvate sheath. Under such circumstances one would naturally expect that the complexing reaction will have a higher activation energy. In general it could be expected that the activation barriers will be lower with the linear than the macrocyclic complexones, even when the corresponding complexes have close stability constants. Thus, Ammann *et al.* [12] have recently shown that N,N'-dimethyl-N,N'-di-(ω-carbethoxydecyl)-diamide of ethyleneglycoldiacetic acid (340)

$$
\begin{array}{l}
\qquad\quad CH_3 \\
\qquad\quad | \\
CH_2OCH_2CON(CH_2)_{10}COOC_2H_5 \\
| \\
CH_2OCH_2CON(CH_2)_{10}COOC_2H_5 \\
\qquad\quad | \\
\qquad\quad CH_3
\end{array}
\qquad (340)
$$

forms quite stable complexes with calcium ions in methanol solution ($K \approx 10^3$ l·mole^{-1}). However, according to the NMR data such complexes have a much shorter lifetime than calcium complexes of the cyclic polyethers (cf. [402]).

It has recently been shown by relaxation techniques that the complexing rates of nonactin and of cyclic polyether (305) approach those typical of diffusion controlled processes ($> 10^8$ l·mole^{-1}·sec^{-1}) [143, 202]. One might, therefore, imagine that in these compounds local motion of the individual ligand-carrying segments is largely an independent process. With valinomycin and the enniatins, complexing occurs in two steps (see Part II.A.3). The first is very rapid; conversion of the primary product of interaction with the cation to the "normal" complex then occurs at a rate which, while depending upon the nature of the cation, lies within the range of conformational transition rates for these cyclodepsipeptides (see Table 41 on p. 100). One might, therefore, conclude that it is interaction with the cation which induces the concerted conformational rearrangement in the molecule.

Conformational changes which may occur during formation and dissociation of complexes, by themselves do not as yet prescribe the existence of
conformational differences between the free and complexed molecules.
However, such differences are usually observed and for the following reason.
As has been shown in Part III.A, in the complexes the dipoles of the polar
ligand groups are more or less directed with their negative poles towards the
molecular cavity. Although such a conformation could be preferable on
steric grounds, it is unfavorable in the absence of the bound cation owing to
the electrostatic repulsion of these dipoles. Therefore, in non-polar media,
the conversion from complex to free molecule is accompanied by a sort of
"inversion" of the dipoles, particularly striking in valinomycin and the
enniatins (see Parts III.A.1.a and III.A.1.c).

In polar media the conformational equilibrium of the free complexone
molecules is determined by the solvation of the polar, including ligand,
groups. If these groups are accessible to the solvating molecules in the
complex or near-complex conformation, interaction with the molecules will
stabilize that conformation, mainly due to a resultant decrease in the ligand
dipole—dipole repulsion. Such is the case with the enniatins, their complexing in polar media being unaccompanied by any significant spatial
rearrangements in the molecules. A similar picture can be also discerned in
antamanide. As Wieland *et al.* [1069] have shown, the addition of minute
amounts of water or methanol to a solution of this compound in non-polar
solvents leads to spectral shifts similar to those observed in complex
formation. With valinomycin, on the contrary, as the polarity of the medium
is increased, the "bracelet" changes to the "propeller" conformation, the
driving force of the transition being solvation of the amide groups with
concomitant rupture of the intramolecular hydrogen bonds. When complexing occurs, reversion to the "bracelet" conformation (although of
opposite chirality), in which the ester carbonyls are pointing towards the
internally accommodated cation, takes place. Thus, in this case polar group
solvation destabilizes the complex conformation and it becomes energetically preferable only due to cation—ligand interactions.

Hence, interaction with the cation can be the cause of ordered changes in
the molecular conformations of the macrocyclic complexones. These not
only determine the displacement mechanism of the solvent molecules from
the cation solvate sheath, i.e. the complexing kinetics, but also the
thermodynamics of the process.

The formation in solution of a complex by compound L with cation M^+
can be schematically represented as shown below. The molecule L,
transferred into the gas phase in the same conformation (or population of

conformations) as in solution, reacts with the desolvated ion M^+. The resultant complex M^+L is then returned from the gas phase into the solution.

$$
\begin{array}{ccc}
M_{gas}^+ \;+\; L'_{gas} & \xrightarrow{\;\Delta F'_{ML}\;} & (M^+L)_{gas} \\[2pt]
\parallel\!\!\parallel\!\!\parallel & \Big\uparrow \Delta F_{conf} & \parallel\!\!\parallel\!\!\parallel \\[2pt]
M_{gas}^+ \;+\; L_{gas} & \xrightarrow{\;\Delta F_{ML}\;} & (M^+L)_{gas} \\[2pt]
\Big\uparrow -\Delta F_M^s \quad \Big\uparrow -\Delta F_L^s & & \Big\downarrow \Delta F_{ML}^s \\[2pt]
M_{solv}^+ \;+\; L_{solv} & \xrightarrow{\;\Delta F_{compl}\;} & (M^+L)_{solv}
\end{array}
$$

Obviously the free energy of complexation in solution, $\Delta F_{compl} = -RT \cdot \ln K$, is determined by Eqn 20.

$$
\begin{aligned}
\Delta F_{compl} &= \Delta F_{ML} - \Delta F_M^s + \Delta F_{ML}^s - \Delta F_L^s \\
&= \Delta F_{ML} - \Delta F_M^s + \Delta(\Delta F^s)_L^{ML}
\end{aligned}
\tag{20}
$$

where ΔF_{ML} is the interaction energy of the complexone molecule with the ion in the gas phase, ΔF_M^s is the energy of solvation of the free ion, and $\Delta(\Delta F^s)_L^{ML}$ is the difference in solvation energies of the complex and the free molecule. The quantity ΔF_{ML} is thus defined as the sum of the transition energy (ΔF_{conf}) of the complexone molecules from the conformation L into the complex conformation L' and the interaction energy ($\Delta F'_{ML}$) of the ion with the molecule in this conformation

$$
\Delta F_{ML} = \Delta F_{conf} + \Delta F'_{ML}
\tag{21}
$$

The relative contribution of ΔF_{conf} to the free energy of complexation depends upon the conformational mobility of the molecules. The macrocyclic molecules often carry bulky radicals and their backbones possess rotationally hindered bonds (amide, ester, the bonds common with fused rings, etc.). Moreover, a frequent occurrence in these complexones is intramolecular hydrogen bonding. If, disregarding dipole–dipole interactions, the energy of formation of such molecules is represented as a function of all the torsional angles, several minima will appear corresponding to preferable conformations. Obviously the barrier heights about these minima determine the convertibility of one stable conformation into another and their steepness — the distortion energy of the corresponding conformation, i.e. a parameter which can be termed the conformational rigidity.

Electrostatic interactions and hydrogen bonding, while changing the depth of the minima, little affects their positions.

Theoretical analysis shows that the complexed molecules of the cyclo-depsipeptides such as valinomycin and the enniatins are in conformations closely coinciding with one of these minima (see Part III.A.1). One might conjecture that this is also the case for other efficient macrocyclic complexones, especially of natural origin. Designating the conformation corresponding to this minimum the complexing conformation [897] (cf. [666, 938]), let us consider the interaction energy of the cation with the complexone molecule $\Delta F_{M\,L}$ as a sum of: (a) the energy of transition from the solution conformation to the complexing conformation discussed above; (b) the distortion energy of the complexing conformation peculiar to the given cation (for details see next Part); and (c) the ion—ligand interaction energy in the complex and also the change in the interatomic and inter-group interaction energies in the complexone molecule caused by polarization of the ligand groupings accompanying the incorporation of the cation.

Such an approach is convenient in analyzing the thermodynamics of complex formation, since it separates the cation—complexone interaction into a number of processes of which the energy can, in principle, be calculated.

The energy of the conformational transitions, as has been mentioned earlier, can be obtained by a theoretical conformational analysis. Much more difficult is the evaluation of the energy of the ion—ligand interaction. For the macrocyclic complexes discussed in this book, in conformity with all said above, this interaction is essentially equivalent to cation solvation by the solvent molecules. However, the coordination number of the cation and the orientation of the ligand groups, being determined by the structure and conformation of the macrocyclic compound, can differ considerably in the complexes from those in the solvate sheaths. It is therefore very difficult to compare metal ion binding by the nactins, say, with solvation of the given ion in a tetrahydrofuran—methyl acetate mixture, despite the same nature of the ligands. Even knowledge of the complex conformations has as yet yielded no tangible results in this respect, because we still do not know the dependence of the energy of interaction between the cations and each individual group upon their distance and mutual disposition. In their calculations some authors account for only the electrostatic interaction of the point charge with the fixed dipoles, or, more correctly, with the partial charges on the ligand group atoms [230, 511]; others take into account also the polarization of the cation and ligand groups and make an overall estimate of the ion, dipole and quadrupole interactions and the effect of dispersion and repulsive forces [666, 938] (cf. [466]). However, the choice of pertinent values for the permanent and induced dipole and quadrupole moments, polarizabilities, etc., is rather arbitrary so that all these calculations bear a qualitative character. Moreover, in the electrostatic approximation it is too great a simplification to regard the ligand atoms as spherically symmetric, since the cation—ligand interaction energy apparently depends on the orientation of the cation relative to the lone electron pairs. Thus, judging from X-ray data, in cyclic polyether complexes the cations are preferentially in a direct line with the tetrahedral direction of the lone pairs or are trigonally oriented with respect to both tetrahedral directions of the lone pairs [260].

The contribution by donor—acceptor interactions is still unclear. Recourse is often made to such interactions in treating cation solvation mechanisms (see, for example, [443, 445]). There is no doubt as to the existence of the donor—acceptor electron exchange in Tl^+ cryptates, for here the coordinative character of the cation—ligand bond follows from the spin—spin coupling of the thallium nucleus with the CH_2N and CH_2O protons as revealed in the NMR spectra (see Part II.A.1). Direct proof of donor—acceptor interactions has not yet been obtained for the macrocyclic complexes of the alkali ions, but their existence is predicted by quantum-chemical calculations [508, 997] and is in agreement with a number of experimental findings. Thus, analysis of the NMR spectra of sodium salt solutions under conditions precluding ion pair formation showed the ^{23}Na chemical shift to be linearly dependent upon the solvent donor number, many of the solvents containing amide, ether and ester groupings typical of the macrocyclic complexone ligands [238]. The donor number is an energy parameter reflecting the ability of the solvent to form coordinate bonds by means of the lone pairs of the polar group atoms [338, 339]. This relation is in accord with the linear dependence of the ^{23}Na chemical shift upon the free energy of complexation found for Na^+-macrocyclic complexes [384].

The difficulties related to the calculation of the interaction energy of cations with macrocyclic complexone molecules are, to a considerable extent, reflections of the all too approximate character of modern concepts of solvation mechanisms in non-aqueous solutions. At the same time data obtained in the study of complexes such as those considered here, will undoubtedly be of significant help in the further elaboration of solvation theory, because these complexes can be regarded as "frozen" solvate sheaths.

The stability of complexes in solution can vary considerably, depending upon the properties of the solvent. Thus, with increasing water content of water—methanol solutions the stability constants of the Rb^+ and K^+ complexes of valinomycin decrease by 10^4 times (see Fig. 146). On the contrary, with decrease in the solvent polarity the stability of the complexes

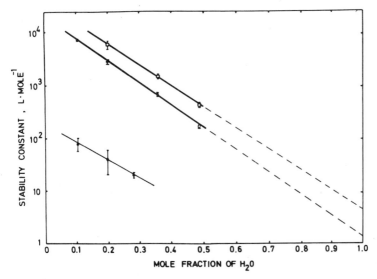

Fig. 146. Stability constants of valinomycin complexes in aqueous methanol solutions [248]. □, Rb⁺; ○, K⁺; ●, NH₄⁺. Measured potentiometrically with the aid of "valinomycin" (see p. 265) and glass cation-sensitive electrodes*.

increases and the formation of a Na^+ complex of valinomycin can be observed in a 1:2 mixture of CH_3CN—CCl_4, whereas it is not detectable under similar conditions ($c_{MX}/c_L \approx 20$) in ethanol solution. Sometimes the solvent effect can be largely ascribed to change in the solvation energy of the free cations (ΔF_M^s). For instance, in the binding of potassium ions by valinomycin in ethanol and methanol solution the value of $RT \cdot \Delta \ln K$ differs by not more than 20% from the difference in solvation energies of the potassium ions in these solvents (cf. Tables 1 and 63). In the majority of cases, however, the factors determining the influence of solvents on the complexing reaction are not so obvious. Thus, it turned out that the stability constants of K^+·valinomycin differ considerably in acetonitrile, methanol and in a 7:3 dioxane—water mixture ($1.5 \cdot 10^5$, $2.7 \cdot 10^4$ and 115 l·mole⁻¹, respectively) despite the close solvation energy values of potassium in these solvents [3]. Other such examples can be cited. In these cases, apparently of considerable importance is the solvent-dependent shift in conformational

---

* The K_w values found by extrapolation to pure water for Rb^+ and K^+ complexes are close to those obtained by Felsenfeld and Feinstein with the aid of ANS (see p. 5). A similar value of $(K_w)_{K^+}$ has been obtained by the same extrapolation to pure water of dependence of the conductimetrically determined stability constant in aqueous ethanol. In this way the $(K_w)_{K^+}$ values of the valinomycin analogs (53) and (79) were found to be 100 and 6 l·mole⁻¹, respectively [3].

TABLE 63

SOLVATION ENERGIES FOR ALKALI METAL IONS (kcal/gram ion) [443]

Cation	Solvent				
	H_2O	CH_3OH	C_2H_5OH	CH_3COCH_3	CH_3CN
Li^+	117	115	113	115	114.5
Na^+	94	92	89	80	90.5
K^+	77	75	72.5	68	75
Rb^+	72.5	69.0	66.5	62	70.5
Cs^+	63	59.5	58.5	57	61.5

equilibrium of the free complexone molecules, and also the more or less associated changes in $\Delta(\Delta F^s)_L^{M\,L}$, which reflects the difference in solvation energies of the complexed and free molecules. Since the complex conformations are usually solvent-independent (see Part III.A) the shift in the conformational equilibrium of free molecules should inevitably affect the magnitude of ΔF_{conf}. At the same time the magnitude and direction of the changes in $\Delta(\Delta F^s)_L^{M\,L}$ should depend upon the degree of shielding of the bound cation. The shielding can be characterized by the effective radius of the complex cation, determining its interaction with solvating molecules and counter-ions. This radius reflects on the one hand the thickness of the "overcoat" of polar groups and hydrophobic radicals surrounding the ion and, on the other, the presence of "rents" in this coat through which the ion can directly interact with the anions and polar molecules. Naturally, the effective radius can differ significantly from the geometrical or the Stokes' radius of the complex.

Let us assume, as a limiting approximation, that the effective radius of the complex cation is so large that its charge has no noticeable effect on its solvation. Then one can expect that in polar, particularly, aqueous, media the solvation energy of the free complexone molecules will be more than that of the complexes since the latter will have a smaller number of solvent-exposed polar groups (cf. p. 217). Thus, changes in $\Delta(\Delta F^s)_L^{M\,L}$ with increased solvent polarity will in this case favor destabilization of the complex. On the contrary, if the effective radius of the complexed cations is small, electrostatic interaction will play an important part, so that the changes in $\Delta(\Delta F^s)_L^{M\,L}$ with increase in solvent polarity will now promote stabilization of the complex [666, 938].

By their properties valinomycin and the nactins seem to approach the former of the two limiting cases, and small-ring cyclic polyethers, the latter. Indeed, according to the results of conductimetric and potentiometric measurements (cf. Table 1 and Fig. 146), the stability constant of the valinomycin·K^+ complex falls to about 100th on passing from ethanol to methanol and, as we have already pointed out, the magnitude of this change is in good accord with the difference in $\Delta F^s_{K^+}$ between these solvents. The stability constant of the K^+ complex of the cyclic polyether (270), however, diminishes only to 1/11 (see Table 24) and it is natural to ascribe the contrasting behavior of the two types of compound to lesser potassium ion shielding in the cyclic polyether complex than in the bulky valinomycin complex.

The difference in shielding of the bound cation can be very readily followed by comparing the crystal structures of salts of the complexes. In the case of valinomycin, nonactin and enniatin B, the $M^+ \cdots X^-$ distance is very large (5—8 Å, see Table 62). Hence, in this case the nature of the crystal lattice is determined not so much by the electrostatic interaction between the ion and counter-ion as by the dispersion forces between the hydrocarbon radicals of the contacting complex cations. It is characteristic that, in general, in the crystal salts of the K^+·nonactin and K^+·enniatin B complexes the anions do not assume a definite position in the lattice, being statistically distributed in the inter-complex spacings [208, 209]. Similar packing is displayed by the salts of larger ring cyclic polyether complexes such as compound (305) [105, 108] which effectively shield the bound cation by chain folding (see Part III.A.3.c).

On the contrary, the cations bound to the small cyclic polyethers, judging from the interatomic distances in the crystals, are accessible to direct interaction with counter-ions and solvating water molecules [88, 105—107, 182, 260]*. These additional ligands can be on either one or both sides of the complexone molecule, their interaction with the cation being capable of forcing it out of the plane of the ethereal oxygen atoms. It is noteworthy that the magnitude of this deviation from the plane and the number of additional ligands are apparently determined mainly by the relative sizes of the cation and the molecular cavity. Thus, in the Na^+ complex of cyclic polyether (270) the inter-oxygen spacing (~2.6 Å) is larger than the sodium ion (cf. Table 64). As a result the steric conditions are favorable for interaction of bound Na^+ not only with one, but with two, additional ligands in *trans* position to each other (see Fig. 147B). The location of the cation is determined by the balance of the forces acting between it and the "proper" and extra ligands. In fact, deviation from the oxygen atom plane is maximum if the sodium ion is linked with one extra ligand (NCS⁻, 0.54 Å

---

* Concerning the possible participation of water molecules in solvation of the sodium ion complexed by antamanide see Part III.A.1.g.

TABLE 64

CRYSTALLOGRAPHIC RADII OF CATIONS [80]

Cation	Å	Cation	Å	Cation	Å
Li^+	0.68	Be^{2+}	0.34	Tl^+	1.49
Na^+	0.98	Mg^{2+}	0.74	Ag^+	1.13
K^+	1.33	Ca^{2+}	1.04	Hg^{2+}	1.12
Rb^+	1.49	Sr^{2+}	1.20	Pb^{2+}	1.26
Cs^+	1.65	Ba^{2+}	1.38	NH_4^+	1.46

[88]); it is less in the case of differing *trans* ligands (Br^-, H_2O, 0.27 Å [106]) and is insignificant if the *trans* ligands are of similar properties ($Br\cdots H_2O\cdots Na^+\cdots H_2O\cdots H_2O$, 0.07 Å [106]).

From this standpoint of interest are the ionic triplets $[(271)\cdot Ba^{2+}]\,Fl_2^-$ (Fl^- = fluorenyl carbanion). According to their ultraviolet spectra in tetrahydrofuran, only one of the anions contacts the bound barium ion. Such contact apparently requires the cation to be so far from the plane of the oxygen atoms as to become inaccessible for effective interaction with the second Fl^- anion. Hogen Esch and Smid [402], who discovered this phenomenon, believe that such ionic triplets possess a sandwich structure (see Fig. 147D) which is partly symmetrized by inversion of the macrocyclic compound or by oscillations of the barium ion.

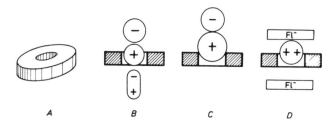

Fig. 147. Disposition of the cation and extra ligands in cyclic polyether complexes with planar arrangement of the oxygen atoms. A, Schematic representation of a cyclic polyether molecule; B, interaction of a small cation with counter-ion and solvent molecule; C, interaction of a large cation with counter-ion; D, probable structure of the ion triplet: Ba^+ complex of cyclic polyether (271) and fluorenyl carbanions (Fl^-) [402].

If the cation, bound by the cyclic polyether with planar complex conformation, is larger than the central hole, then, while remaining in contact with the oxygen atoms, it will deviate considerably from the plane they form. In the Rb^+ complex of compound (270), for instance, the deviation amounts to 0.94 Å (see Part III.A.3.b). Evidently, this asymmetric position of the cation makes it available for one extra ligand and excludes contact interaction with the *trans* counterpart of the latter (see Fig. 147C). Noteworthy from this viewpoint are the data of Smid's group [952, 995, 1093] obtained in a study of the interaction of cyclic polyether complexes with Fl⁻ anions in tetrahydrofuran. The Na⁺ complexes of 18-*crown*-6 type compounds (270), (271), (274), etc., form "separated" ion pairs (see p. 92); however, the proportion of contact ion pairs increases with both increase in cation radius and decrease in size of the ring. The effect of ring size can be clearly seen in the example of the contact pair-forming Na⁺ complexes of compound (247a) belonging to the 15-*crown*-5 type cyclic polyethers (see Table 39). X-ray analysis of the salt (246)·NaI·H₂O showed that the central hole of such cyclic polyethers (approx. 1.5 Å) is smaller than the sodium ion, so that the latter, while remaining in contact with the oxygen atoms, projects out of their plane (see Part III.A.3.a). Hence, here too, as in the case of the (270)·Rb⁺ complex, the metal ion is in an asymmetric position apparently due, not so much to extra ligand interaction, as to steric restrictions.

Interestingly, under similar conditions the Na⁺ complex of the linear polyether tetraglyme ($CH_3O(CH_2CH_2O)_4CH_3$) forms only "separated" ion pairs with Fl⁻ anions. As well as compound (247a), tetraglyme has five ligand oxygen atoms; but the arrangement of these atoms about the cation in its complexes is apparently of a higher order of symmetry (cf. p. 89). The foregoing example clearly illustrates the role of the cyclic structure, restricting the flexibility of the complexone molecule. The forced planar arrangement of the oxygen atoms in complexes of compound (247a) and other such cyclic polyethers necessarily lowers the shielding of the bound cation.

The role of extra ligand can be assumed by a second polyether molecule. The first, but in essence indirect, evidence that two cyclic polyether molecules may take part in the cation binding was obtained by Pedersen [740], who under special conditions was able to prepare crystalline substances of the compositon $L_2M^+X^-$ and $L_3(M^+X^-)_2$ (see Table 26). Pedersen suggested that these substances are the salts of ordinary or double ("club") sandwich complexes (see Fig. 148). This viewpoint was supported, for example, by the fact that the salts could be obtained only with cations commensurate with or larger than the hole in the cyclic polyether molecules. It is difficult to say which of Pedersen's preparations are actually sandwich complexes and which are the result of cocrystallization of the "normal" complexes with free cyclic polyether molecules. In any case, X-ray analysis performed in Truter's laboratory showed that compound (246) is in fact capable of sandwiching potassium ions, whereas the crystalline salt of the

(270)·Rb$^+$ complex can contain a stoichiometric amount of the free cyclic polyether [593] (see also [144, 744, 995]).

Subsequently, cyclic polyether complexes of the composition M$^+$L$_2$ were also detected in solutions (see Tables 25 and 39)* when the cation was larger than the molecular hole. This gives certain grounds for attributing a sandwich structure to such complexes, although no direct proof has as yet been obtained of the participation of both molecules in the cation coordination.

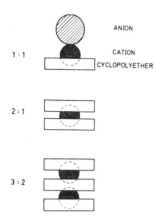

Fig. 148. Cation coordination in the crystalline salt of planar cyclic polyether complexes with large ions. Below, hypothetical structure of a "club" sandwich according to Pedersen [740].

It is of interest that diastereomeric dicyclohexyl-18-*crown*-6 (274A) and (274B) differ in stability of their sandwich complexes with cesium ions (see Table 25). In compound (274A) the cyclohexyl radicals are located cisoidally to the polyether ring, whereas in compound (274B) they are transoidal (see p. 192). One might therefore conclude that in the case of cyclic polyether (274B) steric hindrances in forming a junction are a serious obstacle to sandwich complexing and that they are absent in the case of isomer (274B) because in its sandwich complex the molecules can be contiguous with their non-substituent-carrying sides.

With cyclic polyethers the equilibrium M$^+$L + L $\leftrightarrows$ M$^+$L$_2$ shifts to the left on passing from tetrahydrofuran to methanol. The main reason for dissociation of the sandwich complex is apparently that methanol competes

---

* Eisenman *et al.* [235, 616, 618] who have studied bimolecular lipid membranes modified by cyclic polyethers arrived at the conclusion that in non-polar media these compounds can form complexes of the composition M$^+$L$_3$. For a more detailed discussion see p. 289.

more effectively than tetrahydrofuran for the second cyclic polyether molecule as an extra ligand of the cation in the M^+L complex. The equilibrium shift in methanol is also promoted by hydrogen bond formation with the oxygen atoms of the free molecules.

Enniatin B complexes of the composition M^+L_2 apparently also have a sandwich structure, the cation interacting with three carbonyl ligands which project out from a single side of each ring (see Part III.A.1.c). If free carbonyls in the sandwich are not shielded, its dissociation to the M^+L complex and free molecule is not accompanied by an increase in the number of solvated polar groups. Analogically, on dissociation of enniatin club sandwich $M_2^+L_3$ into two M^+L complexes and a free molecule the number of solvated polar groups does not change. One could, therefore, plausibly expect that with enniatin B the contribution of the polar group solvation to sandwich destabilization should be less than in the case of cyclic polyethers. In aqueous solutions one of the factors promoting sandwich formation can be a decreased area of contact of the hydrocarbon radicals with the medium on junction of the molecules.

Competition between the solvent and the second complexone molecule as extra ligands of the bound cation can apparently take place also in the complexes of nigericin-like compounds. Thus, the antibiotic X-537 A in aqueous alcohol binds alkaline earth metal ions into $M^{2+}L$ type complexes, but in extraction experiments it is an $M^{2+}L_2$ complex which passes into the organic phase [122].

The stability of sandwich complexes can considerably increase due to entropy factors, if the molecules potentially capable of forming such complexes are joined together by bridges. As a typical example can serve compound (226) described in Parts I.B.6 and III.A.1.j, the molecules of which comprise two cysteine-containing cyclopentapeptides joined by a disulfide bond. Contrary to the monomeric cyclopentapeptide, compound (226) complexes alkali metal ions in solution. A similar effect apparently takes place in the case of polymer (316) containing benzo-15-*crown*-5 residues. As compared with methylbenzo-15-*crown*-5 (246) this polymer has an enhanced affinity to large cations, the limiting ratios of the number of bound ions to the number of polyether residues being 1/2 in the case of K^+, Rb^+ and Cs^+, whereas, in the case of Na^+, it is close to unity [504, 995].

It is to be noted that, for the formation of sandwich complexes, it is unessential whether the ligand groups are attached to the macrocyclic compound or to some other relatively rigid matrix, if only the steric requirements are complied with on junction. As a sort of intermediate case, one may mention here *syn*-tris-epoxycyclohexane (benzene trioxide) which can be regarded as a 9-*crown*-3 polyether fixed in a conformation optimal for cation coordination. This compound forms crystalline salts of sandwich complexes with various metals [874] (cf. [1039]).

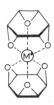

Many properties of the macrocyclic complexes are only to a small extent dependent on the nature of the cation. For instance, in alcoholic solutions the mobilities (or the Stokes' radii) of the complexes of various cyclo-depsipeptides, compound (270) and antamanide change by not more than

TABLE 65

BULK EXTRACTION CONSTANTS OF ALKALI ION—MONACTIN COMPLEXES [233]

Conditions	Solvent		Anion
A	CH_2Cl_2		2,4-dinitrophenolate
B	CH_2Cl_2		picrate
C	$CH_2Cl_2 - n$-hexane (64 : 36)		picrate

Cation	$K_{w/o}$ $(l \cdot mole^{-1})$			$(K_{w/o})_{M^+}/(K_{w/o})_{K^+}$		
	A	B	C	A	B	C
Na^+	0.11	8.0	$4.8 \cdot 10^{-5}$	$8.2 \cdot 10^{-3}$	$9.4 \cdot 10^{-3}$	$4.6 \cdot 10^{-3}$
K^+	13.4	850	$1.04 \cdot 10^{-2}$	1.0	1.0	1.0
Rb^+	4.0	290	$2.3 \cdot 10^{-3}$	0.30	0.34	0.32
Cs^+	0.23	25	$1.1 \cdot 10^{-4}$	$1.7 \cdot 10^{-2}$	$2.9 \cdot 10^{-2}$	$1.1 \cdot 10^{-2}$
NH_4^+	240	$1.6 \cdot 10^4$	0.42	18	18.8	40.4

10% on passing from Na^+ to Cs^+ (see Table 40). Eisenman and coworkers [233] found that the ionic selectivity of nactins, as determined by extraction experiments, is practically independent of the polarity of the organic phase and of the anion species (see Table 65). The extraction selectivity for ions M_1^+ and M_2^+ can be described by Eqn 22.

$$\frac{K_{w/o}^{M_1}}{K_{w/o}^{M_2}} = \frac{K_w^{M_1}}{K_w^{M_2}} \cdot \frac{k_{M_1L}}{k_{M_2L}}$$

(22)

Since the experimental partition ratios were found to be close to the stability constant ratios of the respective complexes in methanol (and probably in aqueous solution) the authors believe that, in this case, complexes with differing cations possess practically coinciding partition coefficients.

The similar behavior of complexes with different cations can be plausibly interpreted by assuming that the bound ions are in the internal cavity of the bulky organic molecule and more or less isolated from the surrounds. As a result the properties of the complexes are largely determined by the structure and conformation of the macrocyclic compound. In Part III.A it has been shown for a number of examples that complexes with differing ions possess similar conformations and to a first approximation can be regarded as isosteric. The isostericity of the complexes, by which is meant not only geometric equivalence but also similarity of other properties, has been used by Eisenman et al. as the basis for a theoretical treatment of the cationic extraction selectivity and also of the effects exerted by macrocyclic complexones on artificial membranes (see Parts II.B, V.B and V.C). Quite obviously, the nature of the bound ion has lesser bearing on the properties of the complexones, the larger the complexone molecule and the more effectively it shields the cation. These conditions are best fulfilled by such compounds as valinomycin and the nactins.

Evidently least sensitive to the cation species should be the size-controlled properties of the complexes. Such, for instance, are the diffusion coefficients, or electrophoretic mobilities. On the other hand, parameters depending on the solvation energy of the complexes (partition coefficients, dissociation constants of type $(M^+L)X^-$ ion pairs, etc.) can differ significantly on going from one cation to another. Such differences are particularly manifest in cyclic polyether complexes with planar arrangement of the oxygen atoms, being exhibited in the solvent- and anion-dependence of the cationic complexing selectivity. This problem comprises one of the topics in the next Part.

III.C. The essence of the ion selectivity of macrocyclic complexones

It is natural to define complexing selectivity for the ions M_1^+ and M_2^+ as the stability constant ratios of their respective complexes in solution. By means of Eqn 20 one may readily obtain Eqn 23, from which it follows that the complexing selectivity depends upon the difference in discriminating capacities of the complexone and solvent.

$$-RT \cdot \ln \frac{K_{M_1}}{K_{M_2}} = (\Delta F_{M_1L} - \Delta F_{M_2L}) - (\Delta F^s_{M_1} - \Delta F^s_{M_2}) + (\Delta F^s_{M_1L} - \Delta F^s_{M_2L})$$

$$= \Delta(\Delta F_{ML})^{M_1}_{M_2} - \Delta(\Delta F^s_M)^{M_1}_{M_2} + \Delta(\Delta F^s_{ML})^{M_1}_{M_2} \tag{23}$$

The first term on the right-hand side of Eqn 23, $\Delta(\Delta F_{ML})_{M_2}^{M_1}$, reflects the difference in the free energy of interaction of the complexone molecule with the M_1 and M_2 ions and is independent of the solvent. Taking into consideration that on passing from ion M_1 to ion M_2 both the complex conformation and the ion–ligand and ligand–ligand interaction energies may change, this term, in conformity with Eqn 21, can be represented as follows.

$$\Delta(\Delta F_{ML})_{M_2}^{M_1} = \Delta(\Delta F_{conf})_{M_2}^{M_1} + \Delta(\Delta F'_{ML})_{M_2}^{M_1} \tag{24}$$

The second term of Eqn 23, $\Delta(\Delta F_M^s)_{M_2}^{M_1}$, the solvation energy difference between the free ions in the given medium, is independent of the complexone species. The third term reflects the solvation energy difference of the complexes with the different cations. In this way the factors responsible for the complexing selectivity are precisely defined and we can now proceed to discuss each one separately.

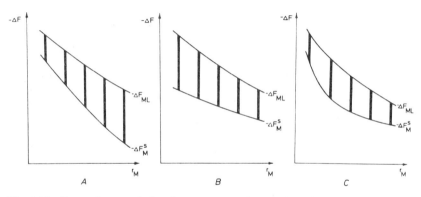

Fig. 149. Dependence of the free energy of complexation on the cation radius. The heights of the vertical bold faced lines correspond to ΔF_{compl} values. For further details see text.

It was shown in the previous Parts of this Chapter that in the macrocyclic complexes the ligand groups surround the cations, forming a sort of solvate sheath, and that complex formation can therefore be likened to change of ligands in the coordination sphere of the free cation when it is transferred from one solvent to another. It is well known that the solvation energy of alkali and alkaline earth ions diminishes monotonously with increasing ion size (see Table 63), the steepness of the ΔF_{ML} vs. ionic radius r_M curve serving as measure of the solvent's discriminating power. Hence, in this approximation the complexing selectivity will be determined by the differences in steepness of the ΔF_{ML} and the ΔF_M^s curves. If the solvent discriminates between the cations better than the complexone does, the complex stability will increase with increasing r_M; in the opposite case, the same complexone will preferably bind the smaller ions (see Figs 149A and B).

As a rule, metal-binding macrocyclic compounds in solution form the most stable complexes with medium-sized ions. This can mean that they distinguish between the smaller ions to a worse extent and between the larger ions to a better extent than the solvent does. In other words, in the region of small r_M values the ΔF_M^s curve will be steeper than the ΔF_{ML} curve, whereas the inverse holds with increase in r_M (see Fig. 149C). Krasne and Eisenman [234, 511] proposed that such curves be called "asymmetric" and showed that the condition for the asymmetry is a non-linear relation between ΔF_{ML} and ΔF_M^s. Analyzing the nature of the cation selectivity of complexation, these authors discussed in the electrostatic approximation the interaction of the cation with a single diatomic molecule (or ligand grouping).

If one assumes a certain fixed charge distribution between the atomic constituents of the molecule, the energy of their interaction with a monovalent cation can be represented by Eqn 25.

$$U = 332e \left(\frac{\delta^-}{r_M + r_n} + \frac{\delta^+}{r_M + r_p} \right)$$

$$= -332 \cdot \delta \cdot e \frac{r_p - r_n}{(r_M + r_n)[(r_M + r_n) + (r_p - r_n)]} \tag{25}$$

where $\delta = \delta^+ = |\delta^-|$ are the partial charges on the atoms P and N; r_M is the cation radius; r_p and r_n are the distances from the cation "surface" to the centers of atoms P and N, respectively (see Fig. 150).

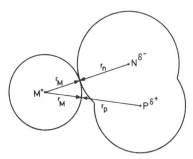

Fig. 150. Mutual disposition of a cation and diatomic ligand group (or molecule). For further details see text.

When the difference $r_p - r_n = (r_M + r_p) - (r_M + r_n)$ considerably exceeds $r_M + r_n$, the energy of interaction with the positively charged atom can be neglected and then U will be inversely proportional to $r_M + r_n$. If, however,

$r_M + r_n \gg r_p - r_n$, the interaction energy U will then be inversely proportional to $(r_M + r_n)^2$. In other words, in these limiting cases the ligand molecule or group can be regarded as a point charge or as a dipole, respectively. Since the value of $r_M + r_n$ increases with increase in cation radius, in principle the steepness of the $-U$ vs. r_M dependence should be variable, depending on both the charge and size of the atoms forming the dipole and the orientation of the latter with respect to the cation. This creates conditions for "asymmetry" in the interaction of the cations with the solvent molecules containing various ligand groupings. A characteristic example of this are the energies of alkali ion transfer from water to dimethylformamide, given in Fig. 151. It is therefore natural to expect that in the formation of the macrocyclic complexes also, structural differences in the ligand groups of the complexone and the solvent may be one of the factors determining the preferential binding of medium-sized ions.

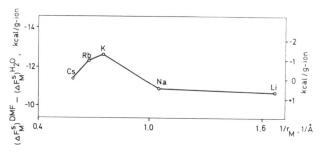

Fig. 151. Free energies of transfer of alkali metal ions from water to dimethyl formamide [511]. The right-hand ordinate axis represents the transfer energy values calculated relative to the value for Cs$^+$.

It would be an error, however, to overestimate the part played by the nature of the ligand groupings in determining the macrocyclic complexing selectivity in the alkali and alkaline earth metal series. Indeed, among the cyclic polyethers, of which each contains ligand oxygen atoms, compounds are encountered with the most varied cationic selectivities. On the other hand, replacement of ester ligands by amide ligands in valinomycin, although causing a certain fall in the K/Na selectivity (see Table 2), in no way imparts to the complexone the Na-specificity of antamanide or compound (225B), whose ligand groups are also amides. The nature of the ligand groups has a much greater effect on the relative affinity of the macrocyclic complexones for the transition metal ions. It has been shown on the example of cyclic polyethers that stability of the complexes with transition metal ions increases and with alkali metal ions falls on replacing some of the oxygen ligand atoms by sulfur or NH groups (see Part I.D). This effect is in accord with the general principles of coordination compound chemistry and cation solvation theory (see, for instance, [338, 339, 873]).

Since alkali and alkaline earth metal ions are spherically symmetric "hard" cations, the ion—ligand interaction energy is not crucially dependent on the ligand disposition symmetry in the complexes (but, of course, not on the ligand orientation with respect to the cation). The dispositions of the ligands do, however, play an important part in determining the affinity for ions of non-spherical symmetry. Among these are, for example, ammonium, where the protons are at the vertices of a tetrahedron in which the nitrogen atom occupies the center. The ammonium ions are bound particularly strongly by the nactins (see Table 65) in complexes in which the cations display a cubic type of coordination. The probable orientation of the NH bonds in the NH_4 complexes of these antibiotics are represented in Fig. 152. Judging from infrared data these complexes contain no NH···O hydrogen bonds [794].

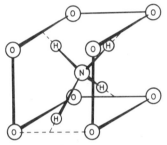

Fig. 152. Possible position of the NH_4^+ ion inside the nactin "tennis ball seam" structure. Alternatively one can assume that NH bonds in the ammonium are oriented towards four tetrahedrally disposed ligand O atoms.

The binding of the polyatomic organic cations by macrocyclic complexones takes place providing insertion of the charge-carrying group into the molecular cavity is sterically possible. This can readily be seen for the example of compounds with a protonated primary amino group, these being bound exclusively by cyclic polyethers, the oxygen atoms of which assume a planar arrangement in the complex. Interestingly, these cyclic polyethers, for instance compound (270), also form complexes with hydrazinium, hydroxyl-ammonium and guanidinium ions. The inter-oxygen space in these complexes is apparently occupied solely by the NH_3^+ or NH_2^+ group. It would be worthwhile to see what effect the electron density distribution in the guanidinium cation has on such selective "solvation" of an amino group.

Pressman [788] has shown that a correlation exists between the amine binding ability of nigericin antibiotics and their ability to complex alkaline earth metal ions. Since the M^{2+} complexes may contain two molecules of the antibiotic, one might suppose that the reason for such correlation could be accessibility of the internal cavity to bulky groupings (extra ligands belonging either to the second antibiotic molecule or to solvent molecules, or the radical on the N atom). The affinity for divalent ions and amines is manifested most strongly in antibiotic X-537 A and dianemycin. These antibiotics effectively discriminate between primary and secondary amines of closely related structure. Thus, X-537 A forms a complex with nor-adrenaline which is about 20 times more stable than that with adrenaline.

The analogy discussed above between changing the solvate sheath of an ion and complexation has its limitations however. The specificity of the "mono-molecular cation solvation" in the macrocyclic complexes is due first of all

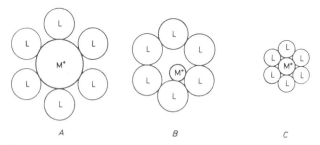

Fig. 153. Cation in the inter-ligand cavity: A, $r_M > r_m$; B, $r_M < r_m$; C, the decrease of r_M with decreasing the ligand atom radius.

to independence of the number of ligands from the complexed cation species and to their spatial arrangement being unequivocally determined by the structure and conformation of the complexone molecule. The part played by these factors in complexing selectivity has been discussed in detail by Simon and Morf [665, 666, 938, 939].

Let the conformational characteristics of the macrocyclic complexone be such that the inter-ligand distances can be widely variable while the ligand groups retain their orientation with respect to the complexed cation. With diminution of the cation radius the ligand groups will be drawn closer together, tending to a position of maximum ion—ligand interaction. Under such conditions the distances between the cation and its ligand atoms approach the sum of their crystallographic radii (see Fig. 153A). Naturally, in the series of alkali and alkaline earth metals, decrease in ion size should be expected to be accompanied by a monotonic increase in $-\Delta F_{M\,L}$. However, at a certain cation radius ($r_M = r_m$) repulsion between the ligand atoms becomes very strong. Further diminution of the inter-ligand gap will thus become energetically unfavorable and cations for which $r_M < r_m$ will not be able to interact equally with all the ligand groups (see Fig. 153B). It is not difficult to see that r_m will increase with increasing number of ligand atoms about the complexed cation and diminish with their decreasing size (cf. Figs 153B and C). The augmentation of ligand—ligand repulsion and withdrawal of part of the ligands from contact with the cation flattens the $-\Delta F_{ML}$ vs. r_M curve (Fig. 154). This effect will apparently be weaker the less symmetric the disposition of the ligands and the greater the differences in their cation interaction energy. As a limiting case one can imagine that the small cations will "stick" to those ligands which make the largest contribution to ΔF_{ML}.

It is accepted that the lowered discriminating capacity in the region of $r_M < r_m$ is a consequence of the independence of the number of ligand

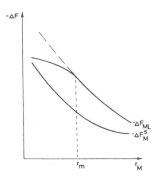

Fig. 154. Flattening of the $-\Delta F_{ML}$ vs. r_M curve in the region $r_M < r_m$.

groups from the cation species. For small ions, therefore, there is more or less deviation of the coordination from the optimal. On the other hand the cation—solvent interactions occur at maximum solvation energy, the "optimization" being attained primarily through decrease in the number of solvent molecules in the inner layer of the solvate sheath on passing to the smaller cations. Owing to this, in the region $r_M \approx r_m$ one might expect asymmetry in the interactions of the cation with complexone and solvent. Moreover, due to flattening of the ΔF_{ML} curve in the region $r_M < r_m$, the complexing selectivity will be determined mainly by the discriminating power of the solvent which can be very high, particularly for the smaller ions.

In most of the complexes discussed here, the bound cations are surrounded by oxygen atoms ($r = 1.40$ Å). From a comparison of Tables 64 and 66 it can be seen that, for certain types of coordination, the r_m values lie within the range of the alkali and alkaline earth ionic radii. Thus, lithium ions are smaller than the minimal inter-oxygen cavity for any arrangement of the oxygen atoms except tetrahedral, square or octahedral. Apparently, it is here wherein lies the reason for the high Na/Li selectivity of the nactins, in the complexes of which the cations are of cubic coordination (see Part

TABLE 66

MINIMUM RADII OF THE INTER-OXYGEN
CAVITIES FOR DIFFERENT TYPES OF
CATION COORDINATION [938]

Coordination number	r_m (Å)
4 (square)	0.58
4 (tetrahedron)	0.31
6 (hexagon)	1.40
6 (octahedron)	0.58
7 (C_{3v} symmetry)	0.83
8 (cube)	1.02
8 (square antiprism)	0.90
9 (D_{3h} symmetry)	1.02
12 (cubo-octahedron)	1.40

III.A.2). It is evidently this factor which is responsible for the low stability of the complexes with magnesium and beryllium ions in the alkaline earth metal series.

While the number and nature of the ligands determine the minimal cation radius at which effective complexing can occur, it is the macrocyclic structure of the complexones which lays the natural upper limits to the complexable cation radius. This is most strikingly exemplified by cyclic polyether (243), the diameter of the inter-oxygen hole of which does not exceed 1.2—1.5 Å. Of the alkali metal ions only Li^+ can be accommodated in such a hole, thus explaining the Li^+ specificity of this cyclic polyether [739]. Characteristically, a molecule of compound (243) has only four oxygen atoms, so that in this case the condition $r_M > r_m$ for Li^+ is satisfied.

The range of r_M values within which effective cation binding is observed cannot always be correlated so simply with the type of coordination and with the size of the ring, mainly because of the limited conformational mobility of the complexone molecules. Complexes with differing ions usually have a macrocyclic spatial structure which corresponds to one and the same minimum in the energy maps so that the effect of the cation species on the molecular "conformational energy", expressed by the $\Delta(\Delta F_{conf})$ term, depends upon the rigidity of the complex conformation (see p. 218). If it is low, then for $r_M > r_m$ the complexone may "adapt" its intramolecular cavity to fit the cation. If, on the contrary, the complex conformation is very rigid, the minimum radius r_M at which all ligands will be in contact with the cation will be close to the maximum value, because change in the cavity size will require considerable energy. In other words, such a molecule will in a sense be "tuned" to cations of a given size.

Unfortunately, theoretical calculation of the conformational rigidity of complexones encounters considerable obstacles. Among these are above all the difficulty in predicting the cation-specific differences in orientation of the ligand groups. It is also difficult to account for the resultant change in interactions between groups, which although considerably separated along the chain, are yet neared enough through space for such interaction to take place. Present methods for estimating the conformational rigidity are therefore confined to studies of molecular models and to comparison

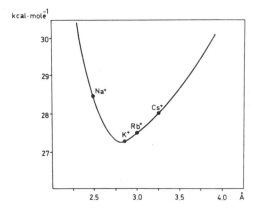

Fig. 155. The dependence of the valinomycin "conformational energy" on the inter-oxygen cavity radius according to Mayers and Urry [607]. The abscissa gives the distances of the oxygen nuclei to the cavity center. The points on the curve correspond to sums of the cation and oxygen radii.

between spectral differences and cation selectivities. As shown in Part III.A, despite its shortcomings such an approach has proved very fruitful in rationalizing the ionic selectivities of valinomycin, the enniatins and antamanide. It should be noted that a theoretical evaluation by Mayers and Urry [607] of the dependence of the "conformational energy" of valinomycin upon the molecular cavity size has led to the same conclusion; namely, that the Rb^+ and K^+ complexes are the least sterically hindered (see Fig. 155). This conclusion had been arrived at earlier from spectral data.

As well as conformational rigidity, factors determining the ion complexing selectivity should include cation-induced orientational changes of the ligand groupings and the accessibility of the bound cation to counter-ions and solvent molecules. The relative contribution of each of these factors is determined by the complexone structure. Thus, in the case of the cryptates, the decisive role is apparently played by the limited variability of the molecular cavity, in good accord with the dependence of the ion selectivity of bicyclic diaminopolyethers on ring size. If in compound (324) specific to potassium and barium ions the number of CH_2CH_2O units is diminished, the specificity will first shift to sodium and strontium (compound 323) and then to lithium and calcium (compound 322) (see Tables 27 and 28).

A peculiar situation arises in the case of valinomycin. According to the spectral data all six ester carbonyls interact equally with the cation in the K^+, Rb^+ and Cs^+ complexes of this cyclodepsipeptide. However, in the valinomycin·Na^+ complex these carbonyls are not equivalent, despite the fact that $r_{Na} > r_m$ in the approximately octahedral arrangement of the ligand oxygens characteristic of valinomycin. As mentioned earlier (see p. 130) this effect is evidently due to the high rigidity of the "bracelet" conformation which makes it energetically unfavorable to bring all six oxygen atoms into contact with the Na^+ ion. The decreased number of ligands in the coordination sphere of the bound Na^+ apparently leads to a significant fall in $-\Delta(\Delta F_{ML})_K^{Na}$. In this case the K/Na complexing selectivity will be mainly determined by the discriminating power of the solvent, i.e. by $-\Delta(\Delta F_M^s)_K^{Na}$. Using the data of Table 63, it can readily be seen that in the limit, when $\Delta(\Delta F_{ML}^s)_K^{Na} = 0$, the attainable K/Na selectivity can exceed 10^{10}. Hence, the cause of the high K/Na selectivity of valinomycin ($> 10^4$ in alcoholic solutions; see Table 1) is the rigidity of the "bracelet" conformation.

Cation-dependent changes in ligand group orientation can be manifested most readily in complexes with low conformational rigidity. Since the orientational changes of these groups are associated with either decrease or increase in steepness of the $-\Delta F_{ML}$ vs. r_M dependence, a sort of positive or negative "feedback" can arise with increase in cation radius. Reorientation of the ligand groups will lead to increasing relative affinity in the first case for large ions and in the second case for small ions. As mentioned on p. 143, the latter case is apparently realized in the enniatins, where with growth of cation radii in the complexes the center of the cations shifts away from the axes of the carbonyl ligands. The orientation effect is of a somewhat different nature in the planar cyclic polyether complexes. Here the change in ion—ligand interaction can be due to shift of the cation with respect to the oxygen lone pairs when the cation is forced out of the plane of the oxygen atom.

Of considerable interest is the question of the contribution to the ion complexing selectivity by the solvation energy differences of the complexes as reflected by the $\Delta(\Delta F_{ML}^s)_{M_2}^{M_1}$ term in Eqn 23. If we revert to the analogy between complexation and solvation, then for well-shielding complexones the contribution of this term is in a certain sense equivalent to that of the external solvate sheaths for the free cations, which is usually insignificant, especially for bulky solvent molecules. However, when the counter-ions or the solvent molecules can penetrate the cation's coordination sphere, the differences in the interaction energies of differing cations with these extra ligands could become quite large. The importance of this factor is as yet hard to assess because of the sparsity of experimental data. An additional difficulty is that the change in selectivity on changing the solvent is associated with small differences in the free energy of complexation, such that, as a rule, these are commensurate with the ambiguity of the solvation energy data for the free ions. This considerably complicates distinction between the effects on $\Delta(\Delta F^s_M)^{M_1}_{M_2}$ and $\Delta(\Delta F^s_{ML})^{M_1}_{M_2}$ of changes in solvent properties.

It is natural to expect that solvation of the complexing cations should play an important role in the ability of macrocyclic compounds to discriminate between alkali and alkaline earth metals. If it be assumed that the size of the complexes is independent of the nature of the bound ion, the difference in solvation energies of $M^{2+}L$ and M^+L can roughly be represented by Eqn 26 (cf. [444]).

$$-\Delta U = \frac{A \cdot \mu}{(r_e + a)^2} + \frac{B}{(r_e + b)} \cdot \left(1 - \frac{1}{\epsilon}\right) \qquad (26)$$

where A and B are constants; r_e is the effective radius of the complex; μ is the dipole moment of the solvating molecules; a and b are parameters, dependent upon the dimensions of the solvating molecules; and ϵ is the solvent dielectric constant.

The first term of this equation reflects solvate sheath formation and the second, Born's solvent polarization energy. From this equation it follows that the difference in solvation energies of $M^{2+}L$ and M^+L will increase with increase in dielectric constant of the solvent and will decrease with increase in effective radius of the complexes. According to Simon and Morf [666, 938], $\Delta(\Delta F^s_{ML})^{M^{2+}}_{M^+}$ plays a decisive part in the high Ba/K selectivity displayed in the formation of cryptates, in whose molecules the bound cations have relatively thin "overcoats". In fact, one can see from Table 67 that some monocyclic and bicyclic polyethers display significantly higher specificity to divalent cations than valinomycin, the nactins and enniatin B, in whose complexes the ion and the ligand groups are surrounded by bulky hydrocarbon radicals. However, it is most probably not so much the thinness of the "overcoat" which causes the enhanced Ba/K selectivity of polyethers (at least in aqueous solutions) as the presence in it of "rents"; in other words, accessibility of the complexed cations to direct binding of the solvent

TABLE 67

THE Ba/K SELECTIVITY OF MACROCYCLIC COMPLEXONES (cf. TABLES 1, 4, 21, 27 and 28)

Complexone	$K_{Ba^{2+}}/K_{K^+}$	Solvent
Nonactin	$1.4 \cdot 10^{-2}$	methanol
Valinomycin	$2.8 \cdot 10^{-2}$	methanol
Enniatin B	1	methanol
Cyclic polyether (274A)	25	water
Bicyclic polyether (324)	$1.3 \cdot 10^4$	water
	~ 100	methanol—water (95:5)
Bicyclic polyether (324a)	$<5 \cdot 10^{-3}$	methanol—water (95:5)
Bicyclic polyether (324c)	~ 1	methanol—water (95:5)

molecules and counter-ions as extra ligands. The cryptates apparently differ from the planar cyclic polyether complexes in that only divalent cations in the former take on extra ligands. At any rate X-ray analysis shows that in the salts $(324) \cdot Ba(NCS)_2 \cdot H_2O$ and $(326) \cdot Ba(NCS)_2 \cdot 2H_2O$, contrary to the salts of alkali metal cryptates, the bound cation is neared to the counter-ion and to the molecule of solvate water so that the distance between them is close to the sum of the crystallographic radii [633] (cf. Part III.A.4). If this difference is maintained on passing from the crystalline state to solution it may be a source for augmentation of the M^{2+}/M^+ selectivity in cryptate formation.

Recently, Dietrich *et al.* [206] showed that incorporation of two aromatic rings into the bicyclic polyether (324) sharply reduces the absolute and relative stability of its complexes with alkaline earth metal ions and particularly its Ba/K selectivity (see compound (324C) in Tables 27, 28 and 67). It would be natural to attribute this effect to the appearance of hindrances to solvation of the bound Ba^{2+} cations, since the Na^+ and K^+ complexes undergo no significant stability changes. It is noteworthy that a still greater drop in the Ba/K selectivity is observed on replacing some of the ethereal oxygen ligands by methylene groups (see compound (324a) in Table 67), i.e. on diminishing the cation coordination number.

Differences in the solvation energies of the complexes are evidently of some importance in the alkali ion complexing selectivity of monocyclic polyethers with planar disposition of the oxygen atoms. A typical example is compound (270), the molecular cavity of which is considerably larger than the sodium ion (see p. 223). On this basis one could have expected a small $-\Delta(\Delta F_{ML})_K^{Na}$ value for this compound. In fact, this is in agreement with the conclusions of Tsatsas *et al.* [1021] reached from an infrared study of the

Na^+ and K^+ complexes. At the same time, in methanol and ethanol solutions the K/Na selectivities are only 4.4 and 7.3, respectively (see Tables 15 and 24); i.e. the values are much less than could have been attained had they been determined using $\Delta(\Delta F_M^s)_K^{Na}$ (see Table 63). The low selectivity is in this case apparently due to the fact that the solvation energy of the complexes increases with decrease in the bound cation radius. Such an assumption is not contradictory to X-ray data revealing a close contact between the cations and the solvating molecules and counter-ions (see p. 191).

As has been mentioned in Part III.A.1.g, a number of data point to the participation of water molecules in the sodium coordination of antamanide complexes. If this is so, the Na^+ specificity of antamanide should be more or less due to incomplete desolvation during the complexing reaction.

It is to be noted that solvent participation in the bound cation coordination is equivalent to complexing of a partially solvated cation. Clearly, whatever the thermodynamic cycle chosen for describing the complexing process, partial solvation of the complexed small cations promotes the decreased relative specificity to the larger ions, the more this is so the less its effect on ΔF_{ML}*.

It is still obscure to what extent differences in the solvation energies of the complexes can affect the salt extraction selectivity by complexone solutions in such organic solvents as CH_2Cl_2, $CHCl_3$, etc. One can see from Eqn 22 (p. 228) that this selectivity is dependent on the ratio of the partition coefficients of the complexes which reflects the relative discriminatory power of the organic solvent and water.

$$-RT \cdot \ln \frac{k_{M_1L}}{k_{M_2L}} = (\Delta F_{M_1L}^s - \Delta F_{M_2L}^s)_{org} - (\Delta F_{M_1L}^s - \Delta F_{M_2L}^s)_{water} \qquad (27)$$

From the data of extraction experiments it follows that the partition coefficients of nactin complexes with various alkali metal ions are close to

---

* Above we have made the tacit assumption that the nature of the bound cation has no effect on the solvation of this or that polar group in the complex molecule. Such an approximation is undoubtedly permissible if all the polar groups are effectively shielded as, for instance, in the nactin complexes or the K^+, Rb^+ and Cs^+ complexes of valinomycin. However, it is not difficult to imagine a situation where the complexes with different cations have conformations differing in the number of polar groups accessible for solvation, or when bound cation and solvent molecules compete for the individual ligand groupings. In the latter case, the solvent species will affect the magnitude not only of $\Delta(\Delta F_{ML}^s)$, but also of $\Delta(\Delta F_{ML})$. It is interesting to note that the CD spectra of the Na^+ complexes of valinomycin differ from those of K^+, Rb^+ and Cs^+ in that they depend on the polarity of the solvent (Th. Funk, private communication). This may be due to the availability for solvation of part of the ligands in the Na^+ complexes.

each other. This is in good accord with the aforementioned effective shielding of nactin-bound ions. It could be surmised that the partition coefficients of the planar cyclic polyether complexes would depend on the nature of the cation. Regrettably, no studies are being carried out in this direction. Of course, with the cyclic polyethers the partition coefficients should increase on passing from the simple to. the sandwich complexes. Therefore, as the cyclic polyether to metal salt ratio is increased, the extraction selectivity will shift towards the cation forming the more stable sandwich complexes. The effect could possibly have practical implications, since in the formation of sandwich complexes cyclic polyethers now and then manifest higher ionic selectivity. Thus, on extraction of alkali metal picrates by a CH_2Cl_2 solution of benzo-15-*crown*-5 (246) approximately equal amounts of Na^+ and Rb^+ pass into the organic phase. If, however, under similar conditions the polymer (316) containing benzo-15-*crown*-5 residues but which is more prone to form sandwich complexes is used as complexone, a threefold amount of Rb^+ with respect to Na^+ passes into the organic phase [504].

In weakly polar media free cations and macrocyclic complexes associate into ion pairs. Under such conditions complexation can in the limit be represented as $L + M^+X^- \rightleftharpoons (M^+L)\dot{X}^-$.

The ion pairing affects not only numerical values of the equilibrium constants but even the cation selectivity sequence. Thus, in tetra-hydrofuran solution the equilibrium $[(271) \cdot K^+] \| Fl^- + Na^+Fl^- \rightleftharpoons [(271) \cdot Na^+] \| Fl^- + K^+Fl^-$ (where Fl^- is fluorenyl carbanion) is shifted to the right [1093], whereas in methanol, as we have noted above, the structurally close compound (270) preferably binds potassium ions. The change in free energy in the course of this reaction can be represented by Eqn 28.

$$\Delta F = \Delta(\Delta F_{ML})_K^{Na} - \Delta(\Delta F_M^s)_K^{Na} + \Delta(\Delta F_{MX})_K^{Na} + \Delta(\Delta F_{MLX})_K^{Na} + \Delta(\Delta F_{sp})_K^{Na}$$

(28)

where ΔF_{MX} and ΔF_{MLX} are the free energies of dissociation of the ion pairs $M^+\|X^-$ and $(M^+L)\|X^-$, respectively; and ΔF_{sp} is the free energy change in the separation of M^+X^- by the solvent.

Let us make the quite plausible supposition that the energy of dissociation of the solvent-separated ion pairs $M^+\|X^-$ and $(M^+L)\|X^-$ is independent of the cation species and that $\Delta(\Delta F_{ML})_K^{Na}$ in the case of compound (271) is very small (cf. p. 240). Then, judging from the equilibrium position of the exchange reaction, one can conclude that the Na^+Fl^- pairs undergo separation much more readily than the K^+Fl^- pairs. In agreement with this conclusion is that a slight increase in polarity within the limits of a single solvent type, viz. on passing from tetrahydrofuran to trimethylene oxide, will cause separation of the Na^+Fl^- but not of the K^+Fl^- pairs. The trimethylene oxide molecules successfully compete with cyclic polyether (271) as ligands for the sodium ion and the equilibrium in this solvent is shifted in the direction of formation of K^+ complexes [952, 1093].

Complexes of compound (247a) in tetrahydrofuran form not only solvent-separated, but also contact ion pairs and the equilibrium position of $(M^+L)Fl^- \rightleftharpoons (M^+L)\|Fl^-$ in this case apparently reflects the competition of anions and solvent molecules as extra ligands for the bound cations. It is to be expected that the anions should possess a lower discriminating power than the tetrahydrofuran molecules, since the formation of contact ion pairs of the complexes proceeds with less Na/K selectivity than for the "separated" ion pairs [995].

It has recently been shown that $(M^+L)\|Fl^-$ "separated" pairs including planar cyclic polyether complexes associate in solutions [996]. Such association should diminish the contact pair to "separated" pair ratio with increasing complex salt concentration and, consequently, affect the apparent cation complexing selectivity.

The differences in properties of the ion pairs of type $(M^+L)X^-$ is manifested also in extraction of the complex salts by organic solvents. For instance, in the extraction of alkali trinitrophenolates in the presence of valinomycin, the K/Na selectivity fell considerably on passing from methylene chloride to hexane or decane. Tosteson et al. [1011, 1014], describing this phenomenon, found that the ultraviolet spectra of decane (but not of CH_2Cl_2) solutions of trinitro-m-cresolates of the valinomycin complexes differ in the position of the anion absorption bands. The band at approx. 350 nm present in the spectra of the trinitrophenol salt solutions in non-polar solvents is known to undergo a hypsochromic shift with decrease in cation radius [343] (cf. p. 95). As the solvent polarity increases the ion pairs dissociate and the position of the absorption band ceases to depend upon the cation species. Hence the shift of this band in decane solutions from 355 nm for the potassium complexes to 340 nm for the sodium complexes is evidence of ion pairing on the one hand, and of a lesser effective radius of the Na^+ complexes on the other. The fall in K/Na selectivity of valinomycin on extraction with decane may be the result of augmented stability of the ion pairs formed by the Na^+ complexes. In fact, the cationic selectivity of the extraction by non-polar solvents, in conformity with Eqn 16 (see p. 106), is determined in the limit by the ratio $(K_{w/o} \cdot K')_{M_1} / (K_{w/o} \cdot K')_{M_2}$, so that the high K' values can compensate the low $K_{w/o}$ values. The higher stability of ion pairs with Na^+ complexes of valinomycin is apparently due to their lesser effective radii. Characteristically, the CD spectra of a decanic solution of Na^+· but not K^+·valino-mycin trinitrocresolate complex salt displays a band corresponding to the aromatic chromophore situated in an asymmetric environment (private communication by D. Davis). This bears witness to closer contact of the Na^+ complex with the anion. The enhanced stability of ion pairs is in this case in accord with the aforementioned solvent sensitivity of the Na^+ complex spectra (see p. 241). The common cause of these effects is probably the

TABLE 68

THE CALCULATED SELECTIVITY SEQUENCES
FOR THE CATION TRANSFER FROM
UNIMOLECULAR AQUEOUS "SHELL" TO
ANIONIC SITES OF DIFFERING EFFECTIVE
RADII [230]

I	Cs > Rb > K > Na > Li
II	Rb > Cs > K > Na > Li
IIa	Cs > K > Rb > Na > Li
III	Rb > K > Cs > Na > Li
IIIa	K > Cs > Rb > Na > Li
IV	K > Rb > Cs > Na > Li
V	K > Rb > Na > Cs > Li
VI	K > Na > Rb > Cs > Li
VII	Na > K > Rb > Cs > Li
VIIa	K > Na > Rb > Cs > Li
VIII	Na > K > Rb > Li > Cs
IX	Na > K > Li > Rb > Cs
X	Na > Li > K > Rb > Cs
XI	Li > Na > K > Rb > Cs

more "open" structure of the sodium complex, owing to which the polar groups and cation are accessible to solvent molecules and the counter-ion.

One should not think that the stability of the complexes depends upon such a large number of independent factors that, in principle, a combination of macrocyclic complexone with solvent and anion could be found which could give rise to any desired cation selectivity sequence of alkali metal ions. In fact, all the observed variations fall into a set of selectivity sequences found by Eisenman [227—230] in a theoretical treatment of cation binding by cation exchangers of different effective anion radii (cf. Tables 68 and 69). Since in his calculations Eisenman took into account only electrostatic interactions, this coincidence may be regarded as evidence that electrostatic interactions also play a decisive part in the binding of alkali metal ions by macrocyclic complexones.

It is significant that these (and only these) selectivity sequences are observed also with other classes of cation exchangers. In order to compare the different classes of ion exchangers, Eisenman proposed the use of so-called selectivity isotherms, depicting the dependence of the free energy of complexation upon the "anionic strength", a parameter equivalent in a sense to the effective radius of the anionic site. The isotherms are

TABLE 69

THE CATION SELECTIVITY SEQUENCES OF MACROCYCLIC AND PSEUDOCYCLIC COMPLEXONES

Complexone	Eisenman's selectivity sequence*	Determined in: A, solution; B, two-phase system	References
X-537 A	IIa	B	[791]
Valinomycin	III	A, B	Table 1, [235]
Enniatin B	III	A	Table 49
Enniatin C	III	A	Table 49
Dicyclohexyl-24-crown-8 (300)	IIIa (IV)	B	[739]
X-206	IV	B	[791]
Dibenzo-18-crown-6 (270)	IV	B	[739]
Triton X-100	IV (IIIa)	B	[91]
Bicyclic polyether (324)	V	A	Table 27
Nigericin	V	B	[791]
Enniatin A	V (VI)	A	Table 49
Dibenzo-18-crown-6 (270)	VI	A (ethanol)	[15]
Octa-enniatin B (141)	VI	A	Table 49
Dianemycin	VII	B	[791]
Dibenzo-18-crown-6 (270)	VII	A (THF)**	[952]
tert-Butylcyclohexyl-15-crown-5 (250)	VII (VIIa)	B	[739]
Monensin	VII (VIII)	B	[791]
Bicyclic polyether (323)	VIII	A	Table 27
Cyclopeptide (225b)	VIII (IX)	A (DMSO)***	[191, 192]
Antamanide	X	B	[1068]
Bicyclic polyether (322)	XI	A	Table 27
Di-tert-butylcyclohexyl-14-crown-4 (243)	XI	B	[739]

* The alternative sequence in parentheses is given in cases of an uncertain position of one of the cations.
** THF, tetrahydrofuran.
*** DMSO, dimethyl sulfoxide.

constructed by laying off on an arbitrary curve known $\lg K_K/K_{M_i}$ values (M_i is any given cation, say Na^+) for a wide variety of exchangers (see Fig. 156). Vertical lines are drawn through these points and on these lines are laid off to scale other cation $\lg K_K/K_M$ values of the respective exchanger. The points in the direction along the arbitrary curve corresponding to each given ion assume regular positions such that lines, called selectivity isotherms, can

246

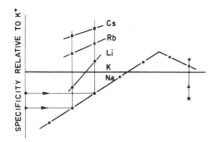

Fig. 156. Eisenman's cation selectivity isotherms (for details see text). The branches on the arbitrary sodium curve have been drawn to avoid intermingling of two sequences with the same K/Na selectivities, bearing in mind the single-value character of the selectivity dependence on the exchanger properties.

be drawn through them. A family of such curves has been obtained, for instance, for ion-exchanging glass and collodion films (see Fig. 157). Since the very existence of the selectivity isotherms points to a definite relationship between the complexing constants for the different cations, it follows that one cannot arbitrarily change the stability of complexes of only

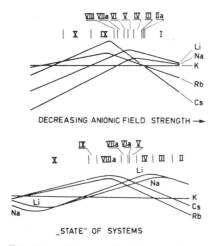

Fig. 157. Cation selectivity isotherms of "inorganic" (above) and biological (below) cation exchangers according to Eisenman [227—230].

one cation without changing the stabilities of all the others. Moreover, for each class of ion exchangers, if one knows the binding selectivities for two or three ions, one can with the aid of the isotherms determine the cation selectivities for the entire sequence.

It turned out that such systems of isotherms can also be plotted for a wide range of biological objects (Fig. 157). The selectivity data corresponding to these isotherms are values obtained in studies of the ion exchange properties of proteins, in determinations of the cation dependence of enzymic activities, in permeability studies for different biological membranes, etc. The biological isotherms differ markedly from the "inorganic" ones, although in both cases the selectivity sequences are practically within the set obtained in the theoretical calculation. The main difference between the two systems lies in the difference between numerical selectivity values (for the

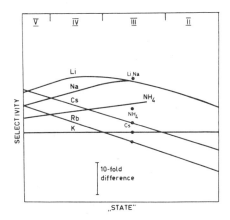

Fig. 158. A segment of the "biological" isotherms (see Fig. 157) on which points have been laid off corresponding to experimental lg $K^{pot}_{M_iK}$ values for valinomycin-treated bovine brain lipid bilayers [235, 541].

same sequences). In other words among the silicate glasses there are not, and apparently cannot be, such that could pertain to, say, sequence III and have the same K/Na selectivity as biological objects corresponding to this sequence. Such a difference points to the apparent existence of certain specificities in the structure of the ion exchange sites of biological cation-dependent systems. Unfortunately, despite the important part played by sodium and potassium ions in many physiological processes (see, for example, review [1078]), we still know practically nothing about the molecular organization of the binding sites for these ions in excitable membranes, K,Na-dependent transport ATPase, etc. It is therefore not surprising that the attention of investigators was attracted by the almost ideal correspondence of selectivity data on the valinomycin-modified lipid bilayers with the biological isotherms (see Fig. 158; cf. Part V.C). Can one propose, on these grounds, that there must be structural similarity between valinomycin and the ion exchange sites of those biological objects pertaining to the given region of the isotherms? This question was considered by Krasne and Eisenman [234, 511] who proposed that the selectivity of multidentate cation-binding macrocyclic compounds and biological objects be compared not only with respect to alkali ions, but also with ammonium and transition metal (Tl^+, Ag^+) ions. To a certain approximation it can be said that selectivity in the alkali metals is determined by the size and accessibility of the inter-ligand cavity; the relative affinity for the transition metals by the electronic structure of the ligand groups (ligand "softness" [338, 339, 873]); and the NH_4^+ specificity to a considerable degree by the geometry of their disposition. Therefore, these more comprehensive spectra of ionic selectivity better reflect the "individuality" of the cation-binding molecule and they can be used as a sort of fingerprint for comparative purposes. Such a comparison of macrocyclic complexones and the resting neural membrane is presented as an illustration in Fig. 159, which shows Krasne and Eisenman's calculations of the cation binding energies in aqueous solutions with reference to the Cs^+ ions. It can be clearly seen that with respect to the Rb^+/NH_4^+ and Rb^+/Tl^+ selectivities both valinomycin and nonactin differ quite definitely from the neural membrane and that the ion exchange properties of the latter are most closely approximated by cyclic polyether (274). On this basis the authors suggested that six ether or "ether-like" oxygen atoms take part in the coordination of ions in the "K^+ channel" of the neural membrane. Such ether-like ligands they were inclined to see in the amide carbonyls of protein molecules [234]. In this connection it is interesting that, judging from the data of Wieland et al. [1069], the amide complexone, antamanide, displays the same selectivity sequence $Tl^+ > K^+ > Rb^+ > NH_4^+$ as cyclic polyether (274). It should be mentioned, however, that the set of multidentate complexones which has been investigated is still rather limited and one cannot be sure that no other system of ligands with similar "fingerprints" exists. Interestingly, in their ionic selectivity, cyclic polyethers whose complexes possess a planar arrangement of the oxygen atoms disclose a certain similarity also with the "Na^+ channel" of the neural membrane. It is well known that the tetrodotoxin-sensitive "Na^+ channels" are capable of transmitting hydra-zinium, hydroxylammonium and guanidinium ions as well as alkali metal ions (see, for example [485]). At the same time, as has been mentioned above, among the known macrocyclic complexones only the monocyclic polyethers form complexes with these cations. However, one could hardly consider this to be more than an outward resemblance. Indeed, on interaction with cyclic polyethers the above-named cations behave as NH_4^+ analogs, whereas in the "Na^+ channel" they imitate to all appearances the partially hydrated sodium ion.

Unfortunately, one cannot state as yet whether the alkali metal ion binding mechanism of the macrocyclic complexones holds for membranes, cation-dependent enzymes, etc., or if it is characteristic of only such specific metabolic products as antibiotics or antamanide. It would, however, have been strange if Nature had not used more widely this mechanism in which

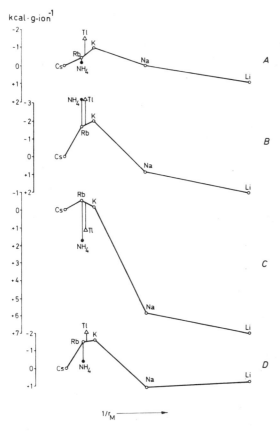

Fig. 159. The free energies of cation binding in aqueous solutions calculated by Eisenman and Krasne [511] for the cyclic polyether (274) (A), nonactin (B), valinomycin (C), and for the resting membrane of a frog nerve (D).

the properties of the cation exchanging site are so strongly dependent upon the conformational characteristics of the entire molecule. It is very easy to picture a protein whose structure provides the conditions for the drawing together of appropriately oriented ligand carbonyl or other groupings. The binding of the cation in such a multidentate center could change the conformation of the protein globule and thus allosterically affect the enzymic and other properties of the protein. On the other hand, conformational changes accompanying, say, phosphorylation, protonation or deprotonation of the protein ionogenic groups, could have a strong bearing on the affinity of the multidentate center for cations. For all that, in order to change the cation selectivity of such a center, in principle, significant conformational rearrangements are not even necessary because the selectivity is determined not only by the number and spacing of the ligand groupings but also by the rigidity of the entire molecule. In the case of a protein globule the rigidity can vary within wide limits, as, for instance, by changing the degree of dissociation of the side groups. It is therefore quite natural to expect the occurrence of multidentate binding centers in systems responsible for the active ion transport (for instance, in K,Na-dependent ATPase) or for the controlled passive transport across excitable membranes; in short, in those cases where there are grounds to assume that the specific binding of alkali metals is determined by (or vice versa determines) the state of the macromolecules.

Finally, one can also picture to oneself a situation such that part of the ligands filling the cation coordination sphere belong to the macromolecule and part to a low molecular weight substrate. In this case complexation will serve as the driving force for binding such a substrate. Possibly this type of interaction occurs in systems responsible for the coupled transmembrane transport of sodium ions and monosaccharides (see, for instance [643]). An interesting hypothesis has recently been advanced by Urry [1026, 1032] according to which the binding of calcium ions in the polypeptide helices of elastin molecules initiates the calcification process (cf. [1040]).

Ion—dipole interactions in the complexes of macrocyclic peptides and depsipeptides give rise to the notion that the chemically inert peptide groups of the protein skeleton can play an important part in the functioning of the active centers of enzymes. Indeed, these groups form a system of dipoles which produce a microheterogeneous electrical field within the protein molecule. Such a field can be likened to a mountainous terrain with its peaks, its valleys and its gulfs. Naturally, the pK values of ionogenic groups surrounded by the amide dipoles will depend on the magnitude of the local potential (cf. pp. 214 and 315). Moreover, translocation energy of the ions participating in the catalytic act should depend on the electrical potential profile along their way. This equally holds for charge separation, i.e. for displacement of the electron density in the substrate molecule or in the transition complex. However, until recently, it was quite unknown whether the "dipole" potential profile in the enzymic active center is, in fact, in conformity with the catalytic mechanism and what contribution this factor makes to the energetics of its elementary stages. A first step in the elucidation of this problem has been made by Johannin and Kellershohn [457]. On the basis of the known three-dimensional chymotrypsin structure, these authors showed that the electrical potential at the site occupied by the N_2 atom of the His-57 residue is lower by 200—600 mV* than that at the sites of the Asp-102 carboxyl and the Ser-195 hydroxyl, and that, consequently, on transferring a proton from any of these groups to the imidazole grouping 5—14 kcal·mole^{-1} are gained. Thus the electrical field induced by the amide dipoles is undoubtedly an essential factor in the proton transfer system of chymotrypsin.

---

* Depending on the selected value of the dielectric constant of the protein (3 and 1, respectively).

Chapter IV

CURRENT TRENDS IN THE APPLICATION OF MACROCYCLIC COMPLEXONES IN CHEMISTRY AND TECHNOLOGY

Although few papers have been devoted to the practical application of macrocyclic complexones, there are no doubts as to their wide potentialities in various areas. From this standpoint, of particular interest are the readily available synthetic cyclic polyethers, characteristic features of which are outstanding chemical and thermal stability [744]. The complexones can be varied to give a wide range of cation selectivities and complex lipophilicities. Moreover, practically any functional groups can be incorporated into their molecules, especially if the latter contain aromatic rings. Some of the derivatives can be used as monomers for polymerization or poly-condensation, or can be grafted onto other polymers.

Of course, it is difficult at present to predict all possible uses of the macrocyclic complexones in organic and analytical chemistry, in extraction technology and in other fields of pure and applied science. In this chapter we shall therefore limit ourselves to a consideration of only a few of the most promising trends, with the hope that acquaintance with them will stimulate the interest of the reader in further work in this area.

The first and most evident application of complexones is for selective extraction of alkali and alkaline earth metal salts by organic solvents. Among the compounds suitable for this purpose are cyclic polyethers whose molecules contain hydrophobic radicals. The distinguishing feature of extraction by neutral complexones is that the anion is carried into the organic phase along with the cation. This provides fundamentally new ways for controlling multistage extraction processes, as the degree of extraction depends upon the lipophilic anion content of the aqueous solutions. The neutral complexones can also be utilized reversely, namely, in selective extraction of lipophilic anions from alkaline or neutral solutions.

Another important application of neutral macrocyclic complexones is based on their ability to solubilize salts of the complexable cations in weakly polar solvents. Its importance to preparative organic chemistry is hard to overestimate. Numerous reactions which must currently be carried out in hardly appropriate or not readily available solvents, and sometimes under heterogeneous conditions can, by means of the complexones, be carried out homogeneously in non-polar media. For instance, Sam and Simmons [847] have found that 30—60 mM $KMnO_4$ can be dissolved in benzene with the aid of cyclic polyether (274). The solution is sufficiently stable at temperatures below 50°C (its half-life at 25°C is about two days). When used in stoichiometric amounts the solubilized $KMnO_4$ practically quantitatively oxidizes various olefines, arylaliphatic compounds, alcohols and aldehydes. Such a procedure is not only very convenient from a preparative standpoint, but can apparently be used for the quantitative determination of the corresponding functional groups (cf. [133]).

As another example can serve Fraenkel and Pechhold's method of preparing carbanions in homogeneous solutions of weakly polar solvents [269], based on the ability of cyclic polyether (274) to solubilize potassium phenylazoformiate (reaction A) and potassium methoxide (reaction B).

$$C_6H_5-N=N-COO^-[(274) \cdot K^+] \xrightarrow[\text{boiling}]{\text{THF}} C_6H_5^-[(274) \cdot K^+] + N_2 + CO_2 \tag{A}$$

$$C_6H_5CO-N=N-tert\text{-}C_4H_9 \xrightarrow[\text{benzene}]{CH_3O^-[(274) \cdot K^+]} tert\text{-}C_4H_9^-[(274) \cdot K^+]$$

$$+ C_6H_5COOCH_3 + N_2 \tag{B}$$

Thus the macrocyclic complexones can successfully compete with lipophilic organic cations of which the salts also possess enhanced solubility in organic solvents and are employed in reactions such as permanganate oxidation [287], borohydride reduction [981], nucleophilic substitution by phenoxide ions [1022], etc.

The salt solubilities in cyclic polyether-containing non-polar solvents considerably increase in the presence of some polar additives, particularly methanol (see Table 70). According to Pedersen and Frensdorff [744] it is possible to obtain, for example, molar solutions of KOH or NaOH in a 99:1 benzene—methanol mixture containing compound (274). This effect, which is more strongly pronounced with small "hard" anions, is undoubtedly caused by methanol participation in the solvation of both anions and complexed metal ions.

Also, by augmenting the concentration of relatively hydrophilic anions in the non-aqueous phase of two-phase water—organic solvent systems (see Part II.B), macrocyclic complexones, like lipophilic cations, such as triethylbenzyl-ammonium, can undoubtedly be used as so-called phase transfer catalysts which promote reactions between anions and water-insoluble compounds by conveying the former from the aqueous into the organic phase. Such catalysts can be employed in low concentrations if the anionic reactants are more lipophilic than the anionic products and the system does not contain highly lipophilic foreign anions. In recent times phase transfer catalysis is gaining more extensive application in multifarious reactions such as generation of dihalo- and thiohalocarbenes [465, 585, 586, 588, 591, 964], alkylation of compounds with active methylene or methine groups [35, 36, 389, 453, 464, 523, 580—584, 587, 589, 590], permanganate oxidation [1048], benzoine condensation [957], borohydride reduction, hydrolysis of esters and chlorides, deuterium exchange in ketones [964], etc.

It should be stressed that the reactivity of anions is determined to a large extent by whether they are in the free state or in ion pairs, and also depends

TABLE 70

THE EFFECT OF METHANOL (250 mM) ON ALKALI HALIDE SOLUBILITY IN ORGANIC SOLVENTS CONTAINING 50 mM COMPOUND (274) [744]

Solvent	Methanol	Solubility in mM*				
		NaCl	NaBr	KCl	KBr	KI
C_6H_6	—	0.01	1.8	0.03	2.3	9.2
	+	0.48	24	8.7	30	46
CCl_4	—	0.03	2.7	0.6	4.1	0.8
	+	1.1	28	8.8	34	15
$CHCl_3$	—	1.8	37	21	41	43
	+	5.7	41	34	44	44
CH_2Cl_2	—	1.8	35	17	41	43
	+	5.8	42	33	42	44
Tetrahydrofuran	—	0.02	1.2	0.1	3.6	45
	+	0.04	5	0.4	13	50

* The solubility values in these experiments could not exceed 50 mM, being limited by the initial amounts of the salts.

on the structure of these ion pairs. Therefore, the presence of solubilizing cations (whether tetra-alkylammonium ions or macrocyclic complexes) affects the reaction rate in non-polar solvents, not only because of increased concentration of the reacting anions but also due to changes in the corresponding rate constants.

Since ion pairs dissociate more and more as the effective cation radius increases, complexation may considerably accelerate reactions which proceed more rapidly with the free anions. Thus, according to the data of Smith and Hanson [955], on introducing a stoichiometric amount of cyclic polyether (270) to a solution of the sodium salt of fluorenone oxime in a 2:1 acetonitrile—tert-butanol mixture the rate of its alkylation by methyl iodide increases 14-fold, the alkylation rate constant becoming independent of the Na salt concentration, whereas the yield ratio of the resulting O- and N-methyl derivatives approaches that found by extrapolation to infinite dilution in the absence of cyclic polyether. It seems as if the increase in dissociation of the ion pairs (or possibly higher associates) can also be mobilized for explaining the ability of cyclic polyether (259) to prevent inactivation of potassium *tert*-butoxide as catalyst, which takes place in its concentrated dimethylsulfoxide solutions.

In other cases the complexone-induced reaction acceleration is due to increased reactivity of the anions which remain in ion pairs but are now coupled to the bulky complexes. Indeed, on interaction of ion pairs with electrophilic centers of organic molecules the energy spent in charge separation on forming the activated complex is less the larger the effective cation radius* [733]:

$$M^+X^- + \underset{|\ \ |}{Y-Z} \rightarrow M^+X^{\delta-} \cdots \underset{|\ \ \ \ |}{Y} \cdots Z^{\delta-} \rightarrow \rightarrow \rightarrow$$

This can be illustrated by the data of Zaugg *et al.* [1106] who investigated the effect of cyclic polyether (274) on the rate of alkylation of the sodium salt of *n*-butylmalonic acid diethylester by *n*-butyl bromide in tetrahydrofuran. In this and other such reactions the activating effect of the complexones resembles the effect of adding efficiently cation-solvating solvents that promote transition of the contact ion pairs to solvent-separated pairs (cf. p. 93). Clearly the reason for the increased anion reactivity is in both cases one and the same, viz. the increased distance from the metal ion.

The peculiarity of the complexones is here manifested in that "separation" of the ion pairs occurs when the former are added in stoichiometric amounts, and the properties of the complexone-separated ion pairs are standard and solvent-independent, since the effective radius of the complexes is unilaterally determined by the structure of the macrocyclic compound. Besides aprotic polar additives, cations of large radius, for instance cesium or tetra-alkylammonium, are frequently used as catalysts of reactions occurring in non-polar media with the participation of ion pairs (see, for instance [219, 475, 492, 510, 1108]). Doubtlessly, in many cases macrocyclic complexones could successfully compete with these cations as catalysts since they would greatly simplify and standardize the reaction procedures. They are of considerable interest, in particular as regulators of anion polymerization rates (cf. [993, 994]).

The above-numerated factors—solubilization of salts in non-polar media and decrease in the ion pairing of the reacting anions or increased reactivity of the pairs—make possible the achievement of truly amazing results with the

---

* This holds for the cases when the cation does not take part in polarization of the Y—Z bond.

complexones. Thus, Pedersen [737, 738] has observed for $[(274) \cdot K^+] OH^-$ in solution in benzene a rapid acyl—oxygen cleavage of the esters of 2,4,6-trimethylbenzoic acid, which are distinguished by their outstanding resistance to saponification.

An especial reservation must be made for anions with greatly delocalized charge. Such anions are usually highly reactive in the form of contact ion pairs, the reactivity diminishing with increase in cation radius and on passing to solvent-separated pairs. From the above it is therefore not surprising that complexing in weakly polar solvents is sometimes accompanied by a fall in anion reactivity. For instance, in tetrahydrofuran, where the alkali metal salts of fluorene form contact ion pairs, cation binding by cyclic polyethers (270) and (271) inhibits proton transfer from 3,4-benzofluorene to the fluorenyl anions [402, 1093].

Not only can the increase in the effective cation radius and displacement of the anion from the metal ion coordination sphere inhibit or accelerate carbanion reactions, but they can also change the direction and mechanism of the reactions. With respect to cation complexes of cyclic polyethers this has been demonstrated by Cram and coworkers [10] on the example of deuterated (—)-1-methyl-3-*tert*-butylindene, which isomerizes to (+)-1-*tert*-butyl-3-methylindene (see Fig. 160). Concurrently with this reaction,

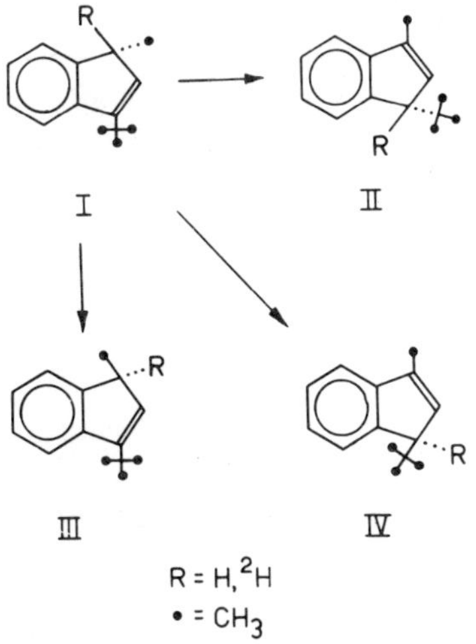

R = H, ^{2}H
• = CH$_3$

Fig. 160. The transformation pathways of (—)-1-deuterio-1-methyl-3-*tert*-butylindene (I, R = ^{2}H) on heating its solutions in potassium phenoxide-containing benzene—phenol (3:1) mixture. (I, R = ^{2}H) → (I, R = H) hydrogen—deuterium exchange with retention of configuration; (I, R = ^{2}H) → (I, R = H) + (III, R = H), hydrogen—deuterium exchange with racemization; (I, R = ^{2}H) → (II, R = H, ^{2}H), stereospecific isomerization coupled with partial hydrogen—deuterium exchange; (I, R = ^{2}H) → (II, R = H) + (IV, R = H), non-stereospecific isomerization coupled with deuterium loss.

isotopic exchange takes place with retention of the asymmetric center configuration. The stereospecificity of these processes is due to the fact that both exchange and proton (deuteron) migration proceed within the confines of the contact ion pair coordinated with phenol and phenolate molecules (see Fig. 161; cf. [169]).

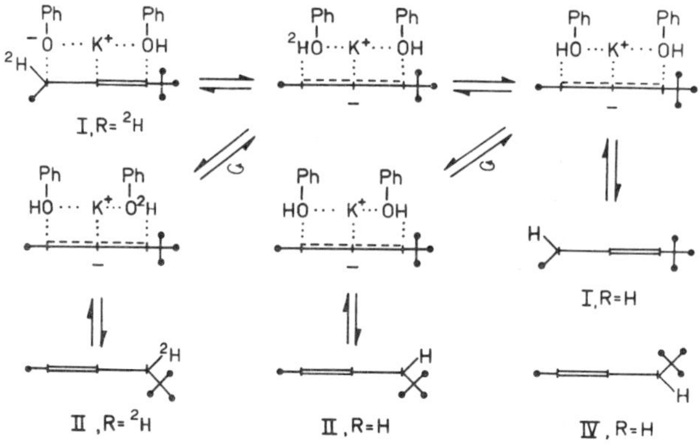

Fig. 161. The probable mechanisms of the reactions shown in Fig. 160 (view from the aromatic ring side). Compound (IV, R = H) could be produced exclusively by proton attachment to the free carbanions.

In the presence of compound (274) not only the rate but also the stereospecificity of these reactions diminishes, because the formation of an ensemble of reacting particles with the complex cation, similar to that shown in Fig. 161, becomes impossible. An effect of similar nature is observed with (−)-phenyl-4′-diphenylmethoxydeuteromethane

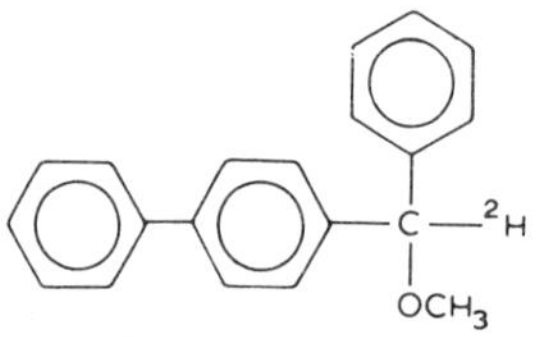

which in potassium tert-butoxide-containing tert-butanol solution exchanges its deuteron for a proton. At 70°C the rate of exchange is about 50-fold that of racemization. Both processes are accelerated in the presence of cyclic polyether (274), the ratio of their rate constants being close to unity. Roitman and Cram [820] who studied this reaction believe that the increased racemization is here also due to the impossibility of forming a reaction ensemble like the one shown in Fig. 161 (cf. [1090]).

Cyclic polyethers or other macrocyclic complexones can be utilized not only as tools for probing into the finer details of carbanion reaction mechanisms. In a number of cases the ability of these compounds to change the stereochemistry of the products is of certain preparative value. Thus, elimination of the acid elements from the 7-bromide or 7-tosylate of 1,1,4,4-tetramethylcyclodecane leads to the formation of a mixture of *cis* and *trans* olefines, the relative proportions of the geometrical isomers depending upon both the nature of the cation and the solvating capacity of the medium.

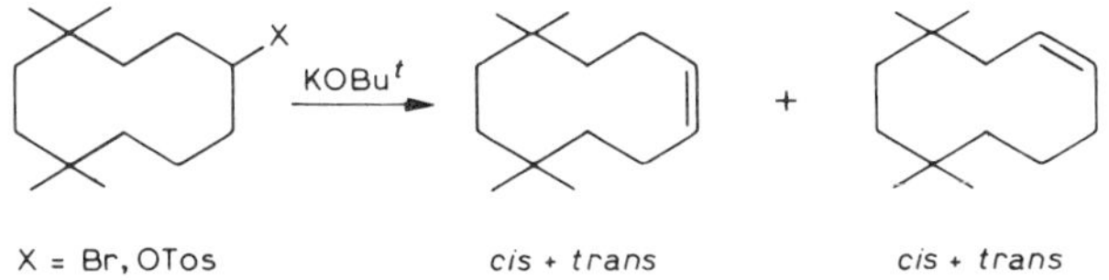

X = Br, OTos   *cis + trans*   *cis + trans*

The Czechoslovak workers who studied this and similar reactions arrived at the conclusion that the formation of *trans*-olefines proceeds via a transient complex with a contact ion pair [986, 1107]:

In weakly polar media, where potassium *tert*-butoxide is in the form of a contact ion pair, *trans* olefines are predominant, their relative yield decreasing with increase in solvent polarity. The *trans/cis* yield ratio dramatically diminishes on addition of equimolar amounts of cyclic polyether (274), the relative growth in the *cis*-isomer yield increasing with increase in solvent polarity (see Table 71) [985]. Evidently the cause of these effects is again the complexing-induced "separation" of the ion pairs. This principle of stereochemical control can apparently be extended to other reactions with participation of contact ion pairs.

If the conversion of carbanions proceeds in several parallel directions, the presence of macrocyclic complexes (as well as of large cations; see, for instance, [176]) can affect the relative yields of the not only stereo-chemically, but also structurally differing products. Thus, Bartsch and Wiegers [49] have recently studied the effect of cyclic polyether (274) on toluene sulfonic acid elimination from 2-phenylcyclopentyltosylate in a potassium *tert*-butoxide containing *tert*-butanol solution. They found that in the presence of stoichiometric (with respect to butoxide) amounts of complexone the reaction is greatly accelerated, the main product (90%

TABLE 71

THE EFFECT OF CYCLIC POLYETHER (274) ON THE YIELD OF *trans*-OLEFINES IN THE REACTION OF 7-BROMO (X = Br) AND 7-TOSYLOXY (X = OTos) DERIVATIVES OF 1,1,4,4-TETRAMETHYLCYCLODECANE WITH POTASSIUM *tert*-BUTOXIDE [985]

X	Solvent	Compound (274)	*Trans*-isomer formation in % of total olefine yield	*trans/cis* ratio
Br	Benzene	—	97	32
		+	9.9	0.1
	tert-Butanol	—	88.2	7.3
		+	10	0.1
	Dimethylformamide	—	8	0.09
		+	3.8	0.04
OTos	Benzene	—	97.6	40
		+	12.5	0.15
	tert-Butanol	—	81.4	4.4
		+	25	0.3
	Dimethylformamide	—	43	0.77
		+	6.1	0.06

yield), 1-phenylcyclopentene, giving way to 3-phenylcyclopentene (70% yield). As another example can serve the data of Staley and Erdman [959] who studied the rearrangement and cleavage of 6-methyl-6-phenylcyclo-hexadienyl anions in liquid ammonia. The ratio of the conversion products depends on the nature of the cation and changes noticeably when a stoichiometric amount of cyclic polyether (274) is added (see Fig. 162). In particular, the yield of 2-methylbiphenyl diminishes from 24 to 6% on passing from Li^+ to K^+, but rises again to 22% in the presence of the complexone. This can apparently be explained by assuming that, like lithium

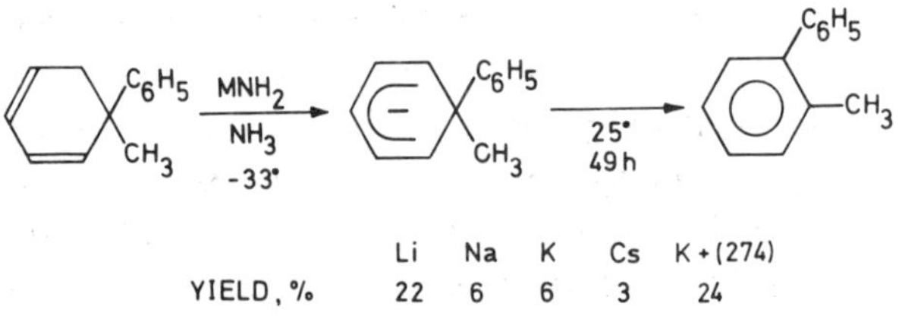

Fig. 162. The effect of cyclic polyether (274) on isomerization of 1-methyl-1-phenyl-cyclohexa-2,4-diene [959]. Other reaction products than those shown here include toluene, biphenyl and isomeric 1-phenylethylcyclopentadienes.

ions, complex cations form "separated" ion pairs with carbanions, whereas under the same conditions potassium ions form contact ion pairs.

Interesting possibilities arise from the ability of cyclic polyethers to complex protonated primary amines. In particular, it is highly tempting to employ the ethers for amino group protection from acylating or alkylating reagents in non-aqueous solutions. One may expect that in the presence of cyclic polyethers an apparent increase in basicity of the amines would take place, similar to the apparent decrease in pK_a of acids in the binding of lipophilic anions by cyclodextrins [170].

In some cases metal ion complexing by neutral macrocyclic compounds can be used for shifting chemical equilibria. For instance, Dye $et\ al.$ [220] found that in the presence of cyclic polyether (274) a considerable increase in concentrations of M^- anions and solvated electrons occurs in diethyl ether or tetrahydrofuran solutions contacting with potassium or cesium mirrors, owing to a shift to the right of the equilibria: $M_{2n}^\circ \rightleftharpoons nM^+ + nM^-$ and $M^\circ \rightleftharpoons M^+ + e^-$. Reactions which are usually carried out in solutions of alkali metals in liquid ammonia could apparently be performed under such conditions.

Shchori and Jagur-Grodzinsky [892] have recently shown that cyclic polyethers are strong bases in non-polar media. For instance, compound (274) when added in stoichiometric amounts to a chloroform solution of HBr causes its almost complete dissociation with the formation of the complex salt $[(274){\cdot}H^+]\,Br^-$ (in the presence of Br_2 the salt $[(274){\cdot}H^+]\,Br_3^-$ is formed). Thus, cyclic polyethers can sometimes be used for increasing the anion concentration in non-polar media also in the absence of complexable metal ions. Moreover, owing to their basicity cyclic polyethers are capable of inhibiting reactions proceeding with the participation of non-dissociated acid molecules, as for instance, the HBr catalyzed bromination of $trans$-stilbene [891].

Interesting possibilities are opened up by the synthesis of polymers containing macrocyclic complexone residues. Such polymers constitute an entirely new type of ion exchanger. The binding of cations by the complexing polymers converts the latter into a kind of polycation. As a result, these materials are actually "salt exchangers", that is they combine the properties of both cation and anion exchangers. Indeed, a polymer containing macrocyclic complexone residues is capable of extracting the salt of the complexable cation from solution, acting as a mixture of cation and anion exchangers in the H^+ and OH^- forms, respectively. Being "charged" with the $M_1^+X^-$ salt, on equilibration with a solution of the salt of another complexable cation $M_2^+X^-$ it can exchange M_1^+ ions for M_2^+ ions, i.e. it can function as a cation exchanger. Finally, on contact with a solution of the salt of a non- (or poorly) complexable cation $M_3^+ Y^-$, such a "charged" polymer would carry out anion rather than cation exchange. Particularly attractive is the preparation of the salt exchangers based on bicyclic nitrogen-containing polyethers of the type of compound (324), because the cryptates they form are stable in aqueous solutions. Furthermore, complexation in that case depends upon the pH of the solution, providing additional possibilities for ion exchange control. Undoubtedly, polymers carrying monocyclic poly-ether residues will be more readily available, but they will apparently be suitable for functioning mainly in non-aqueous (for instance, alcoholic) solutions.

Complexing polymers are of interest not only as ion exchangers. One may cite, for instance, the work of Feigenbaum and Michel [257] who showed that in films from cyclic polyether-containing polyamides (see p. 57) a considerable portion of the complexone residues can bind alkali metal chlorides without the films losing their mechanical or dielectric properties. One may calculate on finding among such "salt filled" polymers new materials with enhanced thermostability and valuable optical and ablational properties.

This naturally far from complete enumeration of the possible applications of macrocyclic complexones in science and technology already gives a clue to the fruits to be gained from further development of this field. In this chapter we have not touched upon the practical aspects of the membrane activity of these compounds. These questions will be elaborated in the following chapters.

Chapter V

THE ACTION OF COMPLEXONES ON ARTIFICIAL MEMBRANES

V.A. Introduction

The rapid progress in the structural and conformational elucidation of naturally occurring macrocyclic and pseudocyclic complexones has been largely the outcome of interest in the mechanism of their membrane activity. Indeed, valinomycin, nigericin and other antibiotic complexones were among the first compounds in which the ability to increase the cation permeability of both biological and artificial membranes was revealed. The action of these antibiotics springs from the high lipophilicity of their complexes with alkali metal ions so that in the bound form the latter can be conveyed across the hydrophobic membrane. Particularly enticing to researchers has been the fact that these complexones could induce alkali cation transport across such popular models of biological membranes as lipid bilayers (in the form of both planar films and liposomes).

At present complexone-mediated cation transport has become an intensively developing field of biophysics, part of a wide front of investigations into biologically active compounds whose role consists of increasing the permeability of membranes. Among these compounds there are, for instance, proton transporting agents or protonophores (dinitrophenol, 2-trifluoro-methyltetrachlorobenzimidazole, etc.), polyene antibiotics (nystatin, amphothericin B, etc.), organomercury compounds (dibarenyl mercury) promoting transport of anions, etc. It should be noted that no other membrane-modifying agents have as yet been found that can impart to the membranes such exceptional ability to discriminate between alkali metal ions as the complexones discussed here.

An important role in these studies belongs to model lipid membranes. For the first time the experimentalist has received into his hands systems whose properties approach those of biomembranes and at the same time are amenable to quantitative study of transport processes and contain as the permeability inducing component molecules of known structure. This has stimulated the elaboration of theoretical concepts of induced ion transport on the one hand and has made possible direct experimental proof of these concepts on the other. Real perspectives have now been opened up for the use of membrane-active compounds, including complexones, in systems simulating specific biological membrane functions such as excitability and active ion transport. Still another, no less important, direction in the model studies is detailed analysis of the factors determining the behavior of membrane-active molecules in the membrane (for instance, the effect of lipid structure and packing) so that the permeability induced by these molecules could serve as a parameter reflecting given properties of the membranes. An important asset is the convenience with which the relation between molecular structure and transport-inducing activity can be studied on model lipid membranes, so that these membranes may serve as a proving ground, as it were, in the preparation of substances with predetermined biological action.

Whilst bimolecular lipid membranes modified by complexones are of interest mainly in connection with biological problems, liquid membranes containing these substances are suitable for use in cation-sensitive electrodes. Such electrodes displaying high ion selectivity will undoubtedly find wide application for the potentiometric determination of alkali, alkaline earth and ammonium ions in hydrochemical investigations, in analysis of biological fluids, etc.

Following Pressman [790], valinomycin, nigericin, gramicidin and other antibiotics promoting cationic permeability of membranes are often referred to as ionophores, i.e. ion carriers. Despite its popularity we will refrain from using this term mainly because it does not always give a justifiable impression of the mode of induced cation transport.

V.B. Liquid membranes

The liquid membrane is a layer of water-immiscible solvent separating two aqueous solutions. As such layers various workers have used high-density organic solvents in U-tubes and similar contrivances [23, 751, 784, 790, 791], the solvent-filled space between two cellophane films [17, 540, 542, 543, 1010, 1012], or, finally, solvent-impregnated porous films of inert material (glass, polythene, polyvinylchloride, millipore filters, thixotropic gels, etc.) [248, 569, 753, 811, 966, 967, 1081—1084]. Such membranes possess a very high electrical resistance and, as a rule, low ionic selectivity (see, however, [248, 419]).

The liquid membranes containing neutral macrocyclic complexones display a noticeably higher conductance providing the aqueous solutions contain complexable cations. Moreover, such membranes have a cation function, their selectivity approaching that for the complexing reaction of the macrocyclic compounds. Studies in this direction were begun in 1966 by Simon *et al.* [966] and have led to the production of cation-sensitive electrodes suitable for practical purposes [167, 270, 477, 478, 515, 755, 866, 937] (see also reviews [102, 150, 165, 940]).

Valinomycin-containing membrane electrodes have an exceptionally high K/Na selectivity ($K_{KNa}^{pot} \leqslant 10^{-4}$), much higher than the best glass electrodes ($K_{KNa}^{pot} \geqslant 6 \cdot 10^{-2}$, [227]). With their aid, one can measure in, say, 0.1 M NaCl solution, potassium ion concentrations of the order of 10^{-5} M. The application of "valinomycin" electrodes for determination of potassium in soils, sea water, biological fluids, etc., is described in the papers [18, 44, 651, 652, 754, 812, 865, 1020, 1085]. Of considerable interest are electrodes for determining the intracellular activity of potassium ions. For such micro-electrodes a finely drawn out end of a glass capillary is made hydrophobic and filled with a valinomycin solution in a water-immiscible organic solvent [486, 487, 1041]. In ordinary macroelectrodes, the membrane as a rule is a porous film impregnated with a valinomycin solution. The potential lg $vs.$ a_{K^+} curves of freshly prepared electrodes in $10^{-4}-10^{-1}$ M KCl solutions are straight lines, the slopes of which approach the thermodynamic value [145, 270, 522, 755], diminishing, however, in the course of time [110, 652]. As a result the lifetime of such an electrode is limited to several weeks. The degraded electrode can easily be restored by replacing the membrane. A diminishing of the slope is also observed in solutions containing lipophilic anions (I^-, NCS^-, Ph_4B^-) [145, 522, 755]. The pH sensitivity of the electrode is evidently dependent upon the nature of the porous support. Thus, according to published data, a valinomycin-containing membrane in an Orion electrode Ser. 92 displays a relatively small K/H selectivity [270, 522], whereas when millipore filters are used the K_{KH}^{pot} value can attain a value of $5 \cdot 10^{-5}$, permitting measurements in the pH range 1—10 [755] (cf. [543]).

An electrode, the porous membrane of which is impregnated with nactin solution, while displaying considerably less K/Na selectivity, has been shown by Scholer and Simon [866] to be very convenient for determining ammonium ions. For solutions of ionic composition close to blood serum (150 mM sodium, 5 mM potassium) this electrode permits measurements to be made in the pNH_4^+ range of 1—2.4 so that it can be used for clinical analysis of urea [866] and arginine [680], following treatment of the sample with urease or urease plus arginase, respectively.

The Beckman company manufactures cation-sensitive electrodes with "solid" membranes which, judging from their properties, apparently contain valinomycin and nactins [167, 515]. Possibly of similar type are the polymer films recently proposed by Petranek and Ryba [749]. Such films are obtained by evaporating a cyclohexanone solution of polyvinylchloride containing as additive a dipentyl phthalate solution of the macrocyclic complexone. According to these authors the selectivity of the valinomycin-

or nactin-containing films does not differ significantly from that of the impregnated porous membranes, but the former considerably surpass the latter in the rate at which the potential is established.

Membrane electrodes specific for alkaline earth metal ions can also be made with the aid of neutral complexones. As Ba^{2+}-sensitive electrode Levins [545] has employed a porous membrane impregnated with a p-nitroethylbenzene solution of the complex of $Ba(Ph_4B)_2$ with the

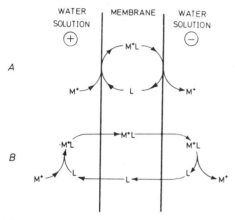

Fig. 163. Possible mechanisms for the electrophoretic transport of cations coupled with circulation of the neutral carrier molecules. Complex formation and breakdown are taking place: A, at the membrane—aqueous solution interfaces ("small carrousel"), or B, in the non-stirred pre-membrane layers of the aqueous solution ("big carrousel").

non-ionic detergent Igepal CO-880 (nonylphenoxypolyethyleneglycol, molecular weight approx. 1500). A Ca^{2+}-specific electrode has been proposed by Ammann et al. [11, 12], utilizing for the membrane filter paper impregnated with a neutral acyclic complexone (340) solution (see p. 216).

Some properties of complexone-containing membrane electrodes are summarized in Table 72.

The mode of ion transport across the liquid membrane containing neutral complexone is still very unclear. Very likely cations are transported within such membranes in the form of complexes, whereas on their surface the mediating factor is the heterogeneous reaction $M_w^+ + L_o \rightleftharpoons (M^+L)_o$. It is also possible that the cations may become attached to the complexone molecules in the non-stirred pre-membrane layer and will penetrate the membrane already in the complexed form. In the first of these limiting cases cation transport is accompanied by circulation of the complexone molecules within the membrane (see Fig. 163A), and this should be more expressed the weaker the convection and the slower the intercomplexone exchange of

TABLE 72

CATION SELECTIVITIES ($K_{M_jM_j}^{pot}$) OF COMPLEXONE-CONTAINING LIQUID MEMBRANES

Complexone (concentration)	Solvent	Porous film	M_j												References
			Li⁺	Na⁺	K⁺	Rb⁺	Cs⁺	NH₄⁺	H⁺	Mg²⁺	Ca²⁺	Sr²⁺	Ba²⁺	
Valinomycin ($9 \cdot 10^{-3}$ M)	diphenyl ether	millipore filter*	$2 \cdot 10^{-4}$	$2.5 \cdot 10^{-4}$	1	1.9	0.4	$1 \cdot 10^{-2}$	$5 \cdot 10^{-5}$	$<2 \cdot 10^{-4}$	$2 \cdot 10^{-4}$			[755] (cf. [248])
Valinomycin (5–10%)	nitrobenzene, bromobenzene, diphenyl ether	membrane electrode Orion (ser. 92)†		$7.5 \cdot 10^{-5}$	1		~1	$2 \cdot 10^{-2}$	$~1 \cdot 10^{-2}$	$<2 \cdot 10^{-4}$				[270]
Biosynthetically prepared mixture of nactins (saturated solution)	tri-(2-ethylhexyl) phosphate	millipore filter	$3.5 \cdot 10^{-2}$	$1.7 \cdot 10^{-2}$	1	0.36	$4 \cdot 10^{-2}$	8.3	0.13	$<2 \cdot 10^{-3}$	$2 \cdot 10^{-3}$			[866]
Benzo-15-crown-5 (246)‡	nitrobenzene	membrane electrode		$3 \cdot 10^{-2}$	1	0.32	0.14	$5.15 \cdot 10^{-2}$						[811]
Dibenzo-18-crown-6 (270)‡	nitrobenzene	membrane electrode		$4.15 \cdot 10^{-2}$	1	0.35	0.27	$6.25 \cdot 10^{-2}$						[811]
Dicyclohexyl-18-crown-6 (274)‡	nitrobenzene	membrane electrode		$4 \cdot 10^{-2}$	1	0.36	0.12	0.17						[811]
Dibenzo-30-crown-10 (303)‡	nitrobenzene	membrane electrode		$1.75 \cdot 10^{-2}$	1	1.15	0.48	0.21						[811]
(Igepal CO-880)·[Ba(Ph₄B)₂]₇	p-nitro-ethylbenzene	membrane electrode§	$<1 \cdot 10^{-4}$	$<1 \cdot 10^{-4}$	$1 \cdot 10^{-3}$			$1 \cdot 10^{-4}$	$<1 \cdot 10^{-4}$	$<1 \cdot 10^{-4}$	$<1 \cdot 10^{-4}$	$<1 \cdot 10^{-3}$	1	[545]
Compound (340)	p-nitro-ethylbenzene	filter paper	$3 \cdot 10^{-3}$	$7 \cdot 10^{-3}$	$9 \cdot 10^{-2}$	0.2	$6 \cdot 10^{-2}$			$<3 \cdot 10^{-5}$	1	$1 \cdot 10^{-2}$	$8 \cdot 10^{-2}$	[12]

* Membrane resistance approx. 10^6 ohms; time of establishment of the potential, 10 sec (± 1 mV), 60 sec (± 0.1 mV).
† Time of the potential establishment, 30 sec.
‡ Mean values.
§ Time of establishment of the potential, 10–180 sec; potential drift, 1–2 mV per day.

cations. Markin *et al.* [594] have termed such complexone movement within the membrane the "small carrousel". If, however, the charge transport across the membrane surface is mediated by the complexes and the fluxes of free molecules between the aqueous solutions and the non-stirred pre-membrane layers are less than their fluxes between these layers and the membrane, conditions arise for the functioning of the "big carrousel", the circulation of the complexone molecules now extending beyond the membrane into the non-stirred pre-membrane layers (see Fig. 163B).

Sometimes the participation of complexes in transmembrane cation transport can be directly demonstrated. Thus Wipf *et al.* [1081, 1082], using

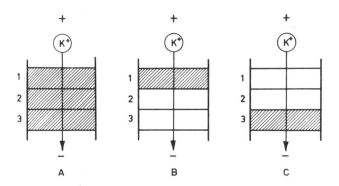

Fig. 164. Transfer of a labeled complexone in liquid membranes consisting of three contingent films of porous polyvinylchloride impregnated with a 6 mg/ml octanol-2 solution of nactins preliminarily equilibrated with a dilute aqueous solution of HCl (pH 3.5) [1082]. The hatched areas represent films that initially contained the ^{14}C-labeled complexone. Under these conditions the potassium ion transport number is 0.42 for an applied membrane voltage of 30 V. Experiment A: Label transfer from film 1 to film 3 corresponds to the number of transported potassium ions. Experiment B: Label transfer from film 1 to film 3 corresponds to $\frac{1}{4}$ to $\frac{1}{6}$ the number of transported potassium ions. Experiment C: Reverse transfer of the label from film 3 to film 1 corresponds to about $\frac{1}{10}$ the number of transported potassium ions.

labeled nactins and valinomycin, showed that electrophoretic transport of potassium ions under conditions excluding intramembrane convection is accompanied by generation of an oppositely directed concentration gradient of the complexone in the membrane. This bears evidence of potassium current in the membrane phase as the flow of complex cations. Experiments with initial asymmetric label distribution across the membrane indicate that potassium is exchanged between the complexone molecules (see Fig. 164). This can be the consequence of direct interaction of the complexes with free molecules or can result from reversible dissociation of the former in the membrane phase. The latter process seems to be the more probable, since in these experiments the membranes contained a relatively polar solvent (octanol-2, preliminarily equilibrated with dilute hydrochloric acid). As indirect evidence in favor of the diffusion mechanism of intramembrane ion transport can also serve the data of Eyal and Rechnitz [248] who worked with porous membranes impregnated with a solution of valinomycin in diphenyl ether. Cooling such membranes to below the solvent melting point strongly increased their resistance and caused a sharp fall in cation selectivity. At the same time, the functioning of complexone-containing "solid" membranes of plasticized polymers is difficult to explain in terms of such a mechanism. It would be interesting to determine the diffusion coefficients of the complexed and free macrocyclic molecules in such membranes. Apparently they should not differ significantly from values for the usual solutions. In any case, films in which the macrocyclic metal-binding compounds are wittingly restrained by incorporating them into the polymer chain, retain a low conductivity even for rather high degrees of complexation [257] (see p. 57).

Based on the supposition that mobile complexes are the only current carriers within a membrane and that exchange currents at the membrane—water interfaces are much stronger than the transmembrane current, Ciani, Eisenman and Szabo [148, 230] derived Eqn 29 for the membrane potential in the presence of neutral complexones. This equation was obtained without any special assumption regarding electroneutrality of the membrane phase.

$$\Delta V = \frac{RT}{F} \ln \frac{\sum\limits_{i=1}^{n} \frac{P_i}{P_j} \cdot a'_{M_i}}{\sum\limits_{i=1}^{n} \frac{P_i}{P_j} \cdot a''_{M_j}} + \frac{RT}{F} \ln \frac{(c'_L)_{tot}}{(c''_L)_{tot}} +$$

$$+ \frac{RT}{F} \ln \frac{1 + \sum\limits_{i=1}^{n} (K_w)_i \cdot a''_{M_i} + \sum\limits_{i=1}^{n} \sum\limits_{x=1}^{m} (K_w)_i \cdot K'_{M_iLX} \cdot a''_{M_i} \cdot a''_X}{1 + \sum\limits_{i=1}^{n} (K_w)_i \cdot a'_{M_i} + \sum\limits_{i=1}^{n} \sum\limits_{x=1}^{m} (K_w)_i \cdot K'_{M_iLX} \cdot a'_{M_i} \cdot a'_X} \tag{29}$$

where $P_i = u_{M_iL} \cdot k_{M_iL} \cdot (K_w)_i$; u_{M_iL} is the mobility of the complex cation M_i^+L in the membrane phase; $'$ and $''$ are indices for solutions on opposite sides of the membrane; R is the gas constant; F is the Faraday number; and T is temperature in $°K$.

If the complexone concentration is the same on both sides of the membrane and there is little complexation in the aqueous solution, the membrane potential is determined by the first term of Eqn 29. If, moreover, the solutions contain only two complexable cation species, Eqn 29 simplifies to Eqn 30:

$$\Delta V = \frac{RT}{F} \ln \frac{a'_{M_i} + \frac{P_j}{P_i} \cdot a'_{M_j}}{a''_{M_i} + \frac{P_j}{P_i} \cdot a''_{M_j}} = \frac{RT}{F} \ln \frac{a'_{M_i} + K^{pot}_{M_i M_j} \cdot a'_{M_j}}{a''_{M_i} + K^{pot}_{M_i M_j} \cdot a''_{M_j}} \tag{30}$$

Since these are all grounds to consider the complexes of different cations with valinomycin or nactin-like macrocyclic compound as practically isosteric (i.e. $u_{M_i L} \cdot k_{M_i L} \approx u_{M_j L} \cdot k_{M_j L}$; cf. p. 229), the membrane selectivity determined as $K^{pot}_{M_i M_j} = P_{M_j}/P_{M_i}$ should, in accord with Eqn 30, be close to the complexing selectivity $K^{M_j}_w / K^{M_i}_w$ in aqueous solutions. This prediction is usually confirmed when comparing the $K^{pot}_{M_i M_j}$ values and the corresponding stability constant ratios [235, 248, 811]. From Eqn 29 it follows that the slope of the ΔV vs. $\lg a_M$ curves should diminish with increase in complexation and may even become negative*. In fact, in solutions with high concentrations of well complexing cations the membrane potential is markedly below the thermodynamic value [110, 270, 755]. Hence, regardless of the assumptions made by Ciani et al., the equations they have derived for given conditions satisfactorily describe the properties of liquid membranes containing neutral complexones. Consideration of Eqn 29 shows that the selectivity of such membranes in dilute solutions calculated with the aid of Eqn 30 should be independent of the cation concentration ratios. A study of the properties of some "valinomycin" electrodes has shown, however, that the $K^{pot}_{K N a}$ value increases with increasing $a_K / a_{N a}$ ratio (and is also dependent on the absolute values of a_K and $a_{N a}$ [110, 145, 522]). A similar behavior of $K^{pot}_{K M_i}$ has been observed for the Rb^+, Cs^+, NH_4^+, Tl^+ and Ag^+ ions.

---

* The reason of the decrease and even sign reversal of the slope for the dependence of ΔV on $\lg a_M$ with increase in concentration of the transported cation is the enhanced complex formation. In the range of $a_M < 1/K_w$ the concentration of the complexes and cations in the aqueous solutions are linearly correlated so that ΔV is described by Eqn 30, independent of whether cations in the membrane are transferred in the form of complexes or by the heterogeneous complexing reaction. At $a_M > 1/K_w$ the contribution of the heterogeneous reaction diminishes (cf. p. 284), and the ratio of the complex concentrations on either side of the membrane approaches the ratio of the complexone concentrations. At the same time the formation of $(M^+L)X^-$ type ion pairs in the aqueous solutions brings about a fall in activity of the complexes, which is more the higher the concentration of the respective anions.

Within the framework of the theory of Ciani *et al.* the cation selectivity of the membranes is determined by the ratio of the fluxes of the respective ion complexes induced therein. However it is difficult to reconcile this conclusion with the results of studies on the electroconductivity of complexone-containing membranes. Such studies were initiated by Tosteson and Andreoli [17, 1010, 1012] who have investigated the electric parameters of phospholipid-containing liquid hydrocarbon (heptane, decane) membranes. In order to determine the specific resistance of the bulk organic phase (R_b) and the "surface" resistance (R_s) associated with charge transfer across the interfaces these authors made use of the empirical linear membrane resistance *vs.* thickness (*l*) relation in the form of $R = R_b \cdot l + 2R_s$. The value of R_s is determined by extrapolation to zero thickness, and of R_b from the slope of the R *vs.* l curve. By means of this method it could be demonstrated that valinomycin and other neutral complexones significantly promote potassium ion transfer across the membrane—water interfaces. These findings were subsequently confirmed by Tosteson [1011], who showed that valinomycin considerably augments the potassium self-diffusion coefficient at the interface between a heptanic phospholipid and KCl aqueous solutions. Valinomycin was, however, found to have a relatively small bearing on the "bulk" resistance R_b of the membrane. At first sight this result appears quite surprising. In fact, if one assumes in conformity with extraction data that a large portion of the valinomycin molecules is present in the membrane as complexes, then R_b should be about 10^5—10^6 ohm · cm, which is about two—three orders of magnitude below the experimental values. The weak effect of valinomycin on the "bulk" resistance is apparently due to a large proportion of the complexes being associated into electrically neutral ion pairs. Similar results have been obtained by Lev *et al.* [540, 542, 543] who studied the properties of valinomycin-containing liquid hydrocarbon membranes in the absence of phospholipids. These authors paid particular attention to the cation selectivity of such membranes. They found essential differences between the $1/K_{KNa}^{pot}$ values and the conductivity ratios g_K/g_{Na}*. With increase in valinomycin concentration to 10^{-4} M, $1/K_{KNa}^{pot}$ increases to approx. $1.5 \cdot 10^3$†, whereas the membrane conductivity ratios for

---

* Data on the current—voltage characteristics of liquid membranes containing neutral macrocyclic complexones are to be found in [17, 542, 1012]. Under symmetrical conditions (0.1 M KCl/0.1 M KCl) valinomycin-containing heptane membranes possess ohmic characteristics up to 0.3 V. A saturation effect is observed at higher voltages. The conductivity of such membranes is little dependent on the KCl concentration (slight maximum at 0.1 M) and is independent of the NaCl concentration. Under asymmetric conditions (0.1 M KCl/0.1 M NaCl) a very faintly expressed rectification appears.

† In the valinomycin concentration range of 10^{-8}—10^{-4} M the K/Na selectivity of the membranes is described by the expression $1/K_{KNa}^{pot} = 1.8 \cdot 10^4 \cdot (c_L)^{0.43}$.

KCl and NaCl solutions differ only very little from unity (1.4—3.0). Utilizing Tosteson and Andreoli's procedure, Lev *et al.* found that the presence of valinomycin tells mainly on the magnitude and cation selectivity of the "surface" resistance of membranes, whereas their "bulk" resistance is practically independent of the nature and concentration of the cations in the aqueous solution ($R_b^{NaCl}/R_b^{KCl} \approx 2$; see Fig. 165, left) or of the valinomycin concentration in the membrane. It thus seems as if the cation selectivity of

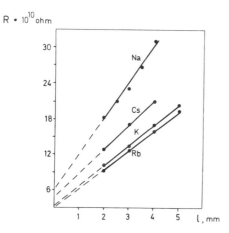
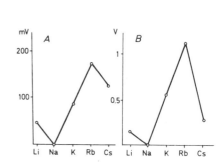

Fig. 165. Properties of heptanic valinomycin-containing membranes. Left, Electrical resistance—thickness relation for membranes from a 10^{-4} M heptanic valinomycin solution, separating 0.1 M aqueous alkali metal chloride solutions [542]. Right, Correlation between bi-ionic potentials (*A*) and interfacial potentials (*B*) in the system 10^{-6} M valinomycin in heptane: 10^{-4} M alkali metal chlorides in water [276].

"valinomycin" membranes is determined by the stage of ion exchange between membrane and water solutions. In this connection of considerable interest are data on the effect of the neutral macrocyclic complexones on the interphase potential between the non-polar organic solvent and aqueous solutions of alkali metal salts. Studies in this direction were first reported by Simon and coworkers [756, 757, 1083] on electrochemical cells of the type Ag ∥ complexone in CCl_4 ∥ 0.1 M MCl, 0.1 M NH_4NO_3, $KCl_{sat.}$, Hg_2Cl_2, Hg. The authors determined the emf changes in such a cell on passing from LiCl to the other alkali metal chlorides and showed that the magnitude of these changes correlates with the complexing selectivity. Recently Frumkin *et al.* [276] investigated the effect of the cation species on the emf of the cell Au ∥ air ∥ valinomycin in heptane ∥ aqueous solution MCl, $KCl_{sat.}$, Hg_2Cl_2, Hg. With heptane containing 10^{-6} M valinomycin and the 0.1 M KCl solution the potential jump at the heptane/water interface attains a value of 3 V (*plus* in heptane). The dependence of the interphase potential on the cation species correlates well with the bi-ionic potentials (see Fig. 165, right). It was found that the potential with KCl solution increases in the presence of heptanoate ions (pH $>$ 4) and decreases in the presence of non-dissociated heptanoic acid (pH $<$ 4). These data lead to the speculation that the main contribution to the measured potential difference is made by adsorption on the interface of the positively charged K^+·valinomycin complexes. The adsorption of these complexes is facilitated by surface-active anions whereas neutral surface-active molecules displace the complexes from the interface.

In view of the above, it is natural to assume an insignificant diffusion component of the membrane potential, the latter being determined mainly by ion exchange equilibria at the membrane—water interfaces. From this standpoint the data of Lev *et al.* [542] are of interest. They found that the K/Na selectivity of valinomycin-containing heptane membranes strongly diminishes in the presence of multivalent Th^{4+} ions. This effect is apparently due to adsorption of Th^{4+} on the membrane, consequently to decrease in the surface concentration of valinomycin complexes.

The question of why the "bulk" resistance of complexone-containing liquid membranes is only weakly dependent on the nature and concentration of the cations and just what particles are the current carriers obviously deserves further study. Within the framework of the theory of Ciani *et al.* the membrane conductivity is independent of the rate of processes on the membrane surface, so that the ratio of the "bulk" conductivities in KCl and NaCl solutions should be close to the permeability ratios:

$$\frac{g_K}{g_{Na}} \approx \frac{P_K}{P_{Na}} \tag{31}$$

If, however, the requirement of electroneutrality of the organic phase and the anionic contribution to the "bulk" conductivity are taken into consideration the conductivity ratios are described by Eqn 32.

$$\frac{g_K}{g_{Na}} \approx \frac{u_{KL} + u_X}{u_{NaL} + u_X} \cdot \left(\frac{k_{KL} \cdot K_w^K}{k_{NaL} \cdot K_w^{Na}} \right)^{1/2} \approx \left(\frac{K_w^K}{K_w^{Na}} \right)^{1/2} \tag{32}$$

But, as has been noted above the g_K/g_{Na} values of valinomycin-containing non-polar membranes do not differ significantly from unity, i.e. are much smaller than the values predicted by Eqns 31 and 32 ($\sim 10^4$ and $\sim 10^2$, respectively). If, as is generally accepted, the current is considered to be mainly due to mobile complexes one of the possible reasons for the divergence is that the intramembrane concentration of complexes not associated into ion pairs cannot be described by the theory of Eisenman *et al.* (see Part II.B). In fact, assuming that on equilibrating an aqueous solution of radioactive alkali metal chlorides with a heptane solution of valinomycin, the concentration of the label in the latter corresponds to the overall concentration of the free and ion-paired complexes, Lev *et al.* [889] were able to show that this overall concentration is proportional to $(a_{MCl})^{1/2}$ rather than $(a_{MCl})^n$, where n varies from 1 to 2 (cf. Eqn 16 on p. 106). With increase in polarity of the organic phase and lipophilicity of the anions, the complex concentration in the organic phase and the electrochemical properties of the corresponding liquid membranes begin to approach the values predicted by Eqns 16 and 29 [542].

However, the divergences could possibly have been due to formation of water-containing micelles by valinomycin in decane or heptane equilibrated with the aqueous solutions. Grounds for such a view have been furnished by Tosteson *et al.* [1011] who showed that this antibiotic increases the solubility of water in heptane. If this is so then a certain and perhaps even considerable proportion of the ions in such a solution might be not so much in the form of complexes as enclosed within the micelles and solvated by water molecules (cf. [698]). It is also of interest to ascertain whether current carriers in heptane solutions too are micelles in which anions are surrounded by complexes. It is well known that micelles of this kind are formed in non-polar media if the cations are much larger than the anions [746].

The complexone-induced ion fluxes across liquid membranes are coupled either electrically (valinomycin-type compounds) or directly via a common transport mechanism (nigericin-type compounds). If lipid-soluble permeant anions (X^-) are present in solutions bathing the neutral-complexone containing membrane, the equilibrium distribution of the anions and complexable cations (M^+) in the absence of an external electrical field is given by Eqn 33 [23, 751].

$$\frac{a'_M}{a''_M} = \frac{a''_X}{a'_X} \quad \text{or} \quad \Delta pM + \Delta pX = 0 \tag{33}$$

where

$$pM = -\lg a_M \,;$$
$$pX = -\lg a_X \,.$$

Obviously, at sufficiently high initial M^+ or X^- gradients, the corresponding counter-ion can be carried through the membrane against its own concentration gradient. In liquid membranes of high dielectric constant, wherein the $(M^+L)X^-$ ion pairs are dissociated, transmembrane MX transport proceeds as parallel, electrically coupled flows of anions and complexed cations. Very likely, with non-polar membranes electrically silent flow of the ion pairs plays a considerable part in the complexone-induced salt transport*.

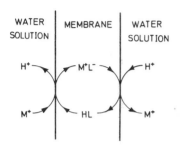

Fig. 166. Probable M^+–H^+ exchange mechanism across liquid membranes containing antibiotic complexones of the nigericin group.

The nigericin antibiotics induce M^+–H^+ or M_i^+–M_j^+ exchange across liquid membranes, the rate of which is independent of the electrical field or of the anion species in the solutions. This effect can be rationalized in terms of only neutral complexes of the type M^+L^- and non-dissociated HL molecules being present in the membrane. One of the possible exchange flow diagrams is represented in Fig. 166 (cf. Fig. 163). Here the equilibrium distribution of the complexable cations M_i^+ and M_j^+ or cations and protons is described by Eqns 34 and 35 respectively [23, 790].

$$\frac{a'_{M_i}}{a''_{M_i}} = \frac{a'_{M_j}}{a''_{M_j}} \quad \text{or} \quad \Delta pM_i - \Delta pM_j = 0 \tag{34}$$

$$\frac{a'_M}{a''_M} = \frac{a'_H}{a''_H} \quad \text{or} \quad \Delta pM - \Delta pH = 0 \tag{35}$$

The data on the rate of M^+–H^+ exchange permit assessment of the complexing selectivity. The selectivity sequences thus obtained for nigericin,

---

* Such ion pair participation probably occurs in the case of salt transport across chloroformic membranes containing di-(4-aminophenyl)-methane. This compound forms crystalline complexes with sodium salts, the structure of which is very similar to that of salts of the macrocyclic complexes (cf. p. 213). The membranes are permeable to NaCl but not permeable to NaBr [304]. Since bromide is a more lipophilic anion than is chloride it would have been natural to expect the reversal relation. Therefore, it must be assumed that NaCl traverses the membrane in the non-dissociated forms $(M^+L_n)Cl^-$ or, perhaps, $(M^+L_n \cdot H_m^+)Cl_{m+1}^-$ which owing to the lesser anionic radius is more stable than NaBr counterparts.

monensin and dianemycin (see Fig. 167) coincide with those found in extraction experiments.

The question of whether L^- anions and/or M^+LH complexes are present in membranes containing nigericin-like antibiotics, however, cannot be considered to have been unequivocally answered. Lutz *et al.* [569] in a

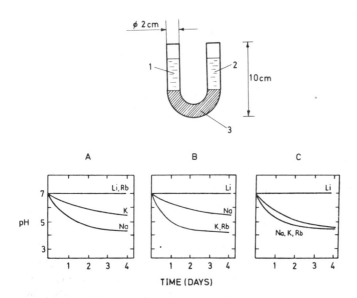

Fig. 167. Kinetics of M^+-H^+ exchange across liquid membranes containing antibiotics of the nigericin group (A, monensin; B, nigericin; C, dianemycin) [23]. The U-tube contains 1 M and 0 M aqueous solutions of the salt MCl (solutions 1 and 2, respectively), separated by a $2 \cdot 10^{-7}$ M solution of the antibiotic in chloroform (3), which is stirred with a magnetic stirrer. The curves represent changes in the pH of solution 1.

study of the properties of nigericin and monensin found that in the electrochemical cell Ag, AgCl; 10^{-2} M NaCl ‖ $3 \cdot 10^{-2}$ M antibiotic in decanol-1 ‖ 10^{-2} M MCl ‖ decanol-1 ‖ 10^{-2} M NaCl; AgCl, Ag the emf correlates with the M/Na selectivity of the antibiotic (*plus* to the left if M^+ forms more stable complexes). It is as yet not very clear whether this effect is directly due to antibiotic-induced charge transport or is a consequence of the establishment of a pH gradient. If the latter is true one must assume that decanol membranes possess a proton function and that the buffer capacity of the aqueous solution in those experiments was of insufficient magnitude*.

---

* The aqueous solutions contained imidazole—HCl buffer (10^{-4} M, pH 6.5).

V.C. Bimolecular lipid membranes

V.C.1. The mode of action of macrocyclic complexones

Bimolecular membranes from phospholipids and certain other surface-active compounds are frequently used as models for studying the mode of action of substances affecting the permeability of biological membranes. Such bimolecular membranes or bilayers, as they are often called, are formed from films of a solution of the lipid in a non-polar solvent (usually heptane,

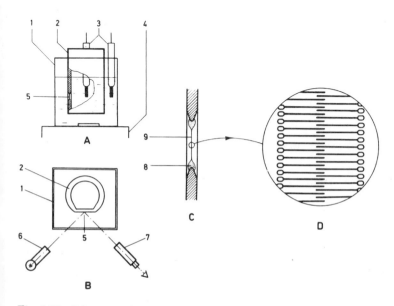

Fig. 168. Scheme of simplest cell for measuring the electrical characteristics of bimolecular membranes. A, Side view: 1, exterior glass beaker; 2, interior teflon beaker with specially made orifice; 3, AgCl—Ag electrodes; 4, magnetic stirrer; 5, hole closed by membrane. B, View from above: 6, light source; 7, microscope for observing the "blackening" of the membrane. C, Enlarged view of the membrane covering the hole: 8, torus of the lipid solution; 9, bimolecular membrane. D, Schematic representation of the orientation of the lipid molecules in the bilayer. The circles designate the polar heads; bold lines, the hydrocarbon chains.

decane or a chloroform—methanol mixture) pipetted or brushed on so as to close a specially made orifice in a lucite, polythene or teflon plate separating the aqueous solutions (see Fig. 168). In the spontaneous thinning out of these initially thick films a membrane is formed first appearing colored because of light interference and then giving rise to so-called secondary blackening, evidence of the formation of a 60—80 Å bilayer. A distinctive feature of the bilayer membrane from the standpoint of its serving as a model is the relative simplicity of measurement of its electrical parameters (membrane potential, resistance, capacitance, etc.) considerably facilitating study of ion transport mechanisms. In the absence of membrane-modifying agents the lipid bilayers possess a high specific resistance (up to 10^8 ohm · cm^2) and low ion selectivity. The structure and properties of the lipid bilayers and their use in the elucidation of induced ion transport mechanisms are discussed in detail in a number of reviews (see, for instance, [41, 387, 548, 671, 1007]).

The first reports on the ability of neutral macrocyclic antibiotic complexones to increase selectively the cationic permeability of bimolecular lipid membranes were published practically simultaneously in the beginning of 1967 by Lev and Buzhinsky [541] and by Mueller and Rudin [669]. These studies attracted the attention of numerous workers because the effects observed were extraordinarily large and specific*. In fact, the K^+ permeability of a membrane noticeably increases already in the presence of 10^{-11} M valinomycin and in concentrations of $10^{-7}-10^{-6}$ M the antibiotic lowers the resistance 10^4-10^5-fold. In subsequent years the action of macrocyclic complexones on bilayers became the object of research in many laboratories.

Despite the fact that many problems associated with this phenomenon remain unresolved to this day, it is doubtless that the underlying cause of the membrane activity of valinomycin, the nactins and other such compounds is their ability to bind alkali metal ions into positively charged lipophilic complexes which participate in cation transport across the hydrophobic interior of the bilayer. Not one of the synthetic analogs of valinomycin and the enniatins which was not able to form complexes exerts a noticeable effect on the cation permeability of the bilayers [897, 904]. At the same time, there is an evident similarity and, under certain conditions, even identity in the selectivity sequences of induced permeability and of complexation. Certain compounds possessing complexone properties increase the cation permeability of membranes only to a very slight extent (for instance, cyclic polyethers containing aromatic rings [232]) or are completely inactive (tri-N-desmethylenniatin B (135) [904]). The most probable reason for this is the poor membrane solubility of their free molecules or their complexes.

The formation of complexes exerts a very significant effect on the behavior of the lipid films. Thus, secondary blackening of the colored

---

* It is practically unknown whether macrocyclic complexones affect transport of solutes other than complexable metal ions, in particular non-electrolytes. Lippe [557] observed that enniatin A increases the permeability of lipid bilayers to thiourea.

membrane is inhibited by valinomycin and the nactins if the bathing solutions do not contain complexable ions. In this case the formation of bilayer regions on the membrane can be achieved only by the application of a direct or alternating electric field (100—500 mV). In solutions containing complexable cations the presence of these antibiotics does not have any bearing on the ease of formation of the bilayers [550, 990]*.

The action of macrocyclic complexones had been studied on bilayers prepared from highly varied lipids: lecithin [393, 669, 960, 962, 990], sphingomyelin [669], phosphatidylethanolamine [617], phosphatidyl-inositol [617, 960], phosphatidylglycerol and phosphatidylglycero-phosphate [617], lipids of the white matter of bovine brain [111, 198, 199, 541, 550, 626, 669], mitochondrial lipids [669], erythrocyte lipids [16, 960] and also on bilayers from oxidized cholesterol [669], 7-dehydro-cholesterol [617], oleylamine [991], mono- and diesters of glycerol with long chain fatty acids [147, 512]. As a rule, the increase in cation permeability is observed independently of the lipid species but the magnitude of the effect and its dependence on pH, ionic strength and on the transportable cation concentration is determined to a great extent by the molecular structures of the bilayer components. Resistance measurements of the modified membrane as a rule give well reproducible results. This enables one, after pertinent calibration, to use the bilayers for estimating the concentration of valinomycin, the nactins, etc., in the aqueous solutions (see, for instance, [461]).

On adding valinomycin, the nactins or enniatins (usually in methanol or ethanol) to aqueous solutions the membrane conductivity acquires practically steady-state values in the course of several minutes (in the case of valinomycin up to 40 minutes) and is independent of whether the active substance is on one or on both sides of the membrane [541, 669]†. Andreoli *et al.* [16] observed that in the absence of an ionic gradient the asymmetric introduction of valinomycin in NaCl solutions causes the appearance of a membrane potential (*minus* on the opposite side); however, this observation has so far not been confirmed. If the membrane-active complexone is added

---

* The inhibition of bilayer formation by valinomyc·n is possibly due to the ability of the compound to change the molecular packing in the water—lipid systems [263]. The incorporation of cation complexes into the membrane apparently counteracts this effect. Interestingly, in the presence of valinomycin analog (79) (see Table 2), the complexes of which are more surface active (see p. 136), the spontaneous blackening appears at lower KCl concentrations than in the case of valinomycin itself [625].

† The very slow fall in conductance which subsequently takes place is sometimes attributed to transition of the complexone from the aqueous solution to the lipid torus (see [961]). The part played by the torus in the complexone-induced conductance is treated in detail in [380].

directly to the lipid solution, stable conductivity values for the bilayer are established immediately after its formation. A comparison of the bilayer conductivity on adding valinomycin and nonactin in aqueous and lipid solutions (0.3% lecithin in decane) permitted determination of the partition coefficients of the antibiotics between these two phases ($2.5 \cdot 10^4$ and $6 \cdot 10^3$, respectively) [960].

The complexable cation gradient-induced membrane potential and the bi-ionic potentials depend on the complexone concentration and on the absolute cation concentrations in the aqueous solutions [541, 550]; however, the modified bilayers can be utilized as membranes with practically ideal cation function over a wide range of concentrations*. High K/Na selectivity and low resistance of the valinomycin modified bilayers makes convenient their use as electrodes for studying fast valinomycin-induced processes associated with potassium uptake or release by cells, mitochondria, etc. (see, for instance, [549]).

Table 73 lists the $K_{KM_i}^{pot}$ values for some complexones calculated from bi-ionic potential data for 0.1 M salt solutions when the selectivity sequences of the modified bilayers usually coincide with the complexing selectivity sequences for solution. It can be seen from the table that, as in the case of liquid membranes, the K/Na selectivity diminishes in the order valinomycin—nactins—enniatins; but in contrast to the liquid membranes the modified bilayers are usually characterized by close and sometimes even coinciding values of $K_{KM_i}^{pot}$ and of the respective zero current conductance ratios g_{M_i}/g_K [147—149].

When valinomycin-modified bilayers separate relatively dilute solutions of potassium and sodium salts the current—voltage characteristics display a distinct rectification effect [541, 669] which is only very weakly expressed in liquid membranes [540]. If experimental conductivity values obtained under the corresponding symmetrical conditions are used these current—voltage characteristics are well approximated by curves calculated according to Goldman's equation [111]. This indicates the absence of ion flux coupling in the membrane and is in accord with the close ratios for the conductivities and for the permeabilities.

The experimental data on the properties of macrocyclic complexone-

---

* The potassium transport numbers determined for the valinomycin-modified bilayers are usually equal to unity [16]. Recently Tosteson [1016, 1017] has shown that under certain conditions the current through valinomycin- or monactin-containing erythrocyte lipid bilayers can be threefold that of the potassium ion flux. The difference disappears in the presence of catalase or superoxide dismutase and also on freeing the aqueous solution of O_2. Such an effect is attributed to participation in the charge transport of superoxide anions apparently resulting from peroxide oxidation of lipids.

TABLE 73

COMPARISON OF BULK EXTRACTION CONSTANT RATIOS $(K_{w/o})^{Mi}/(K_{w/o})^K$
WITH RATIOS OF THE PERMEABILITIES P_{Mi}/P_K AND OF THE ZERO CURRENT
CONDUCTIVITIES $(g_0)^{Mi}/(g_0)^K$ CHARACTERISTIC OF LIPID BILAYERS
MODIFIED BY MACROCYCLIC COMPLEXONES [992]

Complexone	Cation	$(K_{w/o})^{Mi}/(K_{w/o})^K$	P_{Mi}/P_K	$(g_0)^{Mi}/(g_0)^K$
Nonactin	Li^+	$2.6 \cdot 10^{-4}$	$<10^{-3}$	$4.2 \cdot 10^{-4}$
	Na^+	$17 \cdot 10^{-3}$	$7.1 \cdot 10^{-3}$	$6.7 \cdot 10^{-3}$
	K^+	1	1	1
	Rb^+	0.47	0.58	0.58
	Cs^+	$6.1 \cdot 10^{-2}$	$3.3 \cdot 10^{-2}$	$3.9 \cdot 10^{-2}$
	NH_4^+	47	8	5
Valinomycin	Li^+	$7 \cdot 10^{-6}$	$4.7 \cdot 10^{-6}$	$<5 \cdot 10^{-3}$
	Na^+	$1.7 \cdot 10^{-5}$	$3.6 \cdot 10^{-5}$	$<5 \cdot 10^{-3}$
	K^+	1	1	1
	Rb^+	1.95	1.8	1.5
	Cs^+	0.62	0.76	0.25
	NH_4^+	$19 \cdot 10^{-2}$	$3.9 \cdot 10^{-2}$	—

modified bilayers can best be treated in terms of modern concepts on
induced ion transport. Cation transport across bilayers can be visualized as
the entirety of the consecutive stages of the cation movement across the
non-stirred pre-membrane layers of the aqueous solution, the membrane—
water interfaces and the internal hydrophobic zone. Evidently each stage can
in principle become rate limiting for the overall transport. The ratios of the
rates of the individual stages and hence the location of the "bottleneck"
depend, as will be shown below, upon the properties of both the membrane
and the environment, in particular on the composition of the aqueous
solutions. In any case, we have the right to assume that the metal ions within
the membrane are present only as lipophilic complexes with the macrocyclic
compound. Let us at first confine ourselves to neutral complexones which
form equimolar complexes with ions of univalent metals in both the
membrane and the aqueous solution.

First of all, two possible types of induced ion transport can be
distinguished. In one the current passage does not cause (nor, under
asymmetrical conditions, change) charged particle concentration gradients in
either the membrane or the non-stirred pre-membrane layers. In other words
the steady-state concentrations of the free cations and the complexes
approach and, in limit, coincide with equilibrium values, so that following
Eisenman et al. [147, 992], this region, independent of the type of
transport mechanism, can be termed the *equilibrium domain*. Deviations

from equilibrium values arise when the membrane current becomes limited by the rate of the reversible complexing reaction or by charge transport across the membrane—water interfaces. It can be seen in Fig. 163 (see p. 266) that, for instance, in the case of the "small carrousel", these conditions coincide. If the rate of cation transport across the membrane proper exceeds the diffusion rate in the non-stirred pre-membrane layers, the steady-state cation and complex concentrations in these layers will also differ from the equilibrium values. However, such concentration polarization of bilayers has been observed only with very dilute KCl solutions (10^{-5} M) containing valinomycin in a very high concentration (10^{-5} M) [550].

Let us first examine the behavior of a membrane in the *equilibrium domain*, assuming it to be a thin plate (thinner than the Debye length) of an isotropic, hydrophobic material in which the only charged particles are the macrocyclic complexes. Under these conditions the electrical properties of the bilayers are determined by the rate of charge transport within the membrane. Such transport can be the result of movement of the cationic complexes in the electric field or can proceed by a series of cation transfers between immovable complexone molecules (relay transfer), or, finally, it can be a combination of the one and the other of the limiting mechanisms, assuming that the complexes and free molecules capable of participating in the transfer interaction are moving about in the membrane. It should be noted that cation transfer between complexone molecules in hydrophobic media can occur only by direct contact and not via reversible dissociation, since the latter case would require exit of the cations into a non-polar environment. Obviously the relay mechanism can make a marked contribution to the conductivity only when the complexone molecules in the membrane are associated in the form of a sort of "conductivity chain", or when the cation displacement in the course of the exchange reaction is greater than the diffusional drift of the complex for the same time.

Issuing from the assumption that the complex concentration of the membrane is proportional to the concentration in the aqueous solution and that the contribution of intermolecular cation transfer is negligible, we should expect that in dilute solution the conductivity of the membrane in the *equilibrium domain* would increase linearly with increase in a_M, attaining a limiting value in the region of $a_M > 1/K_w$. Under such conditions a lowering of the free molecule concentration due to complex formation will have no bearing on the conductivity. For relay transport, on the contrary, the probability of the intermolecular cation transfer is proportional to the product of the free and complexed molecule concentrations. Hence, the conductivity will be proportional to the concentration of transported cations in the low concentration range where c_L^w is practically constant, but for

$a_M > 1/K_w$ the relative diminution of free molecules with increase in a_M outstrips the relative growth in the number of complexes, causing a decline in conductivity, which thus passes through a maximum. Such an effect will also be manifested when the relay contribution is determined by the probability of chain building by the free complexone molecules. We note that such probability, as also the probability of the bimolecular transfer reaction, is superlinearly dependent on the complexone concentration. It can, therefore, be expected that the potential contribution of relay transport and, correspondingly, the slope of the conductivity $vs.$ c_L^w curve would increase as the complexone concentration increases*.

Under certain conditions the transport "bottleneck" can be cation passage across the non-stirred pre-membrane layers or membrane—water interfaces. Such conditions, which Eisenman and his collaborators have termed the *kinetic domain*, are associated with the establishment of a certain concentration gradient of the complexes in the membrane or in the non-stirred layers. Let us examine the various mechanisms for translocating cations across the membrane boundary. First, it can be assumed that the cations penetrate the membrane as complexes which form and break down in the aqueous solutions. In that case the conductance in the *kinetic domain* is determined by the rate of formation of the complexes or their diffusion across the membrane—water interface. However, this mechanism is apparently not the main contributing factor to metal ion insertion into bilayers, at least in the case of valinomycin and monactin. In fact calculations by Ştark and Benz [960] have shown that the transmembrane current in the presence of these antibiotics exceeds the value corresponding to maximum possible complexation rate in aqueous solution without any diffusional limitations $(5 \cdot 10^{11}$ $cm^3 \cdot mole^{-1} \cdot sec^{-1})$. Hence, there are grounds to assume that there is another process which is the main contributor to the cation crossing the membrane boundary. Such is apparently the so-called heterogeneous complexing reaction, in the course of which complexone molecules localized in the membrane seize cations from the aqueous solution. Herein the high complexing rate is due chiefly to the

---

* If the partition coefficient of the complexes is sufficiently large, one may expect that with increase in a_M their intramembrane concentration will attain a limiting value, determined by the maximum attainable density of the space charge [684]. In principle, this effect can lead to the appearance of a plateau on the conductivity $vs.$ a_M curve in the range of $a_M < 1/K_w$. If the contribution of the relay mechanism is small, the height of this plateau will not increase with increase in complexone concentration (cf. [147]). If the contribution by the relay mechanism becomes noticeable, the height of the plateau will increase with increase in complexone concentration, the approach to the plateau being accompanied by concomitant decrease in the slope of the conductivity $vs.$ c_L^w curve.

fact that the complexone concentration in the membrane considerably exceeds its concentration in solution. In the region $a_M > 1/K_w$ the contribution of the heterogeneous complexing reaction naturally diminishes because of decreased free molecule concentration, so that the principal contribution to the trans-boundary charge flows is now due to the complexes.

When can the heterogeneous complexing reaction prove to be the "bottleneck", i.e. to proceed slower than intramembrane cation transport? First, this can, of course, occur when the rate constants of the reaction are relatively low. Also the rate of interfacial complex formation can be limited by supply of the free complexone molecules. Such diffusional limitations in the region $a_M < 1/K_w$ will arise, for instance, when the free molecules are much less mobile than the complexes and only slowly enter the membrane from the aqueous solution. At $a_M > 1/K_w$ the heterogeneous reaction is retarded owing to diminution of the free molecule concentration by complex formation. These factors may be responsible for the appearance of a plateau or maximum on the conductance *vs.* a_M curve, the position and width of which depend on the value of K_w and on the partition coefficient and mobility ratio of the free and complexed molecules in the membrane.

It should be pointed out that most workers now assume diffusion of the free complexone molecules in bilayers to proceed faster than their reversible transition from the membrane into the aqueous solution. This can be due to the relatively low speed of crossing the membrane boundaries or to diffusional restrictions in the non-stirred pre-membrane layers. If indeed this is so, then current passage through the bilayer should be accompanied by establishment of steady-state concentration gradients of the free molecules in either the membrane itself or the pre-membrane layers of the aqueous solution. At the same time, one may conclude from this that in the absence of current no considerable intramembrane free molecule gradient can appear in a bilayer separating solutions differing in the concentration of the complexone.

Let us assume that transfer of the complexes across the membrane—water interfaces is negligible and the cations penetrate the membrane by means of the heterogeneous complexing reaction. Then, in the absence of a complexable cation gradient, diffusion of the complexone through the bilayer can cause the appearance of a membrane potential providing it is accompanied by establishment of a sufficiently large intramembrane free molecule gradient. It is well known that addition of valinomycin or the nactins to one side of the bilayer does not lead to the appearance of an unambiguous potential difference of the expected sign (*minus* from the side of the complexone) (see, for example, [991]). Apparently this can be interpreted in two ways: either transfer of the complexes across the membrane boundaries is slow, or owing to one of the aforementioned reasons the intramembrane free molecule gradient is small.

The ratio of the rates of the elementary transport stages determines the type of mechanism underlying the behavior of the complexone as cation carrier. Besides the above discussed big and small carrousel mechanisms whereby circulation of the complexone molecules occurs, there can also be a non-circulatory mechanism, viz. so-called direct passage of the complexes. In the latter case the current carriers across the membrane boundaries are complexes which form and dissociate in the aqueous solutions, and the bottleneck of the free molecule transport is not diffusion across the non-stirred pre-membrane layers. Hence, on passage of current the generation of a free molecule gradient (which is a necessary condition for circulation of the complexone) becomes possible only by changing their concentration in the bulk aqueous solutions.

Already this purely qualitative consideration shows that contribution of the individual mechanisms and localization of the rate-limiting stages can vary over wide ranges, depending on the active compound and the transportable cation concentration. Therefore, establishment of concrete transport mechanisms for each complexone and quantitative assessment of the parameters of the individual transport stages is a complicated and still unresolved problem. Studies in this direction are based on a number of variously proposed mathematical models whereby the cation transport is attributed to mobile complexes [147—149, 232, 380, 530, 594, 597, 598, 684, 960, 991, 992], to a relay mechanism (see, for instance, [101, 140, 141]) or to a combination of both [6, 239]. In terms of these models one may make definite predictions regarding the membrane behavior in the *equilibrium* and *kinetic domains* and correlate the experimentally measured electrical characteristics of the bilayers (conductivity, bi-ionic potential, current—voltage characteristics, etc.) with the parameters of the individual transport steps. The main difficulty in such an approach is that very often the observed dependencies can be satisfactorily interpreted by alternate models (for instance, localization of the maxima on the zero current conductivity *vs.* a_M plots in the vicinity of $1/K_w$ [550]) and only in rare cases can an *experimentum crucis* be found wherein unequivocal inferences can be made as to the nature of the actual transport mechanism. A detailed discussion of this problem can be found in papers by Eisenman *et al.* [147, 992], Markin and Chizmadjev [596], Haydon and Hladky [380] and Läuger [529].

It should be mentioned that the proposed models of mobile complex-mediated cation transport differ in the manner of describing movement of the complex. The relation between the thickness of the bilayer and the size of the complexes is such that their passage through the membrane interior should involve relatively few diffusional jumps (see Fig. 169). Different authors regard this process either as movement in an isotropic medium, characterized by the corresponding diffusion coefficient, or as passage over a single activation barrier obeyed by Eyring's absolute reaction rate theory (cf., for instance, [597] and [530]). Hall *et al.* [348] showed recently that differences in the models depend upon the proposals concerning the number of activation barriers and the electrical potential profile in the membrane. According to these authors, best accord with experiment is achieved by assuming that the membrane potential barrier is of an approximately trapezoidal shape, the width of the flat part being about 7/10 the thickness of the membrane and its height, about 13 kT^*.

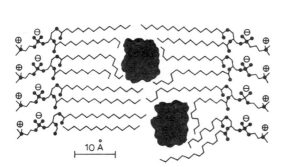 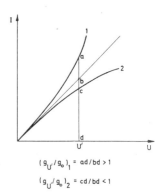

$$(g_U/g_0)_1 = ad/bd > 1$$

$$(g_U/g_0)_2 = cd/bd < 1$$

Fig. 169 (left). Cross section of a valinomycin-containing lecithin bilayer [326].
Fig. 170 (right). Concave (1) and convex (2) current—voltage curves. For describing the curvature of the $I-U$ curve Stark and Benz [960] have proposed the use of the parameter g_U/g_0, i.e. the ratio of the integral conductivity at the voltage U to the zero current conductivity. If $g_U/g_0 > 1$ the $I-U$ curve is concave (1); if $g_U/g_0 < 1$ the curve is convex (2). For ohmic characteristics $g_U/g_0 = 1$ over the entire range of applied voltages.

Analysis of the theoretical transport models which are based on the existence of only M^+L type complexes shows that for direct current the complexone-modified membrane can be represented by an equivalent electrical circuit made up of resistances connected in series, with their magnitude inversely proportional to the rates of cation passage across the membrane—water interfaces and the membrane interior (assuming the resistance of the pre-membrane layers to be relatively low). Classification as

---

* The membrane potential profile proposed by Hall *et al.* is in line with concepts on the dipole-induced boundary potential jump (see p. 301).

to *equilibrium* or *kinetic domains* is evidently determined by the relation between the "surface" and "bulk" resistances.

When the membrane current is determined by the "bulk" resistance, i.e. it is in the *equilibrium domain*, at low voltages the membrane displays concave* current—voltage characteristics ($I \sim ShnU$, where n ($0 < n \leqslant 1/2$) depends on the number and the shape of the activation barriers [348])†. Under these conditions the zero current conductance ratios for the various cations at $a_M < 1/K_w$ coincide with the permeability ratios as determined from data on the bi-ionic potentials. Moreover, if the hypothesis of complex isostericity (see p. 229) holds, these ratios should have values close to the complexing-constant ratios in the aqueous phase (cf. Eqn 28). Judging from the data of Eisenman *et al.* [235, 990, 992] and Stark and Benz [960], these conditions are complied with in valinomycin- or nactin-modified lecithin bilayers in KCl solution. The linear dependence of the conductivity on the antibiotic concentration under such conditions bears evidence of unimportant contribution by exchange reactions.

The transition to the *kinetic domain* occurs when the "bulk" membrane resistance becomes less than the "surface" resistance. In the *kinetic domain* the conductivity ratios differ from the permeability ratios in the region $a_M < 1/K_w$ and the current—voltage characteristics become convex [147, 960]. Such passing over to the *kinetic domain* is observed, for instance, in erythrocyte lipid bilayers modified by valinomycin. In that case, with increase in concentration of potassium ions, sign reversal of the curvature of the current—voltage characteristics occurs (see Fig. 171). In valinomycin-modified bilayers from bovine brain lipids, with increasing potassium ion concentration not only is there a transition from the concave to the convex current—voltage characteristics, but the conductivity passes through a maximum. The fall in conductivity following the maximum is due to a rise in the "surface" resistance connected apparently with diminution of the free valinomycin concentration [550, 625, 627].

Polarization effects in the *kinetic domain* are the cause of relaxation phenomena. If a low-frequency alternating field is applied to membranes operating in this domain then the resulting alternating intramembrane concentration gradients of the complexes will augment the membrane capacitance to above its geometrical value, but if the frequency is increased to beyond the point when the oscillation amplitude of the complexes (in the

---

* The concavity of the current—voltage curve is defined with reference to the current axis (see Fig. 170).

† It is to be noted that even in the *equilibrium domain* the concavity of the current—voltage curves should disappear when the complex concentration in the membrane becomes very large (for details see [596]).

general case, of the current carriers) becomes less than the membrane thickness, the contribution of these gradients will become so small that the membrane capacitance will approach the geometrical, and the measured conductance will approach the "bulk" conductance. In other words, in the *kinetic domain* one should expect membrane conductance and capacitance dispersion [6, 140, 239, 595, 596]. Such dispersion is in fact observed in

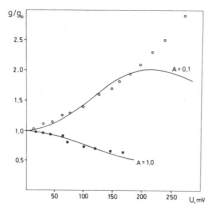

Fig. 171. Curvature sign reversal of current—voltage characteristics with increasing concentration of the transported cations [960]. Erythrocyte lipid bilayers are in 10^{-7} M valinomycin-containing KCl + LiCl solutions of constant ionic strength (1 M). The curves have been calculated according to equation

$$\frac{g_U}{g_0} = \frac{2\,(1 + A)\,\mathrm{Sh}\,\dfrac{U}{2}}{\left(1 + A \cdot \mathrm{Ch}\,\dfrac{U}{2}\right) \cdot U}$$

based on the Eyring model (see p. 286). The parameter A monotonically increases with increase in concentration of the transported ion. It is seen from the equation that at low voltages the I vs. U curves change from concave to convex in the region $0.1 < A < 1$. The observed divergences from the theoretical curves are apparently associated with the fact that in terms of the model the rates of formation and dissociation of the complexes on interfaces are independent of the electrical field strength. A thorough discussion of models which account for this factor is given in [596].

bovine brain lipid bilayers modified by valinomycin and some of its analogs when the current—voltage characteristics display a convex shape [625, 626] (see Figs 173 and 174 on p. 291). If one now measures the high-frequency conductance of the membrane at differing transportable cation concentrations, the character of the a_M-dependence of the "bulk" conductance can be elucidated. This method has been proposed for evaluating the contribution of the relay mechanism. It has already been mentioned above that for relay translocation of cations the "bulk" conductance of a membrane decreases in the region $a_M > 1/K_w$, whereas for carrier transport of the cations it should increase monotonously up to a certain limiting value. Hence the presence of a maximum on the high-frequency conductance $vs.$ a_M curve can be regarded as argument in favor of relay contribution to cation transport in the membrane interior [140, 239]. Regrettably, such measurements still encounter considerable methodological difficulties.

Another approach to the study of polarization effects involves measurement of current relaxation under voltage-clamp conditions. It has enabled Stark et $al.$ [962] to assess the rates of the individual stages of potassium ion transfer across a valinomycin-containing phosphatidylinositol membrane (see footnote to p. 293).

In a number of cases the transport mechanisms discussed above require certain refinements and additions. Thus, the behavior of bilayers modified by cyclic polyether (275) (see Fig. 13 on p. 54) can be explained by assuming that complexes of this compound present in the membrane have the composition M^+L_3 or M^+L_2, whereas in aqueous solutions the usual equimolar complexes M^+L are formed. The observed differences between the permeability ratios and conductivity ratios for various cations, depending on their concentration, could be interpreted, and the data used for calculating the complex stability constants in aqueous solutions. The values obtained were close to those determined by other methods. Moreover, such an assumption is in accord with the presence of a maximum on the conductivity $vs.$ a_M curves in CsCl solutions (at $a_M = 1/2\ K_w$), due to depletion in the membrane of the main current carriers, namely M^+L_3 complexes [149, 616, 618, 992] (cf. [599]).

The concept that in the non-polar zone of the membrane a cation can be attached to more than one cyclic polyether molecule is in good agreement with the formation of sandwich complexes by these compounds (see Parts I.D, III.A.3 and III.B). It should be mentioned, however, that the results obtained can also be rationalized in terms of a relay mechanism and much further study is required to resolve this dilemma. One of the possible approaches to it is measurement of the a_M-dependence of the short-circuited currents (for details see [596]).

According to unpublished data of Melnick, in enniatin-containing membranes complexes of the composition M^+L_2 and possibly $M_2^+L_3$ compete with the usual M^+L complexes for participation in the transmembrane cation transport.

The valinomycin-modified bovine brain lipid bilayers are also distinguished by a number of specific features. Analysis of the experimental data shows that under certain conditions the "surface" conductivity of such bilayers is superlinearly dependent on the antibiotic concentration of the aqueous solution. For describing the properties of these membranes a scheme for the cation transport has been proposed involving a new stage, namely charge transfer from the membrane surface to the interior, kinetically similar to the bimolecular exchange of the cation between the complexone molecules [625—627] (see Fig. 172). The model can be regarded as a sort of combination of the carrier and relay mechanisms. It is as yet unknown whether the model actually reflects the true nature of the non-linear stage, but it permits description of many effects observed not only with valinomycin, but also with its various synthetic analogs, despite

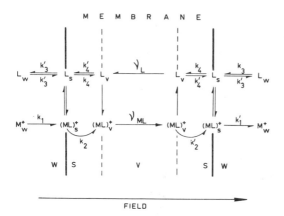

Fig. 172. Diagram of cation transport proposed for bovine brain lipid bilayers modified by cyclodepsipeptides of the valinomycin group [626]. W, S, V are aqueous solutions, surface layers of the membrane and its internal zone, respectively; k_1 and k_1', rate constants of complex formation and dissociation; k_2 and k_2', rate constants of the exchange reaction; k_3 and k_3', rate constants of the free molecule flows into and out of the membrane surface layers; k_4 and k_4', rate constants of free molecule transitions between the surface layers and the internal zone; ν_{ML} and ν_L, probabilities of complex and free molecule transitions across the internal zone of the membranes.

the considerable dissimilitude in the properties of bilayers which they have modified.

The non-linear nature of the "surface" conductivity can be followed particularly well on the example of cyclodepsipeptide (53) (see Table 2). In a 1 M KCl solution containing 10^{-8} M of this compound the bilayer manifests a superlinear conductivity $vs.$ cyclodepsipeptide concentration dependence ($d lg\, g_0 / d lg\, c_L^w = 1.7$), convex current—voltage characteristics and a marked conductance and capacitance dispersion at approx. 2.5 kHz (see Fig. 173). These findings bear witness to the conductance being limited here by the non-linear stage which determines the "surface" resistance on the membrane. It can be expected that with increase in cyclodepsipeptide concentration the "surface" resistance will diminish faster than the "bulk"

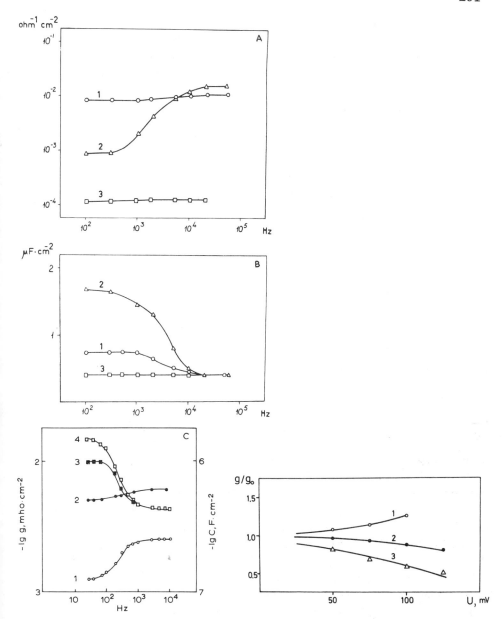

Fig. 173. Conductivity and capacity dispersion of bovine brain lipid bilayers modified by valinomycin and related cyclodepsipeptides [625, 626]. A, B, the bilayers are in 10^{-6} M cyclodepsipeptide-containing 1 M KCl solution: 1, valinomycin; 2, compound (79); 3, compound (57). C, conductivity (1, 2) and capacity (3, 4) curves for the bilayers which are in 1 M KCl solutions containing differing amounts of compound (53): 1 and 4, $2 \cdot 10^{-7}$ M; 2 and 3, $8 \cdot 10^{-7}$ M.

Fig. 174 (bottom right). Curvature sign reversal of the current—voltage characteristics of bovine brain lipid bilayers in 1 M KCl with increase in concentration of cyclodepsipeptide (53): 1, $8 \cdot 10^{-7}$ M; 2, $1.4 \cdot 10^{-7}$ M; 3, $2.25 \cdot 10^{-8}$ M [625].

292

resistance since the latter depends linearly on c_L^w. In fact, with increase in c_L^w a decrease in the conductance and capacitance dispersion and reversal of the sign of curvature of their current—voltage characteristics are observed (see Fig. 174).

This model rests on the assumption that the concentrations of valinomycin and its complexes in the surface layers of the membrane differ from that in the internal zone, transfer of the free molecules and complexes between the two regions requiring the surmounting of an activation barrier. Such a

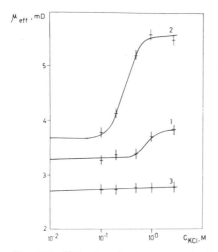

Fig. 175. Effective dipole moment of a cyclodepsipeptide molecule in the monolayer at the air—water interface depending on the concentration of potassium ions in subphase [712]. Calculated from the surface potential data in the plateau on potential *vs.* monolayer area curve: 1, valinomycin; 2, compound (79); 3, compound (57).

proposal appears to be quite plausible if one takes into account the surface activity definitely displayed by cyclodepsipeptides of the valinomycin group and that their conformation at the interface most likely differs from that in the hydrophobic membrane interior [157, 904]. The cation complexes are also surface active. Support for this is the fact that potassium ions are bound by monolayers of valinomycin and its analogs at the air—water interface, as is evidenced by the increase in surface potential [157, 158, 482, 483, 592, 625, 712]. The degree of cation binding depends on the amphiphilicity of the complexes. Thus, it is particularly large in compound (79) (see Fig. 175) in whose complexes the bound cation is accessible to water molecules on one side of the "bracelet", whereas it is effectively screened by the bulky hydrocarbon radicals on the other side (cf. p. 136). On the contrary, in the symmetric compound (57) complex formation in the monolayer is particularly weakly expressed.

Further evidence of localization of the complexes on the membrane surface is the data of Stark *et al.* [962] according to which the equilibrium constant of the heterogeneous valinomycin complexing of K^+ in the lipid bilayer (approx. $1 \ 1 \cdot mole^{-1}$) is close to that for the apparent binding constant of potassium by a valinomycin monolayer. The closeness of the constants to the K_w values is apparently due to the valinomycin polar groups being solvated at the interface by water molecules.

Comparing valinomycin with its synthetic analogs, one cannot but note the amazing adaptation of the molecules of this antibiotic to perform its carrier function. Valinomycin-treated membranes have the highest conductance in the physiological range of potassium ion concentrations. At the same time these membranes possess very small conductance and capacitance dispersion. In other words, valinomycin is equally effective in maintaining potassium ion transport across the boundaries and within the bilayer*. Changes in the valinomycin structure as a rule render compounds less capable of inducing K^+ conductivity since at least one of the stages of the transport mechanism is thereby inhibited. A typical illustration is compound (79) which approaches valinomycin in respect of the high frequency (i.e. "bulk") conductance of the membrane, but is inferior to it in respect of the "surface" conductance (see Fig. 173). The point is that one and the same groups in valinomycin play decisive roles in different properties of this antibiotic such as, say, hydrophobicity, complexing rate, ease of conformational rearrangement, etc. Therefore, practically every modification has a bearing on many parameters determining the rates of the various transport stages. Naturally, all these parameters cannot be simultaneously "improved" from the standpoint of increasing the induced K^+ conductivity. Hence, while the modification accelerates the rate of one of the stages it will inevitably prove detrimental to some other stage, so that in general K^+ transport in the presence of valinomycin analogs will be slower than in the presence of the parent compound. In dilute ($< 10^{-2}$ M) and in concentrated (> 1 M) KCl solution the ratio of the "surface" to "bulk" resistances of valinomycin-modified membranes differs from the optimum. In such solutions some of the analogs surpass valinomycin in the magnitude of the induced conductivity, since in their presence, those stages proceed very quickly, which in the case of valinomycin under the given conditions are rate limiting. Thus, compound (79) turns out to be more effective than valinomycin in dilute

---

* Stark *et al.* [962] have arrived at a similar conclusion while studying current relaxation in voltage-clamp experiments. According to these authors the rate of transfer of free and complexed valinomycin molecules in phosphatidylinositol bilayers is approx. $2 \cdot 10^4 \ sec^{-1}$, whereas the rate of dissociation of the complexes at the interface is approx. $5 \cdot 10^4 \ sec^{-1}$.

KCl solutions, whereas in concentrated solutions it is compound (57). In this connection of interest is the comparison of the parameters of the individual transport stages for various valinomycin analogs which would shed light on the functional significance of the different structural and conformational specificities of the molecules. This undoubtedly would facilitate the rational search for membrane-active compounds of predetermined properties, for instance, of definite conductance vs. cation concentration relationships.

It will be shown in Chapter VI that the effect on biological objects of the complexones discussed here is primarily due to their ability to transport cations across membranes of cells and subcellular organelles. The question naturally arises as to what extent the biological activity of these compounds depends on the mechanism of their induced transport. At low concentrations the highest specific activity should obviously be manifested by those complexones of which the individual molecules while in the membrane perform the largest number of cation transmission acts in the small carrousel. The limiting turning speed of such a carrousel is determined by the mobility of the free complexone molecules in the membrane (cf. [968]; it is to be noted that this limiting transport rate per molecule can in principle be surpassed, if at sufficiently high complexone concentrations the cations in the inner zone of the membrane are transported mainly by the relay mechanism; cf. p. 282). The higher the contribution of the relay mechanism or of complexes of the type M^+L_n ($n > 1$) the steeper should be the dependence of the observed effect on the complexone concentration. Such correlation is observed between the slopes of the c_L^w-dependences of the activity and conductance induced in lipid bilayers on comparing the effects of valinomycin and the enniatins on the energy-dependent potassium uptake by isolated mitochondria. In the respective concentration regions the "enniatin" curves turn out to be steeper, apparently due to contribution of the sandwich complexes to the potassium ion transport (see Fig. 228, p. 384). In those cases when it is possible for cations in the membrane to be transferred from molecule to molecule of the complexones there are grounds to expect a synergism between compounds of which for one the rate-limiting stage of the transport is ion transfer across the membrane—water interfaces and for the other its movement across the internal zone. If the transport mechanism discussed on p. 290 is valid then such synergism should be revealed in the series of valinomycin analogs.

V.C.2. Dependence of induced cation transport on the membrane properties

An important place in studies of the membrane activity of macrocyclic complexones is assumed by investigations of the effect of the structure and properties of lipid bilayers on the induced cation transport. The successes in this field have not only led to better understanding of the mode of action of these compounds, but have laid the foundation for their use as a sort of probe in studies of model and biological membranes. Below we shall discuss the role of such factors as the membrane surface charge, the dipole component of the boundary potential and the packing of the lipid molecules in determining the electrical parameters of complexone-modified bilayers.

V.C.2.a. Surface charge

It is well known that the K^+ conductance induced by valinomycin, nactins, etc., is higher the higher the negative charge density on the bilayer surface. This effect can be naturally explained by dependence of the conductance on the potential of the pre-membrane electrical double layers. In order to evaluate the activity of the transportable cation in the diffuse region of the layer, $\bar{a}_M$, Eqn 36 (taken from classical double-layer theory) can be utilized [617].

$$\bar{a}_M \sim a_M \left[\alpha + \sqrt{\alpha^2 + 1} \right]^{\pm 2} \tag{36}$$

where

$$\alpha = \frac{\sigma}{\sqrt{\sum_i a_i}} \cdot \sqrt{\frac{\pi}{2 \epsilon R T}}$$

a_i is the activity of the ith ion in the bulk solution; σ, surface charge density; ϵ, dielectric constant; $-$ and $+$ refer to negatively and positively charged surfaces, respectively.

Analysis of this equation leads to the following predictions regarding changes of the cation activities in the pre-membrane layer: (i) dependence on the ionic strength should be manifested only in the presence of a surface charge; (ii) effect of change in the surface charge should diminish with increase in ionic strength; and (iii) for negatively charged surfaces increase in charge density and decrease in ionic strength should increase the value of $\bar{a}_M$; for positively charged surfaces the reverse effect should occur.

The quantity $\bar{a}_M$ determines the intramembrane complex concentration since the former reflects the double layer contribution to the potential difference between the membrane and the aqueous solution (cf. p. 301) [617] as well as the rate of complex formation on the membrane—water interface [960]. Hence, when investigating the properties of the bilayer relative to the transportable cation concentration, account must be made of the effect on $\bar{a}_M$ caused by change in the ionic strength. In most cases the ionic strength of the solutions is kept constant by adding salts of non-permeant cations (see, for instance, [990]).

The membrane surface charge is one of the main factors determining the choice of lipid when working with macrocyclic complexone-modified bilayers. Thus, in the case of valinomycin, the maximum on the conductance *vs.* a_M curve due to diminution of the free molecule concentration in the membrane can be clearly noted in the highly negatively charged bovine brain lipid bilayers [540, 550, 625, 626] (see Fig. 176), but it is absent from neutral lecithin bilayers [617, 960].

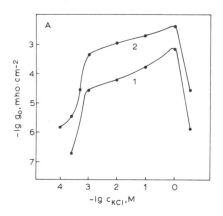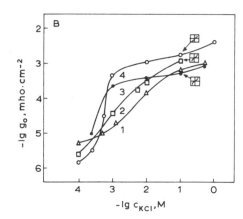

Fig. 176. Dependence of the conductivity of bovine brain lipid bilayers on the KCl concentration in valinomycin-containing bathing solutions [625, 626, 628]. A, the bilayers are in pure KCl solutions containing 10^{-8} M (1) and 10^{-7} M (2) valinomycin. B, effect of foreign cation addition to the 10^{-7} M valinomycin-containing KCl solutions: 1, solutions contain NaCl of the same concentration as KCl; 2, solutions contain 5 mM histidine—HCl buffer (pH 6.5); 3, solutions contain $4 \cdot 10^{-7}$ M cetyltrimethylammonium bromide; 4, control. The plots in squares schematically represent the shape of current—voltage curves.

For the example of valinomycin-modified bovine brain lipid bilayers in KCl solutions ($>10^{-3}$ M) it has been shown that increase in ionic strength leads to effects close to those that are observed on lowering the KCl concentration. In the region to the left of the conductivity maximum (10^{-3}—1 M KCl) the addition of the salt of a non-complexing cation (NaCl) is accompanied by a fall in the conductivity, whereas to the right of the maximum (3 M KCl) the latter markedly increases. On the other hand, such bilayers are characterized by transition from convex to concave current—voltage characteristics when the KCl concentration is lowered to less than 1 M. The region in which this transition takes place was found to shift in the direction of higher KCl concentrations with increase in ionic strength of the solutions [550, 626]. Such effects set certain limitations to the use of "inert salt" additions for maintaining the constancy of the ionic strengths, because the concentration range of the transportable cation in which a given effect is to be observed can thereby be shifted to inaccessible regions [625, 628].

It is to be noted that an ionic strength dependent shift can occur only when the measured parameter is dependent on $\bar{a}_M$. As an illustration may be considered the conductance $vs.$ a_M dependence when the "bottleneck" of the overall transport process is complex formation at the membrane—water interface (see p. 284). If for $a_M < 1/K_w$ this reaction is not accompanied by diffusional restrictions, its rate, and consequently the membrane conductance, should be proportional to the cation concentration near the membrane surface rather than in the bulk of the aqueous solution. Therefore, for charged membranes the a_M dependence of the conductance can be non-linear in some region of a_M values below $1/K_w$, the direction and magnitude of the deviation from linearity being determined by the dependence of the $\bar{a}_M/a_M$ ratio on the concentration of the salt of the transportable cation. Evidently the non-linearity region will shift to higher a_M values if the ionic strength of the solution is increased by "inert salt" additions. In the region $a_M > 1/K_w$, owing to complexation in the aqueous solutions, there will be a fall in concentration of the free complexone molecules. Under these conditions the rate of the heterogeneous complexing reaction will be determined by the flux of the free molecules in the membrane and will therefore be independent of its surface charge and, consequently, of changes in the ionic strength. From this it can be seen that study of the effect of changes in ionic strength may afford valuable information on the nature of the rate-limiting stages of the cation transport.

In the case of amphoteric lipids the sign and magnitude of the surface charge depends on the pH of aqueous solutions. A change in the balance of positively and negatively charged groups in titration of the lipids can be clearly followed for the example of nonactin-modified phosphatidyl-ethanolamine bilayers. The conductivity of these bilayers in KCl solutions increases with increase in pH from 2 to 5 (region of dissociation of the phosphate groups) and from 7 to 11 (region of deprotonation of the ammonium groups), remaining constant at pH values between 5 and 7 (see Fig. 177A). Hence, a conductivity increase is observed with either fall in the positive surface charge or enhancement of negative surface charge of the bilayers. As the ionic strength of the solutions is augmented, the conductivity of positively charged bilayers (pH 2.4) will increase, and that of negatively charged bilayers (pH 10.9) will decrease (see Fig. 177B) [617].

Adsorption of divalent metal ions on the surface of negatively charged bilayers diminishes the complexone-induced conductance, an effect which may be used for quantitative estimation of changes in the double layer potential [619]. A similar effect is exerted by surface-active cations (for instance, cetyltrimethylammonium; see Figs 176 and 178) and certain other organic cations (tris-onium, histidinium; see Fig. 176). This must be borne in mind when using them in buffers.

The insertion of macrocyclic cation complexes into the membrane should also lead to changes in the double layer potential. It is this which is apparently the cause for lessening of the slope of the conductance *vs.* valinomycin concentration curve characteristic for bovine brain lipid bilayers in 1 M KCl solution (i.e. in the neighborhood of $a_M = 1/K$) (cf. Figs 175 and 179) [625, 626] (cf. [615]).

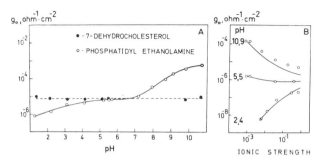

Fig. 177. Dependence of the nonactin (10^{-6} M)-induced conductivity of bilayers on the surface charge [617]. A, effect of pH on the conductivity of bilayers of amphoteric (o) and neutral (•) lipids. The aqueous solutions contain 100 mM KCl. B, effect of the ionic strength of the aqueous solutions on the conductivity of phosphatidylethanolamine bilayers for different pH values. The ionic strength was changed by additions of LiCl.

When the bilayers have a considerable positive surface charge one can expect that in solutions of low ionic strength ($\alpha^2 \gg 1$; see Eqn 35) the conductance will be a quadratic rather than linear function of the complexable cation concentration. Such a dependence has been observed, for instance, in the action of monactin on oleylamine bilayers [991].

It should be pointed out that in some cases the effect of change in ionic strength cannot be ascribed solely to concentration changes of the transportable cations in the double layer. In this connection of interest is the anomalous behavior of valinomycin-modified bilayers of negatively charged bovine brain lipids. In dilute salt solution ($< 5 \cdot 10^{-4}$ M), a very small increase in the ionic strength augments rather than diminishes the conductance (see Fig. 176) [628]. It thus follows that, if for solutions containing insignificant amounts of potassium the NaCl concentration is raised, one may expect a maximum in the neighborhood of 10^{-3} M NaCl.

Apparently this is the reason for the appearance of the Li and Na maxima observed by Liberman *et al.* [550] with complexone-modified brain lipid bilayers in dilute salt solutions.

The source of these anomalies is the non-monotonic dependence of the properties of brain lipids on the ionic strength of the aqueous solutions which can be observed in studies of the surface potential of the monolayers

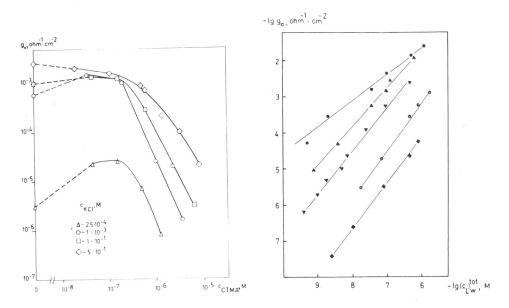

Fig. 178 (left). Effect of cetyltrimethylammonium bromide (CTMA) on the conductivity of 10^{-7} M valinomycin-modified bovine brain lipid bilayers [628].

Fig. 179 (right). Conductivity dependence on valinomycin concentration in KCl solutions for bovine brain lipid bilayers [626]. ○, 3 M KCl; ●, 1 M KCl; ▲, 0.1 M KCl; ▼, 0.001 M KCl; ◆, 0.0025 M KCl.

(see Fig. 180) and the degree of ordering of the hydrocarbon chains in lamellar lipid—water liquid crystals [109]. The binding of cations by the anionic centers of phospholipids (apparently, sphingomyelin; cf. [882]) occurring in dilute solutions causes structural changes in the bilayer which are accompanied by a decrease in potential difference between the membrane interior and the aqueous phase as well as a certain loosening of the lipid chains. As will be shown below, both these effects enhance the induced cation conductance. It is interesting that cetyltrimethylammonium which is practically irreversibly bound by the membrane acts in similar fashion in insignificant concentrations ($< 10^{-7}$ M) (see Fig. 178). With increase in ionic strength or cetyltrimethylammonium concentration another factor is brought into play, namely change in the double layer potential due to screening of the negative charges, or, correspondingly, increase in number of the positive charges attached to the membrane surface. The anomalous properties of the brain lipids are apparently also responsible for the sharp increase in bilayer conductivity in KCl solutions ranging from 10^{-4} to 10^{-3} M (see Fig. 176).

Of considerable interest are the anion effects on the complexone-induced membrane conductance. To all appearances the conductance does not

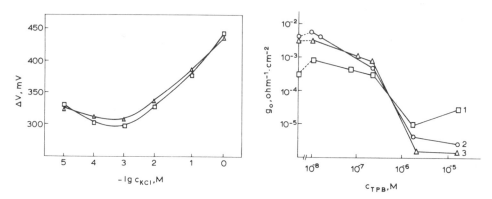

Fig. 180 (left). Dependence of the surface potential of bovine brain lipid monolayers on the water—air interface upon the ionic strength of subphase [628]. The presented values correspond to the plateau segments of potential $vs.$ monolayer area curves: □, NaCl; △, KCl.

Fig. 181 (right). Effect of tetraphenylboron anions (TPB) on the conductivity of valinomycin (10^{-7} M)-modified bovine brain lipid bilayers in solutions of KCl: 1, 10^{-2} M; 2, 10^{-1} M; 3, 1 M [199].

change when hydrophilic anions such as Cl^-, NO_3^- and SO_4^{2-} are interchanged in the solution. Adsorption of the more hydrophobic organic anions on the membranes can augment the conductance. Such anions include, for instance, 8-anilino-1-naphthalene sulfonate (ANS) which, as McLaughlin et $al.$ [619] have shown, in concentrations up to 10^{-4} M increase the nonactin-induced K^+ conductivity of phosphatidylethanolamine bilayers with no sign of saturation. Salicylate and dinitrophenolate possess a similar action but in higher concentrations [615].

Apparently ANS-type amphiphilic anions which are localized in the membrane surface layers (cf. [382]) mainly affect the conductance by changing the double layer potential. This viewpoint has received convincing support in the recent work of McLaughlin [615] who investigated the mechanism of dinitrophenol-induced proton conductivity with the objective of demonstrating that in this case intramembrane charge transport is mediated by dimeric anions of the type $HX \cdot X^-$, attributing the observed deviation from the predictions of the corresponding theoretical model to changes in the double layer potential on adsorption of X^- ions on the membrane. He calculated these changes from the data on the effect of dinitrophenol on the K^+ conductivity induced by nonactin (which considerably exceeds the proton conductivity), assuming that the increase is due only to augmentation of the negative surface charge on the membrane. Taking into account the dependence of the pre-membrane $HX \cdot X^-$ concentration on the double layer potential, McLaughlin achieved complete accord between the experimental results and the dimeric model. The same result has been obtained with ANS.

Anions displaying the highest hydrophobicity, such as tetraphenylboron, possess biphasic activity [199]. It was shown for the example of valinomycin-modified bovine brain lipid bilayers that in very low concentrations ($< 10^{-8}$ M) tetraphenylboron augments the K^+ conductance, particularly in dilute KCl solutions, but inhibits it at higher concentrations (see Fig. 181). A similar effect is manifested by the phenyldicarbaundecaborane anions. The hydrophobic anions apparently penetrate the internal zone of the membrane, thereby compensating the dipole component of the boundary potential jump (see below). A probable cause of conductance inhibition by these anions is their association with the complexes in both the aqueous solutions and the membrane to form electrically neutral ion pairs or micelles (cf. p. 94). Such association decreases not only valinomycin-induced K^+ permeability but also the permeability for the corresponding lipophilic anions. It is tempting to use this phenomenon in titration of membrane-active complexones and possibly for determining the stability constants of their complexes in aqueous solutions.

V.C.2.b. Dipole component of boundary potential

It is well known that monolayers of various lipids at the air—water interface possess a surface potential of the same sign (*plus* in air) independent of the pH or ionic strength of the subphase. This leads to the conclusion that the main contribution to the surface potential is made by the dipoles of the lipid polar groups, unidirectionally oriented with respect to the interface. Since a bilayer can be regarded as two monolayers put together, an electrical potential difference can naturally be expected between its internal zone and the bathing aqueous solutions (positive in the membrane interior) [34]. The principal contribution by the dipolar component to the potential difference is supported by the fact that lipophilic anions independent of the lipid species have a considerably higher partition coefficient in the bilayer—water system than the isosteric or structurally similar cations [552].

It thus follows that the concentration of the macrocyclic cation complexes in the internal zone of the bilayers is determined not only by their hydrophobicity, but also by the potential jump due to presence of the dipoles. Obviously, by changing the magnitude of this jump one may vary the partition coefficient ratios of the free and cation-carrying complexone molecules and thereby exert a directive effect on the rate ratios of the individual stages of the cation transport. Since the lipid molecules differ in structure and in the number of polar groups, this can be achieved in principle by appropriate variations in the lipid composition of the bilayer. The first step in this direction was made by Eisenman and coworkers [147, 231], who showed that on passing from glycerol dioleate to phosphatidylethanolamine the surface potential of the monolayers increases, whereas the trinactin-induced K^+ conductivity of the corresponding bilayers falls.

The boundary potential jump can be lessened by means of surface-active compounds which contain dipolar groupings, such that they orient themselves antiparallel to the lipid dipoles on insertion of the molecules into the membrane. Among these, for instance, are ω-chloropentadecanoic acid and N-methyltetrachlorobenzimidazole (MTCB) complexes with univalent copper ions which form monolayers on the air—water interface with a negative surface potential [198, 200] (cf. [275, 673]). The insertion of these molecules into the lipid monolayers lowers their positive surface potential. Such an effect is undoubtedly due to the "anomalous" orientation of the C—Cl dipoles relative to the interface. The orientation is retained also on adsorption of these compounds onto lipid bilayers. This is evidenced by the ability of these compounds to increase the bilayer conductance in solutions containing such permeant lipophilic cations as, for instance, tetraphenylphosphonium (see Figs 182A and 182B). Since the principal contribution to the resistance is here made by the stage of intramembrane cation transport it is natural to ascribe the growth in conductance to increase in the tetraphenylphosphonium partition coefficient resulting from decrease in the dipole component of the potential jump. Support for this is to be found among others in the similarity of the pH dependencies of the membrane conductances in the presence of MTCB—Cu^+ complexes and of the surface potential of the monolayers on the air—water interface (cf. Figs 182B and 183). We note that in the absence of tetraphenylphosphonium neither chloropentadecanoic acid nor MTCB—Cu^+ increases the membrane conductance. On the other hand, the bilayer conductance in tetraphenyl-phosphonium salt solutions does not change in the presence of either MTCB

itself or hexadecanoic acid which is practically isosteric with chloropentadecanoic acid. It is noteworthy that these inactive compounds are also
unable to lower the surface potential of lipid monolayers.

The effect of ω-chloropentadecanoic acid and MTCB—Cu^+ complexes on
bilayers treated with valinomycin or its analogs can also be explained in

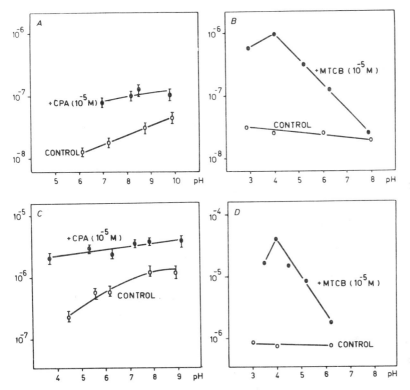

Fig. 182. Effects of ω-chloropentadecanoic acid (CPA) and Cu^+ complexes of N-methyl-
tetrachlorobenzimidazole (MTCB) on the conductivity (in mho · cm^{-2}) of bovine brain
lipid bilayers in solutions containing $5 \cdot 10^{-6}$ M tetraphenylphosphonium bromide (A, B)
and $5 \cdot 10^{-8}$ M cyclodepsipeptide (57) (C, D). The aqueous solutions contain also 0.1 M
NaCl (A, B) and 0.1 M KCl, 10^{-5} M $CuSO_4$ and $5 \cdot 10^{-5}$ M hydroquinone (reducing agent
for Cu^{2+} ions) (C, D) [198].

terms of "bulk" conductance increase due to rise in the intramembrane
cation complex concentration. Thus, with the cyclodepsipeptide (79), the
"surface" conductance of the membrane is less than its "bulk" conductance
(cf. p. 293). One can therefore expect that augmentation of the latter would
increase the conductance and capacitance dispersion and lower the g_U/g_0
parameter reflecting the curvature of the current—voltage characteristic (see
legend to Fig. 170). It is just such effects that have been observed

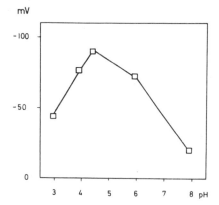

Fig. 183. pH-Dependence of the surface potential of a monolayer formed on solutions containing $5 \cdot 10^{-2}$ M KCl, 10^{-5} M CuSO$_4$, $5 \cdot 10^{-5}$ M hydroquinone and $1.7 \cdot 10^{-6}$ M N-methyltetrachlorobenzimidazole. The monolayer does not form on solutions of low ionic strength or in the absence of copper salts, reducing agents or N-methyltetrachlorobenzimidazole [198].

experimentally (see Fig. 184). Interestingly, the direct current conductance g_0 remains practically constant, showing that the "surface" conductance is most probably limited by flux of the electrically neutral free molecules in the membrane. On the other hand, it is possible that the limiting stage also involves the cyclodepsipeptide complexes on the membrane surface which are, however, beyond the zone of potential profile change caused by the presence of chloropentadecanoic acid or MTCB—Cu$^+$ complexes.

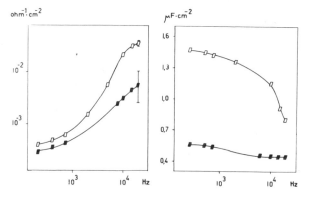

Fig. 184. Effect of N-methyltetrachlorobenzimidazole on the conductivity dispersion (to the left) and capacity dispersion (to the right) of bovine brain lipid bilayers modified by cyclodepsipeptide (79) ($5 \cdot 10^{-7}$ M). Composition of solution: 1 M KCl, 10^{-5} M CuSO$_4$, $5 \cdot 10^{-5}$ hydroquinone (pH 4.0). ■, control; □, $+10^{-5}$ M N-methyltetrachlorobenzimidazole [198].

If the "surface" conductance of the membrane is initially larger than the "bulk" conductance, then with decrease in potential jump this relation may become reversed. Such change in the transport rate determining stages occurs in the action of MTCB—Cu$^+$ complexes on bilayers modified by cyclo-depsipeptide (57). In this case there is an increase in the direct current conductivity (see Figs 182C and 182D), the appearance of conductance and capacitance dispersion (cf. Figs 173 and 185) and transition from concave to convex current—voltage curves (see Fig. 186).

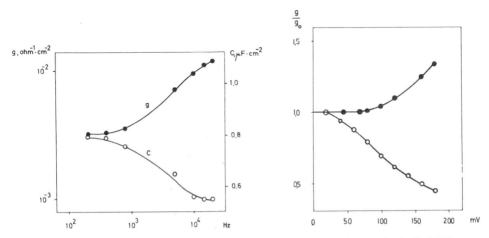

Fig. 185 (left). Conductivity (●) and capacity (○) dispersion of bovine brain lipid bilayers in solutions containing 1 M KCl, $5 \cdot 10^{-7}$ M cyclodepsipeptide (57), 10^{-5} M CuSO$_4$, $5 \cdot 10^{-5}$ M hydroquinone and 10^{-5} M N-methyltetrachlorobenzimidazole (pH 4.0) [198].

Fig. 186 (right). Curvature sign reversal of the current—voltage characteristics of bovine brain lipid bilayers modified by cyclodepsipeptide (57) ($5 \cdot 10^{-7}$ M) in the presence of Cu$^+$ complexes of N-methyltetrachlorobenzimidazole. For composition of solution see legend to Fig. 184. ●, Control; ○, $+10^{-5}$ M N-methyltetrachlorobenzimidazole [198].

The wide possibilities opened up by the use of compounds lowering the dipole component of boundary potential can be illustrated on the example of valinomycin analog (32), whose molecules possess an NH$_2$ group. In principle, this cyclodepsipeptide can be in the membrane in the neutral (L—NH$_2$) or the protonated (L—NH$_3^+$) form. If the main contribution to the cation transport is due to the protonated form then a negative slope region can appear on the steady-state current—voltage characteristics, because in a strong field the L—NH$_3^+$ molecules will be "pressed" onto the cathodically polarized membrane boundary and thereby become excluded from the cation transport cycle [548]. Under ordinary conditions this is not observed even in the low pH range. One could expect that with decrease in the dipole component of boundary potential, there should be an increase in contribution of the protonated form to the K^+ transport on the one hand, and in the contribution of the "surface" resistance on the other. Hence, especially in the range of high potassium ion concentrations, the K^+ transport "bottleneck" can become the flux of $L—NH_3^+$ molecules, a situation favoring the appearance of non-monotonic current—voltage characteristics. In fact, in the presence of MTCB—Cu^+ complexes one can, under certain conditions, observe a declining segment on the current-voltage curve of compound (32)-modified bilayers (see Fig. 187).

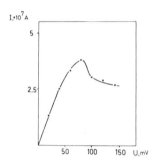

Fig. 187. Current—voltage characteristics of a bovine brain lipid bilayer in a solution containing 1 M KCl, $1.3 \cdot 10^{-6}$ M cyclodepsipeptide (32), 10^{-5} M $CuSO_4$, 10^{-4} M hydroquinone and 10^{-5} M N-methyltetrachlorobenzimidazole [200].

It should be noted that the effect of chloropentadecanoic acid type compounds may apparently be due not only to concentration changes of the lipophilic cations or complexes in the membrane, but to changes in the probability of their being conveyed across the internal zone. This is because the insertion of C—Cl dipoles near the middle plane of the membrane with their negative poles in its direction should give rise to a sort of potential well. Such modification of the potential profile in the membrane can in itself have a bearing on the shape of the current—voltage characterictics (cf. [348]).

V.C.2.c. Packing of the lipid molecules

The significance of this factor is most lucidly illustrated by the data of Krasne *et al.* [512]. These authors showed that cooling of the membrane to below the phase transition temperature of the corresponding lipid—water liquid crystals completely inhibits the valinomycin- or nonactin-induced K^+ conductance. This effect is hard to explain by the antibiotic molecules only being displaced from the membrane, since under similar conditions another hydrophobic agent, gramicidin, retains its activity (see Fig. 188). The most probable interpretation, therefore, is that the phase transition is accompanied by a sharp fall in mobility of the complexes and/or free molecules in the membrane*.

A similar although weaker effect is caused by the introduction of cholesterol into lecithin bilayers. As is well known this compound promotes the immobilization of lipid chains. Szabo *et al.* [990] showed that for a lecithin:cholesterol proportion of 1:1 the nonactin-induced conductivity is diminished to about 1/20 regardless of the nature of the transported cation.

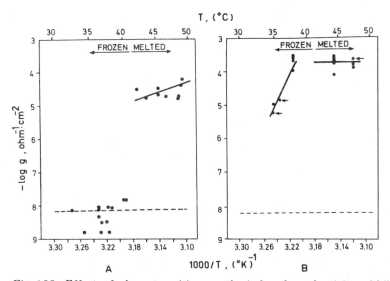

Fig. 188. Effect of phase transition on the induced conductivity of bilayer membranes from a *n*-decane (1 ml) solution of glycerol dipalmitate (20 mg) and glycerol distearate (20 mg). The aqueous solutions contained 10^{-2} M KCl. Dashed lines represent the conductivity of non-modified membranes [512]. A, Nonactin (10^{-7} M). Apparent activation energy as found from the slope of the conductance *vs.* reciprocal temperature plot is 29 kcal · mole^{-1}. B, Gramicidin (10^{-9} M). Apparent activation energy near zero. Arrows designate the results of experiments in which gramicidin was added after blackening of the membrane.

Since under such conditions the membranes are in the *equilibrium domain*, these data can be regarded as argument in favor of isostericity of complexes (cf. p. 287).

The inhibiting effect of cholesterol depends upon which of the transport stages is rate-limiting. For instance, for lecithin bilayers, cholesterol also

---

* In a study of the temperature dependence of valinomycin-induced conductance, Stark *et al.* [961] found that in the case of lipids with saturated chains the activation energy changes at a certain temperature, which is higher than the "melting" temperature of the membrane. The nature of this change is still obscure.

lowers the conductivity induced by cyclodepsipeptide (57) to about 1/20, whereas the valinomycin-induced conductivity is affected very little, apparently owing to the smaller contribution of the "bulk" resistance [625, 626].

Stark *et al.* [961] have studied the dependence of induced conductivity on the hydrocarbon chain length of lipid molecules. They showed that the conductance of bilayers prepared from decane solutions of saturated lecithins (55°C, 1 M KCl, 10^{-8} M valinomycin) falls by about two orders of magnitude as the number of carbon atoms in the chain increases from 12 to 18. This effect, which cannot be ascribed to increased membrane thickness, the authors associate with changes in packing of the lipid polar groups lowering the rate of the heterogeneous complexing reaction.

An interesting effect has been observed by Stark *et al.* [961] in studies of conductance relaxation of valinomycin-modified bilayers when the temperature is raised stepwise. The membrane conductance first rapidly rises and then falls to nearly the original level. The first phase of this process is apparently the temperature-induced changes in the transport rate, whereas the second is due to outflow of valinomycin from the membrane. Thus, in the case of valinomycin there is an automatic compensation, as it were, of the temperature effect on the steady-state conductance.

King and Steinrauf [491] have recently found that unusually stable bilayers (with lifetimes of up to two days) can be obtained by their treatment with polylysine (with a chain length of about 100 residues) in combination with glutaraldehyde. In their opinion such treatment divides up the bilayers, its fragments becoming suspended from the pores of the cross-linked polypeptide framework. The longer lifetime thus attained can be likened to the effect obtained when "after rain a window fitted with a wire screen will support a film of water, while an open window will not" [491]. If valinomycin is added to the solution prior to the treatment of the bilayers by polylysine and glutaraldehyde, superstable membranes selectively permeable to potassium ions can be obtained. This finding is of interest not only from a purely practical standpoint but also in connection with the problem of simulating lipid—protein interactions in biological membranes.

In concluding the discussion on the effect of macrocyclic complexones on bilayer membranes a number of new directions may be indicated. Among these, first and foremost is the utilization of the complexones as probes in studying not only the properties of membranes but also the mode of action of other membrane-active agents (see, for instance, [393]). In order to make full use of its potency further investigation is needed on the effect of the bilayer composition and structure on the parameters of complexone-induced cation transport. It is also intriguing to attempt to use the macrocyclic complexones for making model membranes that would reproduce given functions of their biological counterparts. A possible way of achieving this has been proposed by Liberman [548] who suggested the use of complexones containing cation groups or of bifunctional complexones (for instance, two covalently bound macrocyclic compounds) for imitating the potential-dependent permeability changes of excitable membranes. Schwyzer [875] discussed a model in which a complexone is used for reproducing the

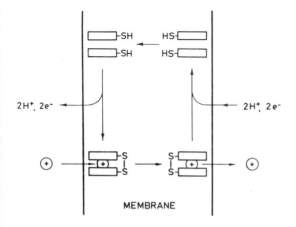

Fig. 189. Possible model of a cation pump driven by oxidoreduction processes (according to Schwyzer [875]). A compound is introduced into the membrane, that is structurally similar to peptide (226) and in the oxidized state forms complexes which can act as cation carriers. The solution to the left contains the oxidizing system and to the right, the reducing system. The driving force of the cation transport is the oxidoreduction supported gradient of the oxidized form in the membrane.

function of primary translocases, the transmembrane ion transport being energetically coupled with the chemical reactions proceeding in the membranes (cf. p. 347; see Fig. 189).

One of the least investigated problems remains the mode of action of nigericin antibiotics on bilayers. Recently there have been indications that these antibiotics not only promote non-electrogenic cation exchange but are also capable of increasing bilayer conductance [261, 671, 788], the reason for the increase, however, is unknown. Probably, the enhanced conductance is associated with the presence of M^+LH complexes or L^- anions in the membrane (cf. p. 276). These data are of interest in connection with the discovery of the uncoupling effect of nigericin on membrane phosphorylation (see pp. 387 and 392).

A first step forward has now been made in the study of cation transport induced by potential helical complexones. Van Zutphen *et al.* [1109] have shown that certain polyether detergents, in particular Triton X-100, in sublytic concentrations increase the conductance of lipid bilayers. The cation selectivity sequence of the effect coincides with that observed in extraction experiments (see p. 215), indirectly pointing to participation of the complexes in the transmembrane ion transport. An interesting feature of the polyether detergents, which as yet has found no satisfactory explanation, is their ability to diminish durably the resistance only when added to both sides of the membrane. If they are added to only one side the resistance, after a rapid fall, increases again attaining in about an hour a constant value only a little less than the original one. The high and lipid-dependent slope of the transient resistance drop *vs.* detergent concentration plots (in double logarithmic scale the slopes are -6 and -10.5 for bilayers from mito-chondrial lipids and egg lecithin, respectively) indicates the participation of molecular aggregates interacting with the membrane.

V.C.3. Substances forming pores in membranes

In the foregoing we have discussed the membrane activity of neutral compounds capable of forming stable complexes with cations in solutions. Apparently, practically all alkali metal-binding macrocyclic compounds, both free and complexed, possessing sufficient hydrophobicity can convey cations more or less effectively across bilayer membranes. At any rate, as has been mentioned above, such properties have been revealed in cyclic polyethers, synthetic compounds which could hardly be regarded as structural analogs of antibiotic complexones selected by Nature as ion carriers. The question now naturally arises as to whether the functioning of ion transport systems in biological membranes might not be connected with the presence of mobile carrier molecules behaving like valinomycin or nigericin. Such carriers have often been postulated, but no-one has as yet been able to provide proof for their existence. Doubtless, searching for them is an exceptionally difficult task since they could be present in the membrane in exceedingly small amounts. On the other hand, it is possible that ions are transported across biological membranes by other means, for instance, by molecules or supermolecular systems that form channels or pores through the membrane. If the pore diameter is commensurate with that of the ion, transport of the latter is associated with reversible dehydration and its probability is determined by the energy of interaction between the ion and the groups lining the pore interior. It is only natural to conjecture that "solvation" by neutral ligand groupings could take place within such narrow pores responsible for translocation of cations. Thus, in recent studies Hille

[392] concluded that the "Na$^+$ channel" of the axonal membrane is in the form of a pore in which the partially hydrated cation is surrounded by six ligand oxygen atoms of which only two belong to ionized carboxyl groups. The insertion of a cation into such a pore can be likened to its binding by a macrocyclic complexone, and its movement therein can be regarded as a process similar in a way to a series of cation transfers between complexone molecules. While such concepts had lacked experimental backing, the cation-selective narrow-pore hypothesis has now found support in the results of a study of the membrane activity of the gramicidins A, B and C. These linear peptides do not form stable complexes with alkali metal ions in

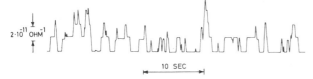

Fig. 190. Conductance oscillations of a bilayer from glycerol monooleate in n-decane. The aqueous solutions (1 M KCl) contain gramicidin A in very low concentrations. Applied voltage, 100 mV [396].

solutions (see p. 31). They form stable monolayers on the air—water interface and are capable of penetrating lipid monolayers. However, in contrast to valinomycin, the surface potential of gramicidin-containing monolayers is independent of the alkali ion concentration in the subphase, providing evidence of the absence of complexation or of low stability of the complexes [482]. Although gramicidins have no detectable complexing properties, they are nevertheless outstanding in their ability to increase selectively the permeability of lipid bilayers to alkali metal ions [550, 669, 1012].

Gramicidin-mediated cation transport differs in some ways from that due to valinomycin and the nactins. First, the induced cation permeability is not so crucially dependent on whether the membrane is above or below the lipid-phase transition temperature (see Fig. 188). This shows that cation movement in the membrane is not coupled with diffusion of the gramicidin molecules [512]. In accord with such an inference are the results obtained in a study of the discrete fluctuations of membrane conductance observed at low gramicidin concentrations. Haydon and Hladky [378, 379, 396, 397, 380] who investigated this effect arrived at the conclusion that each step on the conductance—time curve (see Fig. 190) corresponds to the "switching on" or "switching off" of a single conductance channel, i.e. of the smallest molecular system capable of effecting the transport. Estimating the mobility of potassium ions in such a channel to be approx. 10^{-6} cm$^2 \cdot$ sec^{-1} they showed that it is significantly higher than one could have expected from the values for the diffusion coefficients of the antibiotic complexes.

Secondly, it turned out that the probability of "switching on" a channel and the time it remained in the "switched on" state increased with diminishing thickness of the membrane which depends on the chain lengths of the lipid and of the hydrocarbon solvent (see Table 74) [397, 380] (cf. [262]). Similar phenomena were also observed with increase in the voltage applied to the membrane, apparently because its thickness diminishes under the action of an electric field. This effect provides the key to an

TABLE 74

CONDUCTING SINGLE CHANNELS IN THE GRAMICIDIN A-MODIFIED BILAYERS [397]

Membrane-forming lipid solution	Thickness of hydrocarbon layer of membrane (Å)	Channel conductance in 0.5 M NaCl ($\times 10^{11}$ mho)	Channel lifetime (sec)
Glycerol monopalmitoleate + n-hexadecane	26	1.7	60
Glycerol monooleate + n-hexadecane	31	1.7	2.2
Glycerol monooleate + n-tetradecane	40	1.7	1.3
Glycerol monooleate + n-decane	47	1.7	0.4
Glycerol monooleate + n-decane + polyhydroxystearic acid	64	1.7; 0.8	0.03
Glycerol monooleate + n-decane + cholesterol	47	1.7	0.4

understanding of why gramicidin-modified bilayers have concave current—voltage characteristics in 0.025—1 M salt solutions, whereas the characteristics of the single channels are close to ohmic [380]. The dependence of the number of "switched on" channels on the membrane thickness is probably also the cause of the differences in the kinetics of increase in conductance of bilayers from different lipids, observed by Goodall [300, 302, 303], during their formation in gramicidin-containing solutions. Significantly, the conductance of the single channels is practically independent of the composition of the lipid solution, in particular, of the presence of cholesterol. Only in the case of very thick membranes containing polyhydroxystearic acid do there appear, alongside the ordinary channels, channels of approximately half the conductance (see Table 74), the proportion of the latter diminishing the higher the applied voltage.

These results have led to the natural inference that each conductance channel in the gramicidin-modified membrane corresponds to a pore capable of selectively passing through cations. The gramicidin pores operate independently of each other for, as shown by Hladky and Haydon [397], the relative number of the "switched on" channels is given by Poisson's equation. These authors found that for transportable cations the conductance ratios of a single channel resemble the mobility ratios of the cations in aqueous solutions (see Table 75). The analogy is retained also with respect to the activation energies of the K^+ and Na^+ conductance of a channel (5.4 and 4.9 kcal $\cdot$ mole^{-1}, respectively). From this one may conclude that the cations interact relatively weakly with the functional

TABLE 75

CONDUCTANCE RATIOS (g_{Mi}/g_K) FOR A SINGLE GRAMICIDIN A CHANNEL IN 0.5 M SOLUTIONS, WITH 100 mV APPLIED [397]

	H^{+*}	Li^+	Na^+	K^+	Rb^+	Cs^+	NH_4^+
Channel conductance ratio (g_{Mi}/g_{Na})	14*	0.23	1.0	1.8	2.9	2.9	2.4
Single ion conductance ratio at infinite dilution in aqueous solution $(\Lambda_0^{Mi}/\Lambda_0^{Na})_{25°}$	9.1	0.77	1.0	1.5	1.6	1.5	1.5

* At 0.01 M.

groups lining the pore and apparently cross the membrane in at least a partially hydrated state. The possibility of the incorporation of water molecules into the pore is supported by its high proton permeability. It is as yet unknown whether the protons pass through the pore as mobile hydroxonium cations or whether it is the Grotthus mechanism, i.e. proton transfer along a chain of water molecules by redistribution of hydrogen and covalent bonds, that is operative. The latter mechanism seems to be the more probable. Interestingly, the permeability ratios of the various cations in gramicidin-modified bilayers as determined from the bi-ionic potentials differ from the conductance ratios for the single channel and also display a different dependence on the cation concentration in the solution [684]. Possibly this is due to interaction of the ion fluxes in the pores (cf. [240]).

The molecular organisation of the gramicidin pores is still unknown. Urry suggested that in the membrane gramicidin is in the form of dimers with the molecules folded into one of π_{LD} type helices and joined head to head by hydrogen bonds between the amide groupings (see Part III.A.2.a). Such a dimer can be visualized as a sort of continuous tube piercing the entire membrane and serving as a channel or pore through which the cations can pass. The tubular dimer hypothesis is in accord with the findings of a number of authors that the bilayer conductance is proportional to the square of the gramicidin concentration in the aqueous solution [300, 550, 1012]. Moreover, it is not contradicted by results obtained with malonyl-*bis*-desformylgramicidin A, that is with a compound in which two gramicidin molecules are joined head to head by covalent bonds such as to retain the possibility of forming the continuous helix [1031]. Such a "covalent" dimer possesses activity comparable with that of gramicidin and the membrane conductance is then linearly dependent on its concentration in the aqueous solution. Unfortunately, in the case of gramicidin, determination of the slope of the conductance—concentration dependence is very difficult and may lead to unreliable conclusions (for a detailed discussion of this question see [380]). In fact, the conductance values change with time and differ from membrane to membrane, apparently owing to the instability of the aqueous solutions: at low gramicidin concentrations ($< 10^{-11}$ M) adsorption on the walls of the measuring cell will become significant; at higher concentrations ($> 10^{-10}$ M) association into micelles will take place [482].

Despite the fact that Urry's idea has as yet not received unequivocal experimental backing, it is doubtless very attractive. Of all possible types of π_{LD} helices the most probable appears to be the π_{LD}^6 helix with dimer length 25—30 Å and of channel diameter approx. 4 Å [1031]. Such a pore size is quite sufficient for the passage of ammonium and alkali ions; within the channels the ions can interact with at least two water molecules. It should be noted that the channel diameter is in any case less than the size of tetramethylammonium ions (approx. 7 Å), to which the gramicidin-modified bilayers are practically impermeable [674].

Urry believes that movement of the cations along the intramolecular channel is accompanied by local changes in the helical conformation in the course of which the oxygen atoms of the amide groups are attracted to the cation to form a sort of "transient complex". The distortions should not be large because, as we have seen, cation transport does not proceed by the overcoming of marked activation barriers. Taking into consideration that the conductance and permeability sequences of the gramicidin-modified membranes correspond to the mobility sequences of the cations in water, it is natural to conclude that interaction with the pore walls does not have a significant bearing on the cation selectivity. A similar conclusion can also be reached if the pore is formally considered as an ion exchanger that, in respect of selectivity, corresponds to sequence I in Eisenman's classification (see Table 68). The group comprises complexones with the lowest discriminating power with respect to alkali metal ions (it is to be recalled that the discriminating power of a complexone is estimated from the steepness of the ΔF_{ML} vs. cation radius relation; see p. 230). The position of the proton in the selectivity sequence of the gramicidin pore can be explained from this point of view by considering it to be in the form of a hydroxonium cation.

Despite the fact that the gramicidin pore weakly interacts with the penetrating cations and is only able to differentiate between them weakly, it nevertheless unequivocally discriminates between a cation and an anion. Mayers and Haydon [674] have shown that in alkali metal chloride solutions the cation transport number in the case of gramicidin-modified membranes is unity and only for the weakly permeable Ca^{2+} does the transport number become about 0.8. These authors made the very plausible suggestion that a potential difference exists between the internal space of the pores and the aqueous solution (approx. 60 mV; negative inside the pore) due to specific orientation of the amide dipoles (see pp. 215 and 250). Thus, independent of Urry's hypothesis the conclusion could be drawn that within the interior of the pores there is a weak ion—dipole interaction.

The switching mechanisms of the single channels are as yet unknown. It is not very probable that a "switching on" would correspond to the act of incorporating into the membrane a single molecule or an aggregate of molecules of gramicidin from the aqueous solution. Therefore, one may consider three limiting cases: in the first, the "switching on" of a channel is due to changes in the spatial organization of the membrane transfixing gramicidin molecules, which transform them into a conductive state. The probability of such a transition could depend on the membrane potential, either indirectly, due to decrease in thickness of the membrane, or directly if, for instance, it is accompanied by re-orientation of the dipoles. Such a mechanism is, in particular, discussed by Urry [1027] who assumed that structures of the π_{LD} type could exist in equilibrium with other helices lacking an internal channel (see p. 162). On the other hand, the "switching on" of the channel could result from fluctuational constriction of the membrane at the site of localization of gramicidin in the "conductive" state which could lead to formation of a pore. Finally, one could assume that the "switching on" corresponds to association of gramicidin molecules attached to opposite sides of the membrane, the probability of this act being enhanced by membrane constriction either under the influence of a field or as the result of local fluctuation. As indirect argument for the latter mechanism can serve the current relaxation data of Bamberg and Läuger [40] under voltage-clamp conditions. It turned out that the relaxation kinetics can be described if it is assumed that the rate of "switching on" of the channels is determined by the dimerization reaction. From this standpoint the high activation energy of "switching off" the channels (approx. 19 kcal · mole^{-1} [397]) would correspond to dissociation of the dimers.

A special position among the membrane-active antibiotic complexones is assumed by alamethicin. Essentially the very reference of this compound to neutral complexones is based only on the report of Pressman and Haynes

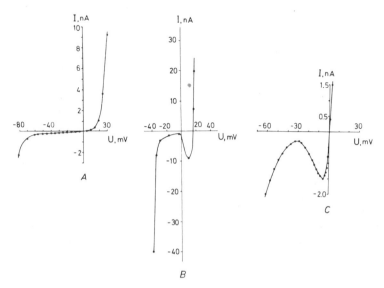

Fig. 191. Current—voltage characteristics of alamethicin-modified membranes. A, Egg lecithin bilayers in 0.1 M KCl solutions, alamethicin (3 · 10^{-6} M) added from the side of the anode [137]. B, Sphingomyelin bilayers interposed between 0.05 M NaCl solution and 5 · 10^{-8} M alamethicin-containing 0.1 M NaCl solution (anolite) [670]. C, Bilayer and solution composition see A; protamine added from the side of the anode in concentration 5 mg/ml [137].

[791] that in its presence alkali metal ions are extracted from aqueous solutions by organic solvents, the degree of extraction being independent of the pH of the aqueous phase. Such behavior is natural when the carboxyl group of the antibiotic does not participate in the cation binding, for instance, if a multidentate complex is formed with the amide groups as ligands. However, these results can hardly be regarded as unequivocal proof of complex formation, the more so that the NMR spectra of alamethicin solutions in water or organic solvents are not affected by the presence of K$^+$, Na$^+$ or Ca^{2+} [377]. Possibly, the cation extraction is due to incorporation of the aqueous phase into alamethicin micelles.

In its effect on lipid bilayers alamethicin differs considerably from both macrocyclic complexones, such as valinomycin or the nactins, and gramicidin. The chief peculiarity of alamethicin is that the bilayer conductance it induces depends strongly on the applied voltage [137, 301, 670, 671] (see Fig. 191). Over a certain range of voltage it obeys the exponential

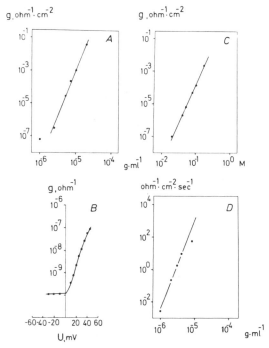

Fig. 192. Conductivity of alamethicin-modified bilayers. A, Effect of alamethicin concentration in the aqueous media on the conductivity of membranes from a 2.5% solution of sphingomyelin in a tocopherol—methanol—chloroform mixture (5:2:3) at 35°C. The aqueous solutions contained 0.1 M NaCl; applied voltage 60 mV [670]. B, Dependence of the conductivity on the NaCl concentration (for conditions see A). Alamethicin concentration 10 μg/ml. C, Dependence of the conductivity on the applied voltage for a membrane from egg lecithin in 5 mM NaCl solution. Alamethicin added from the anode side (3 μg/ml) [137]. D, Effect of alamethicin concentration upon the maximal rate of conductivity increase on applying a rectangular voltage pulse of 100 mV. For conditions see A.

expression $g \sim \exp(neU/kT)$ (see Fig. 192B), the curvature of which increases with increase in cation charge (for bilayers of egg lecithin or of its 1:1 mixture with cholesterol, the value of n in KCl, $CaCl_2$ and $AlCl_3$ solutions is 4.5—6.5, 10—12 and $\geqslant 17$, respectively) and is sensitive to the lipid structure (for dimyristoyl lecithin bilayers in KCl and $CaCl_2$ solutions n has the values 3.0—3.5 and ~ 7, respectively) [137].

Measurements of the membrane potential under salt gradient conditions showed that alamethicin-modified neutral or negatively charged lipid bilayers are predominantly cation permeable. However, if organic polycations (spermin, histones, protamine) are adsorbed on the membrane surface, the bilayers become more permeable to anions, showing that the ionic selectivity is apparently determined by charge of the membrane surface [137, 670].

When alamethicin is introduced from the side of the more concentrated salt solution, a negative slope region will appear on the current—voltage curve, whereas the high differential conductance is retained at zero current, corresponding to a membrane potential in the absence of an external field (see Fig. 191B). This can apparently be accounted for by the conductance increase occurring at lower voltages than the diffusion potential [226, 670]. A negative slope region can also appear at zero diffusion potential, for instance, on adding protamine on one side of the membrane (see Fig. 191C). Mueller and Rudin [670] ascribe this effect to asymmetry of the surface charge arising on adsorption of protamine.

The "switching voltage" at which conductivity begins to increase is linearly dependent on the logarithms of concentrations of alamethicin and of salts. Eisenberg-Grünberg [226], who investigated this relationship, showed that on addition of alamethicin in very high concentration ($\sim 0.5 \cdot 10^{-5}$ M) to one side of the membrane a negative slope region appears on the current—voltage curves, which resemble the curves obtained in the presence of protamine. In both cases the "switching voltage" is shifted to negative values.

The presence on the current—voltage curves of a branch with exponential growth of conductance at the negative voltages (positive on the side of alamethicin; see Fig. 191A) is apparently due to alamethicin diffusing through the membrane. Eisenberg-Grünberg showed [226] that this branch is absent from curves obtained for bacterial lipid bilayers which contain no unsaturated fatty acid residues.

Mueller and Rudin [670] could select a protamine concentration at which the alamethicin-modified bilayers under a salt gradient acquired a "bistability". Such membranes can be in two states, transitions between which are induced by current pulses. In one, the bilayers displayed predominantly cationic permeability and, in the other, predominantly anionic permeability. In some cases the regenerative process on applying a depolarizing current ended with transition of the membrane to the initial state*. On varying the conditions, Mueller and Rudin observed membrane

---

* Boheim [79] recently showed that regenerative transitions in such membranes can be initiated by hydraulic pressure pulses.

potential oscillations resembling sub-threshold responses and the action potentials of excitable biological membranes.

These effects are undoubtedly due to potential-dependent changes in the cationic and anionic conductance of the membranes. However, there is as yet no direct evidence that anion and cation transport across alamethicin-modified bilayers treated by protamine occur by a common mechanism independent of the sign of the ion charge. Boheim [79] believes that alamethicin contribution to the conductance of such bilayers is insignificant, and this antibiotic exerts only regulatory potential-dependent action on the protamine-induced transport.

The membrane conductivity is proportional to a sixth (ninth [226]) power of the alamethicin concentration over the entire voltage range (see Fig. 192A). This indicates participation in the ion transport of an ensemble of interacting antibiotic molecules. A similar dependence is also characteristic for the conductance relaxation rate under voltage clamp (see Fig. 192D). Mueller and Rudin also note the high exponent in the dependence of the alamethicin-modified bilayer conductance on the salt concentration in aqueous solutions (see Fig. 192B).

Important information on the mechanism of the alamethicin activity has been obtained by Gordon and Haydon [309, 310, 380] and also by Eisenberg-Grünberg [226] from studies on the fluctuations of membrane conductance. With decrease in the antibiotic concentration and lowering of the voltage, the fluctuations separate into packets in the intervals between which the conductivity is the same as in the non-modified bilayers. In each packet one may discern several discrete conductance levels (see Fig. 193). The current change on transition between neighboring levels exceeds by approximately one order of magnitude that for the gramicidin, so that there can be no doubt that alamethicin produces ion-conducting pores in the membrane.

The behavior of these pores is much more complicated than those of gramicidin. The packets are independent of each other and, in case of overlap, can be separated by statistical analysis of the fluctuation levels. The frequency of occurrence of the packets increases with increase in voltage and in the concentration of alamethicin and of salts in the aqueous solution*. This leads to the inference that each packet is the result of the "switching on" of a certain self-contained conducting unit which can assume several conductance states. Since the spacings between two neighboring conduc-

---

* The true order of the conductance dependence on the alamethicin concentration may differ from that found experimentally owing to association of the antibiotic in aqueous solution. According to MacMullen and Stirrup [622] the critical micellar concentration of alamethicin at $30°C$ in 0.05 M KCl solution (pH 8) is $2.4 \cdot 10^{-6}$ M.

320

tance levels are not equal (their ratios are $(0 \leftrightarrow 1):(1 \leftrightarrow 2):(2 \leftrightarrow 3):(3 \leftrightarrow 4) = 1.0:1.8:2.1:2.8:2.8$ [309]), this unit cannot consist of a set of functionally equivalent individual pores which, when it is "switched on", have the same probability of becoming permeable. The probability of each of the "switched on" units being on a given conductance level is not the same for all levels and does not depend on the voltage. From this one may conclude that the state of the "switched on" conducting unit is determined by the cooperative inter-

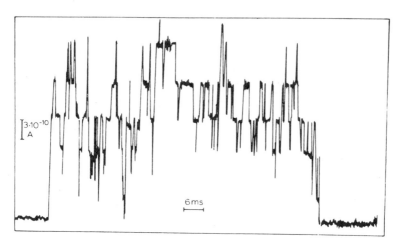

Fig. 193. Current oscillations across a bilayer membrane from a solution of glycerol monooleate in *n*-hexadecane. The alamethicin concentration in the 2 M KCl bathing solutions is very low. Applied voltage, 210 mV [309].

action of its constituents. Clearly each such unit must consist of several alamethicin molecules. At the same time it is still unknown whether each molecule forms a pore or whether the pore comprises all the molecules in the unit. In the former case, transition from one level to another should be accompanied by change in the number of pores, whereas in the latter the number should apparently remain constant, changes occurring only in the conveyance capacity.

Independent of the conductance level of the units they will display near ohmic current—voltage characteristics. Hence the non-linear character of the integral current—voltage characteristics of alamethicin-modified membranes is due to potential-dependent changes of the "switching on" and "switching off" probabilities of the separate units. This has been confirmed in a kinetic analysis of Eisenberg-Grünberg.

As already mentioned, the conductance of the alamethicin pores greatly exceeds that of the gramicidin pores. This leads to the belief that the former must have a larger diameter and that the pore-traversing ions, completely hydrated, interact very weakly with the walls. From Table 76 it can be seen that on passing from the first to the second level the conductance ratios approach the aqueous solution values, the membrane becoming relatively more permeable to the larger ions (here the hydrated cation radii should be compared). These results can be regarded as an argument in favor of the conductance increment being due to increased pore size. A clearer manifestation of this could apparently have been obtained with larger, organic cations.

TABLE 76

CONDUCTANCE RATIOS (g_{Mi}/g_K) CHARACTERISTIC OF THE AQUEOUS SOLUTION AND OF THE ALAMETHICIN CONDUCTING UNIT IN DIFFERENT STATES [226]*

Salt	Aqueous solution	Conducting unit state			
		I	II	III	IV
LiCl	0.48	0.26	0.50	0.48	0.46
NaCl	0.83	0.65	0.83	0.83	0.87
KCl	1	1	1	1	1
CaCl$_2$	0.77	0.58	0.73	0.74	0.76
Tris–HCl	0.49	0.23	0.38	0.38	0.41

* Aqueous solutions contain $2 \cdot 10^{-8}$ g/ml of alamethicin and 1 M of the salt. Bilayers were prepared from a decane solution of the bacterial phosphatidylethanolamine.

Unfortunately, the molecular mechanism of functioning of the conducting alamethicin unit is still unknown and one can merely conjecture as to whether it is due to change in orientation of the alamethicin molecules in the membrane, to changes in their conformation or to some other factors. The part played by cations in the "switching on" of conducting units and whether the membrane activity of alamethicin is at all related to its complexone properties (if it actually is a complexone) also remain unresolved. Eisenberg-Grünberg showed that the number of the "switched on" alamethicin units is proportional to $c_L^9 \cdot c_M^4$. On that basis he assumed that the cations are a necessary component of the latent unit, the field causing them to be ejected from the membrane and thus transforming the unit into a conducting state. The interest that many investigators are showing in the mechanism of alamethicin action on lipid bilayers is quite understandable because we have here a unique example of a compound with known chemical structure, of which the state of the molecules in the membrane should be so highly dependent on the electrical field.

The principles underlying the structure of the pores arising in molecules or supramolecular aggregates of peptide antibiotics incorporated into membranes may also be of interest from the viewpoint of the study of certain non-membrane cellular structures. One may opportunely recall Mitchell's electrophoretic mechanism of the movement of bacterial cells which carry immobile flagella. In terms of Mitchell's hypothesis the flagellum is regarded as a rod with an internal channel through which only certain ions can pass, a circumstance which prompted him to call it a "giant ionophore" [641].

V.D. Liposomes

In studying the properties of biological membranes, frequent use for models is made of the so-called liposomes*, water-filled microvesicles surrounded by a single lipid bilayer or particles consisting of several concentric bilayer membranes separating the water compartments (see, for instance, [135]). Liposomes are ordinarily prepared by shaking or by sonicating aqueous solutions containing phospholipids which had been preliminarily swelled therein. Transfer of molecules and ions from the internal water layers to the surrounding solution and vice versa requires passage of the molecules and ions across the lipid membranes. In the absence of membrane permeability increasing agents such transfer occurs rather slowly, making it possible arbitrarily to change the composition of the environmental solution by dialysis or gel filtration without affecting the liposome content. At the same time the lipid bilayer is usually very permeable to water molecules, so that actually the liposome is a sort of miniature osmometer. One usually follows the changes in the liposome volume which occur with shift of the osmotic equilibrium by measuring light absorption or scattering by the suspensions. Such properties make the liposomes an indispensable object for study of ion transport across lipid membranes, displaying many of the effects observed when membrane-active complexones act on cells and subcellular particles. Moreover, aqueous suspensions of the liposomes are a convenient system for optical and radio spectroscopic study of the interactions of the membrane-active compounds with the lipid membranes.

Liposomes effectively absorb lipophilic complexones from aqueous solutions. According to Johnson and Bangham [461] the partition co-

---

* Besides the term liposomes various authors use the names lipid liquid crystals, smectic mesophase, lipid micelles, etc., for these particles.

efficients of valinomycin between unsonicated lecithin liposomes and a 0.16 M KCl solution is about $4.4 \cdot 10^4$ (with respect to lipid). It has been shown with unsonicated cardiolipin liposomes that gramicidin and valinomycin markedly raise the membrane permeability to complexable cations already at antibiotic:lipid molar ratios of $1:(5—20) \cdot 10^6$. The observed effects are significantly diminished in sonicated liposomes. One of the possible causes for this is the considerable increase in the number of particles in the suspension which can even surpass the number of added antibiotic molecules. Since there can hardly be rapid transfer of antibiotic molecules between the liposomes, it is natural to expect that some of them should retain the initial permeability [842]. The effect of membrane-active complexones on ion transport in liposomes has been extensively studied by Chappell's group (see [386] and references therein). Employing radioactive alkali metal isotopes, these authors have shown that valinomycin, gramicidin, nigericin and allied compounds accelerate cation exchange between the liposomes and the environment, the selectivity sequence of this effect coinciding with that of complex formation. If the liposomes do not contain permeant anions the neutral complexones markedly increase efflux of complexable cations from the liposomes only if it is compensated by counter-fluxes of other cations or protons. Coupling of the flows can well be seen on the example of liposomes from a 9:1 mixture of lecithin and dicetyl phosphate*, containing a solution of alkali metal hydroxyadipates. If such liposomes are suspended in isotonic choline chloride solution, the potassium ion efflux will increase only weakly in the presence of valinomycin, the nactins and enniatin A. The efflux is limited by the low membrane permeability of anions, protons and choline. If compounds (dinitrophenol, FCCP†, etc.) augmenting the proton permeability of lipid bilayers are added, rapid $K^+—H^+$ exchange is observed. It has already been mentioned in Part V.C that gramicidin enhances not only the cationic but also the proton permeability of bilayers. In line with this, at sufficiently large concentrations gramicidin induces very fast $K^+—H^+$ exchange, the rate of which is independent of the presence of the proton carriers mentioned above. $K^+—H^+$ exchange is also considerably accelerated by compounds of the nigericin group (cf. p. 275). Here also the exchange rate does not increase with addition of proton carriers. These data lead to the inference that in the absence of permeant anions neutral complexones of the valinomycin type give rise to a potential difference on the liposome membranes determined by

---

* The addition of dicetyl phosphate confers a negative surface charge to the liposomal membranes, thus enhancing their anion impermeability (cf. p. 295).

† Carbonyl cyanide p-trifluoromethoxyphenylhydrazone.

324

the activity gradients of the complexable cations, whereas compounds of the nigericin group, neutral complexones in combination with proton carriers, and also (in high concentrations) gramicidin transform this cation gradient into a pH gradient.

If the internal or external solutions contain complexable cations and permeant anions, valinomycin-like compounds will stimulate their migration through the membrane along the concentration gradient, transport of the salt being accompanied by osmotic swelling or contraction of the liposomes. Barsukov *et al.* [50] have recently shown that unsonicated egg lecithin

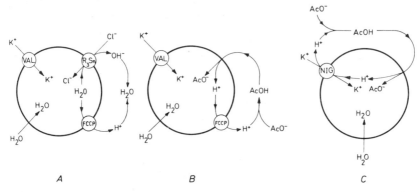

Fig. 194. Diagram of complexone-induced ion fluxes in liposomes undergoing swelling in potassium salt solutions: A, Combined action of valinomycin (VAL), tripropyltin chloride (R_3Sn) and proton carrier (FCCP) in KCl solution. B, Combined effect of valinomycin and proton carrier in potassium acetate (KOAc) solution. C, Effect of nigericin (NIG) in potassium acetate solution.

liposomes containing potassium salt placed in valinomycin-containing isotonic glucose solution increase the rate of potassium release in the following order with respect to the anions: $Cl^- < C_2O_4^{2-} < NO_3^-$. If liposomes containing KNO_3 are placed in isotonic KCl, in the presence of valinomycin they will first rapidly contract and then will slowly recover almost their original volume. This effect is apparently the result of the faster exit of KNO_3 than entry of KCl. A similar, although not so well expressed effect occurs when the surrounding solution contains $NaNO_3$.

The loss of alkali metal ions from liposomes containing the salts of strong acids is not accompanied by pH changes in the medium; this provides evidence of the coupling of the cation and anion fluxes.

The neutral complexone-induced transport of alkali metal chlorides can be accelerated by adding certain organometallic compounds. It has recently been shown that phenylmercury acetate and tripropyltin chloride promote transmembrane Cl⁻—OH⁻ exchange. In combination with proton carriers these compounds markedly increase the KCl transport in valinomycin-treated egg lecithin liposomes [880, 1046]. Fig. 194A schematically shows the correspondingly initiated fluxes.

Many weak acids (for instance, acetic acid) penetrate the lipid membranes as undissociated molecules easier than as anions. Consequently at pH values close to pK_a, transfer of the anions of such acids across the membrane can be regarded as *symport* with protons (see footnote to p. 333). In this case the anion flux is dependent on its concentration gradient and on the pH gradient rather than on the membrane potential. Naturally, transport of the salts of weak acids and of permeant cations will be significant only when protons are transported across the membrane. Hence, liposomes from negatively charged lipids placed in a potassium acetate solution containing valinomycin, the nactins or (in low concentration) gramicidin, swell very slowly. If, however, these substances are used in combination with proton carriers as well as in the presence of nigericin antibiotics the swelling rate becomes very high (for a schematic of the ion flows see Fig. 194B and C)*.

In all the above cases, the driving force of the complexone-induced transport was the artificially created ion gradients. However, induced osmotic flows can also be observed under non-gradient conditions. Singer and Bangham [42, 941] found that liposomes from a mixture of lecithin and dicetyl phosphate prepared in salt solution with which they are in osmotic equilibrium undergo swelling if their membranes become permeable to both cations and anions. Such swelling can, for instance, be observed in the presence of valinomycin in potassium salicylate of iodide solutions (salicylate and iodide are permeant anions) and in potassium acetate solutions on further adding proton carriers. On the contrary, in solutions of KCl or potassium acetate in the absence of proton carriers valinomycin cannot induce swelling, because of the low anion permeability of the liposomes. The nature of this effect is as yet unclear. Possibly the liposome forming process causes excess mechanical strain in the membranes and/or non-balanced

---

* Henderson *et al.* [386] have investigated the effect of the valinomycin antibiotics on the osmotic behavior of KCl-containing liposomes placed in isotonic solutions of other alkali metal chlorides. Under these conditions the antibiotics cause contraction of the liposomes, which is accelerated by proton carriers, proceeding faster the less specific the complexone to the external cation. According to the authors, Cl⁻—OH⁻ exchange takes place. However, the action of the proton carrier is possibly due to the presence of phosphate anions (5 mM Tris buffer) which, like acetate, could be transported across the membrane in the form of neutral phosphoric acid molecules. On the other hand one cannot exclude the possibility that the direct cause of the observed changes in the optical characteristics of suspensions were changes in the liposome structure accompanying the fall in pH of the water compartments.

repulsive forces between the membranes carrying charges of the same sign. An increase in the salt permeability should then disturb the osmotic equilibrium and the liposomes will pass over to a new state with a changed osmotic pressure in the interior.

The effect of the lipid composition of liposomes on the complexone-induced ion transport has only been investigated to a relatively small extent. Van Deenen *et al.* [194, 196] have shown that the rate of valinomycin-promoted $^{86}Rb^+$ exchange in lecithin liposomes falls as the unsaturated fatty acid residue content diminishes and also on introducing cholesterol (see

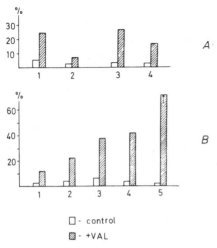

- □ - control
- ▨ - +VAL

Fig. 195. Effect of the lipid species on the rate of valinomycin (VAL)-induced $^{86}Rb^+$ release from lecithin liposomes containing 4% (mol.) of phosphatidic acid. The axis of ordinates represents the percentage of label remaining in the liposomes after incubation for 30 minutes at 40°C. A, Effect of adding cholesterol (30 molar %): 1, egg lecithin; 2, egg lecithin + cholesterol; 3, dioleyl-lecithin; 4, dioleyl-lecithin + cholesterol. B, Effect of the fatty acid structure: 1, dipalmitoyl-lecithin; 2, 1-palmitoyl-2-oleyl-lecithin; 3, egg lecithin; 4, dioleyl-lecithin; 5, lecithin containing up to 70% linoleate residues [196].

Fig. 195). Interestingly, the inhibiting effect of cholesterol is more strongly manifested if lecithin contains only a single mono-unsaturated hydrocarbon chain at the C_2 atom of the glycerol residue. As is well known, only then is the lecithin—cholesterol interaction manifested to the utmost, leading to more ordered chain packing in the hydrophobic zone of the membrane (cf. [197]). Hence, the effect of cholesterol is as in the case of bilayer membranes probably due to a decreased "bulk" conductivity (cf. p. 308). Liposomes from lysylphosphatidylglycerol are distinguished by a very low exchange rate, due most likely to the large positive surface charge of the membranes (cf. p. 295).

Of obvious interest is the possibility of employing membrane-active complexones as tools for the study of liposomal membranes. Some of the first work in this direction was that of Scarpa and De Gier [860]. These authors worked from the assumption that for membranes with low anion permeability the K^+—H^+ exchange rate in the presence of valinomycin is determined by the proton permeability, whereas in the presence of proton carriers (FCCP) it is determined by the potassium permeability. Comparing the rate of valinomycin- and FCCP-induced K^+—H^+ exchange driven by K^+ and H^+ gradients, they were able to show that liposomes display a low

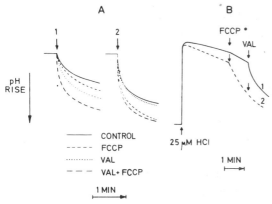

Fig. 196. Use of valinomycin and proton carriers for comparing H^+ and K^+ permeabilities of multilayer liposomes from egg lecithin (1) and lecithin containing up to 70% of linoleate residues (2). A, K^+—H^+ exchange in potassium-poor media (50 mM choline chloride, approx. 3 mM KCl, 0.5 mM choline citrate (pH 5.8)). Additives: valinomycin (VAL), 0.25 μg/ml; FCCP, $3 \cdot 10^{-7}$ M. The arrow indicates the time of addition of the liposomes. B, K^+—H^+ exchange in liposomes induced by acidification of the external solution containing 50 mM choline chloride. Composition of internal phase of liposomes: 40 mM choline chloride, 3 mM KCl and 10 mM choline citrate (pH 5.8) [860].

proton permeability which is only weakly dependent on the degree of unsaturation of the fatty acid residues, whereas their K^+ permeability markedly falls with decreasing unsaturation of the latter (see Fig. 196). The reliability of this conclusion, by the way, depends upon whether or not valinomycin and FCCP affect the H^+ and K^+ permeability of membranes respectively.

Little work published heretofore has concerned the mechanistic aspects of the complexone-induced cation transport in liposomes. Recently Johnson [460] investigated the temperature dependence of potassium transport in unsonicated lecithin liposomes and showed that valinomycin has no significant bearing on the enthalpy of activation of the process (approx. 15 kcal · mole^{-1}) so that its effect can be attributed to increase in the entropy of activation (+35 e.u./mole). The valinomycin-induced K$^+$ transport is augmented by butanol, diethyl ether and chloroform, an effect which is due to summation of the decrement in both the enthalpy and entropy of activation. Comparing the values found for the enthalpy of activation with the values characteristic of diffusion processes (2—4 kcal · mole^{-1}), Johnson proposed as the rate-limiting stage of cation transport the crossing of the membrane boundary. It should be mentioned that transport phenomena in multilayer liposomes are not very amenable to theoretical analysis since such particles comprise a system with several diffusion barriers. These difficulties can apparently be overcome on passing over to microvesicles with a single bilayer membrane [461]. One must take into account, however, that the membrane properties change considerably when the size of such vesicles is diminished to < 200 Å.

It was pointed out earlier that liposomes are a convenient object for spectral study of the properties of various molecules in a lipid environment. As a rule, for this "monolayer" microvesicles about 100 Å in diameter are used, so as to diminish the effects of light scattering and to obtain sufficiently narrow lines in the NMR spectra. Grell et al. [326] made use of spectropolarimetry to investigate the behavior of valinomycin and enniatin B molecules in lecithin membranes. They found that the CD spectra of valinomycin attached to liposomes are similar to its spectra in relatively low polar solvents and do not change on adding up to 0.5 M KCl to the medium. In the case of enniatin B, however, the CD spectra of the attached antibiotic resemble its spectra in polar solvents and change perceptibly in the presence of potassium ions (apparent complex stability constant, 5 l · mole^{-1}). It is interesting to compare these results with those of Haynes [382] who employed the fluorescent 8-anilino-1-naphthalenesulfonate (ANS) anions to determine complex formation. Using liposomes from dimyristoyl and dipalmitoyl lecithins he was able to show that below the lipid-phase transition temperatures ($T < T_t$) the ANS fluorescence is enhanced by alkali metal ions (Rb$^+ \geqslant$ K$^+ >$ Cs$^+ >$ Na$^+$), the effect being much greater in the presence than in the absence of valinomycin. It thus seems likely that the enhanced fluorescence is due to increased ANS binding by the membrane, coupled with complex formation. The magnitude of the quantum yield and the position of the maximum in the fluorescence spectrum permit the assumption to be made that the ANS is in a relatively polar environment, apparently being on the membrane surface or in its nearest layers. A similar effect is observed in the presence of other macrocyclic complexones (valinomycin $\approx$ monactin $>$ enniatin B $>$ cyclic polyethers of the type of

18-*crown*-6). For all the compounds except enniatin B complexing-dependent fluorescence was not observed at $T > T_t$. Since the experiments of Grell *et al.* were performed at $T > T_t$, the data obtained by both methods are not contradictory. Why the phase transition in the lipid membrane should so strongly affect the complexing of valinomycin is as yet obscure. Possibly the concentration of complexes within the hydrophobic zone of the membrane is so low as to be beyond the range of the applied spectral methods, only the surface complexes being detected. Under such circumstances it may be assumed that the ordering of the lipid chains on phase transition causes the valinomycin molecules to be "displaced" into the surface layers of the membrane and hence increases the content of the complexes. On the other hand, it is also possible that the phase transition is accompanied by a change in the interaction of valinomycin with its environment, augmenting the value of the complexing constant. Similar proposals have been made to explain the enhanced binding of potassium ions by valinomycin monolayers at the air—water interface observed at certain degrees of compression of the monolayers [592].

Chapman *et al.* [127, 263, 377] obtained interesting data on the behavior of alamethicin molecules in the lipid membrane. The adsorption of this antibiotic on liposomes from neutral lipids at pH $>$ pK_a of the carboxyl group leads to the appearance of a negative surface charge on the liposomes which can be estimated from their electrophoretic mobilities. It thus follows that in bound alamethicin the carboxyl group is localized on the membrane surface. These authors showed by ESR with spin label and NMR that insertion of alamethicin into the membrane brings about local changes in the lipid packing which is accompanied by decrease in mobility of the hydrocarbon chains and, to a lesser extent, of the choline residues. This effect is of a cooperative nature for, according to the authors' estimate, in liposomes of phosphatidylserine, for instance, a single alamethicin molecule immobilizes about 600 lipid molecules. Possibly alamethicin promotes the formation of a kind of lipid globule in the bilayer. Such globules could in fact be observed by Chapman and coworkers using electron microscopy in conjunction with freeze etching.

Chapter VI

BIOLOGICAL IMPLICATIONS

VI.A. Introduction

Valinomycin, nigericin and other complexones of microbial origin began their "scientific career" as antibiotics, but owing to toxicity and poor water solubility could find no practical application in medicine. However, they attracted the attention of investigators because of a marked ability to affect the energy-transduction processes in mitochondria, chloroplasts and other membrane systems responsible for ATP synthesis coupled with oxido-reductive reactions.

In the mid sixties it was found by Pressman's and Lardy's groups that there are two types of such compounds, which we shall conditionally call the valinomycin antibiotics (valinomycin, enniatins, nactins, gramicidins A—C) and the nigericin antibiotics (nigericin, monensin, dianemycin, etc.) which exert completely different effects on isolated mitochondria, but have in common the unique property that their activity depends upon the presence of alkali metal ions and, what is particularly significant, is linked with the energy-dependent transport of these ions across the mitochondrial membrane.

The key to the mode of action of these antibiotics was found when it was demonstrated that valinomycin-like macrocyclic complexones and gramicidin increase the alkali ion permeability of the membranes of mitochondria, erythrocytes, bacterial cells, etc. Concurrently it was revealed that the nigericin antibiotics stimulate non-electrogenic transmembrane exchange of protons and alkali metal ions.

Many aspects of the biological activity of both these groups of antibiotics were rationalized after establishment of a parallelism between their effect on passive cation transport across biological membranes and across model lipid bilayers, and association of these phenomena with complex formation. It immediately became clear that the ability of membrane-active antibiotic complexones to promote passive cation transport is not the direct outcome of the interaction of these compounds or their cation complexes with the membrane proteins. In all likelihood, interposed between the biomembrane surfaces are topologically continuous lipid-filled regions and in these regions the membrane-active complexones function in just the same way as they do in the lipid bilayers*. This accords well with the fact that the cation selectivity of induced transport depends only to a small extent upon the origin or composition of the biomembrane, being in the main determined by the specificity of the complexing reaction. In the early period of the study a number of workers considered that the effect of membrane-active antibiotic complexones on the energy-dependent cation transport in mitochondria, chloroplasts, etc., was due to stimulation or inhibition of the corresponding translocases. Now hardly anyone would put to doubt that in this case, too, cation transport is due to increase in membrane permeability and that its driving force is the membrane potential or the pH gradient.

In recent years investigators are turning more and more to membrane-active complexones when they wish selectively to increase the cation permeability of biological membranes or stimulate M^+-H^+ (or $M_i^+-M_j^+$) exchange. In fact, an arsenal of methods has been created that permits a wide range of problems in biophysics and biochemistry to be solved with the aid of these compounds. Many examples of the successful use of valinomycin, nigericin and other antibiotic complexones are discussed in the present Chapter and their number is continually increasing.

May one maintain, however, that the biological action of complexones affecting membranes can be completely rationalized in terms of increased cation permeability or induced cation—proton exchange? This question is still largely unanswered. In pertinent parts of this Chapter evidence will be presented, for instance, of a specific effect of valinomycin on electron transport in chloroplasts and chromatophores, which still awaits an unequivocal explanation. Generally speaking, the observation of "side" effects of complexones, particularly when the latter are used in high concentrations, should not be surprising, since it is clear that inclusion of a large number of bulky foreign molecules into the membrane must have its repercussions. The possibility that in certain cases these compounds can have an effect associated not only (or not so much) with increase in cation permeability of the membrane as with the presence in the latter of lipophilic cation complexes can also not be excluded. For example, the incorporation of complexes of the nigericin antibiotics may lead to exchange of the carboxyl protons of the membrane protein for alkali metal ions. The interaction of cation complexes with proteins acquires particular interest in connection with the unusual properties of antamanide. It has already been mentioned, in Part I.B.2, that this cyclopeptide which in vitro selectively complexes sodium and calcium ions and has a very weak augmenting effect

---

* For a detailed discussion of the molecular architecture of biological membranes and, particularly, of the existence of fluid bilayer regions in the latter, see, for instance, [82, 128, 268, 319, 409, 506, 513, 612, 942, 943, 975].

on cation transport through membranes, in vivo neutralizes the toxic effects of phalloidine and α-amanitine*. The mechanism of this protective action is not only completely unknown, but one cannot even be certain that it is at all connected with increase in membrane permeability. Possibly, a still undiscovered mechanism by which complexones can influence biological processes is operative here.

VI.B. Effect of membrane-active complexones on passive ion transport

VI.B.1. Mitochondria

The effect of membrane-active complexones on passive ion transport in resting mitochondria† has been most elaborately studied on the outwardly manifested acceleration of swelling in 100—150 mM alkali metal salt solutions. Under these conditions the response depends to a large extent upon the nature of the anions in the medium. These can be divided into two principal groups. The first comprises the anions of strong acids with $pK_a <$ 4; some of these (nitrate, iodide, thiocyanate) readily penetrate the inner mitochondrial membrane over a wide range of pH values, for others, such as chloride, the membrane becomes permeable when the pH is raised to 8.5—8.8 [27]. To the second group belong the anions of weak acids such as acetate. In most cases the mitochondrial membrane is less permeable to such ions than to molecules of the corresponding undissociated acids. The electrogenic transport of these anions across the membrane therefore occurs at much lower rates than their non-electrogenic symport‡ with protons. Here also belong substrate anions and phosphate which by specific mitochondrial permeases are non-electrogenically transported across the inner membrane, the passage ultimately being coupled with proton transfer (for more details see Part VI.C.4 and, particularly, Fig. 222 on p. 376).

Usually valinomycin considerably accelerates mitochondrial swelling in potassium salt solutions at pH $\geqslant$ 8.5. When the pH is lowered to 7, the rate (but not the ultimate magnitude) of the swelling decreases; the effect of pH is particularly manifest in the case of the low permeant anions of strong acids (for instance, Cl^-) [27, 75, 93, 650]. Similar phenomena are observed in solutions of rubidium, cesium and ammonium salts, whereas valinomycin

---

* Amanitine is an inhibitor of nuclear DNA-dependent RNA-polymerase [264, 450, 661, 662, 685, 690, 878, 1058, 1110]; the mode of action of phalloidine is unclear [869, 1036, 1058].

† The oxidation of endogenic substrates is usually inhibited by rotenone. The effect of mitochondrial ATPase is nullified by antibiotics of the oligomycin group.

‡ By symport is meant transport of two substances (or ions) across a membrane, when their flows are spatially linked, are uni-directed and are related by constant proportion. The term antiport is used for the case of counter-flows [643].

does not accelerate swelling in the case of salts of non-complexable cations (Na^+, $(CH_3)_4N^+$, Tris-onium) or potassium salts of macromolecular, non-permeant anions (ribonucleate) [27, 93, 95]. With the potassium salts of weak acids (such as acetate) swelling in neutral media is synergistically accelerated by valinomycin and proton carriers such as dinitrophenol or FCCP. This effect is also characteristic of other neutral antibiotic complexones [93, 650].

Synergism of valinomycin and FCCP is also observed when the mitochondria are placed in a potassium chloride solution containing compounds stimulating non-electrogenic transmembrane Cl^-—OH^- exchange (tripropyltin chloride, diphenyliodonium salts or phenylmercury acetate) [403, 880] (cf. [7, 189, 879, 881, 1047]; see Fig. 197).

Usually the swelling is assessed from the diminution of light scattering by suspensions of the mitochondria. It has also been shown in a number of studies that the change in the optical characteristics of the mitochondria in the presence of valinomycin-like complexones is accompanied by the absorption of considerable amounts of salt from the medium [27], an increase in the sucrose-inaccessible space* [75] and also morphological changes typical of the so-called high-amplitude swelling [74, 75]. Swelling is inhibited, with increased osmolarity of the medium, by adding sucrose or mannitol [27].

These data show that valinomycin accelerates mitochondrial swelling in salt solutions by selectively increasing the permeability of the inner membrane of the mitochondria to complexable cations [386, 650, 691, 835]. An ion-flux diagram demonstrating the synergism of valinomycin and proton carriers is presented in Fig. 197.

Gramicidin in low concentrations (0.1 μg/mg mitochondrial protein) behaves like valinomycin [413, 681, 999]. Characteristic differences between their action on mitochondria could be unmasked in potassium acetate solution. In that case, when the gramicidin concentration is raised to 1 μg/mg protein, the swelling rate sharply increases but with simultaneous weakening of the synergistic effect of the proton carriers [94, 386]. This is due to the fact that in high concentrations gramicidin increases not only the cationic, but also the proton permeability of membranes (see Parts V.C and V.D).

In contrast to valinomycin and other neutral membrane-active complexones, antibiotics of the nigericin group only insignificantly augment the swelling rate of mitochondria in solutions of the potassium salts of strong

---

* This space is determined by the volume enclosed by the inner mitochondrial membrane [538].

acids (chloride, nitrate, thiocyanate) but their effect is considerably enhanced in the presence of proton carriers. The swelling rate is higher the greater the permeability of the inner membrane to the respective anions ($NCS^- > NO_3^- > Cl^-$). However, these same antibiotics cause very rapid swelling of the mitochondria when the incubation medium is a solution of

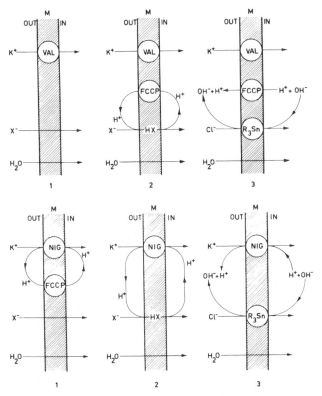

Fig. 197 (above). Valinomycin(VAL)-induced swelling of resting mitochondria in potassium salt solutions. Diagram of ion flows. 1, X^-, permeant anion; 2, synergistic action of proton carriers when X^- is the anion of the permeant acid; 3, synergistic action of tripropyltin chloride and proton carriers in a KCl solution.

Fig. 198 (below). Nigericin(NIG)-induced swelling of resting mitochondria in potassium salt solutions. Diagram of ion fluxes. 1, synergistic action of proton carriers when X^- is the permeant anion; 2, X^-, anion of the permeant acid; 3, synergistic action of tripropyltin chloride and proton carriers in a KCl solution.

the potassium salts of permeant weak acids, the swelling rate then being independent of the presence of proton carriers. Such activity of the nigericin antibiotics can be rationalized in terms of their ability to stimulate transmembrane non-electrogenic M^+–H^+ exchange (cf. Chapter V). From the scheme represented in Fig. 198 it can be seen that augmentation of the mitochondrial swelling rate by these substances is accompanied by cyclic transfer of protons across the membrane [386, 650]. Apparently one can expect nigericin to cause rapid swelling of the mitochondria in KCl solution in the presence of tripropyltin (cf. p. 324; see Fig. 198, 3).

With both the valinomycin complexones and the nigericin-type compounds the cation selectivity sequences for the induced swelling of mitochondria are similar to the corresponding sequences revealed in experiments with model lipid membranes and also in complexing studies in solution (see, for instance, [386]). Owing to its low cation selectivity, gramicidin equally accelerates mitochondrial swelling in solutions of the salts of all alkaline metals [75].

Thus, the effect of antibiotics on mitochondrial swelling in solutions of the salts of complexable cations can be fully rationalized in terms of increase in osmotic ion fluxes across the inner membrane (see, for example, [835]). The same conclusion is arrived at from direct studies of alkali metal ion fluxes with the aid of isotopic labels [27]. The absence of qualitative differences in the behaviors of resting mitochondria and liposomes renders improbable the hypothesis of Green et al. [74, 75], whereby under such conditions the driving force of the swelling process is the specific conformational transitions in subunits of the inner membrane when they bind the permeant ions (cf. [318, 320, 1104, 1105]).

In isotonic media which contain very small amounts of complexable cations (solutions of sucrose, choline chloride, etc.) the membrane-active complexones induce exchange of intramitochondrial potassium for protons (see Fig. 199A and B), proceeding either without significant change in the mitochondrial volume or with some contraction due to exit of endogenic anions [119, 386, 691, 723, 790]*. If in the preincubation period the mitochondria are enriched with permeant anions such as thiocyanate, the valinomycin-induced potassium ion outflow is not accompanied by proton uptake [153]. The rate of the K^+-H^+ exchange induced by valinomycin, nonactin or (in low concentrations) gramicidin is considerably augmented in the presence of proton carriers [30, 386, 603, 649, 690] or on lowering the pH of the medium [119]. With the nigericin antibiotics or with gramicidin in high concentrations K^+-H^+ exchange is very rapid and undergoes no acceleration when proton carriers are added.

If in a potassium-free medium sodium salts of permeant weak acids (for

---

* The valinomycin- and nigericin-induced cation transport across the mitochondrial membranes is inhibited by nupercaine [30, 727]. This basic amphipatic substance, which is a potent local anaesthetic, affects the transport apparently by changing the membrane surface charge (cf. Part V.C.2.a).

instance, acetate) are present causing the mitochondria to swell*, valino-mycin will inhibit this process [691]. This is because the valinomycin-induced K^+–H^+ exchange lowers the matrix pH, inhibiting the accumulation of weak acid anions in the mitochondria (cf. p. 375 and Fig. 199C). For the same reason in potassium-poor media the rapid K^+–H^+ exchange induced by the nigericin antibiotics, or by neutral complexones in combination with proton carriers, is accompanied by considerable depletion of the mito-chondria in phosphate and substrate anions (see, for example [241, 398, 459, 691, 723–725, 729, 802]). As will be shown in Part VI.C.4, the drop in the matrix pH due to K^+–H^+ exchange determines to a considerable extent the effects exerted by the membrane-active complexones on the metabolic processes in mitochondria.

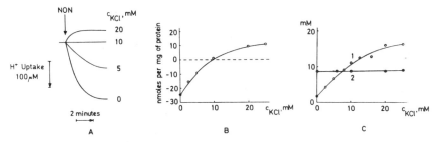

Fig. 199. Effect of K^+ concentration in the medium on the proton and permeant weak acid transport in rat liver mitochondria [723]. The mitochondria were suspended in a solution containing sucrose and KCl (total osmolarity 250 mosm), 2 mM Tris–HCl, rotenone and oligomycin. A, Nonactin(NON)-induced change in pH of the suspension (NON, 0.02 µg per mg protein); B, proton transport (positive values refer to H^+ uptake by mitochondria); C, label uptake from the media containing 1 mM [^{14}C]malate: 1, intramitochondrial malate concentration 90 sec after adding nonactin; 2, control.

Many authors have observed acceleration of Ca^{2+}–K^+ exchange by valinomycin [33, 119, 691, 825, 826, 858]. As is well known, calcium ions easily pass through the inner mitochondrial membrane (see, for instance, [539]), so that the above effect is evidently due to increase in the K^+ permeability. In potassium-free media valinomycin induces potassium outflow from the mitochondria accompanied by uptake of calcium ions. This uptake is inhibited with increase in potassium ion concentration in the medium or in the presence of dinitrophenol or La^{3+} ions [33, 858]. For similar data on gramicidin see [130, 826].

---

* In that case swelling is determined by the presence in the mitochondrial membrane of a system providing for non-electrogenic Na^+–H^+ exchange [650].

Azzone *et al.* [603, 858], studying the kinetics of $Ca^{2+}-K^+$ and K^+-H^+ exchange in the presence of valinomycin and proton carriers, arrived at the conclusion that the mitochondrial membrane contains a common antiporter for Ca^{2+}, K^+ and H^+, effecting non-electrogenic translocation of the ions across the internal zone of the mitochondrial membrane. In their opinion the internal zone is filled with protein molecules and is not very permeable to either potassium or proton carriers, the latter functioning only in the peripheral lipid layers of the membrane. Thus, potassium transfer by the hypothetic antiporter and by valinomycin are of necessity consecutive stages in the induced transmembrane potassium transport. This assumption is, however, completely invalidated by the fact that under the conditions of a potassium gradient, the action of valinomycin leads to the appearance of a membrane potential in the mitochondria (see p. 402).

VI.B.2. Erythrocytes

It has been shown by isotope methods that neutral membrane-active complexones stimulate alkali metal ion exchange between erythrocytes and the environment, the relative rate of this process being determined by the position of the ions in the selectivity sequence of a complexone [386]. The ratio of the unidirected valinomycin-induced isotopic potassium fluxes over a wide range of concentrations is equal to the ratio of potassium activities in the cells and environment.

Independently of the initial cytoplasmic potassium and sodium contents (which are different in erythrocytes of different animal species, and, sometimes, even of subspecies) valinomycin and the nactins promote leveling of the K^+ gradient in the plasma membrane [794]*. The outflow of potassium ions from erythrocytes treated by these compounds is accompanied by no noticeable proton transport and its rate is apparently limited by the membrane permeability to chloride, the major component of the low molecular anions of the cytoplasm. Supporting evidence for this is the considerable acceleration of potassium outflow and the absorption of stoichiometric amounts of protons on further adding proton carriers of the type of FCCP [367, 386, 783] (see Fig. 200B).

Gramicidin in concentrations up to 1 μg/ml has the same effect on the ion flows in erythrocytes as do neutral macrocyclic antibiotic complexones, the only difference being its low ion selectivity [367, 386]. As the gramicidin concentration increases the erythrocyte membrane acquires marked proton

---

* The metabolism-independence of ion flows induced by valinomycin and the nactins in erythrocytes has been demonstrated by inhibition of glycolysis, K,Na-dependent ATPase and carbonic acid anhydrase [859, 1013].

permeability with concomitant acceleration of K^+–H^+ exchange in salt-free media which ceases to depend upon the presence of proton carriers [386]. Bielawski [65] and Scarpa *et al.* [859] showed that in potassium or sodium chloride-rich media erythrocytes swell in the presence of gramicidin; with sodium chloride a transient shrinkage associated with potassium outflow is first observed. The swelling rate depends upon the nature of the external ion, decreasing in the order chloride $\approx$ acetate $>$ succinate $>$ nitrate $>$ ribonucleate. By varying the tonicity of the medium by addition of sucrose,

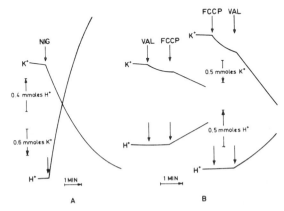

Fig. 200. Effect of nigericin (A) and valinomycin (B) on K^+–H^+ exchange in erythrocytes [369]. The cells (0.042 ml per 1 ml) were suspended in a medium containing 300 mM sucrose, 10 mM NaCl, 20 mM Tris–HCl. Additives: 0.5 μg/ml nigericin (NIG), 0.4 μg/ml valinomycin (VAL), $5 \cdot 10^{-7}$ M FCCP.

conditions can be found under which the gramicidin-induced changes in the cationic composition of the erythrocytes are not accompanied by volume changes. Subsequent washing of the gramicidin restores the initial permeability of the membrane, thus permitting cells to be obtained with arbitrary potassium and sodium contents [65].

In isotonic media of low alkali ion content nigericin antibiotics induce exchange of the internal potassium and sodium for protons, the exchange rate being independent of the presence of proton carriers [367, 386] (see Fig. 200A). In media enriched by alkali metal salts (150 mM) these compounds cause erythrocytes to exchange their potassium ions for environmental cations. The cation exchange is accompanied by transient changes in the environmental pH, the direction and magnitude of which depend upon the relative positions of potassium and the external cation in the sequence of the cation complexing selectivity of the antibiotic. Thus with K^+-specific nigericin in NaCl medium transient basification is observed,

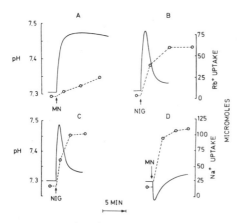

Fig. 201. M$^+$—H$^+$ exchange in erythrocytes in the presence of monensin and nigericin [386]. Composition of the medium in experiments A and B: 77.5 mM RbCl (^{85}Rb), 70 mM choline chloride, 16 mM Tris—HCl (pH 7.3). Composition of the medium in experiments C and D: 150 mM NaCl (^{22}Na), 16 mM Tris—HCl (pH 7.4). Additives: 32 µg/ml monensin (MN), 32 µg/ml nigericin (NIG). Solid lines represent the changes in pH; dashed lines, accumulation of the label in the cells.

whereas with Na$^+$-specific monensin transient acidification occurs [386] (see Fig. 201). Evidently this phenomenon reflects the competition of protons and cations for the complexone.

VI.B.3. Bacterial cells

Owing to the work of Harold's group [354, 357—359, 361, 362, 735] the effect of membrane-active complexones on resting *Streptococcus faecalis* cells has been elucidated in detail. It was shown by tracer experiments with ^{86}Rb$^+$ that in concentrations of $10^{-8}—10^{-6}$ M the valinomycin and nigericin antibiotics stimulate alkali ion exchange between the cytoplasm and the medium in native cells and in protoplasts without changing the membrane permeability to phosphate, amino acids and such non-electrolytes as mono- and disaccharides. Contrary to valinomycin and gramicidin, monactin and nigericin can be washed from the cell surface by water or buffer solutions. Thus cells can be obtained with varying cationic composition, but with the original permeability. The rate of Rb$^+$ exchange for ions of the other alkali metals depends upon the position of the latter in the complexing selectivity sequence of the antibiotic and decreases in the presence of Ca^{2+}, Mg^{2+} and organic polycations (for instance, spermin). Valinomycin and monactin also induce exchange of intracellular potassium for lipophilic permeant cations such as dimethyldibenzylammonium or triphenylmethylphosphonium [361].

It is interesting that in the presence of valinomycin the rate of $*Rb^+-Rb^+$ exchange falls sharply on lowering the temperature from 37 to 0°C, whereas in the case of gramicidin exchange is rapid also at low temperatures. Thus, in respect to the temperature dependence of the antibiotic-induced cation permeability bacterial membranes reveal a striking resemblance to lipid bilayers (cf. p. 306). One might therefore assume that the difference in the mode of action of valinomycin and gramicidin described in Part V.C is characteristic not only of model, but also of biological, membranes.

In media with low alkali metal ion content, nigericin and gramicidin promote exchange of intracellular potassium for protons. Such K^+-H^+ (or

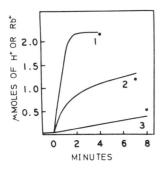

Fig. 202. $^{86}Rb^+-H^+$ exchange in *Streptococcus faecalis* cells in the presence of gramicidin and valinomycin [357]. The curves show the amount of dilute HCl required to maintain the pH of the cell suspension (0.7 mg/ml) constant after adding 1 µg/ml of gramicidin (1), 1 µg/ml of valinomycin (2) and in control (3). Black points designate $^{86}Rb^+$ leakage from the cells.

Rb^+-H^+; see Fig. 202) exchange, proceeds very rapidly, also in the presence of valinomycin and monactin. Because of this Harold et al. [357, 358] assumed that macrocyclic antibiotics augment not only the cation, but also the proton permeability of the plasma membranes. However, doubts arise as to the validity of this hypothesis since it contradicts the sum total of the accumulated data on the membrane activity of valinomycin and the nactins. More plausible seems to be the assumption that the membranes of non-metabolizing bacteria possess a certain inherent proton permeability and/or that the potassium ion outflow is electrically coupled to the exit of anions. This refers particularly to lactate anions, which in the resting *Streptococcus faecalis* cells usually occur in large amounts, being the end products of glycolysis. Evidence in support of this assumption can be seen in the biphasic character of K^+-H^+ exchange at large valinomycin concentrations, the second, slower, stage being considerably accelerated on adding proton carriers [313]. Since, as Harold *et al.* [360—363] have shown (cf. Part VI.C.6), glycolysis in *S. faecalis* is a main source of ATP for active electrogenic proton transport (negative within the cells) there should be no accumulation of lactate in actively glycolyzing cells of *S. faecalis*, the membranes of which are permeable to lactate. Of course such an assumption requires experimental verification.

Faust *et al.* [256] recently investigated the effect of valinomycin, gramicidin and the nactins on passive cation flow in *Pseudomonas fluorescens* cells. The results obtained are in qualitative agreement with the findings of Harold's group. Strikingly differing is the effect of valinomycin on the K^+-H^+ exchange in bacteria capable of intensive respiration, the membranes of which display low proton permeability. Thus, when *Staphylococcus aureus* 209P cells in potassium-free media are treated with this antibiotic, K^+-H^+ exchange becomes significant only on further adding dinitrophenol [313].

An interesting observation has been made by Pavlašova and Harold [735] who found that valinomycin has no significant effect on the permeability of the membranes of *E. coli* cells (which are resistant to the bacteriostatic action of this antibiotic; see Table 77 on p. 398). The cells can be sensitized to valinomycin, however, by treating them with a solution of EDTA in Tris buffer (pH 8). If within a short time after such treatment the cells are placed in a valinomycin-containing medium, the K^+ permeability of their membranes is considerably augmented (cf. [1053]). It seems as if the resistance of the bacterial membranes to the action of valinomycin depends in some way on the presence of calcium ions.

The effect of the lipid composition of the bacterial membrane on the magnitude of the valinomycin-induced K^+ permeability has been investigated by Haest *et al.* [342]. These authors showed that the rate of K^+ and Rb^+ transport in *S. aureus* depends upon the lysylphosphatidylglycerol: phosphatidylglycerol ratio in the membrane, the ratio being varied by changing the pH of the incubation medium. With increase in proportion of the former lipids, the rate of the induced cation transport falls, apparently owing to increase in the positive surface charge density on the membrane (cf. Part V.C.2.a). Earlier, in Part V.D, the similar behavior of the liposomes has been mentioned. This analogy bears witness to the fact that the structures of the ionic heads of the lipid molecules determine the barrier properties not only of purely lipid, but also of biological, membranes. At the same time these experiments are a good illustration of the applicability of complexones as indicators of changes in the surface charge of biological membranes.

VI.C. Effect of membrane-active complexones on energy-linked processes in biological membranes

VI.C.1. General considerations

As is well known, cells contain specialized membrane structures (mito-chondria, chloroplasts, the membranes of respiring and photosynthesizing bacteria) where the exergonic oxidoreductive reactions are energetically coupled to the endergonic synthesis of ATP. Thus the energy liberated in the electron transfer from donors to acceptors is converted into energy of the macroergic pyrophosphate bonds. Numerous experimental data bear evidence of the fact that prerequisite to the coupling is that the membrane be topologically closed. On the other hand an intimate relationship exists between energy transformations in such systems (which we shall further call coupling systems) and transport of ions and molecules across the sur-rounding membrane (see, for instance, [299, 385, 525, 538, 539, 799, 944, 1042].

Beginning mainly with the studies of Pressman's [366, 663, 778—780] and Lardy's groups [314, 527], the effects of valinomycin, nigericin and other antibiotic complexones on the energy-dependent processes in coupling systems have been extensively investigated by numerous workers. A characteristic feature of the observed effects was their alkali-ion dependence, the degree of which closely follows the complexing selectivity for the given ion. Moreover, the antibiotic complexones were found to stimulate or inhibit the energy-dependent transmembrane transport of the complexing cations. In their time these findings were the starting point for the hypothesis that the inner mitochondrial membrane contains primary or secondary cation translocases capable of interacting in some way or other with antibiotics and/or their complexes (see, for instance, [297, 364, 526, 603, 785, 791, 829]). However, as the experimental data began to accumulate it became more and more evident that such translocases had to be ascribed very unusual and sometimes mutually contradictory properties. Insurmountable obstacles arise as well in attempts to bring into accord the hypothetical complexone-induced activation or inhibition of cation translocases with the well-known ability of the complexones to transport cations across mem-branes, i.e. to act as ionophores.

At the same time the ionophoric properties of the antibiotics served as a basis for rationalizing their effect on the coupling systems in terms of the transmembrane proton electrochemical potential differences ($\Delta \bar{\mu}_H$) as a form of the conservation of energy and mediating factor in its transformation. The most consistent and elaborated development of this concept is to be found in Mitchell's chemiosmotic theory of energy transformation in coupled systems [640, 642—646]*. The distinguishing feature of the chemiosmotic theory is that $\Delta\bar{\mu}_H$ is regarded as the sole form of energy exchange between the electron transfer chains and the ATP synthesizing system (see Fig. 203).

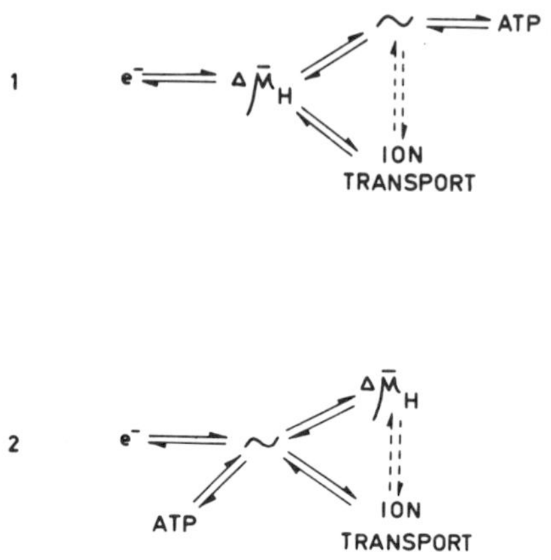

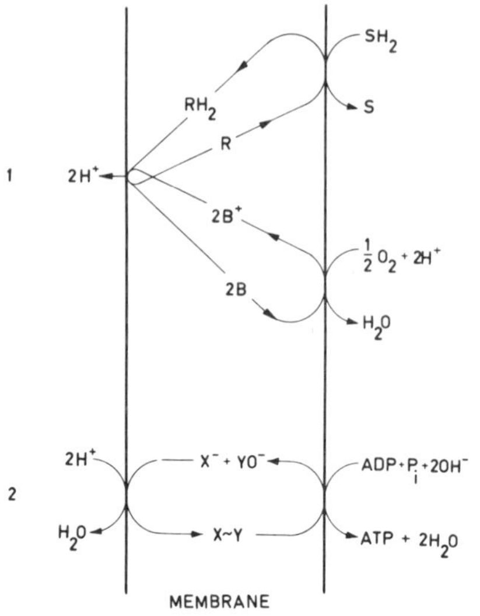

Fig. 203 (left). Energy transduction in coupling systems. 1, *Chemiosmotic hypothesis.* The electron transfer chain (e^-) is in itself a primary proton translocase generating a proton electrochemical potential difference ($\Delta\bar{\mu}_H$) on the coupling membrane. The potential difference serves as energy source for ATP synthesis which possibly proceeds via formation of macroergic bonds ($\sim$) other than pyrophosphate. Ion transport is carried out by $\Delta\bar{\mu}_H$-driven secondary translocases. The existence of other than proton primary translocases in coupling membranes is not very probable. 2, *Chemical hypothesis.* The energy liberated in exoergonic oxidoreductive reactions is directly accumulated in the form of macroergic intermediates ($\sim$) utilized in ATP synthesis and for driving proton and/or other ion pumps. Ions could also be transported by secondary translocases using membrane potential or ΔpH as the energy source.

Fig. 204 (right). Scheme of coupling between oxidoreductive and ATP-synthesizing units. 1, respiratory chain; 2, reversible ATP-driven proton translocase (transport ATPase).

According to Mitchell, electron transfer in respiratory or photosynthetic chains is linked to directed transmembrane proton transport. Such coupling is due to the vector nature of the flows of electrons and hydrogen equivalents in a membrane and their union into consecutive "loops". The

---

* A detailed account of the chemiosmotic theory can be found in the reviews by Greville [328] and Skulachev [949]; see also [31, 118, 126, 292, 356, 385, 551, 819, 832, 945—948, 950, 1088].

underlying principle for their operation is schematically presented in Fig. 204. Thus, electron transfer chains play the part of electrogenic proton pumps transforming the energy liberated in the oxidoreductive reactions into a proton electrochemical potential difference across the membrane. In mitochondria the functioning of such pumps leads to proton ejection from the matrix into the surround (negative inside), whereas in chloroplasts, bacterial chromatophores and in ultrasonic submitochondrial particles with opposite membrane polarity, the protons are "pumped" from the environment into the interior (positive inside) (see Fig. 205).

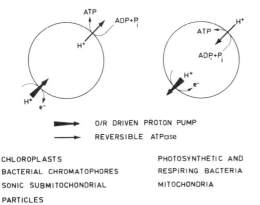

O/R DRIVEN PROTON PUMP

REVERSIBLE ATPase

CHLOROPLASTS

BACTERIAL CHROMATOPHORES

SONIC SUBMITOCHONDRIAL

PARTICLES

PHOTOSYNTHETIC AND

RESPIRING BACTERIA

MITOCHONDRIA

Fig. 205. Polarity of different phosphorylating membrane systems.

According to Mitchell, $\Delta\bar{\mu}_H$ is the driving force of ATP synthesis by reversible membrane ATPase, the formation of each ATP molecule being accompanied by proton transport across the membrane in a direction opposite to that in which protons are "pumped" by the respiratory or photosynthetic chain (see Figs 204 and 205). Evidently when ATPase is functioning under conditions of ATP hydrolysis it also fulfils the role of an electrogenic proton pump.

Such a coupling mechanism can, of course, exist only when the membrane is topologically closed, possessing low proton permeability. It is also quite clear that the phosphorylating capacity in such systems depends upon the ratio of the free energy of ATP hydrolysis to the value of $\Delta\bar{\mu}_H$. The latter is determined by the electrical potential differences ($\Delta\psi$) and the pH gradient (ΔpH) existing on the coupling membrane and can be represented by Eqn 37.

$$\Delta\bar{\mu}_H = F \cdot \Delta\psi - 2.303 \, RT \cdot \Delta\text{pH} \qquad (37)$$

Within the framework of Mitchell's theory the number of "loops" in the electron transfer chain necessarily corresponds to the number of so-called coupling sites, i.e. loci in the chain where electron transfer is accompanied by ATP synthesis. The hypothetical scheme for the mitochondrial respiratory chain is represented in Fig. 206. It can be seen from this scheme that the individual "loops" are the proton pumps operating in parallel.

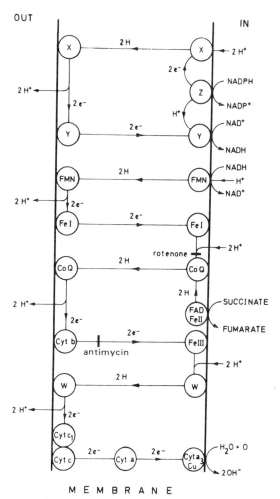

Fig. 206. Hypothetical respiratory chain lay-out in the inner mitochondrial membrane. At the top: one of the possible schemes for the functioning of the transhydrogenase system. In the other "loops" the sequence of the chain components are according to Skulachev [946]. X, Y, Z and W are unidentified components.

By means of selective inhibitors, electron transfer can be blocked at different loci of the chain and then, by choosing appropriate donors and acceptors of electrons and hydrogen equivalents, one can observe $\Delta\bar{\mu}_H$ generation in the operation of a shortened chain or even of single "loops" (for further details see, for instance, [946, 949]). Thus, according to Mitchell, the electron transfer chain and ATPase operating under conditions of ATP hydrolysis are *primary translocases*, i.e. enzyme systems which make direct use of chemical energy for driving the transport processes [643]. The transmembrane proton electrochemical potential difference arising from the action of such translocases is utilized as the energy source for the ionic or molecular transport. The driving force of the transport serves the membrane potential or the pH (or $\bar{\mu}_H$) gradient. In the former case, ion translocation is of an electrophoretic nature and in the latter, it is ultimately the result of

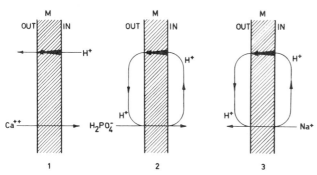

Fig. 207. Energy-dependent ion transport in mitochondria (the black arrow indicates the proton pump). 1, electrophoretic Ca^{2+} transport; 2, symport of protons and phosphate anions; 3, antiport of protons and sodium ions.

ion symport or antiport with protons or OH^-. Thus, for instance, calcium ions are transported across the inner mitochondrial membrane electrophoretically [539, 946], phosphate ions in symport with protons (or in antiport with OH^- ions) [398, 459, 614, 729, 732], and sodium ions in antiport with protons [647, 650] (see Fig. 207).

Despite the fact that some of the assumptions of Mitchell's theory have not yet been experimentally verified, it has not only afforded a unified explanation of a wide number of phenomena brought to light in studies of various coupling systems, but has demonstrated considerable predictive powers, stimulating an enormous number of new, original investigations.

An attempt will now be made to examine some of the underlying principles of the effect of membrane-active complexones on energy transformation processes in the coupling systems. In order for this to be done a number of the characteristics of energy-dependent ion transport in such systems and its connection with $\Delta\bar{\mu}_H$ will first be briefly discussed. It should particularly be stressed that although further discussion will be in terms of chemiosmotic theory, the treatment will in many aspects be valid for any coupling mechanism involving an electrogenic proton translocase for which the primary energy source is oxidoreductive reactions and ATP hydrolysis.

Let us assume for simplicity that in the initial state a proton and alkali-ion impermeable coupling membrane is interposed between two volume-restricted aqueous solutions of the same composition and that the ATP, ADP and phosphate concentrations as well as the oxidation substrates and O_2 (in oxidative phosphorylation) or electron acceptor (in phosphorylation under constant light intensity) remain constant during functioning of the membrane.

In such a system first let phosphorylation be inhibited by removing ADP or phosphate or by treatment with specific ATPase inhibitors. Then, after some time, this system will pass into a quasi-equilibrium state wherein the net rate of electron transfer (and, consequently, of active proton transport) will be zero and the difference $\bar{\mu}_H$ corresponding to the change in free energy of the oxidoreductive reaction will be established in the membrane. We shall refer to the coupling system in such a state as "energized".

Let us consider two limiting cases. If the membrane is ion impermeable, the number of protons crossing it is determined by the electrical capacitance and in the quasi-equilibrium state the main contribution to $\Delta\bar{\mu}_H$ will be made by the membrane potential. At this point there will be very little change in pH. It can thus be shown by means of Eqn 38

$$\Delta pH = \frac{C \cdot \Delta\psi}{F \cdot B} \tag{38}$$

(where C is electrical capacitance of the membrane; B, the buffering capacity of the solution; and F, Faraday's number) that, for instance, for mitochondria ($C/F = 4\ \mu$ equiv $\cdot$ V^{-1} per g of protein, $B = 12\ \mu$ equiv of protons per pH unit calculated per g of protein [644]), the maximum pH change in the internal phase would all in all amount to only 0.07 pH unit for $\Delta\psi = 200$ mV.

If, on the contrary, the membrane is ion permeable (say to the anions X$^-$) then during the energizing process parallel pH and pX gradients will be established, the pH gradient being higher the higher the X$^-$ ion concentration. Indeed, under such conditions $\Delta\psi = -2.303RT/F \cdot \Delta pX$ and, consequently, $\Delta\bar{\mu}_H = -2.303RT(\Delta pH + \Delta pX)$. Since both phases must remain electrically neutral ($\Delta a_H = \Delta a_X$), clearly the equilibrium ΔpH value will increase with decrease in pX. It is also evident that the relative ΔpH contribution to $\Delta\bar{\mu}_H$ will be less and the number of protons transported across the membrane in the transient process will be more the higher the buffering capacity of the solutions.

Now, what effect will the presence of membrane-active antibiotic complexones exert on the ion and proton distribution during energization of our ideal systems?

Let the aqueous solutions contain large amounts of potassium ($a_X \geqslant a_K \gg a_H$) as the only complexable cation. If, in the first of the aforementioned limiting cases, valinomycin can impart K^+ permeability to the membrane of the energized system, a situation will obtain similar to that for a system with an anion-permeable membrane. Under such conditions $\Delta\psi$ should diminish and ΔpH increase and this should be accompanied by the establishment of a pK^+ gradient opposite to the pH gradient. In the transient process the electrophoretic potassium ion flow will be directed oppositely to the active proton transport, and, as Mitchell [644] showed, the fall in $\Delta\psi$ will then precede the increase in the pH gradient. As a result a transient fall in $\Delta\bar{\mu}_H$ will occur due to energy consumption for establishing the ion gradients. The final state will be characterized by the quantities: $\Delta\psi = 2.303\, RT \cdot \Delta pK^+/F$ and $\Delta\bar{\mu}_H = 2.303\, RT\, (\Delta pK^+ - \Delta pH)$.

However, when nigericin is acting on an anion-permeable coupling membrane, non-electrogenic $K^+ - H^+$ exchange develops and the system tends to a state such that alongside the energy-dependent pH gradient a parallel and equal pK^+ gradient will appear on the membrane (see Eqn 35). In the transient process proton migration will be of a cyclic character and the potassium ions will move in parallel with active proton transport. At the same time in the transient process $K^+ - H^+$ exchange lowers the pH gradient (and consequently $\Delta\bar{\mu}_H$).

It should be stressed that, if in our idealized system no processes are accompanied by electrogenic proton transport along the pH gradient, all ion flows should be non-steady state. On the contrary, in the presence of proton carriers (such as FCCP), valinomycin in combination with nigericin and also (in high concentrations) gramicidin (cf. p. 334) will cause the idealized system to pass into a new functional state characterized by a definite steady-state operating rate of the proton pump which, under such conditions, is the driving force for the cyclic proton (and, sometimes, other ion) transport across the membrane (Fig. 208). With such cyclic transport the energy liberated from oxidoreductive reactions is completely transformed into thermal energy. Passage into this state is accompanied by diminution of both $\Delta\psi$ and ΔpH.

Let us now assume that our ideal systems can carry out phosphorylation. Then, in the steady state, the rate of electron transfer parallels the phosphorylation rate; in other words these processes are coupled. By treatment with membrane-active complexones we force the coupled system to generate ion gradients; the resulting osmotic work should therefore be accompanied by a decrease in $\Delta\bar{\mu}_H$ and, correspondingly, by inhibition of ATP synthesis. Concurrently, the diminution in $\Delta\bar{\mu}_H$ activates the proton pump. Hence, during establishment of the ion gradients, electron transfer will precede phosphorylation, which is equivalent to an uncoupling of these processes. If the membrane becomes proton permeable, the uncoupling is not accompanied by ion gradients.

In real coupling systems electron transfer proceeds under steady-state conditions at a certain definite rate even in the absence of phosphorylation. As example may be cited the so-called respiratory control state in

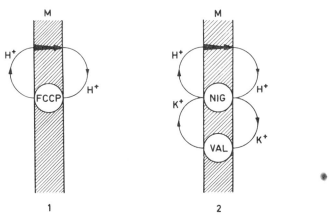

Fig. 208. Short-circuited proton translocase (designated by black arrow) in uncoupling by proton carriers (1) and by valinomycin (VAL) in combination with nigericin (NIG) in the presence of potassium ions (2).

mitochondria, when electron transfer is limited by the absence of phosphorylation substrates (Green's fourth state [538]), and the respiratory rate is only a few percent of the maximal. In any case the relative contribution of ΔpH to $\Delta\bar{\mu}_H$ in the steady state depends upon the ratios of the proton and ion fluxes, increasing with increase in the ion permeability of the membrane. Since the effect of valinomycin is associated with electrical shunting of the membrane and is manifested in a fall in $\Delta\psi$ (with nigericin it is the ΔpH which is lowered), low ion-permeable coupling systems can be expected to decrease $\Delta\bar{\mu}_H$ considerably in the presence of valinomycin, whereas highly ion-permeable systems should be especially affected in this respect by nigericin.

It should be stressed that for purposes of simplification it was tacitly assumed that energy transformation in coupling systems takes place under conditions close to equilibrium. In many cases such treatment can lead to erroneous conclusions since, in fact, phosphorylation proceeds on a background of intensive fluxes of electrons, ions and reacting molecules and the work of the coupling systems can, in principle, be described only in terms of non-equilibrium thermodynamics. This refers especially to evaluation of the various gradients as the driving force for the one or the other process. As Lehninger, Carafoli and Rossi [539] have pointed out ". . . the chemiosmotic hypothesis actually does not require that ⟨for phosphorylation⟩ a very large gradient of H^+ exists across the mitochondrial membrane during active electron transport. The gradient can be quite small providing the flow of electrons through the system is high. The potential work that can be performed by a pH gradient in a non-equilibrium system is a function of the product of the flow and the gradient." This also fully refers to cases where operating values of ΔpH and $\Delta\psi$ are assessed from the standpoint of driving forces of the complexone-induced ion flows. Unfortunately up to now this highly important aspect of the functioning of coupled systems has usually remained outside the attention of investigators. Undoubtedly many unclear features arising in treatments of the effect exerted by antibiotic ionophores on the energy transformation and ion transport processes will be explained as mathematical models of the ion flow and chemical reaction dynamics in coupling systems are developed. The only attempt we know of to analyze this problem theoretically is that of Kaplan and Essig [118] (see also [831]).

Coupling organelles can be regarded as microvesicles whose volume is incomparably smaller than that of the environment. The volume asymmetry manifests itself primarily in the ion gradients being practically completely due to concentration changes of the organelle content. Moreover, the small volume of the internal phase makes for rapid completion of the transient processes. As a rule, establishment of steady-state ion gradients when proton pumps are switched on (by supplying substrates or O_2, exposure to light or the addition of permeability-promoting substances) takes no more than 0.5—2 min.

Up to now we have intentionally not touched upon the question of osmotic equilibrium. However, as we have seen, ion transport in energized coupling systems may change the osmolarity of the internal phase. Since biological membranes possess high permeability towards water molecules, this should stimulate swelling or contraction. The direction of the volume change depends on the polarity of the membrane, i.e. on whether the protons are "pumped" into or out of the organelle. At the same time the membrane polarity determines to a large extent both the number of ions carried across the membrane during the complexone-induced transition to a new stationary state and the uncoupling effect.

As a matter of fact both swelling and contraction counteract the generation of the ion gradient and increase the amount of transported ions. But the absolute values of volume changes during organelle contraction are, as a rule, rather small and, moreover, the exit of ions cannot exceed their initial content in the organelles. These are the reasons for a relatively weak uncoupling effect of complexones when they induce energy-dependent outflow of cations. On the other hand, swelling of the organelles, particularly those in which the membrane forms numerous folds (mitochondria), can go quite far without impairing the membrane integrity. Correspondingly, in cases when the complexones induce cation influx-driven swelling their uncoupling action will be much more strongly pronounced. Some authors (see, for instance, [803]) assume that swelling leads to growth in the proton permeability of the inner mitochondrial membrane. This should inevitably increase the degree of uncoupling and inhibit ion transport. In the limit, swelling may lead to rupture of the membrane and hence to complete uncoupling.

Certain other important features of real coupling systems, distinguishing them from the above model should be mentioned:

(i) In some of these systems, mitochondria, for instance, substrate oxidation and ATP synthesis are carried out on the inner side of the membrane. This means that in the steady state the substrates, ADP and phosphate must be passed across the membrane to the interior of the mitochondria and ATP in the opposite direction. Since, as we shall see, these transport processes depend upon $\Delta\psi$ and ΔpH, the observed change in rate of respiration and phosphorylation in the mitochondria on treatment with membrane-active complexones can ultimately be caused by changes in the steady-state concentration of the reactants and products in the intramitochondrial space.

(ii) The coupling systems may function under conditions of considerable ion gradients. This refers particularly to isolated organelles when they are placed in media of high or low pH, or give rise to a high concentration gradient of the complexable alkali metal ions on the membrane. In these conditions the complexone-induced effects display a number of specific features. Thus nigericin in salt-free media will induce energy-independent K^+-H^+ exchange which will lead to considerable acidification of the internal phase and its depletion of potassium ions. The action of valinomycin in these circumstances can lead to generation of such a large membrane potential that it will spark off the ATP-synthesizing system in the absence of "normal" energy sources. From this standpoint an important, but as yet little explored, factor is the activity coefficients of the alkali ions within the organelles. In particular, practically nothing is known about the effect of pH on the degree of binding of these ions.

(iii) The effect of membrane-active complexones on energy transformation processes is sometimes determined by the dependence of the activity of the electron transfer chain and the ATP-synthesizing system upon the internal pH.

The action of membrane-active complexones on coupling systems will be considered for chloroplasts, bacterial chromatophores, mitochondria and ultrasonic submitochondrial particles. It should be noted that the majority of data in favor of the participation of $\Delta\bar{\mu}_H$ in the energy transformations of the coupled systems was obtained just by studying the action of valino-mycin, nigericin and other antibiotic complexones. At present, however, much independent information has been accumulated regarding the sign, value and mode of generation of $\Delta\bar{\mu}_H$ and in particular of the trans-membrane electrical potential difference (see, for instance, [945—949, 1087, 1088]). This permits one to break through the vicious circle when explaining the effects of these compounds in terms of chemiosmotic theory.

Not setting ourselves the goal of a comprehensive survey of the literature, we shall consider only the most characteristic phenomena, giving special attention to studies in which membrane-active complexones have been directly used to elucidate various functional aspects of coupling systems.

VI.C.2. Chloroplasts

Transformation of light energy in the chloroplasts of green plants takes place in the coupling membranes forming thylakoids — flat, closed vesicles stacked into compact piles called grana*.

The nigericin antibiotics diminish both the initial rate and the limiting value of the light-induced proton uptake by isolated chloroplasts, providing alkali ions are present in the incubation medium [195, 717, 718, 884]. This leads to decrease in the light-induced generation of a pH gradient in the thylakoid membrane [833]. For all the antibiotics investigated the depen-dence of the inhibition on the cation species coincides with the cation complexing selectivity (see, for instance, [884]). Inhibiting proton uptake, nigericin-like complexones simultaneously stimulate uptake of complexable alkali ions [195, 717, 718, 884] (see Fig. 209). The proton uptake-inhibiting effect can thus be qualitatively explained as due to K^+—H^+ exchange.

Taking into account the properties of this group of antibiotics it would

---

* For recent data on the structure and function of chloroplasts see [451, 1012, 1087, 1088].

have been natural to expect that the amount of absorbed cations, ΔM, will coincide in magnitude with the decrease in amount of absorbed protons, $\Delta(\Delta H)$. However, according to Shavit *et al.* [195, 884] the $\Delta M/\Delta(\Delta H)$ ratio is usually much higher than unity during the entire light-induced uptake of protons and cations (cf. [718]). The cause of the apparent deviation from stoichiometry of K^+–H^+ exchange is still unclear. Interestingly, under the same conditions (1 mM KCl) the $\Delta M/\Delta(\Delta H)$ ratio is higher in the presence of nigericin than of monensin. As is well known, monensin only weakly

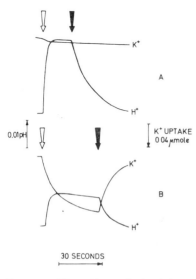

Fig. 209. The effect of nigericin on the light-induced ion fluxes in spinach chloroplasts [884]. Chloroplasts (130 μg chlorophyll) were suspended in 6 ml of a medium containing 100 mM choline chloride, 1 mM KCl and 7 μM pyocyanine (pH 6.3). The white and black arrows indicate the moments of, respectively, switching on and off the light. A, Control; B, the medium contains $7 \cdot 10^{-7}$ M nigericin.

interacts with potassium ions (see Parts I.F.3, V.B and V.D). Moreover, the above ratio increases with increase in nigericin concentration and light intensity. One may thus infer that $\Delta M/\Delta(\Delta H)$ is at least partially dependent on kinetic factors. On the other hand, the nigericin-induced fall in ΔpH leads to a change (in comparison with the control) of the buffering capacity of the thylakoids and, possibly, of the intrathylakoid potassium activity co-efficient.

Illumination of chloroplasts suspended in nigericin-containing KCl solutions causes their swelling, which proceeds markedly slower than potassium uptake [885]. The swelling rate is apparently limited by anion influx because it is considerably retarded on changing the Cl^- in the medium for the less permeant SO_4^{2-} anions [870]. Chloroplasts treated with glutaraldehyde are incapable of swelling. On treating them further with nigericin the light-induced proton uptake is inhibited, but no potassium uptake can be detected [718]. These findings most probably reflect the connection between the swelling ability of organelles and the amount of ions they take up in the presence of the ionophores (cf. p. 352).

If the chloroplasts are equilibrated with a solution of the salts of weak, permeant acids (for instance, alkali acetates), they contract on exposure to light owing to diffusion of undissociated acid molecules in the direction of higher pH, i.e. into the environment [174, 719, 720] (cf. p. 337). Under such conditions nigericin inhibits chloroplast contraction by diminishing the light-dependent pH gradient [719].

Besides lowering the steady-state pH gradient, in light-exposed chloroplasts the nigericin antibiotics induce acceleration of electron transfer, inhibition of both photophosphorylation and the light-triggered ATPase activity, and also of ATP—P_i exchange [677, 717, 884—886]. The most probable explanation of the uncoupling action of these antibiotics is that the pH gradient is the principal contributor to $\Delta\bar{\mu}_H$ in the energized chloroplasts. As we have seen above, such a situation may arise in coupled systems either when the membrane is permeable to ions (but not to protons) or the electrical capacitance is large in comparison with the buffering capacity. It is often assumed that the chloroplast membrane is permeable to Cl^- (see, for instance, [190, 394]; cf. [449]). However, in the dark the swelling rate of chloroplasts in concentrated KCl solution (0.1 M) in the presence of valinomycin is relatively low, increasing considerably on passing over to well-known permeant anions such as iodide or thiocyanate [653, 870]. Moreover, in KCl solutions valinomycin stimulates dark K^+—H^+ exchange, the direction of which is determined by the direction of the K^+ gradient. Thus, under such conditions the Cl^- efflux turns out to be less than the proton influx. Schuldiner and Avron [870] proposed that the Cl^- permeability increases in the light, but it has recently been shown that the conductance of thylakoid membranes appears to undergo practically no change on exposure to light [311]. Some authors assume that chloroplasts are permeable mainly to potassium ions, the conjecture sometimes being made that the buffering capacity of thylakoids decreases with increase in their potassium ion content (see, for instance, [653, 654]). If that is so, nigericin-induced potassium ion uptake should lead to proton release within the thylakoid so that the fall in ΔpH caused by the transmembrane K^+—H^+ exchange should be retarded. Possibly it is this effect which explains why nigericin in low concentrations (approx. 3 ng/mg chlorophyll) causes marked potassium uptake in the light but still does not uncouple photophosphorylation [884]. In general, the question as to whether the predominant part played by ΔpH in chloroplast energization is the result of the high ratio of electrical capacitance to the buffering capacity or is due to the considerable ion permeability of the membrane can as yet not be considered to have been completely clarified*.

Karlish and Avron [472, 473] have found that, in contrast to nigericin, valinomycin causes a noticeable increase in the uptake rate of protons by illuminated chloroplasts. These authors showed that stimulation of proton transport depends on the potassium ion concentration in the medium, increasing with increase in pH from 6.5 to 9. In media of relatively high potassium content, the increase in pH also leads to increase in the limiting value of the proton uptake (by 2.5 times in a medium containing 30 mM KCl and diquat (1,1'-ethylene-2,2'-dipyridinium dibromide) as electron acceptor). Concurrently, valinomycin induces potassium ion efflux from illuminated chloroplasts. The counter direction of the proton and potassium ion flows suggests that the latter ions are carried across the membrane electrophoretically. However, according to Karlish and Avron, the amount of potassium ions crossing the membrane is independent of the pH of the incubation medium. This results in a fall in $\Delta H/\Delta K$ with increase in pH so that the ratio may even become less than unity. Such divergence from stoichiometry indicates, in any case, transport across the membrane of either other ions or undissociated weak acids or bases. No rational explanation has as yet been advanced for this phenomenon. It should be mentioned, by the way, that the very fact of the induction by valinomycin of light-dependent K^+-H^+ exchange in chloroplasts has as yet found no confirmation in the work of other authors (cf., for instance, [654]).

Much also still remains unclear concerning the underlying mechanism whereby valinomycin affects electron transfer and ATP synthesis in the chloroplasts. This antibiotic inhibits both cyclic and non-cyclic photophosphorylation, but this effect bears no evidence of uncoupling. According to Karlish and Avron [472], inhibition of ATP synthesis in the presence of diquat or pyocyanine is due to inhibition of electron transfer, which is

---

* A number of authors issue from the concept that the main contribution to the buffering capacity of the thylakoids is made by the acid groups located in the internal surface of the membrane, considering the number of these groups to be variable depending on the changes in the membrane structure. Hence, on the one hand the degree of ionization of these groups (depending on the cation concentration) determines the structural and functional characteristics of the membrane and, on the other hand, the energy-dependent conformational transitions in the membranes lead to changes in the buffering capacity which may serve as the driving force for the ion transport. For a more detailed discussion of these questions see [653] (cf. [321]).

observed also in the absence of ADP. On the other hand, Keister and Minton [479] were able to observe inhibition by valinomycin of the photoreduction of ferricyanide only in the presence of phosphorylation substrates. The cause for this discrepancy is unclear.

Karlish and Avron [472] have found that inhibition of electron transfer increases with increase in pH in parallel with stimulation of proton uptake. It is thus likely that suppression of electron transfer is associated with increase

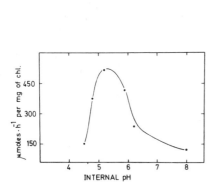
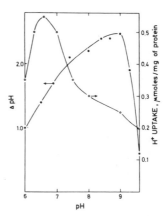

Fig. 210 (left). Dependence of the electron transfer rate in chloroplasts upon the pH of the internal phase [834]. The chloroplasts (75 μg chlorophyll per ml) were suspended in a medium containing 25 mM KCl, 1.5 mM potassium ferricyanide and 10 mM tricine buffer. Light intensity $1.6 \cdot 10^5$ lx. The ΔpH values were calculated from data on the distribution of NH_3.

Fig. 211 (right). Effect of pH on the energy-dependent pH gradient and the proton uptake by isolated chloroplasts [834]. The ΔpH values were calculated from data on the $^{14}CH_3NH_2$ distribution, when chloroplasts (400 μg chlorophyll per ml) were suspended in a solution containing 100 mM KCl, 10 μM pyocyanine, 1 μCi/ml $^{14}CH_3NH_2$ and 10 mM tricine buffer. When measuring proton uptake the chloroplasts (75 μg chlorophyll per ml) were suspended in a medium of lower buffer capacity and higher pyocyanine content. Light intensity $1.6 \cdot 10^5$ lx.

in the transmembrane pH gradient, or, what in the present case amounts to the same thing, with lowering of the pH of the thylakoid interior.

The possible mode of action of valinomycin considered by these authors is based on the fact that electron transfer in chloroplasts is slowed down on decrease of the internal pH to below 5.5 (see Fig. 210). When chloroplasts are illuminated in a medium of pH 8.5, the internal pH is close to 6 (cf. Fig. 211). The fall in the internal pH which under these conditions is due to the valinomycin-stimulated proton uptake leads to inhibition of electron transfer. When the environmental pH is lowered to 6.5 the steady-state internal pH approaches the point of maximum buffer capacity; as a result valinomycin only slightly affects ΔpH and so now only weakly inhibits electron transfer. It should be noted, however, that there is as yet no direct proof of the ability of valinomycin to lower significantly the internal pH of thylakoids.

Possibly there are two independent mechanisms for the valinomycin inhibition of electron transfer in chloroplasts. In support of this is the weaker inhibiting effect of valinomycin at low concentrations ($< 10^{-7}$ M) when the medium contains agents promoting the leveling of the energy-dependent pH gradient. Such agents are, for instance, methylamine, the well-known proton carrier S-13 ($2',5$-dichloro-$4'$-nitro-3-*tert*-butylsalicyloyl-anilide) [479] or gramicidin (in 10^{-6} M concentrations this antibiotic completely dissipates ΔpH) [836]. Thus, valinomycin at low concentration inhibits the electron transfer most likely due to pH decrease within the thylakoids. With increase in the valinomycin concentration, inhibition of the electron transfer becomes increasingly noticeable when proton uptake is also inhibited. When valinomycin is in high concentrations its effect on electron transfer is independent of the presence of potassium ions [479, 654]*. It is noteworthy that potassium-independent inhibition of electron transfer is observed also on treating bacterial chromatophores with valinomycin (see Part VI.C.3). To what extent this effect is specific for valinomycin is still unknown. Highly interesting is therefore Bachofen's investigation of the effect on chloroplasts of another neutral antibiotic complexone, dinactin [37]. In contrast with valinomycin, dinactin uncouples photophosphoryl-ation, especially in media of high sodium and low potassium content, and also induces contraction of the chloroplasts concurrently with their release of potassium. As the relative potassium content of the medium increases, the degree of uncoupling and of proton uptake diminishes. If these findings are confirmed, they can be regarded as evidence of an essential difference between the effects of valinomycin and dinactin on energy transformations in chloroplasts.

The effect of gramicidin on light-induced processes in chloroplasts is in good accord with its well-known ability in low concentrations to increase the membrane permeability mainly to cations and in high concentrations also to protons (see Parts V.C, V.D and VI.B.1). In line with this in 1—100 nM concentrations gramicidin is an effective phosphorylation uncoupler. In low concentrations (0.1—1 nM) it behaves like valinomycin, i.e. inhibits electron transfer and augments proton uptake (H/e) (see Fig. 212).

Both valinomycin and gramicidin (the latter in low concentrations) inhibit

---

* Molotkovsky and Dzyubenko [654] recently reported another potassium-independent effect of valinomycin. They showed that this antibiotic inhibits energy-dependent swelling of chloroplasts in both KCl and NaCl solutions.

light-dependent proton and potassium ion uptake in the presence of nigericin and enhance the uncoupling effect of this antibiotic [195, 683, 718, 884, 885]. This effect is in all likelihood due to proton shunting of the membrane, coupled with cyclic potassium ion transport (cf. Fig. 208, 2). Interestingly, valinomycin increases the ammonium salt-induced uncoupling of photophosphorylation in chloroplasts [96, 609, 683]. Uncoupling is here supposed to be due to decrease in ΔpH resulting from proton binding in the thylakoids by NH_3 molecules penetrating through the membrane. Hence, on exposure to light, chloroplasts accumulate ammonium ions [173]. Since valinomycin augments the NH_4^+ permeability of membranes (cf. p. 281), in its presence ammonium ions function as proton carriers (see Fig. 213) and the pH gradient dissipates at much lower ammonium salt concentrations in the medium*.

Witt and coworkers obtained interesting results from a study of the action of valinomycin on non-steady-state ion flows on exposing the chloroplasts to short ($< 2 \cdot 10^{-5}$ sec) flashes of light. These studies are of particular significance for the information they give on the magnitude and relaxation kinetics of the photo-induced electrical potential gradient in chloroplast membranes. To measure the potential gradient these authors made use of a "molecular voltmeter", viz. the electrochromism of the membrane carotenoids, owing to which changes in $\Delta\psi$ affect the chloroplast absorption in the 515 nm region (see [467—470, 863, 1087, 1088] and references therein).

A single flash exposure of the chloroplasts leads to rapid ($< 2 \cdot 10^{-8}$ sec) generation of a potential gradient†, which subsequently diminishes with a time constant of approx. 100 msec [470]. Using umbellipherone as membrane-impermeant pH indicator, it was found that parallel with

---

* Some authors (see, for instance, [479, 870]) believe that the chloroplast coupling membrane is also somewhat permeable to ammonium ions in the absence of valinomycin. This assumption explains, among other effects, the uncoupling action of ammonium ions in media containing such non-permeant anions as aspartate, i.e. under conditions when the relative contribution of ΔpH to $\Delta\bar{\mu}_H$ is possibly small. However, it becomes correspondingly difficult to understand why chloroplasts do not swell in hypertonic NH_4Cl solutions, assuming Cl^- to be permeant (see p. 355). As we have seen above, a similar problem arises when chloroplasts are placed in valinomycin-containing hypertonic KCl solutions. The validity of the assumption of ammonium ion permeability of thylakoid membranes raises misgivings also in connection with the data of Neiman et al. [683], which show that ammonium salts have no effect on the dark relaxation times of the electrochromic shift of chloroplasts absorption in the 520 nm region. At the same time this parameter is sensitive to permeant cations such as sodium ions in gramicidin-containing media (cf. p. 362).

† Here the potential gradient is generated much faster than any ion gradients can be established. Its appearance is most likely due to charge separation in the membrane as a result of the primary photochemical act [335, 1068] (cf. [449]).

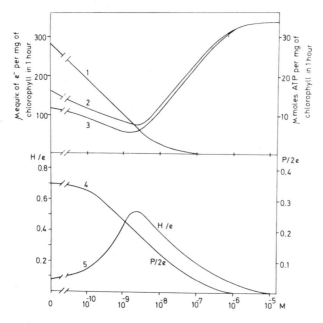

Fig. 212. Effect of gramicidin on phosphorylation and electron transfer in chloroplasts [472]. Chloroplasts (44 μg chlorophyll per ml) were suspended in a medium containing 30 mM KCl, 0.8 mM phosphate, 0.8 mM Mg^{2+}, 0.5 mM azide and 10 μM diquat (pH 8.0); ADP was added to a concentration of 0.2 mM. Actinic light from a 500 W tungsten lamp. 1, ATP synthesis; 2, rate of electron transfer in the presence of ADP; 3, rate of electron transfer in the absence of ADP; 4, ratio of phosphorylation and electron transfer rates (P/2e); 5, ratio of proton transport and electron transfer rates (H/e).

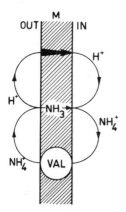

Fig. 213. Transmembrane ion fluxes in chloroplasts and bacterial chromatophores on phosphorylation uncoupling with valinomycin (VAL) and ammonium ions (the black arrow marks the proton pump).

diminution of the potential gradient a pH gradient is generated across the membrane which also then diminishes with a time constant $\tau_{1/2}^{H} \approx 1$ sec.

The pH gradient may be regarded as resulting from cation and anion displacement of protons and OH^{-} ions that are produced on the internal and external membrane surfaces, respectively, by the light-induced oxido-reductive reactions. Alongside this, electrophoretic transmembrane ion transport develops, dissipating the membrane potential. Subsequent ΔpH diminution is due to proton release by the thylakoids coupled with influx of cations or efflux of anions [335].

Valinomycin ($5 \cdot 10^{-7}$ M) and gramicidin (10^{-9} M) at pH 8 affect neither the amplitude of ΔpH nor the value of $\tau_{1/2}^{H}$. This means, first of all, that at the stage of ΔpH decrease, the membrane proton conductance is low relative to that for ions. Indeed, if proton release which is electrically coupled to transport of other ions is not at the same time the rate-limiting stage, an increase in cation permeability should markedly lower the $\tau_{1/2}^{H}$ value. Interestingly, $\tau_{1/2}^{H}$ sharply decreases under the combined action of valinomycin and dinitrophenol ($5 \cdot 10^{-7}$ and $2 \cdot 10^{-4}$ M, respectively), although the latter is inactive by itself. According to Grünhagen and Witt [335] the inability of dinitrophenol to stimulate proton release from chloroplasts is due to this process being driven by a special system in the membrane of non-electrogenic proton symport and/or antiport with other ions. Should this be true, the separate increase in the H^{+} or M^{+} permeability of membranes should not accelerate proton efflux. However, another proton carrier, ClCCP (the chlorine analog of FCCP), in $4 \cdot 10^{-6}$ M concentration diminishes $\tau_{1/2}^{H}$ in a similar way to the combination of valinomycin and dinitrophenol. One therefore comes to the conclusion that the synergism of valinomycin and dinitrophenol is in fact of a different nature. It should be mentioned that in the case of chloroplasts valinomycin synergistically augments the uncoupling effect of dinitrophenol, and also of FCCP in suboptimal concentrations ($< 10^{-7}$ M) [471, 473, 677]. The most probable explanation of this effect has been advanced by Mitchell (cited according to [328]) who assumed that insertion of $K^{+} \cdot$ valinomycin complexes into the membrane facilitates in turn its penetration by the anionic form of the proton carrier, thereby augmenting the proton permeability. According to Karlish et al. [473] nonactin also displays such synergism with proton carriers. In contrast to valinomycin, however, it also promotes inhibition of proton uptake and uncoupling of photophosphorylation by proton carriers in potassium-free Na^{+}-containing media. This is in accord with the lower K/Na selectivity of nonactin. Contrary to valinomycin and nonactin, gramicidin in 10^{-8} M concentration has no effect on the light-induced proton uptake in the presence of proton carriers. Such differences might possibly be due to the molecules of this antibiotic forming cation-conducting pores in the membrane, instead of positively charged mobile complexes (cf. Part V.C).

Although valinomycin and gramicidin do not affect pH relaxation in single flash exposures of chloroplasts, there is a noticeable decrease in the membrane potential gradient relaxation time ($\tau_{1/2}^{\psi}$) [470, 683]. This effect depends on the alkali metal ions present in the incubation medium and increases as their concentration is increased, being thus undoubtedly due to

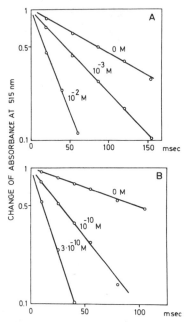

Fig. 214. Acceleration of the relaxation of the spectral shift at 515 nm in chloroplasts in the presence of gramicidin and potassium ions [470]. Exposure to actinic light (630—680 nm) in $1.5 \cdot 10^{-5}$ sec flashes. Medium composition: 200 mM sucrose, $5 \cdot 10^{-5}$ M benzyl viologen, 10^{-3} M tricine—NaOH (pH 7.4). A, Effect of the KCl concentration in 10^{-9} M gramicidin-containing medium; B, effect of the gramicidin concentration in 40 mM KCl-containing medium.

augmented cation conductance of the membrane* (see Figs 214 and 215A). Assuming the electrical capacitance and the antibiotic-independent ion permeability of the thylakoid membrane to be constant, Junge and Schmid [469] used the relaxation curves for calculating the dependence of the

---

* Neiman *et al.* [683], exposing chloroplasts to single 0.1 msec flashes, observed a decrease in $\tau_{1/2}^{\psi}$ in the presence of nigericin. This effect is difficult to rationalize in terms of stimulation of non-electrogenic K^+—H^+ exchange which usually occurs in the presence of this antibiotic (cf. pp. 335 and 339).

valinomycin-induced K^+ current on the potential gradient in the membrane. The current—voltage characteristic thus obtained could be well approximated by a hyperbolic sinus (see Fig. 215B), the zero current conductivity being proportional to the antibiotic concentration. These findings show that in some parameters the valinomycin-modified thylakoid membranes strongly resemble bilayers (see Part V.C). On the other hand, the sequence found by these authors for the valinomycin-induced conductivity ($Rb^+ > Cs^+ > K^+$) differs from the order of both the complexing selectivity and induced

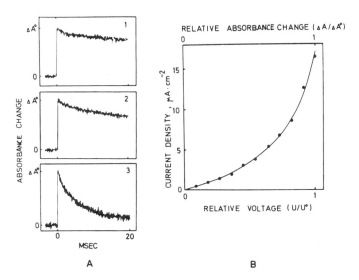

Fig. 215. Effect of valinomycin and potassium ions on relaxation of the spectral shift at 515 nm in chloroplasts [469]. Exposure light as described for Fig. 214. Medium composition: 20 mM NaCl, 10^{-5} M benzyl viologen, $3 \cdot 10^{-4}$ M tricine—NaOH (pH 8.5). A, Relaxation curves (light flashed at zero time): 1, control (the same curve obtained in the presence of 10 mM KCl); 2, in the presence of $1.1 \cdot 10^{-8}$ M valinomycin; 3, in the presence of $1.1 \cdot 10^{-8}$ M valinomycin and 10 mM KCl. B, Current—voltage characteristic of a thylakoid membrane in the presence of $3 \cdot 10^{-8}$ M valinomycin and 20 mM KCl. The current density calculated from a relaxation curve of the chloroplast absorption at 520 nm. Solid line plotted from the equation $i = 1.36 \cdot Sh (3.18 \, U/U_0)$.

conductivity of lipid bilayers. In the case of gramicidin, on the contrary, the cation selectivity of the induced conductivity of thylakoid membrane ($Cs^+ > K^+ > Na^+ > Li^+$) is likewise observed for modified bilayers (cf. p. 313).

As had been mentioned in Part V.C, of the known antibiotic ionophores gramicidin induces the highest cation permeability per molecule. This antibiotic is so hydrophobic that it should be absorbed practically completely from the aqueous solution by the chloroplast membranes, the sojourn of its molecules in the membranes apparently exceeding $\tau_{1/2}^{\psi}$. Both these factors are responsible for the results of the experiments carried out by Junge and Witt [470] who showed that in the region of very low gramicidin concentrations part of the $\Delta\psi$-sensitive chromophores in the chloroplast preparation retain their initial relaxation time, whereas in the rest it is lowered to a certain value equal to this fraction of the particles. The contribution by the latter increases as the number of antibiotic molecules becomes more and more commensurate with the number of thylakoids.

This example clearly illustrates one of the most characteristic features of the action of antibiotic ionophores. For most inhibitors or activators of biological processes the targets are individual, independently functioning biopolymers (proteins, nucleic acids, etc.) or their complexes which together form a single unit (for instance, ribosomes or the electron transfer chain). Such a minimal functional unit, in the case of the membrane-active complexones, which can be modified as a whole by a single active molecule (or by their aggregate) is the entire membrane-enclosed organelle and in a number of cases even the cell itself. Indeed, owing to the high electro-conductivity of the internal phase and minuteness of the organelles, local drops in the membrane resistance should decrease the operating potential difference at all points of the topologically closed membrane.

The use of valinomycin has shed considerable light on the part played by the electrical field in ATP synthesis in chloroplasts. It was shown that single flashes or flashes interrupted for long intervals do not induce ATP synthesis. This is apparently because the number of protons carried across the membrane in a single, short flash is too small for the generation of a $\Delta\bar{\mu}_H$ sufficient for phosphorylation to take place. Moreover, owing to complete dissipation of $\Delta\bar{\mu}_H$ in the inter-flash periods the effects of the flashes are not cumulative. If, however, the inter-flash period is less than $\tau_{1/2}^H$ and/or $\tau_{1/2}^{\psi}$ then ΔpH and/or $\Delta\psi$ will not have time to dissipate and under such conditions a constant $\Delta\bar{\mu}_H$ component will appear which will increase the number and intensity of the flashes [335]. Consequently, when chloroplasts are subjected to a volley of short flashes, phosphorylation develops with the appearance of a rapid phase on the dark potential relaxation curve after each volley [468]. Hence it may be assumed that for ATP synthesis use is made, not only of the energy stored in the form of ΔpH, but also of the electrical energy accumulated in each flash by charging of the membrane capacitance. If this is true then electrical shunting of this capacitance should lead to inhibition of phosphorylation. Such an effect has indeed been observed, the degree of inhibition in the presence of valinomycin being close to that calculated from data on the magnitude of the induced K^+ conductivity of the thylakoid membrane. This study is undoubtedly of considerable importance for proof of the applicability of Mitchell's concept in its general form to energy transduction in the chloroplasts.

While chloroplast absorption in the 515 nm region may be used for estimating the membrane potential (see p. 372), certain other optical parameters of the chloroplasts (delayed light emission and light-induced quenching of fluorescence) reflect the energization level of the thylakoid membrane, i.e. the magnitude of $\Delta\bar{\mu}_H$. Wright and Crofts [1094, 1095] who made a detailed investigation of these parameters used valinomycin, nigericin and dianemycin in order to reveal and study separately the $\Delta\psi$- and ΔpH-dependent processes; this can be illustrated by the determination of the pH-dependences of the rapid and slow phases of delayed light emission in chloroplasts during the initial seconds following exposure to light*. The rapid phase ($\tau_{1/2} \leqslant 0.1$ sec) turned out to be selectively inhibited by valinomycin; nigericin did not affect it, although it completely inhibited the slow phase ($\tau_{1/2} \approx 0.3$ sec), evidently associated with the establishment of a pH gradient. In this way the two phases could be distinctly separated and characteristic pH relationships found for each. It is noteworthy that the amplitude of the rapid phase and the electrochromic shift at 515 nm (30 msec after switching on the light) have the same pH dependencies. This can be regarded as further support for the proposal that the action of valinomycin is directed on the $\Delta\psi$-dependent process [1095] (see Fig. 216).

In a theoretical treatment of delayed light emission, Wright and Crofts [1095] (see also [175]) have found that its intensity I_L is related to $\Delta\bar{\mu}_H$ by the exponential expression

$$I_L \sim e^{\Delta\bar{\mu}_H} \tag{39}$$

Evidently for a constant pH gradient this equation can be represented in the form

$$I_L \sim e^{\Delta\psi} \tag{40}$$

Connection between the delayed light emission intensity and the membrane potential has been demonstrated by Barber [45] (cf. [48]), who was also able to calibrate this "voltmeter" so that it could be used for measuring absolute values of $\Delta\psi$. In creating a KCl concentration gradient on the membrane of valinomycin-treated chloroplasts one could expect the initial $\Delta\psi$ value to correspond to the diffusion potential which can be estimated from the well-known Goldman equation [295]:

$$\Delta\psi = \frac{RT}{F} \ln \frac{a'_K + \rho \cdot a''_{Cl}}{a''_K + \rho \cdot a'_{Cl}} \tag{41}$$

---

* In these experiments the so-called millisecond delayed light emission has been measured, viz. the emission level 1 msec after the beginning of a short interruption in the illumination.

366

where $\rho = P_{Cl}/P_K$, permeability ratio; a'_K and a'_{Cl} are the K$^+$ and Cl$^-$ activities in the medium immediately after KCl addition; a''_K and a''_{Cl} are the K$^+$ and Cl$^-$ activities in the interior of the thylakoids.

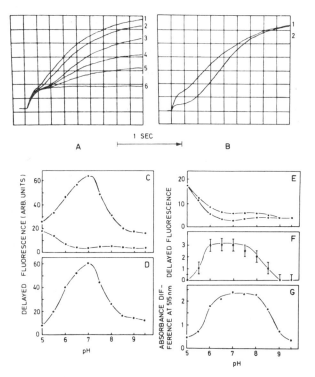

Fig. 216. Effect of valinomycin and nigericin on the delayed fluorescence in chloroplasts [1095]. Medium composition: 100 mM KCl, 20 mM tricine buffer (pH 7.8), 0.2 mM potassium ferricyanide. A, Inhibition of the slow phase in the presence of nigericin (concentration in ng/ml): 1, control; 2, 0.2; 3, 0.4; 4, 2; 5, 4; 6, 20. B, Inhibition of the rapid phase in the presence of valinomycin: 1, control; 2, in the presence of 0.2 ng/ml valinomycin. C, pH-Dependence of the onset of maximal fluorescence in the absence of nigericin (upper curve) and in the presence of 40 ng/ml nigericin (lower curve). D, Difference profile for the upper and lower curves in C. E, pH-Dependence of the rapid phase increase in fluorescence: upper curve, in the presence of 40 ng/ml nigericin; lower curve, in the presence of 40 ng/ml valinomycin. F, Difference profile for the upper and lower curves in E. G, pH-Dependence of the spectral shift at 515 nm on continuous (30 msec) illumination by red light.

In such cases the value of ΔI_L can be described by Eqn 42.

$$\Delta I_L \sim \frac{a'_K + \rho \cdot a''_{Cl}}{a''_K + \rho \cdot a'_{Cl}} \tag{42}$$

Analysis of the experimental data has shown that the change of I_L in the presence of 10^{-6} M valinomycin as a function of the magnitude of the KCl pulse is well approximated by Eqn 42 for ρ = 0.03—0.04* (see Fig. 217). The calibration permitted the evaluation of $\Delta\psi$ for energized chloroplasts in the absence of phosphorylation substrates, the results lying within the limits of 75—105 mV.

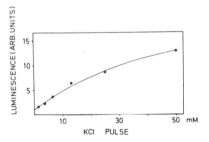

Fig. 217. Initial intensity of the delayed fluorescence of chloroplasts measured after KCl addition in the presence of 10^{-6} M valinomycin [45]. Medium composition: 330 mM sucrose, 5 mM N-Tris-(hydroxymethyl)-methyl-2-aminoethanesulfonic acid, 2 mM KOH. The curve has been plotted from the equation $I_L = c_K (2 + 0.04\,c_{Cl})^{-1}$, where c_K and c_{Cl} are the potassium and chloride concentrations immediately after KCl addition.

VI.C.3. Bacterial chromatophores

Photosynthesizing microorganisms of the *Rhodospirillum*, *Rhodopseudomonas*, etc., families are capable of both photophosphorylation and oxidative phosphorylation, cells grown under aerobic conditions having predominantly developed a respiratory system, whereas those anaerobically grown in the light possess a predominant photosynthetic system [867, 1035]. Ultrasonic treatment of these bacteria yields microvesicles called chromatophores. Chromatophores obtained *lege artis* retain the ability to carry out electron transfer-coupled phosphorylation in which the relative contributions of the respiratory chain and the light-stimulated electron transfer chain are the same as in the original cells [867].

Jackson *et al.* [449] have shown that valinomycin and (in low concentrations) gramicidin cause dark $K^+ - H^+$ exchange in *Rhodospirillum rubrum* chromatophores, the direction of the exchange being determined by the pK$^+$ gradient. Since under such conditions the $\Delta K/\Delta H$ ratios were only somewhat higher than unity one could expect the ionic conductance other than that of potassium of the chromatophore membrane to be commensurate with the proton conductance.

Similar to the chloroplasts, the action of nigericin and dianemycin on chromatophores diminishes the light-dependent proton uptake and causes

* These data appear to show a rather large initial Cl$^-$ permeability of the thylakoid membrane. However, Barber has raised doubts as to the validity of this conclusion [47].

inflow of potassium ions, both effects augmenting with increase in the potassium ion content of the medium [449, 687, 688, 887, 1002]*.

On the other hand in contrast with the chloroplasts the ratio $\Delta K/\Delta(\Delta H) = 1$ so that the nigericin-induced $K^+ - H^+$ exchange is of normal stoichiometry. In the presence of nigericin the initial rate of the light-induced proton uptake exceeds that of potassium uptake. This is a convincing argument in favor of the energy-dependent pH gradient being the driving force of the $K^+ - H^+$ exchange [449].

In contrast to nigericin, valinomycin increases the rate and magnitude of proton uptake and causes potassium ion outflow, the effects becoming stronger as the initial potassium ion concentration in the chromatophores is increased [447, 449, 687, 719, 887]. Similar effects are displayed by valinomycin if the energy source is the oligomycin-sensitive hydrolysis of ATP or the electron transfer in the respiratory chain [867]. In the latter case the valinomycin-induced proton uptake is especially large if the chromatophores are prepared from cells grown under aerobic conditions. As for gramicidin, in low concentrations it stimulates and in high concentrations it inhibits the light-dependent uptake of protons [449, 965].

The identical direction of cation flows which membrane-active complexones induce in illuminated chromatophores and in chloroplasts is undoubtedly due to the same polarity of the coupling membrane of these particles. It is interesting that in the unsonicated cells valinomycin stimulates the energy- (light, oxidation substrates + O_2)-dependent efflux of protons. According to Scholes, Mitchell and Moyle [867], this indicates that in the cell disintegration the coupling membrane reverses its polarity. These authors proposed that the chromatophores resulting from the sonication of the bacteria originate by a closure of the folds or invaginations of the cell membrane so that it is projecting inside out (see Fig. 218). We shall see further on that such membrane inversion also takes place in the ultrasonic fragmentation of mitochondria (see Part VI.C.5). However, in the case of photosynthesizing bacteria, sonication, besides forming vesicles by closure of the folds, apparently also liberates chromatophores which had already been formed in the cell by detachment from the membrane in some natural way [421, 747].

---

* With increase in potassium (or other alkali metal ion) concentration, cation-sensitive electrodes suffer a fall in sensitivity ($\Delta V/\Delta a_M$). Since the inner volume of the membrane particles is much less than that of the incubation medium, the absolute amounts of the absorbed and released ions are very small so that reliable potentiometric measurements of ion flows can usually be carried out in media containing not more than 10 mM of the salt of the corresponding cation. Under these conditions a 1 mV potential shift occurs in the release or uptake of ~ 0.4 microequiv of cation per ml.

In contrast to chloroplasts, phosphorylation in chromatophores is not uncoupled in the presence of nigericin and potassium ions [449, 887]. A similar difference has been observed with ammonium salts, which like nigericin give rise to uncoupling in chloroplasts by decreasing the energy-dependent pH gradient [609]. This leads to the inference that under steady-state conditions the main contribution to $\Delta\bar{\mu}_H$ in chromatophores is made by the membrane potential as compared with the pH gradient for chloroplasts. Such a viewpoint, first expressed by Jackson *et al.* [449] is in accord with the low Cl^- permeability of chromatophore membranes.

On this basis it is natural to expect that photophosphorylation in the chromatophores can be effectively uncoupled by electrical shunting of the

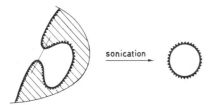

Fig. 218. Reversal of membrane polarity after sonication of bacterial cells and mitochondria according to Scholes, Mitchell and Moyle [867].

membrane, i.e. by increasing its ion permeability. However, such uncoupling cannot be observed in the presence of neutral membrane-active complexones since ATP synthesis is depressed by inhibition of the electron transfer chain (cf. p. 358). In the presence of phenazine methosulfate or tetramethyl-*p*-phenylenediamine phosphorylation is restored to its initial level providing the electrons by-pass the impaired sites of the chain [479]. The uncoupling of phosphorylation and stimulation of proton uptake in chromatophores can be induced by adding permeant anions such as NO_3^-, SO_4^{2-}, Ph_4B^-, picrate, etc., to the medium [420, 660]. Uncoupling is increased in the presence of nigericin and potassium ions [660]. Possibly, nigericin eliminates the lipophilic anion-induced rise in ΔpH, thereby promoting still further decrease in $\Delta\bar{\mu}_H$.

Photophosphorylation in chromatophores is also uncoupled in the presence of valinomycin or nonactin in combination with nigericin antibiotics [449, 887] or with ammonium salts [96, 479, 660], i.e. under conditions favorable to transmembrane proton migration linked with cyclic cation transport (see Figs 208 and 213). The uncoupling action of the combination of neutral and acid antibiotic complexones increases with increase in external potasssium ion concentration, although the potassium content in chromatophores changes insignificantly. The increase is therefore most likely attributable to augmented cyclic potassium ion flux across the membrane. On the contrary, the valinomycin-induced uncoupling in the presence of ammonium ions (2 mM) completely disappears when the potassium ion concentration in the medium is increased to 300 mM, apparently due to competitive displacement of the ammonium ions from the complexes. It is noteworthy that in the presence of methylamine and other alkylamine salts valinomycin does not uncouple phosphorylation in chromatophores although like ammonia these amines penetrate biological membranes. The reason for this is doubtless the inability of valinomycin to complex alkylammonium cations. Interestingly, nonactin, a compound possessing particular affinity for ammonium ions, displays very strong synergism with ammonium salts.

Keister and Minton [479] have shown that in combination with NH_4Cl valinomycin and nonactin not only uncouple photophosphorylation, but increase the ATPase activity of chromatophores. In terms of Mitchell's theory this can be ascribed to shunting of the proton pump, which plays the part of the reversible phosphorylating system operating under conditions of ATP hydrolysis. The level of the induced ATPase activity, as well as the degree of uncoupling, decreases with increased complexable cation concentration of the medium. The ATPase activity inhibiting sequence of the alkali metal ions ($Rb^+ > K^+ > Na^+$) coincides with the order of complexing selectivity. This is consistent with the above-described competition between ammonium and alkali metal ions.

Thus, we see that in combination with ammonium salts valinomycin and the nactins have a proton carrier-like effect on chromatophores. The similarity of these combinations to proton carriers can also be seen in the inhibition of the photoreduction of NAD^+ by both types of uncouplers [330] (cf. [1035]).

It has already been noted that when membrane-active complexones act on coupling systems, the energy liberated in electron transfer is utilized for ion transport, i.e. for osmotic work. If cyclic proton passage across the membrane takes place (see, for instance, Figs 208 and 213) the energy is thereby completely transformed into heat. However, when the action of complexones is not accompanied by cyclic proton transport, during complexone-induced transition of the coupling system to a new steady state, the energy is accumulated in the form of transmembrane ion gradients. If the electron transfer abruptly ceases, this energy is gradually dissipated by ion fluxes across the membrane, although it can be partly used, at least in principle, in the synthesis of ATP. The less the inherent membrane permeability to ions the greater will be the valinomycin-induced accumulation of energy. Apparently this is the nature of the effect investigated by McCarty [610] of the dark synthesis of ATP when illuminated chromatophores are rapidly transferred to an ADP- and phosphate-containing medium. It turned out that the dark synthesis of ATP increases threefold if the exposure to light occurs in the presence of valinomycin at pH 6.2. A necessary condition for the stimulation of photophosphorylation was the presence of potassium ions in the medium. Under similar conditions valinomycin did not increase the dark synthesis of ATP in chloroplasts. At the same time under the above conditions digitonin and ultrasonic subchloroplast particles, which behave similarly in many respects to the chromatophores, displayed considerable increase in dark phosphorylation*. Because the amount of synthesized ATP is independent of the light exposure from 5 to 60 sec, one can conclude that the valinomycin-induced ion gradients are very quickly established.

Jackson et al. [446, 449] were able to estimate the ΔpH and $\Delta\psi$ established on the chromatophore membranes by exposure to light. Their experiments clearly demonstrate the wide potentialities in the use of the valinomycin and nigericin-type antibiotics for studying coupling systems.

First, by varying the KCl concentration in the medium these authors [449] found conditions under which in the dark nigericin causes neither uptake nor release of potassium ion from R. rubrum chromatophores. Since the initial pH gradient was zero, the potassium ion concentration they found (a_K^{dark} = 4 mM) corresponded to the intraparticulate concentration. In a parallel experiment the same selection of potassium ion concentration was made after completion of the transient proton uptake (a_K^{light} = 0.4 mM). Since in the absence of nigericin the transmembrane K^+ flows were insignificant, the pH gradient could be very easily calculated: ΔpH = $(pK^+)^{light} - (pK^+)^{dark}$ = 3.4 $-$ 2.4 = 1.

---

* The polarity of the coupling membrane is not reversed by sonication [608, 609]. As to the digitonin fragments prepared by Nelson et al. [677], since they do not take up protons in the light, no definite opinion regarding the membrane polarity can as yet be formed. Both types of particle have the common feature that photophosphorylation is resistant to ammonia, valinomycin or nigericin in the presence of potassium ions but is synergistically uncoupled by valinomycin and ammonium ions [96, 609, 677, 682, 683]. Nelson et al. [677] and Neiman et al. [682, 683] have shown that digitonin particles in potassium-containing media are also subjected to uncoupling by valinomycin in combination with nigericin and proton carriers. The reason for the differences between the chloroplasts and the subchloroplast particles is as yet unclear. The behavior of the latter could perhaps be ascribed to low Cl$^-$ permeability of the membranes (cf. [96]). However, Neiman et al. [683] have found that the dark relaxation time of the electrochromic absorption shift at 515 nm for the digitonin particles in Cl$^-$-containing media is the same as for the chloroplasts. If this parameter is regarded as a measure of ion permeability of the coupling membrane, considerable doubts arise as to the validity of the low Cl$^-$-permeability hypothesis. Possibly the subchloroplast particles possess a low buffer capacity and/or are less subject to swelling.

For evaluating $\Delta\psi$ Jackson and Crofts [446] made use of the electro-chromism of the membrane carotenoids. As with chloroplasts, exposure of the chromatophores to light is accompanied by specific changes in the 500 nm absorption, ΔA (see, for instance, [38, 265]). In order to show that ΔA depends directly on the membrane potential, Jackson and Crofts varied the KCl concentration in a valinomycin-containing suspension of chromato-phores from *Rhodopseudomonas spheroides*. If the inherent ion per-meability of the chromatophore membrane is small relative to the induced K^+ permeability, the membrane potential should have changed in con-formity with the equation $\Delta\psi = \text{const} \cdot \Delta pK^+ = \text{const} \cdot \lg c_{KCl}$ (cf. p. 349). It turned out that establishment of a K^+ gradient leads to spectral shifts similar to those occurring on exposure to light, the $\Delta(\Delta A)$ value being proportional to $\lg c_{KCl}$ and independent of the presence of electron transfer inhibitors [446]*. The data obtained permitted not only the steady-state values of $\Delta\psi$ for illuminated chromatophores (~200 mV) to be calculated, but also the peak values at the moment of switching on the light (~400 mV). As expected, valinomycin considerably diminishes the steady-state value of $\Delta\psi$ especially in combination with nigericin. Nigericin itself not only does not promote a decrease in $\Delta\psi$ but, on the contrary, even increases its value (cf. [687]; see Fig. 219).

It should be mentioned that utilization of the electrochromism of membrane carotenoids for evaluating $\Delta\psi$ gives somewhat ambiguous results. Strictly speaking, such a shift reflects the field gradient within the membrane rather than the magnitude of the membrane potential, i.e. the potential difference between the membrane-interposed aqueous phases. Owing to charge separation in the course of the photochemical process the potential profile across the membrane can differ under the dark and light conditions even if the membrane potential is the same in both cases (cf. [834]). This especially pertains to pulse illumination when ΔA is established much faster than ion gradients can arise.

The ability of membrane carotenoids to display an electrochromic shift depends upon the state of the entire photosynthetic apparatus. Thus, Sherman and Clayton [927] have recently found that chromatophores from the cells of a *R. spheroides* mutant which, lacking reaction centers, cannot carry out photosynthesis, does not display electrochromism either in the light or in the dark (under K^+ gradient in the presence of valinomycin) despite the considerable carotenoid content in the membranes.

The use of carotenoid electrochromism as a molecular voltmeter yielded

---

* A similar "calibration" of the electrochromic shift for chloroplasts has recently been carried out by Strichartz and Chance [977].

interesting information on the non-steady state processes developing in chromatophores illuminated in the presence of membrane-active complexones. As in the case of chloroplasts, a short flash ($\sim$20 nsec) generates a potential difference in the membrane (with $\tau_{1/2} < 10^{-7}$ sec) due to charge separation which diminishes with the relaxation time, $\tau_{1/2} \sim 600$ msec [448]. An augmentation of the potential difference relaxation time relative to that of the chloroplasts is apparently due to the lower ion permeability of the chromatophore membrane. The relaxation time diminishes in the

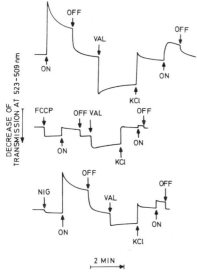

Fig. 219. Effect of valinomycin and nigericin on the spectrum of *Rhod. spheroides* chromatophore carotenoids on exposure to light and under conditions of a K^+ gradient [446]. Medium composition: 100 mM choline chloride, 20 mM 2-(N-morpholino)-ethanesulfonic acid (pH 6.4). Additives: 2 μg/ml valinomycin (VAL), 8 μg/ml nigericin (NIG), $4 \cdot 10^{-6}$ M FCCP, 24 mM KCl. The transmission in the region of 509—523 nm was followed by means of a double beam recording spectrophotometer. ON and OFF refer to the switching on and off of the light.

presence of valinomycin and potassium ions and, as in the case of chloroplasts, it could be shown that the induced K^+ conductance is almost linearly related to the antibiotic concentration (over the region 10^{-8}—10^{-5} M, $d\lg\tau/d\lg c_L = -1.2$; cf. Fig. 220)*.

Certain phenomena observed in the action of valinomycin and nonactin on chromatophores have as yet found no satisfactory explanation. This

---

* For other specific effects of valinomycin and nigericin in the pulse illumination of chromatophores see [448, 687].

refs, first of all, to selective inhibition by these antibiotics of electron transfer through sites of the chain that can be shunted by phenazine methosulfate and tetramethyl-*p*-phenylenediamine [39, 330, 479, 856]. The effect is observed at about the same antibiotic concentrations at which stimulation of light-induced proton uptake occurs ($> 10^{-7}$ M), but it is practically independent of the potassium ion concentration in both the medium (up to 250 mM) and the chromatophore interior (the preparations were made in sucrose, choline chloride or KCl solutions [330, 479]). Inhibition of the electron transfer chain is apparently not due to change in pH of the internal phase, since it occurs also in the presence of ammonium

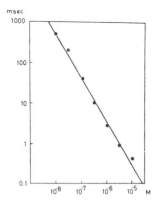

Fig. 220. Decay of the carotenoid shift as function of the valinomycin concentration [448]. Chromatophores from *Rhod. spheroides* suspended in a medium containing 100 mM choline chloride, 20 mM 2-(*N*-morpholino)-ethanesulfonic acid (pH 6.5) were illuminated by a 20 nsec pulse of a ruby laser (693.4 nm).

salts [479]. It is noteworthy that in the joint action of valinomycin and ammonium salts on *R. rubrum* chromatophores the induced ATPase activity (see p. 370) falls as the valinomycin concentration is increased above $7 \cdot 10^{-6}$ M. This parallelism in the action of valinomycin at high concentrations on the ATPase system and electron transfer chain warrants further study.

VI.C.4. Mitochondria

Rat liver and bovine heart mitochondria were the first and are probably still the most popular objects for studying the mechanism by which membrane-active antibiotic complexones affect metabolism-dependent processes in biological membranes. In eucaryotic cells the mitochondria are responsible for the oxidation-coupled ATP synthesis. The role of coupling membrane is played by the inner membranes which are in the form of numerous ridges called cristae (see Fig. 221)*. The polarity of the mitochondrial coupling membrane is opposite to that of chloroplasts and bacterial chromatophores so that electron transfer in the respiratory chain as well as ATP hydrolysis by oligomycin-sensitive, reversible ATPase is accompanied by active proton transport from the internal phase (so called matrix) into the surrounding medium (see Fig. 205). Another important consequence of the reversed polarity is that in mitochondria oxidation of substrates and phosphorylation proceed on the inner side of the coupling membrane. Hence, under steady-state conditions the rate of electron transfer in the respiratory chain is in the limit dependent either on the flux of the oxidation substrates, ATP, ADP and phosphate across the inner membrane, or on their equilibrium concentration in the matrix.

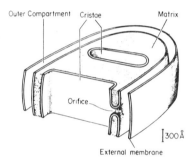

Fig. 221. Schematic drawing of a mitochondrion (reproduced from [1056]).

As is well known, the oxidation substrates in mitochondria are usually anions of the weak acids participating in the Krebs cycle. The entry of these anions and of phosphate into the matrix is provided for by specialized antiport and symport systems located on the inner membrane [129, 131, 398, 459, 498, 614, 723—725, 728, 729, 732, 750, 800—802], some of which are shown in Fig. 222. The distinguishing feature of the substrate anion transport across the membrane is in general its non-electrogenic nature so that the intramitochondrial concentration depends upon the pH of the incubation medium and of the matrix. For instance, at equilibrium, partition of the permeant acid HA is described by Eqns 43 and 44:

$$\Delta pA^- = (pA^-)'' - (pA^-)' = (pH)' - (pH)'' = -\Delta pH \tag{43}$$

$$\Delta p\overline{HA} = (p\overline{HA})'' - (p\overline{HA})' = -\Delta pH + \lg \frac{K_D + a'_H}{K_D + a''_H} \tag{44}$$

---

* A detailed description of the structure and function of mitochondria can be found in the monographs of Lehninger [538] and Skulachev [944, 949] (see also [183, 385, 539, 799, 946]).

where $p\overline{HA} = -lg\,(c_{HA} + c_{A^-})$; K_D is the dissociation constant of HA; a_H is the proton activity; and $''$ and $'$ refer to the matrix and external medium, respectively.

The pK_a of monoanions of the substrate dicarboxylic acids and dianions of the substrate tricarboxylic acids is equal to or less than 6, so that in the physiological pH range (6.5—8.5) an increase in ΔpH is accompanied by an increase in their total intramitochondrial concentration $c_{\overline{HA}}$. Hence, if the

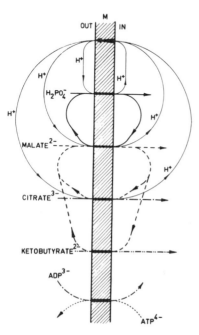

Fig. 222. Anion transport in mitochondria (bold arrow signifies proton pump). From top to bottom: symport of protons and phosphate anions; exchange of phosphate for malate + proton; exchange of malate for citrate + proton; exchange of malate for ketobutyrate and electrogenic ADP^{3-}—ATP^{4-} exchange.

pH gradient changes during functioning of the membrane-active complexones the effect of this change on the respiratory and phosphorylation rate can be mediated through shifts in the steady-state intramitochondrial concentration of substrate anions and/or phosphate. Since, in the general case, these steady-state concentrations are determined by numerous transport equilibria and the rates at which they are attained, many of the complexone-induced effects turn out to be very intricately dependent on the nature of the substrate anions and their accompanying cations (see, for instance, [241, 244]). One must therefore be very careful in the approach selected for their quantitative interpretation. The observed phenomena become still more complicated when phosphate anions take part in the substrate metabolism preceding respiratory electron transfer. Furthermore, the action of the antibiotic complexones is frequently accompanied by changes in the membrane potential which affect the intramitochondrial concentration of ADP and ATP since their transmembrane exchange is of electrogenic nature (see Fig. 222). This in itself should affect the rates not only of "membrane" phosphorylation, but also of ATP synthesis at the so-called substrate level, viz. substrate conversion without direct participation of the respiratory chain (for more details see [538, 944]).

These and other factors highly complicate treatment of the effect of membrane-active complexones on energy transduction in mitochondria, sometimes giving rise to erroneous conclusions and to as yet unsolved problems. Nevertheless, the principal phenomena can be satisfactorily explained in terms of the part played by $\Delta\bar{\mu}_H$ in the functioning of coupling membranes, already discussed earlier.

VI.C.4.a. Respiratory-dependent ion transport

One of the most characteristic phenomena observed in the action of valinomycin and other neutral antibiotic complexones on mitochondria in the state of respiratory control is complexable cation-dependent O_2 uptake [316, 317, 663]. In contrast to proton carriers such as dinitrophenol and FCCP, in high concentrations these antibiotics do not inhibit respiration whatever the substrate anion concentration [368] (cf., for instance, [1079]). Another specific feature of these antibiotics is their induction of complexable cation uptake by the respiring mitochondria [130, 315, 663, 780].

If phosphate, acetate or anions of other weak acids capable of permeating the inner membrane are absent from the incubation medium and respiration is maintained by low concentrations of substrate anions (or endogenic substrates), valinomycin stimulates the uptake of potassium ions and the stoichiometric release of protons (see, for instance, [827]; cf. Fig. 223). Such behavior is characteristic of mitochondria in isotonic media with high mannitol, sucrose or choline chloride content when the potassium ion concentrations are in the 1—10 mM range. Similar effects are manifested by the nactins and enniatins. The activation of respiration and ion transport in mitochondria by the valinomycin and allied antibiotics depend upon the alkali ion species in the medium, its selectivity correlating well with the complexing selectivity (see, for instance, [789]).

The cation dependence of respiratory acceleration is sometimes dependent on the nature of the substrate. Estrada-O et al. [245] have shown that, if

378

the substrate is glutamate, respiration in the presence of beauvericin is promoted most strongly by potassium ions and, if the substrate is succinate or β-hydroxybutyrate, by sodium ions. Such a relation may be assumed to be the result of superposition of the selectivities for two processes, of which one is the beauvericin-mediated cation transport, the contribution of the other being determined by the nature of the substrate and polarity of the membrane. At any rate, for the membrane-inverted submitochondrial particles (see Part VI.C.5) beauvericin-induced respiration displays potassium specificity also with succinate. Such an effect should be manifested with

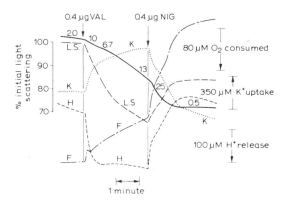

Fig. 223. Effect of valinomycin (VAL) and nigericin (NIG) on respiratory-dependent processes in rat liver mitochondria [790]. Mitochondria (2.4 mg protein per ml) were suspended in a medium (10 ml) containing 200 mM sucrose, 5 mM KCl, 3 mM glutamate, 3 mM malate, 10 mM Tris—HCl (pH 7.4). The curves on the diagram represent the change in the O_2 content of the medium (the numbers alongside the curve refer to the respiration rate in μmoles O_2 absorbed per g protein per minute), K^+ and H^+ content in the medium, fluorescence of pyridine nucleotides (F, upward slope indicates increase in the degree of oxidation) and light scattering (L.S., downward slope indicates mitochondrial swelling).

complexones of comparatively low K/Na selectivity as is characteristic of beauvericin.

Much still remains unclear regarding the effect of antamanide on ion transport. In a number of communications Pressman [787, 791] refers to this compound as stimulating the energy-dependent uptake of potassium ions. Later it was shown that in low concentrations (approx. 10^{-7} M) antamanide noticeably activates respiration [78, 701]. In these experiments sodium ions had no effect on the respiration rate. This appears strange, since, on the one hand, the induced sodium conductance combined with the membrane-inherent system of Na^+—H^+ antiport should promote effective sodium-dependent uncoupling. It was subsequently found, however, that antamanide-treated mitochondrial lipid bilayers possess a higher conductance in potassium-containing solutions than in sodium-containing solutions [78]. It is not known to what extent this applies also to the native mitochondrial membrane.

The data presented above demonstrate that when acting on mitochondria antibiotics of the valinomycin group give rise to energy-dependent proton and complexable cation flows counter-directed to those observed in chloroplasts and bacterial chromatophores. Clearly, this difference is due to the reversed polarity of the mitochondrial coupling membrane.

Since the rate of respiration induced by valinomycin in the presence of potassium ions increases with the substrate concentration in the range 1—10 mM, it is evidently limited under these conditions by the substrate anion influx into the matrix. If one now bears in mind that the inflow of anions into mitochondria is more intense the higher the pH gradient on the inner membrane, it becomes evident that the intramitochondrial pH increase caused by this antibiotic is an extra factor in the activation of respiration [367, 722]. Apparently it is just this circumstance which is the underlying cause of the aforementioned difference between the neutral membrane-active complexones and proton carriers. With increase in concentration of the latter the operating pH gradient and, consequently, the substrate influx decrease ultimately ending in arrest of the primary uncoupled respiration (cf., for example, [1079]). This secondary inhibition diminishes with increase in substrate concentration.

Mitochondria display a characteristic behavior in media containing high concentrations of complexable cations and of the anions of strong acids (chlorides, nitrates, iodides, etc.). If the substrate is added when the valinomycin- and gramicidin-treated mitochondria have already undergone swelling (see Part VI.B.1), energy-dependent contraction is observed, the rate of which falls with decrease in external pH from 8.5 to 7 [28]. Since under these conditions the respiratory rate is maintained at a sufficiently high level, this pH-associated contractility inhibition is naturally attributed to a decrease in the anion permeability of the inner membrane [27]. The driving force of the contractile process must therefore apparently be the electrophoretic anion transport from the mitochondria. If, however, the substrate is added at pH $\leqslant$ 7 and the mitochondria have not had time to swell, a cyclic process is developed involving rapid swelling and slower contraction stages [89, 90, 95]. Both are energy-dependent and the swelling stage is associated with transient outflow of protons (see Fig. 224). The most plausible explanation of this phenomenon [90] is that the neutral antibiotic complexones are capable of stimulating respiratory-dependent increase in the intramitochondrial pH. As a result the anion permeability of the inner membrane increases and the mitochondria undergo rapid swelling due to entry of salt and water (cf. Part VI.B.1). Consequently the situation arises when, as we have noted above, the respiring mitochondria begin to contract. A possible factor in the dissipation of ΔpH on completion of the swelling process is enhancement of the proton permeability of the inner membrane (cf. p. 352).

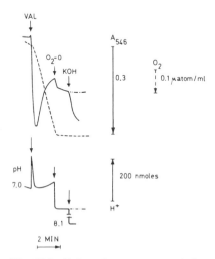

Fig. 224. Relaxation processes induced by valinomycin in bovine heart mitochondria suspended in concentrated KCl solutions [90]. Mitochondria (5 mg protein per 8 ml) were treated with rotenone and suspended in a medium containing 100 mM KCl, 6 mM sucrose, 2 mM Tris—succinate (pH 7.0). Additives: $2.5 \cdot 10^{-7}$ M valinomycin (VAL); KOH up to pH 8.1. Mitochondrial swelling and contraction were followed by measuring the absorption at 546 nm (A_{546}, downward slope refers to swelling). The point $O_2 = 0$ refers to oxygen exhaustion in the cell. Basification of the medium in the absence of energy sources results in swelling of the mitochondria (cf. p. 333).

From this example one can see how in mitochondria relaxational oscillations could arise under the influence of membrane-active complexones (see also [341]). Under certain conditions mitochondria can undergo several swelling and contraction cycles [317, 399] (see Fig. 225) and even pass into an auto-oscillatory state with the pH and pK^+ of the medium as well as the mitochondrial volume oscillating in approximately one-minute periods [125] (see Fig. 226)*. Very likely further study of these phenomena will aid in clarifying the feed-back mechanisms operating in respiring mitochondria.

---

* Such an oscillatory response is characteristic not only of mitochondria; Metlička and Rybova [524] have shown that valinomycin induces light-dependent oscillations of the membrane potential in cells of the alga *Hydrodiction reticulatum*.

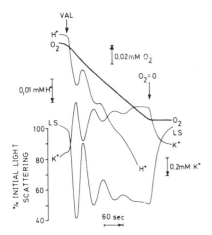

Fig. 225. Oscillatory effects induced by the action of valinomycin on rat liver mitochondria [782]. Mitochondria (45 mg protein) were suspended in 10 ml medium containing 250 mM sucrose, 10 mM KCl, 5 mM Tris—phosphate and 20 mM Tris—HCl (pH 7.2) to which 0.5 µg valinomycin (VAL) was added as an ethanolic solution. The curves represent changes in the O_2, K^+ and H^+ contents of the medium and in the light scattering (LS).

Both the rate and magnitude of the induced complexable cation uptake are augmented in the presence of the anions of permeant weak acids (phosphate, acetate, succinate, etc.) (see, for instance, [368]). Concurrently, these anions accumulate in the mitochondria together with an equivalent decrease in proton efflux. The driving force of the anion accumulating process is apparently increase in the intramitochondrial pH. In fact, a

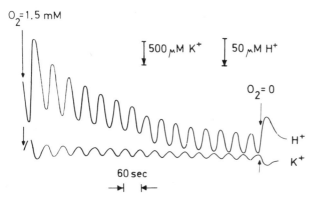

Fig. 226. Oscillatory effects in the action of valinomycin on pigeon heart mitochondria [125]. Mitochondria (2.5 mg per ml) were suspended in a medium containing 225 mM mannitol, 75 mM sucrose, 2 mM Tris—glutamate, 2 mM Tris—malate, 2 mM Tris—phosphate, 2 mM Tris—HCl, 6.7 mM KCl and 83 ng/ml valinomycin (pH 6.25). O_2 pulse was generated by introduction of catalase (17 µg/ml) followed by H_2O_2 addition.

number of data indicate that anion transport begins only after exhaustion of the intramitochondrial buffering capacity. Energy-dependent swelling of the mitochondria is also observed under these conditions [33, 315, 827]*.

It should be mentioned that cation uptake induced by antibiotics of the valinomycin group usually reaches its limit before the substrates or O_2 are exhausted. The mitochondria therefore pass into new steady-state conditions characterized by a higher alkali ion content of the matrix. For the example of valinomycin, Harris et al. [365] have proved that this new steady-state is associated with increased K^+ permeability of the membrane. If measures are taken to prevent mitochondrial swelling, completion of the valinomycin-induced potassium uptake is accompanied by a decreased respiratory rate [603, 649].

Employing potassium-depleted mitochondria, Mitchell and Moyle [649] measured the limiting potassium uptake in the presence of valinomycin under O_2 pulse conditions and used the results for calculating $\Delta\psi$ (cf. p. 349). They estimated that in the state of respiratory control (state 4) the value of $\Delta\bar{\mu}_H/F$ in potassium-free medium is about 230 mV, about 90% of which is due to $\Delta\psi$, but only 40% if the medium contains 10 mM KCl. As we have seen the decreased $\Delta\psi$ contribution to $\Delta\bar{\mu}_H$ is the natural consequence of enhanced ion permeability of the membrane (see p. 348). This approach to determining the operating $\Delta\psi$ values has been subsequently used by several other workers [549, 830]. Characteristically, under O_2 pulse conditions the final mitochondrial steady-state potassium content is dependent to only a small extent on the valinomycin concentration (from 0.025 to 0.3 ng/mg protein), but the duration of potassium uptake diminishes significantly [649] (see Fig. 227).

The rate and magnitude of potassium-ion uptake stimulated by the valinomycin-type antibiotics depends on the transmembrane K^+ gradient. Ordinarily, in experiments with the isolated mitochondria the potassium ion concentration is by 1.5—2 orders of magnitude higher in the matrix than in the incubation medium so that K^+ uptake proceeds against a concentration gradient. If the K^+ gradient is made still higher by decreasing the external concentration or increasing the internal concentration, then instead of inducing potassium uptake by the mitochondria valinomycin will induce its release [154, 779, 828]. Reversal of the induced K^+ flow indicates that the potassium diffusion potential apparently exceeds the energy-dependent component of $\Delta\psi$. An outwardly similar effect is observed by increasing the valinomycin concentration of the medium. Bogolyubova et al. [78] found that under these conditions the rate of aerobic potassium uptake passes

---

* For the effect of valinomycin on the mitochondrial ultrastructure, see [305, 721].

through a maximum, the direction of the induced K^+ flow being reversed in the range of high antibiotic concentrations. Similar effects are observed also with other membrane-active complexing cyclodepsipeptides, inhibition of potassium uptake not being accompanied by arrest of respiration. The nature of this effect is as yet unknown. It is noteworthy that the heights of the maxima can be similar for compounds whose isoeffective concentrations differ by several orders of magnitude, and possibly can depend on their K/Na selectivity. Thus, for valinomycin and its analogs (57) and (79) the maximal

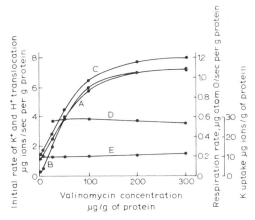

Fig. 227. Valinomycin-induced changes in respiration and proton and K^+ transport in rat liver mitochondria under O_2 pulse conditions [649]. Mitochondria (7.1 mg protein per ml) were suspended in a medium containing 150 mM choline chloride, 25 mM sucrose, 2 mM β-hydroxybutyrate and 3 mM glycylglycine (pH 7.0). The O_2 pulse was generated by additions of catalase and H_2O_2. Curves A, B and C represent the initial rates of proton release, K^+ uptake and O_2 consumption, respectively. D and E curves represent the K^+ uptake and the respiratory rate at the final state of respiratory control.

potassium uptake rate is noticeably higher than for the enniatins (see Fig. 228). It could be assumed that in isoeffective concentrations compounds of similar cation selectivity should increase the K^+ permeability of the mitochondrial membrane to the same extent. This is supported by the fact that in minimal concentrations at which under standard conditions aerobic uptake of potassium is still observed, cyclodepsipeptides increase the K^+ conductivity of lipid bilayers equally. It is noteworthy that the maximal rate of potassium uptake induced by valinomycin and its analogs cannot be surpassed by any of their combinations.

According to the chemiosmotic theory respiratory stimulation in the absence of phosphorylation substrates is inevitably associated with ion transport across the coupling membrane. Hence, within the framework of

384

this theory it is necessary to suppose that inhibition or reversal of K^+ flow in the case of high cyclodepsipeptide concentrations is due to ion competition with potassium for electrophoretic transport. As an example of such competition the data of Harris et al. [365] can be cited, according to which with gramicidin the induced potassium ion inflow diminishes in the presence of sodium salts. Apparently, any permeant cations can be competitors of potassium. Thus, Ogata and Rasmussen [691] have shown that, in a medium containing 0.2 mM Ca^{2+} and 1 mM K^+, valinomycin begins noticeably to affect potassium ion uptake by mitochondria only after the latter have absorbed practically all the calcium ions.

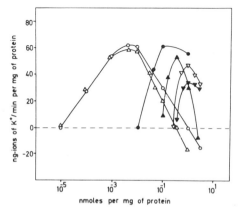

Fig. 228. Effect of cyclodepsipeptides of the valinomycin and enniatin groups on the energy-dependent K^+ transport in mitochondria [78]. Rat liver mitochondria (5 mg protein per ml) were suspended in a medium containing 250 mM sucrose, $5 \cdot 10^{-4}$ M potassium, $1.5 \cdot 10^{-4}$ M sodium, 0.3 mM succinate and 8 mM Tris—HCl (pH 7.2). ○, valinomycin; ●, compound (57); △, compound (79); ▲, enniatin A; ▽, enniatin B; ▼, enniatin C.

Of undoubted interest is the question of whether the action of neutral macrocyclic complexones can be solely ascribed to increased cation permeability of the mitochondrial membrane. One cannot be sure that the very insertion of the molecules of these compounds or their cation complexes into the membrane will not affect in some way or other the functioning of its constitutive enzymic or other systems. From this standpoint, particular importance is attached to such complexones which while not being efficient uncoupling agents are yet able to affect energy transduction and active ion transport. The first paper concerning this subject was published by Lardy in 1968 [524]. He showed that cyclic polyethers (270) and (274) (see Fig. 13) inhibit the valinomycin-induced uptake of potassium. Alone, compound

(274) can only weakly stimulate K^+ absorption, while compound (270) is entirely inactive in this respect. On these grounds, Lardy proposed that the cyclic polyethers or their complexes compete with valinomycin for a certain protein which is taking part in the cation transport. Subsequently, Estrada-O and Carabez [242] confirmed Lardy's observations and, moreover, showed that, in the presence of cyclic polyethers (270) and (274), electron transfer is inhibited in that part of the respiratory chain which includes the flavin-involved coupling site (see Fig. 206). Since such inhibition can also be observed in submitochondrial particles, and these have a reversed membrane polarity (see Part VI.C.5), the action is in all likelihood independent of the direction of the ion flows and hence is straightforwardly directed to some component of the membrane. In contrast to compound (270), cyclic polyether (274) can quite actively carry out transmembrane potassium ion transport so that its electron-transfer inhibiting capacity is combined with a marked uncoupling effect. Noteworthily, the inhibition of glutamate oxidation by the cyclic polyethers is weakened with increase in concentration of monovalent cations ($Cs^+ > Rb^+ > K^+$) if valinomycin and phosphate are present in the medium. These findings have as yet received no satisfactory explanation.

Antibiotics of the nigericin group reverse the effect of valinomycin and other membrane-active complexones. Their presence causes outflow of accumulated cations from the mitochondria, proton uptake and contraction [314, 526, 790] (see Fig. 223). In low concentrations, nigericin and dianemycin accelerate somewhat the respiration induced by valinomycin or nactins, but it is inhibited as the concentration is increased. The inhibition can be counteracted by augmenting the potassium-ion or substrate-anion concentration of the medium (see Fig. 229). The underlying cause of such behavior is the ability of nigericin and related antibiotics to stimulate non-electrogenic K^+–H^+ exchange, lowering (or even reversing) the transmembrane pH gradient established by the respiratory chain. If the fall in intramitochondrial pH is not very significant, dissipation of ΔpH leads to acceleration of respiration; if, on the other hand, it is large, then respiration can be inhibited due to arrest of the substrate anion inflow [367, 722, 790]. In view of this, it is of interest that, if the mitochondria are preliminarily enriched with phosphate, respiration becomes more nigericin-resistant [315], for, on the one hand, many substrate anions can enter by exchange with the phosphate and, on the other, phosphate is necessary for metabolism of such substrates as glutamate and citrate [538, 944]. The combination of these factors determines to what extent respiratory inhibition will depend on the nature of the substrate. Respiration induced by neutral antibiotic complexones is resistant to nigericin-like compounds when β-hydroxybutyrate, succinate, proline or a glutamate—malate mixture are used as substrates.

In their effect on valinomycin- or nactin-induced respiration and ion flows, nigericin-like compounds are rather similar to proton carriers. The latter in combination with valinomycin also causes rapid exchange of intramito-chondrial potassium for protons and inhibition of respiration [386]. The inhibition increases with decrease in pH and diminishes with increase in potassium concentration and, in some cases, substrate concentration [490].

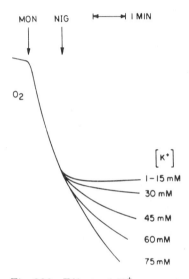

Fig. 229. Effect of K^+ concentration on the nigericin-induced inhibition of glutamate oxidation in rat liver mitochondria [526]. Medium composition: 200 mM sucrose, 12 mM glutamate, 8 mM phosphate, 3 mM $MgCl_2$ (pH 7.4; triethanolamine—HCl). Additives: $2 \cdot 10^{-7}$ M monactin (MON); $6 \cdot 10^{-7}$ M nigericin (NIG).

If, in potassium-free media (containing NaCl and sucrose as the main components) mitochondrial respiration is stimulated by dinitrophenol, its inhibition by valinomycin increases with increase in NaCl concentration from 25 to 100 nM [297]. This is probably due to augmentation of the K^+ gradient on the membrane resulting from contraction of the mitochondria with increased osmolarity of the medium and, consequently, to a greater fall in the intramitochondrial pH during K^+—H^+ exchange. Respiratory inhibition is also observed when gramicidin and ammonium ions act jointly on the mitochondria [386]. The sensitivity of the respiratory process to inhibition caused by valinomycin antibiotics in combination with proton carriers also depends on the nature of the substrate anion, being high in the case of glutamate or citrate and very low in the case of succinate or β-hydroxy-butyrate [386]. As with nigericin, these differences are apparently due to exit of phosphate from the mitochondria. In fact, in the presence of phosphate (but not acetate) glutamate-maintained respiration that had first been stimulated by dinitrophenol and then inhibited by valinomycin is resumed in potassium-free medium.

The effect of the intramitochondrial phosphate content on the rate of glutamate oxidation can be observed also when mitochondria are treated with nigericin in the absence of neutral membrane-active complexones. In phosphate-poor media nigericin inhibits respiration with accumulation in the mitochondria of α-ketoglutarate, an intermediate in glutamate metabolism requiring phosphate for further transformation. The effect weakens with increased concentrations of KCl and phosphate in the medium [261]. On the other hand, it is well known that lowering the phosphate concentration in the medium causes noticeable inhibition of the dinitrophenol-induced respiration if it is supported by glutamate as substrate [32]. Thus, inhibition of respiration in phosphate-poor media is characteristic of all agents which lower or reverse the pH gradient of the inner membrane*.

However, one may not exclude the possibility that decreased substrate anion and/or phosphate content of mitochondria is not the only cause of the observed effects. Many authors have assumed that the metabolic level in mitochondria depends directly on both the volume and pH of the matrix and also on its potassium ion concentration (see, for instance, [296, 490, 790]).

In general, in low concentrations, antibiotics of the nigericin group can inhibit or stimulate respiration, depending on the potassium content of the medium and also on the nature and concentration of the substrates [246, 527, 1086]. In some cases the inhibition can be overcome by adding valinomycin-type compounds to the medium [1086]. Ferguson et al. [261] have recently published an interesting paper in which they showed that in the presence of potassium ions nigericin in high concentrations (about 5 μg/mg protein or $\sim 1.5 \cdot 10^{-5}$ M) practically completely restores respiration in mitochondria brought into a state of respiratory control by means of oligomycin or aurovertin. Stimulation of respiration becomes maximal with increase in potassium ion concentration to 60 mM. A study of the cation selectivity of this effect showed it coincides with the complexing selectivity

* Estrada-O et al. [243] have shown that in a medium containing 15 mM potassium and β-hydroxybutyrate as respiring substrate, the nigericin antibiotics induce over a narrow range of pH ($\sim$6) the coupled uptake of calcium and phosphate ions, which is inhibited by neutral complexones. The underlying cause of this effect might be formation within the mitochondria of calcium phosphate, the solubility of which is dependent on the pH.

of nigericin (see Fig. 230). The removal of respiratory control is observed only with substrates (for instance, succinate or β-hydroxybutyrate) of which oxidation in the absence of ATPase inhibitors is nigericin-resistant.

Generally speaking, the removal of respiratory control in the presence of nigericin might be naturally associated with diminution of $\Delta\bar{\mu}_H$ caused by lowering of the pH gradient. However, strange as it may seem, such activity

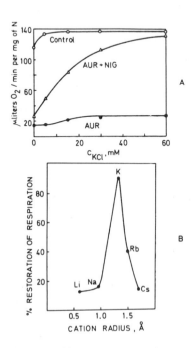

Fig. 230. Nigericin-activated respiration in rat liver mitochondria poisoned by oligomycin antibiotics [261]. A, Dependence on the K^+ concentration. Medium composition: sucrose + KCl (total osmolarity, 250 mosm), 2 mM ATP, 10 mM β-hydroxybutyrate, 3 mM MgCl$_2$ and 13 mM triethanolamine phosphate (pH 7.4). Additives: $1.4 \cdot 10^{-5}$ M nigericin (NIG), 3 μg/ml aurovertin (AUR). B, Effect of cation species. Medium composition: 90 mM sucrose, 45 mM MCl, $1.4 \cdot 10^{-5}$ M nigericin, 3 μg/ml rutamycin; the other components as in A.

is not manifested by other antibiotics of this group: monensin, dianemycin and X-206. As a possible explanation of this anomaly the authors proposed that with nigericin the enhanced antibiotic and potassium ion concentrations lead to the appearance in the membrane of complex cations M^+HL as well as neutral complexes of the type M^+L^-, making possible electrogenic transport of potassium ions across the membrane. In other words, under such conditions nigericin seems to combine within itself the properties of both neutral and acid complexones, providing complete dissipation of $\Delta\bar{\mu}_H$ (cf. Fig. 208B)*. This hypothesis is no doubt of considerable interest, the more so that it is in accord with suggestions made in connection with the similar action of nigericin on other membrane systems (see pp. 276, 309 and 362).

VI.C.4.b. Processes associated with the hydrolysis and synthesis of ATP
If the potassium ion concentration in the medium is not very high (1—5 mM) valinomycin and dinactin augment the rate of oxidative phosphorylation, although as a rule the P/O ratio diminishes (see Fig. 231) [368, 399,

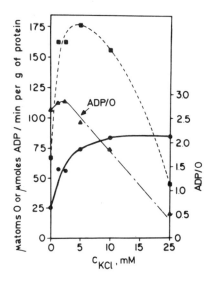

Fig. 231. Effect of dinactin on the rate of oxidative phosphorylation in rat liver mitochondria [368]. Mitochondria suspended in a medium containing KCl in the indicated concentration, dinactin (5 μg per mg protein), 3 mM Tris—glutamate, 3 mM Tris—malate, 5 mM Tris—phosphate, 20 mM Tris—HCl (pH 7.2). ●—●, respiration rate; ■---■, rate of ADP consumption; ▲—·—▲, ADP/O ratio.

781]. This effect is particularly noticeable at low antibiotic concentrations. With potassium-enriched media the phosphorylation rate sharply falls as the antibiotic concentration is raised, resulting in complete uncoupling. The uncoupling effect of valinomycin in K^+-containing media is very strong, being noticeable at antibiotic concentrations as low as 10^{-8} M. Essentially it was this which in its time had attracted the interest of mitochondriologists in

---

* The same result will, of course, be obtained if the appearance of M^+HL complexes in the membrane increases the proton rather than the cation conductance.

this compound [570, 623]. Within the framework of the chemiosmotic concept discussed above, it is only natural that this uncoupling be attributed to dissipation of $\Delta\bar{\mu}_H$ and a switch-over of the energy flux to ion transport. Less evident is the cause underlying stimulation of oxidative phosphorylation at low potassium concentrations. Under the experimental conditions the phosphorylation rate was limited by the influx of the substrate anions. It has been noted above that valinomycin and the nactins, increasing the pH gradient, accelerate the uptake of these anions. Apparently in the low-potassium concentration range this factor has a greater bearing on the oxidative phosphorylation rate than does decrease of $\Delta\psi$. This assumption is indirectly confirmed by the inability, under any circumstances, of gramicidin (which increases not only the cation but also the proton permeability and consequently lowers ΔpH) to stimulate oxidative phosphorylation [368]. Gramicidin is an exceptionally effective uncoupler, but in contrast with valinomycin displays low cation selectivity (see, for instance, [681]).

The valinomycin-induced potassium ion transport and the ATP-synthesizing system compete for the energy liberated in the oxidation of the substrates. Harris *et al.* [368] investigated the effect of ADP additions on oxygen consumption and valinomycin-induced ion flows in the presence of potassium ions and phosphate. These additions were found to accelerate respiration markedly while potassium uptake and swelling were inhibited until all the ADP in the medium was exhausted (see Fig. 232).

On the other hand, in the presence of valinomycin-like antibiotics ATP can be used by the mitochondria as an energy source for the uptake of complexable cations. This can be observed, for instance, in the stimulation of the oligomycin-sensitive cation-dependent ATPase activity, the level of which reflects the ion selectivity of the antibiotic [123]. The most complete summary of data on this effect is contained in the paper by Graven *et al.* [316], who compared the ion selectivity of the ATPase activity induced under identical conditions by valinomycin, the gramicidins and the nactins. In general, the selectivity sequences they found coincide with or are very close to those obtained in studies of complex formation by these compounds or of their action on model membranes. However, it was also shown that the sequences could change with change in antibiotic and cation concentrations in the medium. It is as yet hard to say to what extent such changes are comparable with those observed for bilayers treated with these antibiotics (see Part V.C). Noteworthy is the observation by these authors of the inability of the nactins to promote ATPase activity of mitochondria in the presence of ammonium salts, since these antibiotics are distinguished by high affinity for ammonium ions and are capable of transporting them through biological membranes, mitochondrial membranes included. Evidence of this is the synergistic uncoupling effect which the nactins in combination with ammonium ions exert on oxidative phosphorylation in submito-chondrial particles (see Part V.C.5).

Blocking of ATPase makes the ATP pool in the mitochondria inaccessible to the action of valinomycin. Significant from this aspect are the data of Fang and Rasmussen [252] who studied the incorporation of a phosphate label into the phosphoinositide fraction of mitochondrial lipids. The rate of

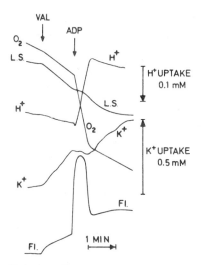

Fig. 232. Effect of ADP on proton and K^+ transport in rat liver mitochondria in the presence of valinomycin [368]. Mitochondria (5 mg protein per ml) suspended in medium containing 250 mM sucrose, 4.5 mM KCl, 0.5 mM $MgCl_2$, 3 mM Tris—glutamate, 3 mM Tris—malate, 3 mM Tris—phosphate and 20 mM Tris—HCl (pH 7.2). Additives: 0.2 μg/ml valinomycin (VAL), 0.4 mM ADP. The curves represent oxygen consumption, changes in the proton and K^+ concentrations in the medium, light scattering (L.S., downward slope indicates mitochondrial swelling) and fluorescence of pyridine nucleo-tide (Fl., upward slope indicates increase in degree of oxidation).

this process parallels the mitochondrial ATP content. On using glutamate oxidation as energy source, valinomycin in the presence of potassium acetate inhibits incorporation of the label in the first minutes of the respiration process. At the same time, in the case of mitochondria preliminarily treated with oligomycin, valinomycin not only inhibits but even accelerates incorporation of the label. This is apparently due to promotion of glutamate metabolism on activation of the respiratory chain and, consequently, to an increase in ATP synthesized at the substrate level.

In low concentrations antibiotics of the nigericin group do not noticeably uncouple oxidative phosphorylation in mitochondria. However, they do inhibit ATP—P_i exchange [527] probably owing to a decrease in the intramitochondrial concentration of phosphate.

In high concentrations (about 10^{-5} M) nigericin (but not dianemycin) increases the oligomycin-sensitive ATPase activity of mitochondria, an effect which is promoted by potassium ions and, to a much lesser extent, by rubidium ions [247]. The increase in ATPase activity in the presence of nigericin is in good accord with the ability of this antibiotic to remove the oligomycin-induced respiratory control (p. 387).

One would naturally assume that nigericin-induced respiratory activation and ATP hydrolysis are caused by a common factor, namely the electrophoretic transport of potassium ions (or protons), as well as non-electrogenic K^+—H^+ exchange. From this standpoint dianemycin should differ from nigericin by its inducing only K^+—H^+ exchange, independent of the concentration.

The effect of nigericin on the mitochondrial ATPase activity has a number of noteworthy features: (a) stimulation decreases with increase in medium tonicity due to sucrose or non-permeating salts; (b) as the potassium ion concentration is increased the activity passes through a maximum, the subsequent decrease being dependent on the anion species, least with acetate, phosphate and formate and considerable with iodide, nitrate and citrate (in the absence of malate); (c) the ATPase activity is inhibited by succinate and β-hydroxybutyrate, the effect of the latter being annulled by rotenone; (d) nigericin (as well as dianemycin) lowers the valinomycin-, dinactin- and gramicidin-induced activity [247].

These effects are apparently due to dependence of the ATPase activity not only upon the operating $\Delta\bar{\mu}_H$ value but also upon pH and the volume of the matrix. If the ATPase displays a bell-shaped activity $vs.$ pH curve [247], its inhibition with increase in non-permeating solutes or potassium salt concentration can be attributed to K^+—H^+ exchange, which respectively leads to excessive acidification or alkalination of the matrix. In the former case, an essential part is probably played by matrix contraction. Swelling of the mitochondria is apparently not so dangerous because the nigericin-induced ATP hydrolysis is not very sensitive to additions of permeating salts such as sodium acetate.

It is interesting to compare the mitochondrial ATPase activation by nigericin with the effect of valinomycin in the presence of dinitrophenol. As is well known, dinitrophenol-induced ATP hydrolysis in potassium-free media is inhibited by valinomycin if the tonicity of the medium is augmented by sucrose or non-permeating salts [297]. No such inhibition is observed in KCl or RbCl solutions of the same tonicity or in solutions of NaCl if valinomycin is replaced by gramicidin. One can therefore postulate here also that the ATP hydrolysis is inhibited by the K^+—H^+ exchange-induced fall in the intramitochondrial pH and/or contraction of the matrix.

However, the effects of nigericin and the valinomycin + dinitrophenol combination have their differences. For instance, synergism of the valino-mycin and dinitrophenol in ATPase stimulation in K^+-depleted mitochondria (down to 15 mM in the matrix) can be detected only on the background of respiration. Why electron transport is required here for ATPase activation is as yet not very clear. A similar effect was earlier observed in the activation of ATP hydrolysis by bovine heart mitochondria in the presence of p-chloro-mercuriphenylsulfonate and potassium ions. In each case respiration is possibly one of the driving forces of ion transport which elevates the mitochondria into a state of enhanced ATPase activity [298].

The quantitative treatment of ATP, ADP and phosphate transport is required for detailed analysis of the kinetics of induced ATP hydrolysis, in which a bottleneck can prove to be, for instance, the electrogenic ATP^{4-}—ADP^{3-} exchange, the rate of which increases if the membrane becomes permeable to potassium ions.

VI.C.5. Submitochondrial particles

Ultrasonic fragmentation of mitochondria leads to the formation of vesicles enclosed by the inner mitochondrial membrane. Electron micro-scopic and other data indicate the polarity of the membrane in these submitochondrial particles to be the reverse of that of the native mito-chondria. The ultrasonic particles can carry out oxidative phosphorylation which is uncoupled in the presence of proton carriers of the type of FCCP. On the other hand, on treatment with inhibitors of the ATP-synthesizing system such as dicyclohexylcarbodiimide or oligomycin, the particles pass into a state of respiratory control which can be removed by proton carriers (see, for instance, [63]). Hence the ultrasonic fragments of the mitochondria possess the properties characteristic of coupling systems. Ordinarily for their energization, use is made of the oxidation of succinate by oxygen, but other exergonic oxidoreductive reactions can also be employed for this purpose; for instance, the oxidation of NADPH by oxygen or fumarate, the oxidation of NADPH by NAD^+, etc. (for further details on the preparation and properties of ultrasonic particles, see [533, 946] and references therein).

Nigericin lowers energy-dependent proton uptake and promotes potassium uptake by submitochondrial particles preliminarily treated with oligomycin and dicyclohexylcarbodiimide*. Both effects increase in the presence of permeant anions and are inhibited by valinomycin and gramicidin [156, 657]. In contrast with nigericin valinomycin stimulates proton uptake [730] and the exit of potassium ions from the particles [657]. Thus, these particles resemble bacterial chromatophores in the effect of the membrane-active complexones on the energy-dependent ion flows, which is reasonable if one bears in mind their possessing the same membrane polarities.

In the presence of potassium ions valinomycin and nigericin, separately, only slightly depress respiratory control [62, 152, 156, 315, 365, 655, 656, 659, 954]. In contrast with valinomycin, nigericin manifests a stronger action the higher the membrane permeability to anions of the medium, the effect being particularly strong in the presence of Ph_4B^- or picrate [656, 657]. When acting together these antibiotics display a synergistic effect independent of the anion composition of the medium [152, 156, 655—657, 659]. Effective removal of respiratory control can also take place in the joint presence of valinomycin and ammonium salts [152, 156, 656]. Gramicidin, like FCCP, removes respiratory control independently of the presence of alkali metal ions [658].

The decrease in respiratory control correlates with uncoupling of oxidative phosphorylation in particles untreated with inhibitors of ATP synthesis. Maximum uncoupling activity is manifested by the following combinations: valinomycin + nigericin + K^+; valinomycin + NH_4^+; and nigericin + K^+ + permeant anion [152, 655, 657]. In the presence of valinomycin alone uncoupling occurs only at high potassium ion concentrations [731] (cf. [62, 146]), whereas with gramicidin in high concentrations the degree of uncoupling is independent of the presence of cations in the medium [368, 658]. These data are additional arguments in favor of similarity of the properties of ultrasonic submitochondrial particles and bacterial chromatophores; underlying their behavior is the low ion-permeability of the membrane. Consequently, on energization of the particles a considerable membrane potential, $\Delta\psi$, is established, the dissipation of which leads to uncoupling of oxidative phosphorylation [657].

---

* Such treatment of the particles defines more clearly the effect of valinomycin, nigericin and other membrane-active complexones on the energy-dependent ion flows (see, for instance, [726]; cf. effect of oligomycin on lipophilic anion uptake [329]). The effect of oligomycin is, at least partially, due to diminution of the proton permeability of the membranes impaired in the fragmentation process [867]. In a similar way, oligomycin increases the valinomycin-stimulated light-induced uptake of protons by bacterial chromatophores [867, 946].

As a typical example of the use of membrane-active complexones for determining coupling membrane polarities, the work of John and Hamilton [458] may be cited. These authors showed that phosphorylating particles obtained from the sonication of *Micrococcus denitrificans* cells are similar to ultrasonic submitochondrial particles. In these bacteria which synthesize ATP by oxidative phosphorylation, the polarity of the coupling membrane is apparently the same as in the mitochondria, since respiration is accompanied by a fall in the pH of the medium [933, 934]. Fragmentation of the cells causes reversal of the membrane polarity. The membrane vesicles obtained retain respiratory control in K^+-containing media in the presence of valinomycin, nigericin and (in 0.1 μg/mg protein concentrations) of gramicidin, but it disappears under the combined action of valinomycin and nigericin in the presence of potassium ions, of gramicidin (0.1 μg/mg protein) and of monensin in the presence of sodium ions and also of valinomycin in the presence of ammonium ions. At high concentrations (10 μg/mg protein) gramicidin annuls respiratory control independently of the presence of alkali ions in the medium. We have already mentioned in Part VI.C.3 that similar properties are displayed by chromatophores resulting from sonication of photosynthesizing bacterial cells which also possess a well developed apparatus for membrane phosphorylation. Apparently the "inside-out" inversion of the membrane during ultrasonic fragmentation is characteristic of many cells and subcellular organelles although it is not a universal property since on similar treatment chloroplasts yield particles with retention of the membrane polarity (see footnote on p. 371). Particles with polarity retention can be prepared from mitochondria by using digitonin as the disintegrating agent. Digitonin particles in the presence of membrane-active complexones behave in many respects similarly to mitochondria (see, for instance, [624, 1050]; cf. [163]).

VI.C.6. Active transport of alkali metal ions across plasma membranes

The ouabain-sensitive system of active potassium and sodium transport widespread in plasma membranes of animal tissue cells is by itself apparently resistant to membrane-active complexones; the latter seem to exert an indirect effect mainly by uncoupling oxidative phosphorylation in the mitochondria and thereby diminishing the cellular ATP pool [306, 547] (cf. [307, 1100]). According to Pressman [781] the activity of K,Na-dependent ATPase preparations obtained from the electrical organs of the eel and bovine or rat brain does not change in the presence of valinomycin and gramicidin (cf. [186]).

The most complete information on the effect of valinomycin and nigericin antibiotics on the bacterial system of active alkali metal ion transport has been obtained for *Streptococcus faecalis*. Harold and associates [354—356, 359—363] have found that the membrane of this microorganism apparently contains ATPase whose functioning is coupled to the electrogenic transport of protons from the cell. The main source of ATP for this primary translocase is glycolysis. The glycolysis-supported pH gradient diminishes in the presence of nigericin and monensin but is insensitive to valinomycin and monactin if the potassium ion concentration in the medium is above 1 mM. In media of pH > 7.5, valinomycin and monactin do not display any significant effect on potassium uptake by glycolizing cells. Since both these antibiotics considerably augment the K^+ permeability of the plasma membrane (see Part VI.B.2) these findings speak in favor of the potassium ions being transported in the cell electrophoretically rather than by a primary K^+-translocase.

Intriguing data have recently been reported on the effect of valinomycin on potassium ion transport in vesicle preparations from *E. coli* cells [64]. It turned out that these membrane vesicles retain the ability for energy-dependent uptake of proline, lactose, etc., but their active potassium-ion transport system is depressed. The vesicles begin to take up potassium ions from the medium under aerobic conditions in the presence of valinomycin and oxidation substrates. Significantly, the kinetics of potassium accumulation by cells of different *E. coli* mutants were found to be similar to those of the valinomycin-treated membrane vesicles prepared from them.

The authors have proposed that valinomycin-induced K^+ transport proceeds by the same mechanism as the native transport and that the antibiotic can replace a certain component of the transport system lost (or possibly inactivated) in the cell fragmentation. Despite the seemingly small probability of this proposal according to the present state of our knowledge, it may in the future perhaps even turn out to be trail-blazing in the application of membrane-active complexones for reconstructing active ion transport systems.

VI.D. Mechanisms of the antibiotic action of the membrane-active complexones

From among the complexone antibiotics discussed here bio-organic chemists have been particularly attracted to valinomycin and the enniatins. The reason is that these antibiotics are characterized by chemical stability and by resistance to enzymes that split amide and ester bonds, so that their activity could hardly be ascribed to covalent bonding with structural components of the cells. On the other hand, the valinomycin and enniatin molecules, built up of α-hydroxy and α-amino acid residues, are to some extent structurally related to peptides and proteins. It was therefore quite enticing to regard the cyclodepsipeptide antibiotics as a kind of model, simulating the stereo-chemistry of the non-covalent interactions of biologically active peptides and proteins with cellular receptors. Such considerations had served as a stimulus for the directed synthesis and comparative activity studies of a large number of valinomycin and enniatin analogs (see, for instance, [897, 900, 904, 922]). During the course of these studies it was found, in particular, that the valinomycin and enniatin A and B enantiomers have the same activity as the naturally occurring antibiotics, a phenomenon exceptionally rare in bio-organic chemistry. As is well known, protein, nucleic acid or mixed biopolymer macromolecules which are usually the targets of optically active effector compounds are distinguished by a very well expressed structural asymmetry so that interaction of these effectors with receptor sites always proceeds with a high degree of stereoselectivity. When enantiomers of antibiotics, hormones and other biologically active compounds became synthetically available, they were either found to be entirely inactive, or their activity differed fundamentally from the natural compound. Hence the identity of the antibiotic action of valinomycin and the enniatins with that of their enantiomers was ascribed to stereochemical (or so-called topo-chemical) similarity of the latter to the naturally occurring forms [902]. The alternative possibility that interaction of these antibiotics with the target was not stereospecific had at that time seemed of little probability.

When, however, the ability of valinomycin and the enniatins to bind alkali metal ions and promote their transport across biological membranes was discovered, the question naturally arose as to what extent the complexing properties and membrane activity of these compounds are responsible for their antibiotic action. It is now clear that these phenomena are intimately related. First, none of the synthetic, non-complexing analogs of valinomycin and the enniatins inhibits microbial growth. Secondly, these antibiotics as well as the nactins, gramicidins A, B and C and nigericins can increase the cation permeability of bacterial membranes (see Part VI.B.3). Finally, one of the principal properties of the valinomycin and other membrane-active complexones is the frequent dependence of their antibacterial action on the cation composition of the media. Thus, addition of potassium removes the growth inhibition of *Streptococcus faecalis* by valinomycin, gramicidin and nigericin [357, 358] (see Fig. 233 and Table 77). For many aerobic microorganisms, on the contrary, the bacteriostatic activity of valinomycin increases with increase in potassium ion concentration of the medium (see, for instance, *Staphylococcus aureus* 209P in Table 77). The antibacterial activity often displays a more complicated dependence on the cationic composition of the medium. For instance, inhibition of *Sarcina lutea* growth by valinomycin increases with increasing potassium concentration of the

398

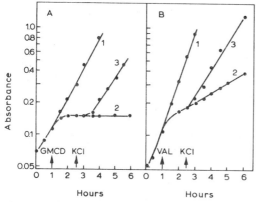

Fig. 233. Gramicidin- (A) and valinomycin- (B) induced inhibition of *Streptococcus faecalis* growth in a potassium-poor medium [357]. Cell population followed by measuring the absorbance of the suspension. Additives: 0.2 μg/ml gramicidin (GMCD); 1 μg/ml valinomycin (VAL); 0.2 M KCl. 1, growth in the absence of additives; 2, growth in the presence of antibiotics; 3, release of growth inhibition by addition of KCl.

medium from 4.5 to 10 mM, but it is suppressed in the presence of 100 mM NaCl. The conditions for manifestation of antibacterial activity apparently depend on the cation complexing selectivity. For instance, among the many synthetic analogs of valinomycin, only some can bind sodium ions to a noticeable degree. One of these — compound (45) (see Table 2) — in contrast

TABLE 77

ANTIMICROBIAL ACTIVITY OF VALINOMYCIN IN MEDIA OF DIFFERING K^+ AND Na^+ CONTENTS [313]*

Microorganism	Minimal growth-inhibiting concentration (μg/ml) for media containing (in mM)			
	K^+ 5 Na^+ 3	K^+ 100 Na^+ 3	K^+ 5 Na^+ 100	K^+ 100 Na^+ 100
Streptococcus faecalis	0.2	**	0.2	**
Staphylococcus aureus 209P	**	0.2	**	0.2
Staphylococcus aureus UV-3	0.1—0.2	0.07	0.1—0.2	0.07
Sarcina lutea	0.1	0.01	0.1	0.01***
Bacillus subtilis	10	1	10	1.5
Mycobacterium phlei	0.3	0.3	0.3	0.3
Candida albicans	0.2—0.4	0.2—0.4	0.2—0.4	0.2—0.4
Escherichia coli B	**	**	**	**

   * Composition of media: 10 g glucose, 5 g peptone and 30 ml Hottinger's broth in 1 l tap water; KCl and NaCl added to required concentrations.
   ** Microbial growth was observed at a valinomycin concentration of 10 μg/ml.
   *** Effect of valinomycin cancelled by addition of 150 mM NaCl.

to valinomycin, inhibits *S. aureus* 209P in potassium-depleted media and *S. faecalis* in potassium-enriched media, i.e. behaves like the low K/Na selective enniatin B.

Collectively, these data are convincing evidence that the antibiotic activity of membrane-active complexones is due to impairment of alkali ion transport. Thus, the most probable cause of their antibacterial activity is the increased cation permeability of the membranes. This is also in accord with the identical activities of enantiomeric cyclodepsipeptide antibiotics, since such enantiomers equally increase the permeability of model lipid bilayers and of bacterial membranes [12]. On this basis it is quite possible that comparison of the biological activities of membrane-active compounds and their enantiomers could serve as a sort of test for substances whose molecules function in the membrane's achiral lipid matrix. This can be illustrated by usninic acid, an antibiotic produced by certain lichens as a mixture of equally active *d* and *l* forms [877]. Usninic acid uncouples oxidative phosphorylation and, being a diphenol, apparently serves as a proton carrier (similar to, say, dicoumarol). On the other hand, Wieland *et al.* [255] have recently shown that antamanide and its enantiomer have markedly different antitoxic activities. One may therefore suppose that in the organism antamanide and its complexes encounter a certain stereospecific target or a stereospecific barrier in their pathway to the target.

What then are the immediate causes of microbial growth inhibition by the antibiotic complexones? In the case of *S. faecalis*, the latter are active under conditions facilitating release of intracellular potassium ions and dissipation of the energy-dependent pH gradient on the plasma membrane (see Part VI.C.5). According to Harold [354] an important factor in the growth inhibition of this microorganism is impairment of its ribosome functioning associated with the fall in potassium concentration; on the other hand, decrease in the cytoplasmic pH can inhibit glycolysis and transport of phosphate across the membrane. The decisive factor in the inhibition by antibiotic complexones of microorganisms with membranes containing an oxidoreduction-coupled ATP-synthesizing system is apparently the lowering of $\Delta\bar{\mu}_H$ and the associated suppression of phosphorylation. In other words, it seems as if the cells are deprived of their energy sources by some process resembling uncoupling of oxidative phosphorylation in mitochondria. Support for such a proposal can be found, for instance, in the valinomycin-induced acceleration of respiration and potassium uptake by *Mycobacterium phlei* and *Azotobacter vinelandii* cells observed by Pressman [781] (for similar data on gramicidin, see [407]).

Little is known about the underlying cause of the resistance of micro-organisms to the action of membrane-affecting complexones. In the case of

E. coli the resistance to valinomycin is apparently due to the inability of the antibiotic to increase the K^+ permeability of the cellular membranes of this species (cf. Part VI.B.3). One may expect that the plasma membrane or mitochondria of the yeasts *Endomyces magnusii*, resistant to valinomycin whatever the sodium or potassium concentration of the medium, should be inaccessible to this antibiotic. However, with the aid of labeled valinomycin it seems to have been shown that this is not so. Moreover, it turned out that in the presence of valinomycin, mitochondria isolated from these yeasts in

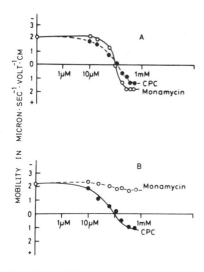

Fig. 234. Effect of cation species on the electrophoretic mobility of monamycin-treated *Staphylococcus aureus* NCTC 6571 cells in potassium (A) and sodium (B) phosphate solutions of ionic strength 0.01 (pH 7.0) [346]. For comparison a plot of the dependence of the cell mobility on the concentration of a cationic surfactant (cetylpyridinium chloride, CPC) has been presented.

many respects behave similarly to mitochondria of the valinomycin-sensitive yeasts *Saccharomyces carlsbergensis*, the differences being only of a quantitative character [78].

Monamycins assume a particular position among the alkali metal-binding antibiotics because their antimicrobial activity is due to lysis of the bacterial cells. Interestingly, being adsorbed on cells (*S. aureus* NCTC 6571) the monamycin alkali metal ion complexes reverse the sign of their surface charge (see Fig. 234). For monamycin and potassium concentrations of the medium corresponding to zero potential, as determined by electrophoretic mobility of the cells, the latter coalesce into clusters [346].

VI.E. Summary and perspectives

When papers dealing with membrane-active complexones published in the past two to three years are considered, it is found that investigations of the mode of action of these compounds on particular biological objects have gradually given place to their directed use for the controlled and selective increase of the cation permeability of membranes. The rationality of this trend becomes quite apparent when it is taken into account that the overwhelming majority of the presently known biological effects of the valinomycin and nigericin antibiotics can be explained by augmentation of the cation permeability of the membrane or stimulation of M^+—H^+ exchange.

The following principal trends in the use of such compounds as tools in biological studies can be discerned.

Firstly, under appropriate conditions one can, by increasing the cation permeability of the membranes, change the ion composition of cells and subcellular particles; the active substance can often be subsequently eluted, restoring, at least partially, the initial membrane permeability (see Parts VI.B.2 and VI.B.3). This affords a convenient way for studying the effect of the cytoplasmic ion composition and of the transmembrane ion gradients on active transport, and metabolic and catabolic processes. Such an approach is exemplified by the work of Harold's group [354, 357—363] on S. faecalis cells. Considerable possibilities are also afforded for study of the regulatory role of alkali metal ions in the functioning of the genetic apparatus.

As has been repeatedly mentioned in the foregoing pages, in the presence of complexable cations the valinomycin antibiotics promote dissipation of the energy-dependent membrane potential, whereas nigericin and allied substances cause a downfall in the pH gradient. Hence these compounds can be utilized for the discrimination and individual study of $\Delta\psi$- and ΔpH-dependent processes. Perhaps the most consistent use of such an approach has been in studies of conditions of phosphorylation uncoupling and also in mechanistic investigations of anion transport in mitochondria (see, for instance, [613, 614, 725, 728, 750, 802, 946, 949]). Recently membrane-active complexones are finding more and more application in studies of the transport processes on a cellular level. With their aid, for instance, the presence has been shown in S. aureus cells of a proton—galactoside (lactose) symport system, and in pigeon erythrocytes symport of sodium ions and amino acids [689, 1000, 1052, 1053]. An interesting area in mechanistic studies of coupling has been opened up by the work of Papa et al. [730] who are examining the effect of valinomycin- or nigericin-induced changes in $\Delta\psi$ and ΔpH on the oxidation kinetics of individual components of the respiratory chain.

Membrane-active complexones cause the energy-dependent redistribution of alkali metal ions between organelles and the environment. The redistribution data serve as a basis for calculating steady-state $\Delta\psi$ and ΔpH values. This method, which had been applied to mitochondria and chromatophores (see pp. 371 and 382), has now been extended to bacterial cells (see, for instance, [455]).

By means of the antibiotic complexones, $\Delta\psi$ and ΔpH can not only be dissipated but also generated. Indeed, by establishing a K^+ gradient on a low ion permeability membrane, the generation of a membrane potential can be achieved with the aid of valinomycin, and of a pH difference with the aid of nigericin. This procedure is especially useful in quantitative studies of $\Delta\psi$-dependent processes in membranes of cells and subcellular particles, of which the direct polarization by external voltage sources is impossible. Its efficiency has been demonstrated, for example, in the investigations mentioned in Parts VI.C.2 and VI.C.3, in which valinomycin was used to elucidate the dependence of the spectral characteristics of chloroplasts and bacterial chromatophores upon the membrane potential. As another example can serve the proof of the electrophoretic nature of 8-anilino-1-naphthalene-sulfonate anions transport across the mitochondrial membrane [454] (see Fig. 235). The diffusion potential which is established by the action of valinomycin on the coupling membrane under K^+ gradient conditions may serve as the driving force for the phosphorylation reaction. Such "ionic phosphorylation", predicted by Mitchell's theory, has been demonstrated in experiments on mitochondria and chloroplasts [155, 816, 828, 871]. The recent years are witnessing ever increasing progress in the reconstruction of the coupling membranes and also of other membranes containing ATP-dependent systems which are carrying out active electrogenic ion transport. The occurrence of "ionic phosphorylation" in such re-assembled systems may become one of the principal criteria for their low ion permeability and the correct assembly of transport ATPase. Still another interesting approach involves the use of the diffusion potential as the driving force of oxidoreductive processes in the electron transfer chains. Thus, Hinckle and Mitchell [395] have observed reversed electron transfer in the respiratory chain of mitochondria in a potassium-poor valinomycin-containing medium.

We see considerable potentialities in using the induced cation permeability as a parameter reflecting the characteristics of biomembranes such as the sign and density of surface charge, microscopic viscosity, intramembrane profile of the electrical potential, etc. Doubtlessly, studies in this direction will advance in line with the elucidation of the properties of complexone-modified model lipid membranes.

One cannot help but note the practically complete absence of information on the effect of valinomycin, gramicidin, etc., upon the generation and propagation of impulses in excitable membranes (almost the only studies in this direction refer to cells of the *Torpedo* electric organs [769, 936]). The unsuccessful attempt to use monactin for modifying the axonal membrane [974] should hardly result in interruption of work in this field.

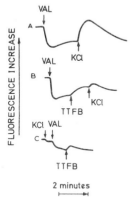

Fig. 235. Electrophoretic transport of 8-anilino-1-naphthalenesulfonate anions (ANS) coupled with valinomycin-induced potassium ion transport in bovine heart mitochondria [454]. Since transport of ANS anions has relatively little effect on their concentration in the medium, but leads to considerable changes in their intramitochondrial concentration (see p. 351), it is these changes which play the decisive part in the rise or fall in ANS content in the hydrophobic zone of the membrane. The quantum yield of fluorescence of the ANS anions considerably increases with their transport into non-polar media, in particular on their uptake by biomembranes [978]. However, the quantum yield of fluorescence of mitochondrial membrane-bound ANS is not changed by valinomycin under potassium gradient conditions [29, 49].Therefore the fluorescence level of the ANS anions is determined mainly by their intramembrane concentration which changes parallelly with the concentration changes of these anions in the matrix. Mitochondria (2.3 mg protein per ml) suspended in a medium containing 200 mM sucrose, 5 mM NaCN, 10^{-5} mM ANS and 50 mM Tris—HCl (pH 7.5). Additives: $2 \cdot 10^{-8}$ M valinomycin (VAL); $2 \cdot 10^{-5}$ M 2-trifluoromethyltetrachlorobenzimidazole (TTFB); 100 mM KCl. Fluorescence was excited by 360 nm light and its intensity was measured at 460 nm. A, Increase in the K^+ permeability of the mitochondrial membrane generates a membrane potential (negative inside) and causes ANS release from the mitochondrion. The subsequent increase in the KCl concentration of the medium reverses the sign of the potential, causing inflow of ANS into the mitochondrion. B, Shunting of the membrane by means of a proton carrier (TTFB) prevents ANS transport. C, Decrease in the K^+ gradient with increase in K^+ concentration of the medium considerably weakens the effect of valinomycin.

Latterly, studies have been initiated which are devoted to the effect of valinomycin and other membrane-active complexones on various highly involved phenomena at the cellular and tissue levels. As an illustration can be cited the works on the antilipolytic activity of valinomycin [249, 518], its effect on hormonal regulation in animals [56, 390], on the circadian movement of plant leaves [112] and also on ion transport and the energetics of normal and tumor cells [70, 306, 308, 547, 770, 771]. Wyssbord [1100] has begun studies on the effect of valinomycin upon the electrical parameters of the turtle bladder. Although as a rule the results obtained do not lead to unequivocal conclusions as to the mechanism of the observed effects, such studies should in the future permit the use of membrane-active compounds for control of various physiological processes.

Mention should also be made of the pharmacological potentialities of the macrocyclic complexones, such as their use for the sequestering and excretion of radioactive and toxic metal ions [672]. Also, of no little interest is the still unknown mechanism by which antamanide counteracts the toxic effects of amanitin and phalloidine [1073].

Evidently we are. at but the threshold of the application of membrane-active complexones to biological studies and the near future is holding in store many a pleasant surprise in this area. Already one of them has turned out to be the amazing similarity between the effect on mitochondria of valinomycin and sublytic concentrations of polyether detergents (Triton X-100, Igepal CO-630) recently discovered by Brierley *et al.* [91, 92] (cf. [123]). In the previous sections (see pp. 215, 266 and 310) data have already been presented showing these compounds to be helical complexones capable of transporting cations across membranes.

Undoubtedly, application of the membrane-active complexones in biology will broaden in scope as compounds with "unusual", for instance, sodium or calcium, specificity will appear. In this connection of considerable interest are the antibiotics X-537 A and A23187*. The former carries not only alkali, but also alkaline, earth metal ions across the membrane, whereas the latter selectively transports calcium and magnesium ions [122, 237, 813, 857, 861]. Reed and Lardy [814] have shown that A23187 uncouples phosphorylation in mitochondria, which, by the nature of the ion flows, resembles the uncoupling caused by valinomycin in combination with nigericin. It is to be remembered that in the latter case non-electrogenic K^+–H^+ exchange is coupled to electrophoretic potassium-ion transport, which is equivalent to increasing the proton permeability of the membrane (see Fig. 208). In the presence of antibiotic A23187 the Ca^{2+}–H^+ exchange it induces proceeds alongside calcium ion transport through the mitochondrial membrane's inherent Ca^{2+} transport system (see p. 347 and Fig. 207). The uncoupling effect of A23187 is competitively inhibited by

---

* Antibiotic A23187 is a carboxylic acid of the composition $C_{29}H_{37}N_3O_6$, which forms salts with magnesium and calcium (but not with potassium) salts, soluble in organic solvents [814].

Ca^{2+}-binding substances (EDTA, ATP), by inhibitors of the mitochondrial Ca^{2+} transport system (La^{3+}, ruthenium red) and by magnesium ions. In the latter case this is apparently due to the magnesium ions displacing the calcium ions from their complexes with the antibiotic, thus lowering the calcium ion contribution to the $M^{2+}-H^+$ exchange, whereas cyclic transport of the magnesium ions is precluded by low Mg^{2+} permeability of the mitochondrial membrane*. Since, contrary to calcium ions, magnesium ions released from the mitochondria are not reabsorbed, the antibiotic rapidly (< 30 sec) depletes the mitochondria of their magnesium ions, the rate of this process being particularly high in the presence of EGTA (up to 2—3 ng ions per mg of protein). Under conditions of maximum magnesium depletion and, correspondingly, maximum matrix acidification, the ATPase activity of the mitochondria is completely inhibited, whereas coupled oxidation of succinate and β-hydroxybutyrate is retarded by not more than 15—20%. A similar effect is exerted by nigericin (see Part VI.C.4), but in the case of antibiotic A23187 the underlying cause is not only decrease in the intramitochondrial concentration of phosphate and substrate anions, but also of magnesium ions. However important these findings may be, the use of membrane-active complexones selective to alkaline earth metal ions for studying active Ca^{2+} transport should open up a field with truly amazing potentialities.

---

* Similar competition for the antibiotic, leading to inhibition of cyclic NH_3 transport, has been described on p. 370.

REFERENCES

[1] Ackmann R. G., Brown W. H. and Wright G. F., J. Org. Chem., 20 (1955) 1147.

[2] Adams R. and Whitehill L. N., J. Am. Chem. Soc., 63 (1941) 2073.

[3] Agranovich A. M., Complexation of cyclic dodecadepsipeptides with potassium and sodium ions in different solvents (Russian), Graduate Thesis, The Moscow State University, Shemyakin Institute for Chemistry of Natural Products, U.S.S.R. Academy of Sciences, 1971.

[4] Agtarap A. and Chamberlin J. W., Antimicrob. Agents Chemother., (1967) 359.

[5] Agtarap A., Chamberlin J. W., Pinkerton M. and Steinrauf L., J. Am. Chem. Soc., 89 (1967) 5737.

[6] Aityan S.Kh., Levich V. G., Markin V. S. and Chizmadjev Yu. A., Dokl. Akad. Nauk. S.S.S.R. (U.S.S.R.), 193 (1970) 1402.

[7] Aldridge W. N. and Street B. W., Biochem. J., 124 (1971) 221.

[8] Alleaume M. and Hickel D., Chem. Commun., (1970) 1422.

[9] Alleaume M. and Hickel D., Chem. Commun., (1972) 175.

[10] Almy J., Garwood D. C. and Cram D. J., J. Am. Chem. Soc., 92 (1970) 4321.

[11] Ammann D., Pretsch E. and Simon W., Anal. Letters, 5 (1972) 843.

[12] Ammann D., Pretsch E. and Simon W., Tetrahedron Lett., (1972) 2473.

[13] Ando K., Murakami Y. and Nawata Y., J. Antibiot., 24 (1971) 418.

[14] Ando K., Oishi H., Hirano S., Okutani T., Suzuki K., Okazaki H., Sawada M. and Sagawa T., J. Antibiot., 24 (1971) 347.

[15] Andreev I. M., Malenkov G. G., Shkrob A. M. and Shemyakin M. M., Mol. Biol. (U.S.S.R.), 5 (1971) 614.

[16] Andreoli T. E., Tieffenberg M. and Tosteson D. C., J. Gen. Physiol., 50 (1967) 2527.

[17] Andreoli T. E. and Tosteson D. C., J. Gen. Physiol., 57 (1971) 526.

[18] Anfält T. and Jagner D., Anal. Chim. Acta, 66 (1973) 152.

[19] Angyal S. J. and Davies K. P., Chem. Commun., (1971) 500.

[20] Antonov V. K., Shchelokov V. I., Shemyakin M. M., Tovarova I. I. and Kiseleva O. A., Antibiotiki (U.S.S.R.), 10 (1965) 387.

[21] Arnett E. M., Ko H. C. and Chao C. C., J. Am. Chem. Soc., 94 (1972) 4776.

[22] Arnett E. M. and Moriarty T. C., J. Am. Chem. Soc., 93 (1971) 4908.

[23] Ashton R. and Steinrauf L. K., J. Mol. Biol., 49 (1970) 547.

[24] Avignon M. and Huang P. V., Biopolymers, 9 (1970) 427.

[25] Avignon M., Huang P. V., Lascombe J., Marrand M. and Neel J., Biopolymers, 8 (1969) 69.

[26] Avron M. and Shavit N., Biochim. Biophys. Acta, 109 (1965) 317.

[27] Azzi A. and Azzone G. F., Biochim. Biophys. Acta, 131 (1967) 468.

[28] Azzi A. and Azzone G. F., Biochim. Biophys. Acta, 135 (1967) 444.

[29] Azzi A. and Santato M., FEBS Lett., 27 (1972) 35.

[30] Azzi A. and Scarpa A., Biochim. Biophys. Acta, 135 (1967) 1087.

[31] Azzone G. F., J. Bioenerg., 3 (1972) 95.

[32] Azzone G. F. and Ernster L., J. Biol. Chem., 236 (1961) 1501.

[33] Azzone G. F., Rossi E. and Scarpa A., in Regulatory Functions of Biological Membranes [BBA Library, Vol. 11], (Järnefelt J., ed.) Elsevier, Amsterdam, 1968, p. 236.

[34] Babakov A. V., Myagkov I. V., Sotnikov P. S. and Terekhov O. P., Biofizika (U.S.S.R.), 17 (1972) 347.

[35] Babayan A. T., Gambaryan N. and Gambaryan N. P., Zh. Obshch. Khim. (U.S.S.R.), 24 (1954) 1887.

[36] Babayan A. T. and Indzhikyan M. G., Zh. Obshch. Khim. (U.S.S.R.), 27 (1957) 1201.

408

[37] Bachofen R., Experientia, 27 (1971) 770.
[38] Baltscheffsky M., Arch. Biochem. Biophys., 130 (1969) 646.
[39] Baltscheffsky H. and Arwidsson B., Biochim. Biophys. Acta, 65 (1962) 425.
[40] Bamberg E. and Läuger P., private communication.
[41] Bangham A. D., Ann. Rev. Biochem., 41 (1972) 753.
[42] Bangham A. D., Chem. Phys. Lipids, 8 (1972) 386.
[43] Bangham A. D., de Gier J. and Greville G. D., J. Mol. Biol., 13 (1965) 238.
[44] Banin A. and Shaked D., Agrochimica, 15 (1971) 238.
[45] Barber J., J. Physiol., 223 (1972) 23P; FEBS Lett., 20 (1972) 251.
[46] Barber J., Biochim. Biophys. Acta, 275 (1972) 105.
[47] Barber J. and Kraan G. P. B., Biochim. Biophys. Acta, 197 (1970) 49.
[48] Barrett-Bee K. and Radda G. K., Biochim. Biophys. Acta, 267 (1972) 211.
[49] Bartsch R. A. and Wiegers K., Tetrahedron Lett., (1972) 3819.
[50] Barsukov L. I., Shkrob A. M. and Bergelson L. D., Biofizika (U.S.S.R.), 17 (1972) 1032.
[51] Bayley P. M., Nielsen E. B. and Schellman J. A., J. Phys. Chem., 73 (1969) 228.
[52] Beck J., Gerlach H., Prelog V. and Voser W., Helv. Chim. Acta, 45 (1962) 620.
[53] Benedict R. G., Bot. Rev., 19 (1953) 271.
[54] Benedict R. G., Pridham T. G., Lindenfelser L. A., Hall H. H. and Jackson R. W., Appl. Microbiol., 3 (1955) 1.
[55] Bennett R. E., Brindle S. A., Ginffre N. A., Jackson P. W., Kowald J., Pansy F. E., Perlman D. and Trejo W. H., Antimicrob. Agents Chemother., (1961) 169.
[56] Bentley P. J., J. Endocrinol., 45 (1969) 287.
[57] Berger J., Kachlin A. I., Scott W. E., Sternbach L. H. and Goldberg M. W., J. Am. Chem. Soc., 73 (1951) 5295.
[58] Bertaud W. C., Probine M., Shannon J. S. and Taylor A., Tetrahedron, 21 (1965) 677.
[59] Bevan K., Davies J., Hall M. J., Hassall C. H., Morton R. B., Phillips D. A. S., Ogihara Y. and Thomas W. A., Experientia, 26 (1970) 122.
[60] Bevan K., Davies J. C., Hassall C. H., Morton R. B. and Phillips D. A. S., J. Chem. Soc. Ser. C., (1971) 514.
[61] Beyer C. F., Craig L. C. and Gibbons W. A., Biochemistry, 11 (1972) 4920.
[62] Beyer R. E., Brinker K. R. and Crankshaw D. L., Can. J. Biochem., 47 (1969) 117.
[63] Beyer R. E., Crankshaw D. L. and Kuner J. M., Biochem. Biophys. Res. Commun., 28 (1967) 758.
[64] Bhattacharjya P., Epstein W. and Silver S., Proc. Natl. Acad. Sci. U.S., 68 (1971) 1488.
[65] Bielawski J., Eur. J. Biochem., 4 (1968) 181.
[66] Birr C. and Lochinger W., Synthesis, (1971) 319.
[67] Bishop E., Griffiths H., Russell D. W., Ward V. and Gartside R. N., J. Gen. Microbiol., 38 (1965) 289.
[68] Bishop E. and Russell D. W., J. Chem. Soc., Ser. C., (1967) 634.
[69] Bissell E. C. and Paul I. C., Chem. Commun., (1972) 967.
[70] Bittar E. E., Experientia, 27 (1971) 657.
[71] Black D. St. C. and McLean I. A., Chem. Commun., (1968) 1004.
[72] Black D. St. C. and McLean I. A., Tetrahedron Lett., (1969) 3961.
[73] Blondin G. A., Decastro A. F. and Senior A. E., Biochem. Biophys. Res. Commun., 43 (1971) 28.
[74] Blondin G. A. and Green D. E., Proc. Natl. Acad. Sci. U.S., 58 (1967) 612.
[75] Blondin G. A., Vail W. J. and Green D. E., Arch. Biochem. Biophys., 129 (1969) 158.
[76] Blount J. F. and Westley J. W., Chem. Commun., (1971) 927.
[77] Blount J. F. and Westley J. W., Chem. Commun., (1972) 112.

[78] Bogolyubova N. D., Oreshnikova N. A., Muravyeva T. I. and Pavlenko I. A., in Biofizika Membran (Alkalene B. D., Zablotskaite D. V. and Narushevichus E. V., eds), Kaunas, 1971, part 1, p. 146.
[79] Boheim G., Erregbarkeit schwarzer Lipid Membranen, D.S. Thesis, Rheinisch-Westfälische Techn. Hochschule, Aachen, 1972.
[80] Bokii G. B., Introduction into Crystallochemistry, Moscow State University Publishing House, Moscow, 1954.
[81] Borgan G. and Dale J., Chem. Commun., (1970) 1340.
[82] Borovyagin V. L., Biofizika (U.S.S.R.), 16 (1971) 746.
[83] Bradbury E. M., Downie A. R., Elliot A. and Hanby W. E., Proc. R. Soc., Ser. A, 259 (1960) 110.
[84] Brändström A. and Junggren U., Acta Chem. Scand., 23 (1969) 2203 and 2204.
[85] Brändström A. and Junggren U., Tetrahedron Lett., (1972) 473.
[86] Brewster A. I. and Bovey F. A., Proc. Natl. Acad. Sci. U.S., 68 (1971) 1199.
[87] Bright D. and Truter M. R., J. Chem. Soc., Ser. B. (1970) 1544.
[88] Bright D. and Truter M. R., Nature, 225 (1970) 176.
[89] Brierley G. P., Biochem. Biophys. Res. Commun., 35 (1969) 396.
[90] Brierley G. P., Biochemistry, 9 (1970) 697.
[91] Brierley G. P., Jurkowitz M., Merola A. J. and Scott K. M., Arch. Biochem. Biophys., 152 (1972) 744.
[92] Brierley G. P., Jurkowitz M., Scott K. M., Hwang K. M. and Merola A. J., Biochem. Biophys. Res. Commun., 43 (1971) 50.
[93] Brierley G. P., Jurkowitz M., Scott K. M. and Merola A. J., J. Biol. Chem., 245 (1970) 5404.
[94] Brierley G. P., Settlemire C. T. and Knight V. A., Arch. Biochem. Biophys., 126 (1968) 276.
[95] Brierley G. P. and Stoner C. D., Biochemistry, 9 (1970) 708.
[96] Briller S. and Gromet-Elhanan Z., Biochim. Biophys. Acta, 205 (1970) 263.
[97] Brockmann H. and Geeren H., Liebig's Ann. Chem., 603 (1957) 216.
[98] Brockmann H. and Schmidt-Kastner G., Chem. Ber., 88 (1955) 57.
[99] Brockmann H., Springorum M., Träxler G. and Höfer I., Naturwiss., 22 (1963) 689.
[100] Brown R., Brennan J. and Kelley C., Antibiot. Chemother., 12 (1962) 482.
[101] Bruner L. J., Biophysik, 6 (1970) 241.
[102] Buck R. P., Anal. Chem., 44 (1972) 270R.
[103] Bugg C. E. and Cook W. J., Chem. Commun., (1972) 727.
[104] Burg K. H., Herrmann H. D. and Rehling H., Makromol. Chem., 111 (1968) 181.
[105] Bush M. A. and Truter M. R., Chem. Commun., (1970) 1439.
[106] Bush M. A. and Truter M. R., J. Chem. Soc., Ser. B, (1971) 1440.
[107] Bush M. A. and Truter M. R., J. Chem. Soc., Perkin Trans. (II), (1972) 341.
[108] Bush M. A. and Truter M. R., J. Chem. Soc., Perkin Trans. (II), (1972) 345.
[109] Butler K. W., Dugas H., Smith I. C. P. and Schneider H., Biochem. Biophys. Res. Commun., 40 (1970) 770.
[110] Butler I. N. and Huston R., Anal. Chem., 42 (1970) 676.
[111] Buzhinskii E. P., Tsitologiya (U.S.S.R.), 10 (1968) 1432.
[112] Bünning E. and Moser I., Proc. Natl. Acad. Sci. U.S., 69 (1972) 2732.
[113] Bystrov V. F., Usp. Khim. (U.S.S.R.), 41 (1972) 512.
[114] Bystrov V. F., Ivanov V. T., Kozmin S. A., Mikhaleva I. I., Khalilulina K.Kh., Ovchinnikov Yu. A., Fedin E. I. and Petrovskii P. V., FEBS Lett., 21 (1972) 34.
[115] Bystrov V. F., Ivanov V. T., Portnova S. L., Balashova T. A. and Ovchinnikov Yu. A., Tetrahedron, 29 (1973) 873.
[116] Bystrov V. F., Portnova S. L., Balashova T. A., Tsetlin V. I., Ivanov V. T., Kostetsky P. V. and Ovchinnikov Yu. A., Tetrahedron Lett., (1964) 5283.

410

[117] Bystrov V. F., Portnova S. L., Tsetlin V. I., Ivanov V. T. and Ovchinnikov Yu. A., Tetrahedron, 25 (1969) 493.

[118] Caplan S. R. and Essig A., Proc. Natl. Acad. Sci. U.S., 64 (1969) 211.

[119] Carafoli E., Rossi C. S. and Gazzotti P., Arch. Biochem. Biophys., 131 (1969) 527.

[120] Carlson R. H., Norland K. S., Fasman G. D. and Blout E. R., J. Am. Chem. Soc., 82 (1960) 2268.

[121] Cary L. W., Takita T. and Ohnishi M., FEBS Lett., 17 (1971) 145.

[122] Caswell A. H. and Pressman B. C., Biochem. Biophys. Res. Commun., 49 (1972) 292.

[123] Cereijo-Santaló R., Can. J. Biochem., 46 (1968) 55.

[124] Chamberlin J. W. and Agtarap A., Org. Mass. Spectrom., 3 (1970) 271.

[125] Chance B. and Yoshioka T., Arch. Biochem. Biophys., 117 (1966) 451.

[126] Chance B., FEBS Lett., 23 (1972) 1.

[127] Chapman D., Cherry R. J., Finer E. G., Hauser H., Phillips M. C., Shipley G. G. and McMullen A. I., Nature, 224 (1969) 692.

[128] Chapman D. and Leslie R. B., in Membranes of Mitochondria and Chloroplasts [ACS Monograph 165] (Racker E., ed.), Van Nostrand Reinhold, 1970, p. 91.

[129] Chappell J. B., Biochem. J., 116 (1970) 2P.

[130] Chappell J. B. and Crofts A. R., Biochem. J., 95 (1965) 393.

[131] Chappell J. B. and Crofts A. R., in Regulation of Metabolic Processes in Mitochondria (Tager J. M., Papa S., Guagliariello E. and Slater E., eds), Elsevier, Amsterdam, 1966, p. 393.

[132] Chappell J. B. and Haarhoff K. N., in Biochemistry of Mitochondria (Slater E. C., Kaniuga Z. and Wojtczak L., eds), Academic Press, London, 1967, p. 75.

[133] Chem. Eng. News, 48, No. 9 (1970) 26.

[134] Cheney J., Lehn J. M., Sauvage J. P. and Stubbs M. E., Chem. Commun., (1972) 1100.

[135] Cheney J. and Lehn J. M., Chem. Commun., (1972) 487.

[136] Cherry R. J., Chem. Phys. Lipids, 8 (1972) 393.

[137] Cherry R. J., Chapman D. and Graham D. E., J. Membrane Biol., 7 (1972) 325.

[138] Chirgadze Yu.N. and Rashevskaya E. P., Biofizika (U.S.S.R.), 14 (1969) 608.

[139] Chizmadjev Yu.A., Markin V. S. and Kuklin R. N., Biofizika (U.S.S.R.), 16 (1971) 230 and 437.

[140] Chizmadjev Yu.A., Markin V. S. and Kuklin R. N., Biofizika (U.S.S.R.), 16 (1971) 437.

[141] Chizmadjev Yu.A., Markin V. S. and Kuklin R. N., Biofizika (U.S.S.R.), 16 (1971) 230.

[142] Chock P. B., Biochimie, 53 (1971) 161.

[143] Chock P. B., Proc. Natl. Acad. Sci. U.S., 69 (1972) 1939.

[144] Christensen J. J., Hill J. O. and Izatt R. M., Science, 174 (1971) 459.

[145] Christian G., Anal. Lett., 3 (1970) 11.

[146] Christiansen R. O., Loyter A. and Racker E., Biochim. Biophys. Acta, 180 (1969) 207.

[147] Ciani S., Eisenman G., Laprade R. and Szabo G., in Membranes — A Series of Advances (Eisenman G., ed.) Vol. 2, Marcel Dekker, New York, 1973, p. 61.

[148] Ciani S., Eisenman G. and Szabo G., J. Membrane Biol., 1 (1969) 1.

[149] Ciani S., Eisenman G. and Szabo G., in Abstracts of papers presented at the Symposium on Molecular Mechanisms of Antibiotic Action on Protein Biosynthesis and Membranes, Granada, 1971, p. 66.

[150] Clerk J. T., Kahr G., Pretsch E., Scholer R. P. and Wuhrmann H.-R., Chimia, 26 (1972) 287.

[151] Cochran G. T., Allen J. F. and Marullo N. P., Inorg. Chim. Acta, 1 (1967) 109.

411

[152] Cockrell R. S., Fed. Proc., 28 (1969) 472.
[153] Cockrell R. S., Biochem. Biophys. Res. Commun., 46 (1972) 1991.
[154] Cockrell R. S., Harris E. J. and Pressman B. C., Biochemistry, 5 (1966) 2326.
[155] Cockrell R. S., Harris E. J. and Pressman B. C., Nature, 215 (1967) 1487.
[156] Cockrell R. S. and Racker E., Biochem. Biophys. Res. Commun., 35 (1969) 414.
[157] Colacicco G., Gordon E. E. and Berchenko G., Abstracts of Reports at 12th Annual Meeting of the Biophysical Society U.S., Pittsburgh, 1968.
[158] Colacicco G., Gordon E. E. and Berchenko G., Biophys. J., 8 (1968) 22a.
[159] Colley C. M. and Metcalfe J. C., FEBS Lett., 24 (1972) 241.
[160] Cook A. H., Cox S. F. and Farmer T. H., Nature, 162 (1948) 61.
[161] Cook A. H., Cox S. F. and Farmer T. H., J. Chem. Soc., (1949) 1022.
[162] Cook A. H., Cox S. F., Farmer T. H. and Lacey M. S., Nature, 160 (1947) 31.
[163] Cooper C. and Lehninger A. L., J. Biol. Chem., 224 (1957) 647.
[164] Corbaz R., Ettlinger L., Gäumann E., Keller-Schierlein W., Kradolfer F., Neipp L., Prelog V. and Zähner H., Helv. Chim. Acta, 38 (1955) 1445.
[165] Coryta J., Anal. Chim. Acta, 61 (1972) 329.
[166] Cosani A., Reggion E., Verdini A. S. and Terbojevich M., Biopolymers, 6 (1968) 963.
[167] Cosgrove R. E., Mask C. A. and Krull I. H., Anal. Lett., 3 (1970) 457.
[168] Craig L. C., Gregory J. D. and Barry G. T., J. Clin. Invest., 28 (1949) 1014.
[169] Cram D., Fundamentals of Carbanion Chemistry, Academic Press, New York and London, 1965.
[170] Cramer F., Einschluss-Verbindungen, Springer-Verlag, Berlin-Göttingen-Heidelberg, 1954.
[171] Cramer F. A., Saenger W. and Spatz H.-Ch., J. Am. Chem. Soc., 89 (1967) 14.
[172] Crippen G. M. and Scheraga H. A., Proc. Natl. Acad. Sci. U.S., 64 (1968) 42.
[173] Crofts A. R., J. Biol. Chem., 242 (1967) 3352.
[174] Crofts A. R., Deamer D. W. and Packer L., Biochim. Biophys. Acta, 131 (1967) 97.
[175] Crofts A. R., Wraight C. A. and Fleischmann D. E., FEBS Lett., 15 (1971) 89.
[176] Curtin D. Y., Crawford R. J. and Wilhelm M., J. Am. Chem. Soc., 80 (1958) 1391.
[177] Czerwinski E. W. and Steinrauf L. K., Biochem. Biophys. Res. Commun., 45 (1971) 1284.
[178] Dale J., J. Chem. Soc., (1963) 93.
[179] Dale J. and Krane J., Chem. Commun., (1972) 1012.
[180] Dale J. and Kristiansen P. O., Chem. Commun., (1971) 670.
[181] Dale J. and Kristiansen P. O., Acta Chem. Scand., 26 (1972) 1471.
[182] Dalley N. K., Smith D. F., Izatt R. M. and Christensen J. J., Chem. Commun., (1972) 90.
[183] Dam van K. and Meyer A. J., Ann. Rev. Biochem., 40 (1971) 115.
[184] Damiani A., de Santis P. and Pizzi A., Nature, 226 (1970) 542.
[185] Dann J. R., Chiesa P. P. and Gates J. W., J. Org. Chem., 26 (1961) 1991.
[186] Dargent-Sallé M.-L. and Wins P., Life Sci., part II, 9 (1970) 1091.
[187] Das B. C., Gero S. D. and Lederer E., Biochem. Biophys. Res. Commun., 29 (1967) 211.
[188] Dashevskii V. G., in High Molecular Compounds. Theoretical Aspects of Macromolecular Conformations (Kitaigorodsky A.I., ed.), VINITI, Moscow, 1970, p. 93.
[189] Dawson A. P., Dunnett S. J. and Selwyn M. J., Biochem. J., 127 (1972) 67P.
[190] Deamer D. W. and Packer L., Biochim. Biophys. Acta, 172 (1969) 539.
[191] Deber C. M., Torchia D. A., Dorman D. E., Bovey F. A. and Blout E. R., in Chemistry and Biology of Peptides (Meienhofer J., ed.), Ann Arbor Science, Ann Arbor, 1972, p. 39.
[192] Deber C. M., Torchia D. A., Wong S. C. K. and Blout E. R., Proc. Natl. Acad. Sci. U.S., 69 (1972) 1825.

412

[193] Deenen van L. L. M., Chem. Phys. Lipids, 8 (1972) 366.
[194] Deenen van L. L. M., de Gier J., Demel R. A., Haest C., van Zutphen H., de Kruyff B. and Norman A. W., in Abstracts of papers presented at the Symposium on Molecular Mechanisms of Antibiotic Action on Protein Biosynthesis and Membranes, Granada, 1971, p. 72.
[195] Degani H. and Shavit N., Arch. Biochem. Biophys., 152 (1972) 339.
[196] De Gier J., Haest C. W. M., Mandersloot J. G. and van Deenen L. L. M., Biochim. Biophys. Acta, 211 (1970) 373.
[197] De Gier J., Mandersloot J. G. and van Deenen L. L. M., Biochim. Biophys. Acta, 173 (1969) 143.
[198] Demin V. V., Babakov A. V. and Shkrob A. M., in Biofizika Membran (Alkalene B. D., Zablotskaite D. A. and Narushevichus E. V., eds), part 1, Kaunas, 1971, p. 337.
[199] Demin V. V., Babakov A. V., Shkrob A. M. and Ovchinnikov Yu. A., Biofizika (U.S.S.R.), in press.
[200] Demin V. V. and Shkrob A. M., IVth International Biophysics Congress, Abstracts of contributed papers, EX-4.
[201] Denisova S. I., Ovchinnikova G. A. and Melnikov G. P., Zh. Obshch. Khim. (U.S.S.R.), 33 (1963) 2058.
[202] Diebler H., Eigen M., Ilgenfritz G., Maass G. and Winkler R., Pure Appl. Chem., 20 (1969) 93.
[203] Dietrich B., Lehn J. M. and Sauvage J. P., Tetrahedron Lett., (1969) 2885.
[204] Dietrich B., Lehn J. M. and Sauvage J. P., Tetrahedron Lett., (1969) 2889.
[205] Dietrich B., Lehn J. M. and Sauvage J. P., Chem. Commun., (1970) 1055.
[206] Dietrich B., Lehn J. M. and Sauvage J. P., Chem. Commun., (1973) 15.
[207] Dobler M., Helv. Chim. Acta, 55 (1972) 1371.
[208] Dobler M., Dunitz J. D. and Kilbourn B. T., Helv. Chim. Acta, 52 (1969) 2573.
[209] Dobler M., Dunitz J. D. and Krajewski J., J. Mol. Biol., 42 (1969) 603.
[210] Dominguez J., Dunitz J. D., Gerlach H. and Prelog V., Helv. Chim. Acta, 45 (1962) 129.
[211] Down J. L., Lewis J., Moore B. and Wilkinson G. W., J. Chem. Soc., (1959) 3767.
[212] Downing A. P., Ollis W. D. and Sutherland J. O., J. Chem. Soc., Ser. B., (1970) 24.
[213] Dreele van P. H., Brewster A. I., Scheraga H. A., Ferger M. F. and du Vigneaud V., Proc. Natl. Acad. Sci. U.S., 68 (1971) 1028.
[214] Duax W. L. and Hauptman H., Acta Cryst., Ser. B., 28 (1972) 2912.
[215] Duax W. L., Hauptman H., Weeks C. M. and Norton D. A., Science, 176 (1972) 911.
[216] Dubos R. J., J. Exp. Med., 70 (1939) 1.
[217] Dubos R. J., J. Exp. Med., 70 (1939) 11.
[218] Dutcher J. D., Antimicrob. Agents Chemother., (1961) 173.
[219] Duynstee E. F. J. and Grunwald E., Tetrahedron, 21 (1965) 2401.
[220] Dye J. L., De Backer M. G. and Nicely V. A., J. Am. Chem. Soc., 92 (1970) 5226.
[221] Edgell W. F., Watts A. T., Lyford J. and Risen W. M., J. Am. Chem. Soc., 88 (1966) 1815.
[222] Edsall J. T., Flory P. J., Kendrew J. C., Liquori A. M., Nemethy G., Ramachandran G. N. and Scheraga H. A., J. Mol. Biol., 15 (1966) 399; J. Biol. Chem., 241 (1966) 1004; Biopolymers, 4 (1966) 121.
[223] Efremov E. S., Kostetsky P. V., Ivanov V. T., Popov E. M. and Ovchinnikov Yu. A., Khim. Prir. Soedin. (U.S.S.R.), (1973) 348.
[224] Efremov E. S., Senyavina L. B., Kostetsky P. V., Ivanova A. N., Zheltova V. I., Ivanov V. T., Popov E. M. and Ovchinnikov Yu. A., Khim. Prir. Soedin. (U.S.S.R.), (1973) 322.

[225] Eigen M. and De Maeyer L., in Techniques of Organic Chemistry (Weissberger A., ed.), Vol. 8, part 2, Interscience, New York, 1963, p. 895.

[226] Eisenberg-Grünberg M., Ph.D. Thesis, Calif. Inst. Technol., Pasadena, U.S., 1972.

[227] Eisenman G., Biophys. J., 2 (1962) 259S.

[228] Eisenman G., Bol. Inst. Est. Medic. Biol., 21 (1963) 155.

[229] Eisenman G., Excerpta Medica, Int. Congr. Ser. No. 87, 1965, p. 489.

[230] Eisenman G., in Ion-Selective Electrodes (Durst R. A., ed.), Natl. Bur. Stand. U.S. Sp. Publ., No. 314, 1969, p. 1.

[231] Eisenman G., private communication.

[232] Eisenman G., Ciani S. M. and Szabo G., Fed. Proc., 27 (1968) 1289.

[233] Eisenman G., Ciani S. and Szabo G., J. Membrane Biol., 1 (1969) 294.

[234] Eisenman G. and Krasne S., Symposium report at IVth International Biophysics Congress, Moscow, 1972.

[235] Eisenman G., Szabo G., McLaughlin S. G. A. and Ciani S. M., in Molecular Mechanisms of Antibiotic Action on Protein Biosynthesis and Membranes (Muñoz E., García-Ferrándiz F. and Vazquez D., eds) Elsevier, Amsterdam, 1972.

[236] Emsley J. W., Feeney J. and Sutcliffe L. H., High Resolution Nuclear Magnetic Resonance Spectroscopy, Vol. 1, Pergamon Press, 1965, p. 26.

[237] Entman M. L., Gillette P. C., Wallick E. T. and Pressman B. C., Biochem. Biophys. Res. Commun., 48 (1972) 847.

[238] Erlich R. H., Roach E. and Popov A. I., J. Am. Chem. Soc., 92 (1970) 4989.

[239] Ermishkin L. N. and Muskhelishvili N. L., Biofizika (U.S.S.R.), 16 (1971) 849; see also typescripts in the publishers lists of VINITI, No. 1491-70 and No. 3947-72.

[240] Ermishkin L. N. and Smolyaninov V. V., Typescript in the publishers lists of VINITI, No. 1416-70.

[241] Estrada-O S. and Calderon E., Biochemistry, 9 (1970) 2092.

[242] Estrada-O S. and Carabez A., J. Bioenerg., 3 (1972) 429.

[243] Estrada-O S., de Caspedes C. and Calderon E., J. Bioenerg., 3 (1972) 361.

[244] Estrada-O S. and Gomez-Lojero C., J. Biol. Chem., 245 (1970) 5606.

[245] Estrada-O S., Gomez-Lojero C. and Montal M., J. Bioenerg., 3 (1972) 417.

[246] Estrada-O S., Graven S. N. and Lardy H. A., Fed. Proc., 26 (1967) 610.

[247] Estrada-O S., Graven S. N. and Lardy H., J. Biol. Chem., 242 (1967) 2925.

[248] Eyal E. and Rechnitz G. A., Anal. Chem., 43 (1971) 1090.

[249] Fain J. N. and Loken S. C., Mol. Pharmacol., 7 (1971) 465.

[250] Falcone A. B. and Hadler H. I., Arch. Biochem. Biophys., 124 (1968) 91.

[251] Falcone A. B. and Hadler H. I., Arch. Biochem. Biophys., 124 (1968) 115.

[252] Fang M. and Rasmussen H., Biochim. Biophys. Acta, 153 (1968) 88.

[253] Faulstich H., Bürgermeister W. and Wieland Th., Biochem. Biophys. Res. Commun., 47 (1972) 984.

[254] Faulstich H., Trischmann H. and Wieland Th., Tetrahedron Lett., (1969) 4131.

[255] Faulstich H. and Wieland Th., Abstracts of communications presented at VIIIth FEBS Meeting (Amsterdam, August 1972), No. 103, North-Holland, Amsterdam, 1972.

[256] Faust M. A. and Doetsch R. N., Can. J. Microbiol., 17 (1971) 183.

[257] Feigenbaum W. M. and Michel R. H., J. Polymer Sci., 9 (A-2) (1971) 817.

[258] Feinstein M. B. and Felsenfeld H., Proc. Natl. Acad. Sci. U.S., 68 (1971) 2037.

[259] Fenton D. E., Mercer M., Poonia N. S. and Truter M. R., Chem. Commun., (1972) 66.

[260] Fenton D. E., Merger M. and Truter M. R., Biochem. Biophys. Res. Commun., 48 (1972) 10.

[261] Ferguson S. M. F., Estrada-O S. and Lardy H. A., J. Biol. Chem., 246 (1971) 5645.

414

[262] Fettiplace R., Andrews D. M. and Haydon D. A., J. Membrane Biol., 5 (1971) 277.

[263] Finer E. G., Houser H. and Chapman D., Chem. Phys. Lipids, 3 (1969) 386.

[264] Fiume L. and Wieland Th., FEBS Lett., 8 (1970) 1.

[265] Fleischman D. E. and Clayton R. K., Photochem. Photobiol., 8 (1968) 287.

[266] Florey H. W., Chain E., Heatley N. G., Jennings M. A., Sanders A. G., Abraham E. P. and Florey M. E., Antibiotics, Vols I and II, Oxford University Press, London, 1949.

[267] Fonina L. A., Sanasaryan A. A. and Vinogradova E. I., Khim. Prir. Soedin. (U.S.S.R.), (1971) 69.

[268] Fox C. F., Sci. Am., 226 (1972) 30.

[269] Fraenkel G. and Pechhold E., Tetrahedron Lett., (1970) 153.

[270] Frant M. S. and Ross J. W., Science, 167 (1970) 987.

[271] French. Pat. 1164181, Ref. Zh. Khim. (U.S.S.R.), (1960) 58393.

[272] Frensdorff H. K., J. Am. Chem. Soc., 93 (1971) 600.

[273] Frensdorff H. K., J. Am. Chem. Soc., 93 (1971) 4684.

[274] Früh P. V., Clerc J. T. and Simon W., Helv. Chim. Acta, 54 (1971) 1445.

[275] Frumkin A. and Gerovich M., J. Chem. Phys., 4 (1936) 624.

[276] Frumkin A. N., Gugeshashvili M. I. and Boguslavskii L. I., Dokl. Akad. Nauk. S.S.S.R. (U.S.S.R.) 198 (1971) 1452.

[277] Fuchs R., Bear J. L. and Rodewald R. F., J. Am. Chem. Soc., 91 (1969) 5797.

[278] Furisaki A. and Watanabe T., Tetrahedron Lett., (1968) 6301.

[279] Gachon P., Kergomard A., Veschambre H., Esteve C. and Staron T., Chem. Commun., (1970) 1421.

[280] Gäumann E., Roth S., Ettlinger L., Plattner Pl.A. and Nager U., Experientia, 3 (1947) 202.

[281] Geddes, A. J., Parker K. D., Atkins E. D. T. and Beighton E., J. Mol. Biol., 32 (1968) 343.

[282] Gerlach H. and Huber E., Helv. Chim. Acta., 50 (1967) 2087.

[283] Gerlach H., Hütter R., Keller-Schierlein W., Seibl J. and Zähner H., Helv. Chim. Acta, 50 (1967) 1782.

[284] Gerlach H. and Prelog V., Liebig's Ann. Chem., 669 (1963) 121.

[285] Giancotti V., Quadrifoglio F. and Crescenzi V., J. Am. Chem. Soc., 94 (1972) 297.

[286] Gibb L. E. and Eddy A. A., Biochem. J., 129 (1972) 979.

[287] Gibson N. A. and Hosking I. W., Austr. J. Chem., 18 (1965) 123.

[288] Gisin B. F. and Merrifield R. B., J. Am. Chem. Soc., 94 (1972) 6165.

[289] Gisin B. F., Merrifield R. B. and Tosteson D. C., J. Am. Chem. Soc., 91 (1969) 2691.

[290] Glickson J. D., Mayers D. F., Settine J. M. and Urry D. W., Biochemistry, 11 (1972) 477.

[291] Glinskii V. P., Samuilov V. D. and Skulachev V. P., Mol. Biol. (U.S.S.R.), 6 (1972) 664.

[292] Glynn I. M., Nature, 216 (1967) 1318.

[293] Go N., Lewis P. N. and Scheraga H. A., Macromolecules, 3 (1970) 628.

[294] Gobillon Y., Piret P. and Van Meersche M., Bull. Soc. Chim. France, (1962) 551.

[295] Goldman D. E., J. Gen. Physiol., 27 (1943) 37.

[296] Gómez-Puyou A., Sandoval F., Chávez E. and Tueña M., J. Biol. Chem., 245 (1970) 5239.

[297] Gómez-Puyou A., Sandoval F., Tueña M., Chávez E. and Peña A., Arch. Biochem. Biophys., 129 (1969) 329.

[298] Gómez-Puyou A., Sandoval F., Tueña M., Peña A. and Chávez E., Biochem. Biophys. Res. Commun., 45 (1971) 104.

[299] Good N. E., Izawa S. and Hind G., in Current Topics in Bioenergetics (Sanadi R., ed.), Academic Press, Vol. 1, 1966, p. 75.
[300] Goodall M. C., Biochim. Biophys. Acta, 219 (1970) 471.
[301] Goodall M. C., Nature, 225 (1970) 1257.
[302] Goodall M. C., Arch. Biochem. Biophys., 147 (1971) 129.
[303] Goodall M. C., Structural effects in the action of antibiotics on the ion permeability of lipid bilayers. IV. Mechanism of "autocatalytic" behaviour, preprint.
[304] Goodman I., Kettle S. J. and Owston P. G., Chem. Ind., (1971) 1300.
[305] Gordon E. E. and Bernstein J., Biochim. Biophys. Acta, 205 (1970) 464.
[306] Gordon E. E. and de Hartog M., Biochim. Biophys. Acta, 162 (1968) 220.
[307] Gordon E. E., private communication.
[308] Gordon E. E., Nordenbrand K. and Ernster L., Nature, 213 (1967) 82.
[309] Gordon L. G. M. and Haydon D. A., Biochim. Biophys. Acta, 225 (1972) 1014.
[310] Gordon L. G. M. and Haydon D. A., Symposium report at IVth International Biophysics Congress (Moscow, August 1972).
[311] Gordon W., J. Membrane Biol., 10 (1972) 193.
[312] Gorman M., Chamberlin J. W. and Hamill R. L., Antimicrob. Agents Chemother., (1967) 363.
[313] Gorneva G. A., Ryabova I. D. and Faizova G., in Membrane Biophysics (Russian) (Alkalene B. D., Zablotskaite D. A. and Narushevichus E. V. eds), Part 1, Kaunas, 1971, p. 318.
[314] Graven S. N., Estrada-O S. and Lardy H. A., Proc. Natl. Acad. Sci. U.S., 56 (1966) 654.
[315] Graven S. N., Lardy H. A. and Estrada-O S., Biochemistry, 6 (1967) 365.
[316] Graven S. N., Lardy H. A., Johnson D. and Rutter A., Biochemistry, 5 (1966) 1729.
[317] Graven S. N., Lardy H. A. and Rutter A., Biochemistry, 5 (1966) 1735.
[318] Green D. E., Proc. Natl. Acad. Sci. U.S., 67 (1970) 544.
[319] Green D. E. and Brucker R. F., Bio Science, 22 (1972) 13.
[320] Green D. E. and Harris R. A., FEBS Lett., 5 (1969) 241.
[321] Green D. E. and Ji S., J. Bioenerg., 3 (1972) 159; Proc. Natl. Acad. Sci. U.S., 69 (1972) 726.
[322] Greene R. N., Tetrahedron Lett., (1972) 1793.
[323] Gregory J. D. and Craig L. C., J. Biol. Chem., 172 (1948) 839.
[324] Grell E., Funk Th. and Sauter H., Eur. J. Biochem., 34 (1973) 415.
[325] Grell E., Eggers F. and Funk Th., Chimia, 26 (1972) 632.
[326] Grell E., Funk Th. and Eggers F., in Molecular Mechanisms of Antibiotic Action on Protein Biosynthesis and Membranes (Muñoz E., García-Ferrándiz F. and Vazquez, D. eds), Elsevier, Amsterdam, 1972, p. 646.
[327] Funk Th., Eggers F. and Grell E., Chimia, 26 (1972) 637.
[328] Greville R. D., in Current Topics in Bioenergetics (Sanadi D. R., ed.), Vol. 3, Academic Press, 1969, p. 1.
[329] Grinius L. L., Jasaitis A. A., Kadziauskas Yu. P., Liberman E. A., Skulachev V. P., Topali V. P., Tsofina L. M. and Vladimirova M. A., Biochim. Biophys. Acta, 216 (1970) 1.
[330] Gromet-Elhanan Z., Biochim. Biophys. Acta, 223 (1970) 174.
[331] Gromet-Elhanan Z., FEBS Lett., 13 (1971) 124.
[332] Gross E. and Witkop B., Biochemistry, 4 (1965) 2495.
[333] Grotens A. M., Smid J. and De Boer E., Chem. Commun., (1971) 759.
[334] Groth P., Acta Chem. Scand., 25 (1971) 725.
[335] Grünhagen H. H. and Witt H. T., Z. Naturforsch., 25B (1970) 373.
[336] Guerilott-Vinet A., Guyot L., Montegut J. and Roux L., Compt. Rend., 230 (1950) 1424.

416

[337] Gunn R. B. and Tosteson D. C., J. Gen. Physiol., 57 (1971) 593.

[338] Gutmann V., Rec. Chem. Progr., 30 (1969) 169.

[339] Gutmann V., Chimia, 23 (1969) 285.

[340] Gyimesi J., Horvath I. and Szentirmai A., Z. Allg. Mikrobiol., 4 (1964) 269.

[341] Hadler H. I. and Falcone A. B., Arch. Biochem. Biophys., 124 (1968) 110.

[342] Haest C. W. M., de Gier J., Op den Kamp J. A. F. and van Deenen L. L. M., Biochim. Biophys. Acta, 255 (1972) 720.

[343] Halban von H. and Szigeti B., Helv. Chim. Acta, 20 (1937) 746.

[344] Hall C. E., Chem. Ind., (1960) 1270.

[345] Hall E. A., Kavanagh F. and Asheshov I. N., Antibiot. Chemother., 1 (1951) 369.

[346] Hall M. J., Biochem. Biophys. Res. Commun., 38 (1970) 590.

[347] Hall M. J. and Hassall S. H., Appl. Microbiol., 19 (1970) 109.

[348] Hall J. E., Mead C. A. and Szabo G., J. Membrane Biol., 11 (1973) 75.

[349] Hamill R. L., Higgens C. E., Boaz N. E. and Gorman M., Tetrahedron Lett., (1969) 4255.

[350] Hamill R. L., Hoehn M. M., Pittenger G. E., Chamberlin J. and Gorman M., J. Antibiot., 22 (1969) 161.

[351] Haney M. E., Hoehn M., Antimicrob. Agents Chemother., (1967) 349.

[352] Harned R. L., Hidy P. H., Corum C. J. and Jones K. L., Proc. Indiana Acad. Sci., 59 (1950) 38.

[353] Harned R. L., Hidy P. H., Corum C. J. and Jones K. L., Antibiot. Chemother., 1 (1951) 594.

[354] Harold F. M., Adv. Microbiol. Physiol., (Rose A. H. and Wilkinson J. F., eds), Vol. 4, Academic Press, 1970, p. 45.

[355] Harold F. M., Biochem. J., 127 (1972) 49P.

[356] Harold F. M., Bacteriol. Rev., 36 (1972) 172.

[357] Harold F. M. and Baarda J. R., J. Bacteriol., 94 (1967) 53.

[358] Harold F. M. and Baarda J. R., J. Bacteriol., 95 (1968) 816.

[359] Harold F. M. and Baarda J. R., J. Bacteriol., 96 (1968) 2025.

[360] Harold F. M., Baarda J. R. and Pavlašova E., J. Bacteriol., 101 (1970) 152.

[361] Harold F. M. and Papineau D., J. Membrane Biol., 8 (1972) 27.

[362] Harold F. M. and Papineau D., J. Membrane Biol., 8 (1972) 45.

[363] Harold F. M., Pavlašova E. and Baarda J. R., Biochim. Biophys. Acta, 196 (1970) 235.

[364] Harris E. J., FEBS Lett., 5 (1969) 50.

[365] Harris E. J., Catlin G. and Pressman B. C., Biochemistry, 6 (1967) 1360.

[366] Harris E. J., Cockrell R. and Pressman B. C., Biochem. J., 99 (1966) 200.

[367] Harris E. J., van Dam K. and Pressman B. C., Nature, 213 (1967) 1126.

[368] Harris E. J., Hofer M. P. and Pressman B. C., Biochemistry, 6 (1967) 1348.

[369] Harris E. J. and Pressman B. C., Nature, 216 (1967) 918.

[370] Hassall C. H. and Magnus K. E., Nature, 184 (1959) 1223.

[371] Hassall C. H., Martin T. G., Schofield J. A. and Thomas J. O., J. Chem. Soc., (1967) 997.

[372] Hassall C. H., Morton R. B., Ogihara Y. and Phillips D. A. S., J. Chem. Soc., Ser. C, (1971) 526.

[373] Hassall C. H., Moschidis M. C. and Thomas W. A., J. Chem. Soc., Ser. B, (1971) 1757.

[374] Hassall C. H., Ogihara Y. and Thomas W. A., J. Chem. Soc., Ser. C, (1971) 522.

[375] Hassall C. H., Sanger D. G. and Thomas J. O., in Peptides — 1968 (Bricas E., ed.), North-Holland, Amsterdam, 1968, p. 70.

[376] Hassall C. H. and Thomas W. A., Chem. Br., 7 (1971) 145.

[377] Hauser H., Finer E. G. and Chapman D., J. Mol. Biol., 53 (1970) 419.

[378] Haydon D. A., Abstracts of papers presented at the Symposium on Molecular Mechanisms of Antibiotic Action on Protein Biosynthesis and Membranes, Granada, 1971.
[379] Haydon D. A. and Hladky S. B., J. Physiol., 210 (1970) 79P.
[380] Haydon D. A. and Hladky S. B., Q. Rev. Biophys., 5 (1972) 187.
[381] Haynes D. H., FEBS Lett., 20 (1972) 221.
[382] Haynes D. H., Biochim. Biophys. Acta, 255 (1972) 406.
[383] Haynes D. H., Kowalsky A. and Pressman B. C., J. Biol. Chem., 244 (1969) 502.
[384] Haynes D. H., Pressman B. C. and Kowalsky A., Biochemistry, 10 (1971) 852.
[385] Henderson P. J. F., Ann. Rev. Microbiol., 25 (1971) 393.
[386] Henderson P. J. F., McGivan J. D. and Chappell J. B., Biochem. J., 111 (1969) 521.
[387] Henn F. A. and Thompson T. E., Ann. Rev. Biochem., 38 (1969) 241.
[388] Herceg M. and Weiss R., Bull. Soc. Chim. France, (1972) 549.
[389] Herriot A. W. and Picker D., Tetrahedron Lett., (1972) 4521.
[390] Hertelendy F., Peak G. and Todd H., Biochem. Biophys. Res. Commun., 44 (1971) 253.
[391] Hewertson W., Kilbourne B. T. and Mais R. H. B., Chem. Commun., (1970) 952.
[392] Hille B., J. Gen. Physiol., 58 (1971) 599; 59 (1972) 637.
[393] Hilton B. D. and O'Brien R. D., Science, 168 (1970) 841.
[394] Hind G., Nakatani H. Y. and Izawa S., Biochim. Biophys. Acta, 172 (1969) 277.
[395] Hinkle P. and Mitchell P., J. Bioenerg., 1 (1970) 45.
[396] Hladky S. B. and Haydon D. A., Nature, 225 (1970) 451.
[397] Hladky S. B. and Haydon D. A., Biochim. Biophys. Acta, 274 (1972) 293.
[398] Hoek J. B., Lofrumento N. F., Meyer A. J. and Tager J. M., Biochim. Biophys. Acta, 226 (1971) 297.
[399] Höfer M. and Pressman B. C., Biochemistry, 5 (1966) 3919.
[400] Hogen Esch T. E. and Smid J., J. Am. Chem. Soc., 88 (1966) 307.
[401] Hogen Esch T. E. and Smid J., J. Am. Chem. Soc., 88 (1966) 318.
[402] Hogen Esch T. E. and Smid J., J. Am. Chem. Soc., 91 (1969) 4580.
[403] Holland P. C. and Sherratt H. S. A., Biochem. J., 121 (1971) 42P.
[404] Horvath G. and Gyimesi J., Biochem. Biophys. Res. Commun., 44 (1971) 639.
[405] Horvath L., Lovrekovich I. and Varga J. M., Z. Allg. Mikrobiol., 4 (1964) 236.
[406] Hotchkiss R. D., J. Biol. Chem., 141 (1941) 171.
[407] Hotchkiss R. D., Adv. Enzymol., 4 (1944) 153.
[408] Hotchkiss R. D. and Dubos R. J., J. Biol. Chem., 141 (1941) 155.
[409] Hsia J. C., Chen W. L., Long R. A., Wong L. T. and Kalow W., Proc. Natl. Acad. Sci. U.S., 69 (1972) 3412.
[410] Huang H. W., J. Theor. Biol., 32 (1971) 351.
[411] Huang H. W., J. Theor. Biol., 32 (1971) 363.
[412] Hunter F. and Schwarz L., in Antibiotics (Gottlieb D. O. and Shaw P. D., eds), Vol. 1, Springer Verlag, Berlin-Heidelberg-New York, 1967.
[413] Hunter G. R., Kamishima Y. and Brierley G. P., Biochim. Biophys. Acta, 180 (1969) 81.
[414] Hybl A., Rundle R. E. and Williams D. E., J. Am. Chem. Soc., 87 (1965) 2779.
[415] Hyman E. S., Fed. Proc., 28 (1969) 284.
[416] Hyman E. S., Biophys. Soc. Abstr., (1970) 72a.
[417] Hyman E. S., IVth International Biophysics Congress (Moscow, August 1972), Abstracts of papers, EXb-4/1.
[418] Iitaka Y., Sakamaki T. and Nawata Y., Chem. Lett., (1972) 1225.
[419] Ilani A., Biochim. Biophys. Acta, 163 (1968) 429.
[420] Isaev P. I., Liberman E. A., Samuilov V. D., Skulachev V. P. and Tsofina L. M., Biochim. Biophys. Acta, 216 (1970) 22.

418

[421] Isaev P. I., Samuilov V. D., Skulachev V. P. and Tsofina L. M., Biokhimiya (U.S.S.R.), 38 (1973) 796.

[422] Isbell B. E., Rix-Evans C. and Beaven G. H., FEBS Lett., 25 (1972) 192.

[423] Ishii S. and Witkop B., J. Am. Chem. Soc., 85 (1963) 1832.

[424] IUPAC Abbreviations, J. Biol. Chem., 247 (1972) 977.

[425] Ivanov V. T., Andreev I. M., Vinogradova E. I., Laine I. A., Malenkov G. G., Ovchinnikov Yu. A., Sanasaryan A. A., Fonina L. A., Shvetsov Yu. B., Shemyakin M. M. and Shkrob A. M., in VIIth IUPAC Symposium on the Chemistry of Natural Products (Riga, June, 1970), Abstracts of papers C-26, Zinatne, Riga, 1970.

[426] Ivanov V. T., Evstratov A. V., Mikhaleva I. I., Balashova T. A., Abdullaev N. D., Bystrov V. F. and Ovchinnikov Yu. A., Khim. Prir. Soedin. (U.S.S.R.), in press.

[427] Ivanov V. T., Kogan G. A., Meshcheryakova E. A., Shilin V. V. and Ovchinnikov Yu. A., Khim. Prir. Soedin. (U.S.S.R.), (1971) 309.

[428] Ivanov V. T., Kogan G. A., Tulchinsky V. M., Miroshinikov A. I., Mikhaleva I. I., Evstratov A. V., Zenkin A. A., Kostetsky P. V., Ovchinnikov Yu. A. and Lokshin B. V., FEBS Lett., 30, (1973) 199.

[429] Ivanov V. T., Kostetsky P. V., Meshcheryakova E. A., Efremov E. S., Popov E. M. and Ovchinnikov Yu. A., Khim. Prir. Soedin. (U.S.S.R.), (1973) 363.

[430] Ivanov V. T., Laine I. A., Abdullaev N. D., Pletnev V. Z., Lipkind G. M., Arkhipova S. F., Senyavina L. B., Meshcheryakova E. A., Popov E. M., Bystrov V. F. and Ovchinnikov Yu. A., Khim. Prir. Soedin. (U.S.S.R.), (1971) 221.

[431] Ivanov V. T., Laine I. A., Abdullaev N. D., Senyavina L. B., Popov E. M., Ovchinnikov Yu. A. and Shemyakin M. M., Biochem. Biophys. Res. Commun., 34 (1969) 803.

[432] Ivanov V. T., Laine I. A., Ovchinnikov Yu. A., Yakovlev G. I. and Chervin I. I., Khim. Prir. Soedin. (U.S.S.R.), (1973) 248.

[433] Ivanov V. T., Laine I. A., Ryabova I. D. and Ovchinnikov Yu. A., Khim. Prir. Soedin. (U.S.S.R.), (1970) 744.

[434] Ivanov V. T., Miroshnikov A. I., Abdullaev N. D., Senyavina L. B., Arkhipova S. F., Uvarova N. N., Khalilulina K. Kh., Bystrov V. F. and Ovchinnikov Yu. A., Biochem. Biophys. Res. Commun., 42 (1971) 654.

[435] Ivanov V. T., Miroshnikov A. I., Kozmin S. A., Meshcheryakova E. A., Senyavina L. B., Uvarova N. N., Khalilulina K. Kh., Bystrov V. F. and Ovchinnikov Yu. A., Khim. Prir. Soedin. (U.S.S.R.), (1973) 378.

[436] Ivanov V. T., Portnova S. L., Balashova T. A., Bystrov V. F., Shilin V. V., Bernat Ya. and Ovchinnikov Yu. A., Khim. Prir. Soedin. (U.S.S.R.), (1971) 339.

[437] Ivanov V. T., Senyavina L. B., Efremov E. S., Shilin V. V. and Ovchinnikov Yu. A., Khim. Prir. Soedin. (U.S.S.R.), (1971) 347.

[438] Ivanov V. T., Shilin V. V., Kogan G. A., Meshcheryakova E. A., Senyavina L. B., Efremov E. S. and Ovchinnikov Yu. A., Tetrahedron Lett., (1971) 2841.

[439] Ivanov V. T. and Shkrob A. M., FEBS Lett., 10 (1970) 285.

[440] Izatt R. M., Haymore B. L. and Christensen J. J., Chem. Commun., (1972) 1308.

[441] Izatt R. M., Nelson D. P., Rytting J. H., Haymore B. L. and Christensen J. J., J. Am. Chem. Soc., 93 (1971) 1619.

[442] Izatt R. M., Rytting J. H., Nelson D. P., Haymore B. L. and Christensen J. J., Science, 164 (1969) 443.

[443] Izmailov N. A., Dokl. Akad. Nauk S.S.S.R. (U.S.S.R.), 149 (1963) 1364.

[444] Izmailov N. A., Electrochemistry of Solutions (Russian), Khimiya, Moscow, 1966, p. 204.

[445] Izmailov N. A. and Kruglyak Yu. A., Dokl. Akad. Nauk S.S.S.R. (U.S.S.R.), 134 (1960) 1390.

[446] Jackson J. B. and Crofts A. R., FEBS Lett., 4 (1969) 185.

[447] Jackson J. B. and Crofts A. R., Eur. J. Biochem., 10 (1969) 226.

[448] Jackson J. B. and Crofts A. R., Eur. J. Biochem., 18 (1971) 120.
[449] Jackson J. B., Crofts A. R. and von Stedingk L. V., Eur. J. Biochem., 6 (1968) 41.
[450] Jacob S. T., Sajdel E. M. and Munro H. N., Nature, 225 (1970) 60.
[451] Jagendorf A. T., Fed. Proc., 26 (1967) 1361.
[452] James A. T. and Synge R. L. M., Biochem. J., 50 (1951) 109.
[453] Jarousse J., Compt. Rend., 232 (1951) 1424.
[454] Jasaitis A. A., Kuliene V. V. and Skulachev V. P., Biochim. Biophys. Acta, 234 (1971) 177.
[455] Jeacocke R. E., Niven D. F. and Hamilton W. A., Biochem. J., 127 (1972) 57P.
[456] Jiroveč O. Schweiz., Z. Allg. Pathol. Bakteriol., 14 (1951) 653.
[457] Johannin G. and Kellershohn N., Biochem. Biophys. Res. Commun., 49 (1972) 321.
[458] John P. and Hamilton W. A., Eur. J. Biochem., 23 (1971) 528.
[459] Johnson R. N. and Chappell J. B., Biochem. J., 116 (1970) 37P.
[460] Johnson S. M., Biochim. Biophys. Acta, 193 (1970) 92.
[461] Johnson S. M. and Bangham A. D., Biochim. Biophys. Acta, 193 (1969) 82.
[462] Johnson S. M., Herrin J., Lin Sh. J. and Paul I. C., J. Am. Chem. Soc., 92 (1970) 4428.
[463] Johnson S. M., Herrin J., Lin Sh. J. and Paul I. C., Chem. Commun., (1970) 72.
[464] Jonczyk A., Serafinova B. and Makosza M., Tetrahedron Lett., (1971) 1351.
[465] Joshi G. C., Singh N. and Pande L. M., Tetrahedron Lett., (1972) 1461.
[466] Julg A., Tetrahedron Lett., (1969) 5155.
[467] Junge W., Eur. J. Biochem., 14 (1970) 582.
[468] Junge W., Rumberg B. and Schröder H., Eur. J. Biochem., 14 (1970) 575.
[469] Junge W. and Schmid R., J. Membrane Biol., 4 (1971) 179.
[470] Junge W. and Witt H.-T., Z. Naturforsch., 23b (1968) 244.
[471] Karlish S. J. D. and Avron M., FEBS Lett., 1 (1968) 21.
[472] Karlish S. J. D. and Avron M., Eur. J. Biochem., 20 (1971) 51.
[473] Karlish S. J. D., Shavit N. and Avron M., Eur. J. Biochem., 9 (1968) 291.
[474] Kashket E. R. and Wilson T. W., Biochem. Biophys. Res. Commun., 49 (1972) 615.
[475] Katyshkina V. V. and Kraft M. Ya., Zh. Obshch. Khim. (U.S.S.R.), 29 (1959) 63.
[476] Kaufman H. P. and Tobschirbel A., Chem. Ber., 92 (1959) 2805.
[477] Kedem O., Furmansky M., Loebel E., Gordon S. and Blodi R., Israel J. Chem., 7 (1969) 87P.
[478] Kedem O., Loebel E. and Furmansky M., Ger. Offen. 2027128 (cl. G01n), 23 XII 1970.
[479] Keister D. L. and Minton N. J., J. Bioenerg., 1 (1970) 367.
[480] Keller-Schierlein W. and Gerlach H., in Fortschritte der Chemie organischer Naturstoffe, (Zechmeister L., ed.), Vol. 25, 1968, S.161.
[481] Keller-Schierlein W., Gerlach H. and Seibl J., Antimicrob. Agents Chemother., (1966) 644.
[482] Kemp G., Jacobson K. A. and Wenner C. E., Biochim. Biophys. Acta, 255 (1972) 493.
[483] Kemp G. and Wenner C. E., Biochim. Biophys. Acta, 282 (1972) 1.
[484] Kendrew J. C., Klyne W., Lifson S., Miyazawa T., Némethy G., Phillips D. C., Ramachandran G. N. and Scheraga H. A., J. Mol. Biol., 52 (1970) 1; Biochemistry, 9 (1970) 3471; Arch. Biochem. Biophys., 145 (1971) 405.
[485] Khodorov B. I., The Problem of Excitability (Russian), Meditsina, Leningrad, 1969, p. 253.
[486] Khuri R. N., Agulian S. K. and Wise W. M., Pflüger Arch., 322 (1971) 39.
[487] Khuri R. N., Hajjar J. J. and Agulian S. K., J. Appl. Physiol., 32 (1972) 419.

420

[488] Kilbourn B. T., Dunitz J. D., Pioda L. A. R. and Simon W., J. Mol. Biol., 30 (1967) 559.

[489] Kimikura H., J. Theor. Biol., 13 (1966) 145.

[490] Kimmich G. A. and Rasmussen H., Biochim. Biophys. Acta, 131 (1967) 413.

[491] King T. E. and Steinrauf L. K., Biochem. Biophys. Res. Commun., 49 (1972) 1433.

[492] Kingsbury C. A., J. Org. Chem., 29 (1964) 3262.

[493] Kiryushkin A. A., Ovchinnikov Yu. A. and Shemyakin M. M., Tetrahedron Lett., (1964) 3313.

[494] Kiryushkin A. A., Ovchinnikov Yu. A. and Shemyakin M. M., Tetrahedron Lett., (1965) 143.

[495] Kiryushkin A. A., Ovchinnikov Yu. A. and Shemyakin M. M., Khim. Prir. Soedin. (U.S.S.R.), (1965) 58.

[496] Kiryushkin A. A., Rozynov B. V. and Ovchinnikov Yu. A., Khim. Prir. Soedin. (U.S.S.R.), (1968) 182.

[497] Kiryushkin A. A., Shchelokov V. I., Antonov V. K., Ovchinnikov Yu. A. and Shemyakin M. M., Khim. Prir. Soedin. (U.S.S.R.), (1967) 267.

[498] Klingenberg M., FEBS Lett., 6 (1970) 145.

[499] Kodaira Y., Agr. Biol. Chem., 25 (1961) 261.

[500] Kodaira Y., Agr. Biol. Chem., 26 (1962) 36.

[501] Köhler W., Thrum H. and Schlegel R., Zbl. Bakteriol. Parasitenkrankh., 194 (1964) 457.

[502] König W. and Geiger R., Liebig's Ann. Chem., 727 (1969) 125.

[503] Konnert J. and Karle I. L., J. Am. Chem. Soc., 91 (1969) 4888.

[504] Kopolow S., Hogen Esch T. E. and Smid J., Macromolecules, 4 (1971) 359.

[505] Kopple K. D., Biopolymers, 10 (1971) 1139.

[506] Korn E. D., Ann. Rev. Biochem., 38 (1969) 263.

[507] Kornberg R. D., McNamee M. G. and McConnell H. M., Proc. Natl. Acad. Sci. U.S., 69 (1972) 1508.

[508] Kostetsky P. V., Ivanov V. T., Ovchinnikov Yu. A. and Shchembelov G. A., in IVth International Biophysics Congress (Moscow, August, 1972), Abstracts of papers, EVIc-4/7.

[509] Kraan G. P. B., Amesz J., Velthuys B. R. and Steemers R. G., Biochim. Biophys. Acta, 223 (1970) 129.

[510] Kraft M. Ya. and Katyshkina V. V., Dokl. Akad. Nauk S.S.S.R. (U.S.S.R.), 109 (1956) 312; Zh. Obshch. Khim. (U.S.S.R.), 29 (1959) 59.

[511] Krasne S. and Eisenman G., in Membranes — A Series of Advances, (Eisenman G., ed.), Vol. 2, Marcel Dekker, New York, 1973, p. 277.

[512] Krasne S., Eisenman G. and Szabo G., Science, 174 (1971) 412.

[513] Kreitz W., Angew. Chem., 84 (1972) 597.

[514] Krigbaum W. R., Kuegler F. R. and Oelschleger H., Biochemistry, 11 (1972) 4548.

[515] Krull I. H., Mask C. A. and Cosgrove R. E., Anal. Lett., 3 (1970) 43.

[516] Kubota T. and Matsutani S., J. Chem. Soc. Ser. C, (1970) 695.

[517] Kubota T., Matsutani S., Shiro M. and Koyama H., Chem. Commun., (1968) 1541.

[518] Kuo J. F. and Dill I. K., Biochem. Biophys. Res. Commun., 32 (1968) 33.

[519] Kuyama S. and Tamura S., Agr. Biol. Chem., 29 (1965) 168.

[520] Lacey M. S., J. Gen. Microbiol., 4 (1950) 122.

[521] Laine I. A., Conformational states of valinomycin and its alkali metal ion complexes in solution (Russian), Ph.D. Thesis, Shemyakin Institute for Chemistry of Natural Products, U.S.S.R. Acad. Sci., Moscow, 1971.

[522] Lal S. and Christian G. D., Anal. Lett., 3 (1970) 11.

[523] Lange L. and Makosza M., Roczn. Chem., 41 (1967) 1303.

[524] Lardy H. A., Fed. Proc., 27 (1968) 1278.

[525] Lardy H. A. and Ferguson S. M., Ann. Rev. Biochem., 38 (1969) 991.

[526] Lardy H. A., Graven S. N. and Estrada-O S., Fed. Proc., 26 (1967) 1355.

[527] Lardy H. A., Johnson D. and McMurray W. C., Arch. Biochem. Biophys. 78 (1958) 587.

[528] Lassigne C. and Baine P., J. Phys. Chem., 75 (1971) 3188.

[529] Läuger P., Science, 178 (1972) 24.

[530] Läuger P. and Stark G., Biochim. Biophys. Acta, 211 (1970) 458.

[531] Läuger P., Stark G., Benz R. and Ketterer B., in Abstracts of papers presented at the Symposium on Molecular Mechanisms of Antibiotic Action on Protein Biosynthesis and Membranes, Granada, 1971, p. 65.

[532] Leach S. J., Nemethy G. and Scheraga H. A., Biopolymers, 4 (1966) 369.

[533] Lee C. P. and Ernster L., in Regulation of Metabolic Processes in Mitochondria (Tager J. M., Papa S., Quagliariello E. and Slater E. C., eds), [BBA Library, Vol. 7], Elsevier, Amsterdam, 1966, p. 218.

[534] Lehn J. M., Angew. Chem., 82 (1970) 183.

[535] Lehn J. M. and Montavon F., Tetrahedron Lett., (1972) 4557.

[536] Lehn J. M. and Sauvage J. P., Chem. Commun., (1971) 440.

[537] Lehn J. M., Sauvage J. P. and Dietrich B., J. Am. Chem. Soc., 92 (1970) 2916.

[538] Lehninger A., The Mitochondrion, W. A. Benjamin, New York — Amsterdam, 1964.

[539] Lehninger A., Carafoli E. and Rossi C. S., Adv. Enzymol., 29 (1967) 259.

[540] Lev A. A., Buzhinskii E. P. and Osipov V. V., VIIth IUPAC Symposium on the Chemistry of Natural Products, Presymposium on Physicochemical Basis of Ionic Transport through Membranes (Riga, June, 1970), Abstracts of papers, III-18, Zinatne, Riga, 1970.

[541] Lev A. A and Buzhinskii E. P., Tsitologiya (U.S.S.R.), 9 (1967) 102.

[542] Lev A. A., Malev V. V. and Osipov V. V., in Membranes — A Series of Advances, (Eisenman G., ed.), Vol. 2, Marcel Dekker, New York, 1973, p. 481.

[543] Lev A. A., Osipov V. V., Buzhinskii E. P., Gotlib V. A. and Shlyakhter T. A., Tsitologiya (U.S.S.R.), 13 (1971) 619.

[544] Levich V. G., Markin V. S. and Chizmadjev Yu. A., Vestn. Akad. Nauk S.S.S.R. (U.S.S.R.), 9 (1969) 61.

[545] Levins R. J., Anal. Chem., 43 (1971) 1045.

[546] Levins R. J. and Ikeda R. M., Anal. Chem., 37 (1965) 671.

[547] Levinson Ch., Nature, 216 (1967) 74.

[548] Liberman E. A., Biofizika (U.S.S.R.), 15 (1970) 278.

[549] Liberman E. A., in Mitochondria. Molecular Mechanisms of Enzymatic Reaction, Nauka, Moscow, 1972, p. 99.

[550] Liberman E. A., Pronevich L. A. and Topaly V. P., Biofizika (U.S.S.R.), 15 (1970) 612.

[551] Liberman E. A. and Skulachev V. P., Biochim. Biophys. Acta, 216 (1970) 30.

[552] Liberman E. A. and Topaly V. P., Biofizika (U.S.S.R.), 14 (1969) 452.

[553] Liberman E. A. and Tsofina L. M., Mol. Biol. (U.S.S.R.), 6 (1972) 386.

[554] Linday L. F. and Bush D. H., Chem. Commun., (1968) 1589.

[555] Lipkind G. M., Arkhipova S. F. and Popov E. M., Zh. Strukt. Khim. (U.S.S.R.), 11 (1970) 121.

[556] Lippe C., Nature, 218 (1968) 196.

[557] Lippe C., J. Mol. Biol., 39 (1969) 669.

[558] Littke W., Tetrahedron Lett., 45 (1971) 4247.

[559] Llines M., Klein M. P. and Neilands J. B., J. Mol. Biol., 52 (1970) 399.

[560] Long R. A., Hruska F., Gesser H. D., Hsia J. C. and Williams R., Biochem. Biophys. Res. Commun., 41 (1970) 321.

[561] Losse G. and Klengel H., Tetrahedron, 27 (1971) 1423.

422

[562] Losse G. and Raue H., Chem. Ber., 101 (1968) 1532.
[563] Losse G. and Raue H., Tetrahedron, 25 (1969) 2677.
[564] Lüttringhaus A., Liebig's Ann. Chem., 528 (1937) 181.
[565] Lüttringhaus A. and Sichert-Modrow I., Makromol. Chem., 18-19 (1956) 511.
[566] Lüttringhaus A. and Ziegler K., Liebig's Ann. Chem., 528 (1937) 155.
[567] Lutz W. K., Früh P. U. and Simon W., Helv. Chim. Acta, 54 (1971) 2767.
[568] Lutz W. K., Winkler F. K. and Dunitz J. D., Helv. Chim. Acta, 54 (1971) 1103.
[569] Lutz W. K., Wipf H.-K. and Simon W., Helv. Chim. Acta, 53 (1970) 1741.
[570] Lynn W. S. and Brown R. H., Biochim. Biophys. Acta, 105 (1965) 15.
[571] MacDonald C. G. and Shannon J. S., Tetrahedron Lett., (1964) 3113.
[572] MacDonald J. C., Can. J. Microbiol., 6 (1960) 27.
[573] MacDonald J. C., Can. J. Microbiol., 15 (1969) 236.
[574] MacDonald J. C. and Slater G. P., Can. J. Biochem., 46 (1968) 573.
[575] Madison V. and Schellman J., Biopolymers, 9 (1970) 569.
[576] Magyar K., Z. Allg. Mikrobiol., 4 (1964) 265.
[577] Maier C. A. and Paul I. C., Chem. Commun., (1971) 181.
[578] Maigret B., Pullman B. and Caillet J., Biochem. Biophys. Res. Commun., 40 (1970) 808.
[579] Maigret B., Pullman B. and Perahia D., J. Theoret. Biol., 31 (1971) 269.
[580] Makosza M., Tetrahedron Lett., (1966) 4621.
[581] Makosza M., Bull. Acad. Polon. Sci. Ser. Sci. Chim., 15 (1967) 165.
[582] Makosza M., Roczn. Chem., 43 (1969) 79 and 333.
[583] Makosza M., Tetrahedron Lett., (1969) 673.
[584] Makosza M., Tetrahedron Lett., (1969) 677.
[585] Makosza M. and Bialecka B., Tetrahedron Lett., (1971) 4571.
[586] Makosza M. and Fedorynski M., Bull. Acad. Polon. Sci. Ser. Sci. Chim., 19 (1971) 105.
[587] Makosza M. and Serafinova B., Roczn. Chem., 39 (1965) 1223, 1401, 1593, 1799 and 1805; 40 (1966) 1647 and 1839.
[588] Makosza M., Serafinova B. and Gajos J., Roczn. Chem., 43 (1969) 671.
[589] Makosza M., Serafinova B. and Jawdosiuk M., Roczn. Chem., 41 (1967) 1037.
[590] Makosza M., Sarafinova B. and Urbanski T., Chimie Ind., 93 (1965) 537.
[591] Makosza M. and Wawrzyniewicz M., Tetrahedron Lett., (1969) 4659.
[592] Malenkov G. G. and Rimskaja V. A., IVth International Biophysics Congress (Moscow, 1972), Abstracts of papers, EXb-4/4.
[593] Mallinson P. R. and Truter M. R., J. Chem. Soc., Perkin Trans. (II), (1972) 1818.
[594] Markin V. S. and Liberman E. A., Dokl. Akad. Nauk S.S.S.R. (U.S.S.R.), 201 (1971) 975.
[595] Markin V. S., Mol. Biol. (U.S.S.R.), 3 (1969) 610.
[596] Markin V. S. and Chizmadjev Yu. A., Induced Ion Transport (Russian), Nauka, Moscow, 1974.
[597] Markin V. S., Krishtalik L. I., Liberman E. A. and Topaly V. P., Biofizika (U.S.S.R.), 14 (1969) 256.
[598] Markin V. S., Pastushenko V. F., Krishtalik L. I., Liberman E. A. and Topaly V. P., Biofizika (U.S.S.R.), 14 (1969) 462.
[599] Markin V. S. and Pastushenko V. F., Typescript in the publishers list of VINITI, No. 1788-70.
[600] Marsch R. E. and Donohue J., Adv. Protein Chem., 22 (1967) 235.
[601] Marullo N. P. and Lloyd R. A., J. Am. Chem. Soc., 88 (1966) 1076.
[602] Maskornick M. J., Tetrahedron Lett., (1972) 1797.
[603] Massari S. and Azzone G. F., Eur. J. Biochem., 12 (1970) 310.
[604] Massari S., Babboni E. and Azzone G. F., Biochim. Biophys. Acta, 283 (1972) 16.

[605] Matsumoto T., Yanagiya M., Maeno S. and Yasuda S., Tetrahedron Lett., (1968) 6297.

[606] Maxey B. M. and Popov A. I., J. Am. Chem. Soc., 89 (1967) 2230.

[607] Mayers D. F. and Urry D. W., J. Am. Chem. Soc., 94 (1972) 77.

[608] McCarty R. E., Biochem. Biophys. Res. Commun., 32 (1968) 37.

[609] McCarty R. E., J. Biol. Chem., 244 (1969) 4292.

[610] McCarty R. E., FEBS Lett., 9 (1970) 313.

[611] McCoy F. C., Reinhard R. R. and Hess H. V., Chem. Ind., (1971) 531.

[612] McFarland B. G., Chem. Phys. Lipids, 8 (1972) 303.

[613] McGivan J. D. and Chappell J. B., Biochem. J., 127 (1972) 54P.

[614] McGivan J. D. and Klingenberg M., Eur. J. Biochem., 20 (1971) 392.

[615] McLaughlin S. G. A., J. Membrane Biol., 9 (1972) 361.

[616] McLaughlin S. G. A., Szabo G., Ciani S. and Eisenman G., J. Membrane Biol., 9 (1972) 3.

[617] McLaughlin S. G. A., Szabo G., Eisenman G. and Ciani S. M., Proc. Natl. Acad. Sci. U.S., 67 (1970) 1268.

[618] McLaughlin S. G. A., Szabo G., Eisenman G. and Ciani S., Abstracts of Reports at U.S. Biophysical Society Annual Meeting, 1970.

[619] McLaughlin S. G. A., Szabo G. and Eisenman G., J. Gen. Physiol., 58 (1971) 667.

[620] McMullen A. I., Biochem. J., 119 (1970) 10P.

[621] McMullen A. I., Marlborough D. I. and Bayley P. M., FEBS Lett., 16 (1971) 278.

[622] McMullen A. I. and Stirrup J. A., Biochim. Biophys. Acta, 241 (1971) 807.

[623] McMurray W. and Begg R. W., Arch. Biochem. Biophys., 84 (1959) 546.

[624] Meisner H. N. and Wenner C. E., Biochim. Biophys. Acta, 223 (1970) 46.

[625] Melnik E. I., Mechanism of cation transport across bilayers induced by the valinomycin cyclodepsipeptides (Russian), Ph.D. Thesis, Physicotechnical Institute, Moscow, 1973.

[626] Melnik E. I., Babakov A. V. and Shkrob A. M., in Biofizika Membran (Alkalene B. D., Zablotskaite D. V. and Narushevichus E. V., eds), Kaunas, 1971, p. 599.

[627] Melnik E. I. and Shkrob A. M., IVth International Biophysics Congress, (Moscow, 1972), Abstracts of papers, EXb-1.

[628] Melnik E. I., Terekhov O. P., Shkrob A. M. and Ovchinnikov Yu. A., Biofizika (U.S.S.R.), 18 (1973) 649.

[629] Menshikov G. P. and Rubinshtein M. M., Zh. Obshch. Khim. (U.S.S.R.), 26 (1956) 2035.

[630] Merrifield R. B., Gisin B. F., Tosteson D. C. and Tieffenberg M., in The Molecular Basis of Membrane Function (Tosteson D. C., ed.), Prentice-Hall, Englewood Cliffs, 1969, p. 211.

[631] Metlička R. and Rybová R., Biochim. Biophys. Acta, 135 (1967) 563.

[632] Metz B., Moras D. and Weiss R., Chem. Commun., (1970) 217.

[633] Metz B., Moras D. and Weiss R., J. Am. Chem. Soc., 93 (1971) 1806.

[634] Metz B., Moras D. and Weiss R., Chem. Commun., (1971) 444.

[635] Meyer C. E. and Reussen F., Experientia, 23 (1967) 85.

[636] Meyers E., Pensy F. E., Perlmann D., Smith D. A. and Weisenborn F. L., J. Antibiot. Ser. A, 18 (1965) 128.

[637] Mikhaleva I. I., Ryabova I. D., Romanova T. A., Tarasova T. I., Ivanov V. T., Ovchinnikov Yu. A. and Shemyakin M. M., Zh. Obshch. Khim. (U.S.S.R.), 38 (1968) 1229.

[638] Ming Keong Wong, McKinney W. J. and Popov A. I., J. Phys. Chem., 75 (1971) 56.

[639] Miroshnikov A. I., Khalilulina K. Kh., Uvarova N. N., Ivanov V. T. and Ovchinnikov Yu. A., Khim. Prir. Soedin. (U.S.S.R.), (1973) 214.

[640] Mitchell P., Chemiosmotic Coupling in Oxidative and Photosynthetic Phosphorylation, Glynn Research, Bodmin, Cornwall, 1966.

424

[641] Mitchell P., FEBS Lett., 28 (1972) 1.
[642] Mitchell P., Nature, 191 (1961) 144; Biol. Rev., 41 (1966) 445; Fed. Proc., 26 (1967) 1370.
[643] Mitchell P., Adv. Enzymol., 29 (1967) 33.
[644] Mitchell P., Chemiosmotic Coupling and Energy Transduction, Glynn Research, Bodmin, Cornwall, 1968.
[645] Mitchell P., Biochem. J., 116 (1970) 5P.
[646] Mitchell P., J. Bioenerg., 3 (1972) 5.
[647] Mitchell P. and Moyle J., Biochem. J., 105 (1967) 1147.
[648] Mitchell P. and Moyle J., Nature, 213 (1967) 137.
[649] Mitchell P. and Moyle J., Eur. J. Biochem., 7 (1969) 471.
[650] Mitchell P. and Moyle J., Eur. J. Biochem., 9 (1969) 149.
[651] Miyada D. C., Inami K. and Matsuyama G., Clin. Chem., 14, (1971) 27.
[652] Mohan M. S. and Rechnitz G. A., J. Am. Chem. Soc., 92 (1970) 5839.
[653] Molotkovskii Yu. G., Structural and metabolic transition of isolated chloroplasts (Russian), D.Sc. Thesis, Institute of Plant Physiology, U.S.S.R. Academy of Sciences, Moscow, 1972.
[654] Molotkovskii Yu. G. and Dzyubenko V. S., Dokl. Akad. Nauk S.S.S.R. (U.S.S.R.), 204 (1972) 1272.
[655] Montal M., Fed. Proc., 28 (1969) 881.
[656] Montal M., Chance B. and Lee C. P., Biochem. Biophys. Res. Commun., 36 (1969) 428.
[657] Montal M., Chance B. and Lee C. P., J. Membrane Biol., 2 (1970) 201.
[658] Montal M., Chance B. and Lee C. P., FEBS Lett., 6 (1970) 209.
[659] Montal M., Chance B., Lee C. P. and Azzi A., Biochem. Biophys. Res. Commun., 34 (1969) 104.
[660] Montal M., Nishimura M. and Chance B., Biochim. Biophys. Acta, 223 (1970) 183.
[661] Montanaro N., Novello F. and Stirpe F., Biochem. J., 125 (1971) 1087.
[662] Montecuccoli G., Novello F. and Stirpe F., FEBS Lett., 25 (1972) 305.
[663] Moore C. and Pressman B. C., Biochem. Biophys. Res. Commun., 15 (1964) 562.
[664] Moras D., Metz B., Herceg M. and Weiss R., Bull. Soc. Chim. France, (1972) 551.
[665] Morf W. E. and Simon W., Helv. Chim. Acta, 54 (1971) 794.
[666] Morf W. E. and Simon W., Helv. Chim. Acta, 54 (1971) 2683.
[667] Mortillaro L., Russo M., Credali L. and DeChecchi C., J. Chem. Soc. Ser. C, (1966) 428.
[668] Möschler H. J., Weder H.-G. and Schwyzer R., Helv. Chim. Acta, 54 (1971) 1437.
[669] Mueller P. and Rudin D. O., Biochem. Biophys. Res. Commun., 26 (1967) 398.
[670] Mueller P. and Rudin D. O., Nature, 217 (1968) 713.
[671] Mueller P. and Rudin D. O., in Current Topics in Bioenergetics (Sanadi D. R., ed.), Vol. 3, Academic Press, New York, 1969, p. 157.
[672] Müller W. H., Naturwiss., 57 (1970) 248.
[673] Myagkov I. V., Sotnikov P. S., Terekhov O. P. and Babakov A. V., in Membrane Biophysics (Russian) (Alkalene B. D., Zablotskaite D. V. and Narushevichus E. V., eds), Part 1, Kaunas, 1971, p. 627.
[674] Myers V. B. and Haydon D. A., Biochim. Biophys. Acta, 274 (1972) 313.
[675] Naider F., Benedetti E. and Goodman M., Proc. Natl. Acad. Sci. U.S., 68 (1971) 1195.
[676] Nawata Y., Ando K. and Iitaka Y., Acta Cryst., 27B (1971) 1680.
[677] Nelson N., Drechsler Z. and Neumann J., J. Biol. Chem., 245 (1970) 143.
[678] Neu R., Arzneim. Forsch., 9 (1959) 585.
[679] Neu R., Fette Seifen Anstrichm., 61 (1959) 980.
[680] Neubecker T. A. and Rechnitz G. A., Anal. Lett., 5 (1972) 653.

[681] Neubert D. and Lehninger A. L., Biochim. Biophys. Acta, 62 (1962) 556.

[682] Neumann J., Arntzen C. J. and Dilley R. A., Biochemistry, 10 (1971) 866.

[683] Neumann J., Ke B. and Dilley R. A., Plant Physiol., 46 (1970) 86.

[684] Neumcke B. and Läuger P., J. Membrane Biol., 3 (1970) 54.

[685] Niessing J., Z. Naturforsch., 25b (1970) 1119.

[686] Nishimura H., Mayama M., Kimura T., Kimura A., Kawamura Y., Tawara K., Tanaka Y., Okamoto S. and Kyotani H., J. Antibiot. Ser. A, 17 (1964) 11.

[687] Nishimura M., Biochim. Biophys. Acta, 197 (1970) 69.

[688] Nishimura M. and Pressman B. C., Biochemistry 8 (1969) 1360.

[689] Niven D. F. and Hamilton W. A., Biochem. J., 127 (1972) 58P.

[690] Novello F., Fiume L. and Stirpe F., Biochem. J., 116 (1970) 177.

[691] Ogata E. and Rasmussen H., Biochemistry, 5 (1966) 57.

[692] Ohnishi M., Fedarco M.-C., Baldeschwieler J. D. and Johnson L. F., Biochem. Biophys. Res. Commun., 46 (1972) 312.

[693] Ohnishi M. and Urry D. W., Biochem. Biophys. Res. Commun., 36 (1969) 194.

[694] Ohnishi M. and Urry D. W., Science, 168 (1970) 1091.

[695] Oishi H., Sugawa T., Okutomi T., Suzuki K., Hayashi T., Sawada M. and Ando K., J. Antibiot., 23 (1970) 105.

[696] Okazaki H. and Arai M., Japan Patent 1391, Sankyo Co. Ltd., 1966.

[697] Olcott H. S., Lewis J. C., Dimick K. P., Fevold H. L. and Fraenkel-Conrat H., Arch. Biochem., 10 (1946) 553.

[698] Osipov V. V., Shchagina L. V. and Lev A. A., IVth International Biophysics Congress (Moscow, 1972), Abstracts of papers, EXb-4/6; see also EXb-4/11.

[699] Ovchinnikov, Yu. A., in Proceedings VIIIth FEBS Meeting (Amsterdam, 1972), Vol. 28, North-Holland, Amsterdam, 1972, p. 279.

[700] Ovchinnikov Yu. A., Ivanov V. T., Antonov V. K., Shkrob A. M., Michaleva I. I., Evstratov A. V., Malenkov G. G., Melnik E. I. and Shemyakin M. M., in Peptides (Proc. IXth European Peptide Symposium, Orsay, France, 1968), (Bricas E., ed.), North-Holland, Amsterdam, 1968, p. 56.

[701] Ovchinnikov Yu. A., Ivanov V. T., Barsukov L. I., Melnik E. I., Oreshnikova N. I., Bogolyubova N. D., Ryabova I. D., Miroshnikov A. I. and Rimskaya V. A., Experientia, 28 (1972) 399.

[702] Ovchinnikov Yu. A., Ivanov V. T., Bystrov V. F., Abdullaev N. D. and Miroshnikov A. I., in Peptides (Proc. XIth European Peptide Symposium, Wien, Austria, 1971), (Nesvadba H., ed.), North-Holland, Amsterdam, 1972, p. 130.

[703] Ovchinnikov Yu. A., Ivanov V. T., Bystrov V. F., Miroshnikov A. I., Shepel E. N., Abdullaev N. D., Efremov E. S. and Senyavina L. B., Biochem. Biophys. Res. Commun., 39 (1970) 217.

[704] Ovchinnikov Yu. A., Ivanov V. T., Evstratov A. V., Bystrov V. F., Abdullaev N. D., Popov E. M., Lipkind G. M., Arkhipova S. F., Efremov E. S. and Shemyakin M. M., Biochem. Biophys. Res. Commun., 37 (1969) 668.

[705] Ovchinnikov Yu. A., Ivanov V. T., Kiryushkin A. A. and Shemyakin M. M., Izv. Akad. Nauk S.S.S.R., Ser. Khim. (U.S.S.R.), (1962) 2046.

[706] Ovchinnikov Yu. A., Ivanov V. T., Kiryushkin A. A. and Shemyakin M. M., Dokl. Akad. Nauk S.S.S.R. (U.S.S.R.), 153 (1963) 122.

[707] Ovchinnikov Yu. A., Ivanov V. T., Kiryushkin A. A. and Shemyakin M. M., Dokl. Akad. Nauk S.S.S.R. (U.S.S.R.), 153 (1963) 1348.

[708] Ovchinnikov Yu. A., Ivanov V. T., Kiryushkin A. A. and Shemyakin M. M., Izv. Akad. Nauk S.S.S.R., Ser. Khim. (U.S.S.R.), (1963) 770.

[709] Ovchinnikov Yu. A., Ivanov V. T. and Mikhaleva I. I., Tetrahedron Lett., (1971) 159.

[710] Ovchinnikov Yu. A., Ivanov V. T., Mikhaleva I. I. and Shemyakin M. M., Izv. Akad. Nauk S.S.S.R., Ser. Khim. (U.S.S.R.), (1964) 1962.

426

[711] Ovchinnikov Yu. A., Ivanov V. T., Miroshnikov A. I., Khalilulina K. Kh. and Uvarova N. N., Khim. Prir. Soedin. (U.S.S.R.), (1971) 469.

[712] Ovchinnikov Yu. A., Ivanov V. T. and Shkrob A. M., in Molecular Mechanisms of Antibiotic Action on Protein Biosynthesis and Membranes (Proceedings of a Symposium, Granada, June 1971) (Muñoz E., García-Ferrándiz and Vázquez D., eds), Elsevier, Amsterdam, 1972, p. 459.

[713] Ovchinnikov Yu. A., Kiryushkin A. A. and Shemyakin M. M., Tetrahedron Lett. (1965) 1111.

[714] Ovchinnikov Yu. A., Kiryushkin A. A. and Shemyakin M. M., Zh. Obshch. Khim. (U.S.S.R.), 36 (1966) 620.

[715] Ovchinnikov Yu. A., Kiryushkin A. A. and Kozhevnikova I. V., Zh. Obshch. Khim. (U.S.S.R.), 41 (1971) 2085.

[716] Pache W. and Chapman D., Biochim. Biophys. Acta, 255 (1972) 348.

[717] Packer L., Biochem. Biophys. Res. Commun., 28 (1967) 1022.

[718] Packer L., Allen J. M. and Starks M., Arch. Biochem. Biophys., 128 (1968) 142.

[719] Packer L. and Crofts A. R., in Current Topics in Bioenergetics, Vol. 2, Academic Press, New York, 1967, p. 23.

[720] Packer L., Deamer D. W. and Crofts A. R., in Energy Conversion by the Photosynthetic Apparatus, Brookhaven Symposia in Biology, No. 19, 1966, p. 281.

[721] Packer L., Wrigglesworth J. M., Fortes P. A. G. and Pressman B. C., J. Cell. Biol., 39 (1968) 382.

[722] Palmieri F., Cisternino M. and Quagliariello E., Biochim. Biophys. Acta, 143 (1967) 625.

[723] Palmieri F. and Quagliariello E., Eur. J. Biochem., 8 (1969) 473.

[724] Palmieri F., Quagliariello E. and Klingenberg M., Biochem. J., 116 (1970) 36P.

[725] Palmieri F., Quagliariello E. and Klingenberg M., Eur. J. Biochem., 17 (1970) 230.

[726] Papa S., Guerrieri F., Bernardi L. R. and Tager J. M., Biochim. Biophys. Acta, 197 (1970) 100.

[727] Papa S., Guerrieri F., Simone S. and Lorusso M., J. Bioenerg., 3 (1972) 553.

[728] Papa S., Lofrumento N. E., Kanduc D., Paradies G. and Quagliariello E., Eur. J. Biochem., 22 (1971) 134.

[729] Papa S., Lofrumento N., Loglisci M. and Quagliariello E., Biochim. Biophys. Acta, 189 (1969) 34.

[730] Papa S., Scarpa A., Lee C. P. and Chance B., Biochemistry, 11 (1972) 3091.

[731] Papa S., Tager J. M., Guerrieri F. and Quagliariello E., Biochim. Biophys. Acta, 172 (1969) 184.

[732] Papa S., Zanghi M. A., Paradies G. and Quagliariello E., FEBS Lett., 6 (1970) 1.

[733] Parker A. J., Chem. Rev., 69 (1969) 1.

[734] Patel D. J. and Tonelli A. E., Biochemistry, 12 (1973) 486.

[735] Pavlašova E. and Harold F. M., J. Bacteriol. 98 (1969) 198.

[736] Payne J. W., Jakes R. and Hartley B. S., Biochem. J., 117 (1970) 757.

[737] Pedersen C. J., J. Am. Chem. Soc., 89 (1967) 2495.

[738] Pedersen C. J., J. Am. Chem. Soc., 89 (1967) 7017.

[739] Pedersen C. J., Fed. Proc., 27 (1968) 1305.

[740] Pedersen C. J., J. Am. Chem. Soc., 92 (1970) 386.

[741] Pedersen C. J., J. Am. Chem. Soc., 92 (1970) 391.

[742] Pedersen C. J., J. Org. Chem., 36 (1971) 254.

[743] Pedersen C. J., J. Org. Chem., 36 (1971) 1690.

[744] Pedersen C. J. and Frensdorff H. K., Angew. Chem., 84 (1972) 16; Int. Ed., 11 (1972) 16.

[745] Pelissard D. and Louis R., Tetrahedron Lett., (1972) 4589.

[746] Peri J. B., J. Coll. Interface Sci., 29 (1969) 6; Ekwall P., ibid., p. 16; Kraus C. A., J. Chem. Educ., 35 (1958) 324; Barlow G. H. and Zaugg H. E., J. Org. Chem., 37 (1972) 2246.

[747] Peters G. A. and Cellarius R. A., J. Bioenerg., 3 (1972) 345.

[748] Petkau A. and Chelack W. S., Biochim. Biophys. Acta, 255 (1972) 161.

[749] Petranek J. and Ryba O., cited in [165]; cf. Additional references no. 127.

[750] Pfaff E. and Klingenberg M., Eur. J. Biochem., 6 (1968) 66.

[751] Pinkerton M., Steinrauf L. K. and Dawkins P., Biochem. Biophys. Res. Commun., 35 (1969) 512.

[752] Pinkerton M. and Steinrauf L. K., J. Mol. Biol., 49 (1970) 533.

[753] Pioda L. A. R. and Simon W., Chimia, 23 (1969) 72.

[754] Pioda L. A. R., Simon W., Bosshard H. R. and Curtius H. C., Clin. Chim. Acta, 29 (1970) 289.

[755] Pioda L. A. R., Stankova V. and Simon W., Anal. Lett., 2 (1969) 665.

[756] Pioda L. A. R., Wachter H. A., Dohner R. E. and Simon W., Helv. Chim. Acta, 50 (1967) 1973.

[757] Pioda L. A. R., Wipf H.-K. and Simon W., Chimia, 22 (1968) 189.

[758] Piret P., Gobillon Y. and Van Meersshe M., Bull. Soc. Chim. France, (1963) 205.

[759] Plattner Pl. A. and Nager U., Experientia, 3 (1947) 325.

[760] Plattner Pl. A. and Nager U., Helv. Chim. Acta, 31 (1948) 665.

[761] Plattner Pl. A. and Nager U., Helv. Chim. Acta, 31 (1948) 2192.

[762] Plattner Pl. A. and Nager U., Helv. Chim. Acta, 31 (1948) 2203.

[763] Plattner Pl. A., Nager U. and Boller A., Helv. Chim. Acta, 31 (1948) 594.

[764] Plattner Pl. A., Vogler K., Studer R. O., Quitt P. and Keller-Schierlein W., Experientia, 19 (1963) 71.

[765] Plattner Pl. A., Vogler K., Studer R. O., Quitt P. and Keller-Schierlein W., Helv. Chim. Acta, 46 (1963) 927.

[766] Plengvidhya P. and Burris R. H., Plant Physiol., 40 (1965) 997.

[767] Pletnev V. Z., Gromov E. P. and Popov E. M., Khim. Prir. Soedin. (U.S.S.R.), (1973) 224.

[768] Pletnev V. Z. and Popov E. M., Khim. Prir. Soedin. (U.S.S.R.), (1973) 220.

[769] Podleski T. and Changeux J.-P., Nature, 221 (1969) 541.

[770] Poole D. T., Butler T. C. and Williams M. E., J. Membrane Biol., 5 (1971) 261.

[771] Poole D. T., Butler T. C. and Williams M. E., Biochim. Biophys. Acta, 266 (1972) 463.

[772] Popov E. M., Lipkind G. M., Arkhipova S. F. and Dashevskii V. G., Mol. Biol. (U.S.S.R.), 2 (1968) 622.

[773] Popov E. M., Lipkind G. M., Pletnev V. Z. and Arkhipova S. F., Khim. Prir. Soedin. (U.S.S.R.), (1971) 184.

[774] Popov E. M. and Pletnev V. Z., Izv. Akad. Nauk S.S.S.R., Ser. Khim. (U.S.S.R.), (1970) 991.

[775] Popov E. M., Pletnev V. Z., Evstratov A. V., Ivanov V. T. and Ovchinnikov Yu. A., Khim. Prir. Soedin. (U.S.S.R.), (1970) 616.

[776] Portnova S. L., Balashova T. A., Bystrov V. F., Shilin V. V., Biernat J., Ivanov V. T. and Ovchinnikov Yu. A., Khim. Prir. Soedin (U.S.S.R.), (1971) 323.

[777] Portnova S. L., Shilin V. V., Balashova T. A., Biernat J., Bystrov V. F., Ivanov V. T. and Ovchinnikov Yu. A., Tetrahedron Lett., (1971) 3085.

[778] Pressman B. C., Biochem. Biophys. Res. Commun., 15 (1964) 556.

[779] Pressman B. C., Fed. Proc., 24 (1965) 425.

[780] Pressman B. C., Proc. Natl., Acad. Sci. U.S., 53 (1965) 1076.

[781] Pressman B. C., Special publication in International Symposium on Mechanisms of Action of Fungicides and Antibiotics, (Castle Reinhardsbrunn, DDR, 1967) Academic-Verlag, Berlin, 1967.

[782] Pressman B. C., in Methods in Enzymology, (Colowick S. P. and Kaplan N. O., eds), Vol. 10, Academic Press, New York, 1967, p. 714.

428

[783] Pressman B. C., in Molecular Mechanisms of Antibiotic Action on Protein Biosynthesis and Membranes (Abstracts of papers presented at the Symposium, Granada, June 1971), p. 62.

[784] Pressman B. C., Mechanism of Action of Transport Mediating Antibiotics, Report in Symposium on Ion Transport and Intramitochondrial pH, New York, 1968.

[785] Pressman B. C., Fed. Proc., 27 (1968) 1283.

[786] Pressman B. C., Antimicrob. Agents Chemother., (1969) 28.

[787] Pressman B. C., in Membranes of Mitochondria and Chloroplasts (Racker E., ed.), (ACS Monograph No. 165), Van Nostrand Reinhold, 1970, p. 213.

[788] Pressman B. C., Symposium report at IVth International Biophysics Congress (Moscow, 1972).

[789] Pressman B. C. and Harris E. J., Abstracts of 7th International Congress of Biochemistry (Tokyo, 1967), Vol. 5, p. 900.

[790] Pressman B. C., Harris E. J., Jagger W. S. and Johnson J. H., Proc. Natl. Acad. Sci. U.S., 58 (1967) 1949.

[791] Pressman B. and Haynes D. H., in The Molecular Basis of Membrane Function (Tosteson, D. C. ed.), Prentice Hall, Englewood Cliffs, 1969, p. 211.

[792] Prestegard J. H. and Chan S. I., Biochemistry, 8 (1969) 3921.

[793] Prestegard J. H. and Chan S. I., J. Am. Chem. Soc., 92 (1970) 4440.

[794] Pretsch E., Vasak M. and Simon W., Helv. Chim. Acta, 55 (1972) 1098.

[795] Pimentel G. C. and McClellan A. L., The Hydrogen Bond, Freeman, San Francisco, 1960.

[796] Prox A., Schmid J. and Ottenheym H., Liebig's Ann. Chem., 722 (1969) 179.

[797] Prox A. and Weygand F., in Peptides (Proceedings of the VIIIth European Peptide Symposium, Noordwijk, Holland, 1966), (Beyerman H. C., van de Linde A. and Maassen van den Brink W., eds), North-Holland, Amsterdam, 1967, p. 158.

[798] Pullman B., Aspects de la Chimie Quantique Contemporaire (Dandel R. and Pullman A., eds), Colloque Internationale du CNRS, Paris, 1971, p. 261.

[799] Pullman M. E. and Schatz G., Ann. Rev. Biochem., 36 (1967) 539.

[800] Quagliariello E., Geuchi G. and Palmieri F., FEBS Lett., 13 (1971) 253.

[801] Quagliariello E., Meijer A. J., Tager J. M. and Papa S., Biochem. J., 116 (1970) 38P.

[802] Quagliariello E. and Palmieri F., FEBS Lett., 8 (1970) 105.

[803] Quitt P., Studer R. O. and Vogler K., Helv. Chim. Acta, 46 (1963) 1715.

[804] Quitt P., Studer R. O. and Vogler K., Helv. Chim. Acta, 47 (1964) 166.

[805] Racker E., J. Membrane Biol., 10 (1972) 221; Hinkle P. C., Kim J. J. and Racker E., J. Biol. Chem., 247 (1972) 1338.

[806] Racker E., J. Biol. Chem., 247 (1972) 8198.

[807] Ramachandran L. K., Biochemistry, 2 (1963) 1138.

[808] Ramachandran G. N., Chandrasekaran R. and Kopple K. D., Biopolymers 10 (1971) 2113.

[809] Ramachandran G. N. and Sasisekharan V., Adv. Protein Chem., 23 (1968) 283.

[810] Rambhav S. and Ramachandran L. K., Ind. J. Biochem. Biophys., 9 (1972) 21.

[811] Rechnitz G. A. and Eyal E., Anal. Chem., 44 (1972) 370.

[812] Rechnitz G. A. and Mohan M. S., Science, 168 (1970) 1460.

[813] Reed P. W., Fed. Proc., 31 (1972) Abstr. No. 432.

[814] Reed P. W. and Lardy H. A., J. Biol. Chem., 247 (1972) 6970.

[815] Roeske R. W., Isaac S., Steinrauf L. K. and King T., Fed. Proc., 30 (1971) Abstr. No. 1340.

[816] Reid R. A., Biochem. J., 116 (1969) 12P.

[817] Rendleman J. A., Adv. Carbohydrate Chem., 21 (1966) 209.

[818] Reusser F., J. Biol. Chem., 242 (1967) 243.

429

[819] Robertson R. N., Endeavour, 26 (1967) 134.
[820] Roitman J. N. and Cram D. J., J. Am. Chem. Soc., 93 (1971) 2231.
[821] Rose I. B., J. Chem. Soc., (1956) 542.
[822] Rosen W. and Busch D. H., J. Am. Chem. Soc., 91 (1969) 4694.
[823] Rosen W. and Busch D. H., Chem. Commun., (1969) 148.
[824] Rosen W. and Busch D. H., Inorg. Chem., 9 (1970) 262.
[825] Rossi C. and Azzi A., in Regulation of Metabolic Processes in Mitochondria (Tager
       J. M., Papa S., Quagliariello E. and Slater E. C., eds), [BBA Library, Vol. 7] Elsevier,
       Amsterdam, 1966, p. 332.
[826] Rossi C., Azzi A. and Azzone G. F., J. Biol. Chem., 242 (1967) 951.
[827] Rossi C., Scarpa A. and Azzone G. F., Biochemistry, 6 (1967) 3902.
[828] Rossi E. and Azzone G. F., Eur. J. Biochem., 7 (1969) 418.
[829] Rossi E. and Azzone G. F., Eur. J. Biochem., 12 (1970) 319.
[830] Rottenberg H., Eur. J. Biochem., 15 (1970) 22.
[831] Rottenberg H., Caplan S. R. and Essig A., in Membranes and Ion Transport (Bittar
       E. E., ed.), Vol. 1, Wiley-Interscience, London, 1970.
[832] Rottenberg H., Essig A. and Caplan S. R., Nature, 216 (1967) 610.
[833] Rottenberg H., Grunwald T. and Avron M., FEBS Lett., 13 (1971) 41.
[834] Rottenberg H., Grunwald T. and Avron M., Eur. J. Biochem., 25 (1972) 54.
[835] Rottenberg H. and Solomon A. K., Biochim. Biophys. Acta, 193 (1969) 48.
[836] Rumberg B. and Siggel U., Naturwiss., 56 (1969) 130.
[837] Russell D. W., Biochim. Biophys. Acta, 45 (1960) 411.
[838] Russell D. W., J. Chem. Soc., (1962) 753.
[839] Russell D. W., J. Chem. Soc., (1965) 4664.
[840] Russell D. W., Macdonald C. G. and Shannon J. S., Tetrahedron Lett., (1964)
       2759.
[841] Sackman E. and Träuble H., J. Am. Chem. Soc., 94 (1972) 4482, 4492, 4499.
[842] Saha J., Papahadjopoulos D. and Wenner C. E., Biochim. Biophys. Acta, 196
       (1970) 10.
[843] Saha J., Shepard D., Jacobson K. and Wenner C. E., unpublished results, cited in
       [842].
[844] Saha S., Izawa S. and Good N. E., Biochim. Biophys. Acta, 223 (1970) 158.
[845] Sallis J. D. and de Luca H. F., J. Biol. Chem., 239 (1964) 4303.
[846] Sallis J. D., de Luca H. F. and Martin D. L., J. Biol. Chem., 240 (1965) 2229.
[847] Sam D. J. and Simmons H. E., J. Am. Chem. Soc., 94 (1972) 4024.
[848] Sanasaryan A. A., Fonina L. A., Shvetsov Yu. B. and Vinogradova E. I., Khim.
       Prir. Soedin. (U.S.S.R.), (1971) 81.
[849] Sandblom J., Walker J. L. and Eisenman G., Biophys. J., 12 (1972) 587.
[850] Sarges R. and Witkop B., J. Am. Chem. Soc., 86 (1964) 1861.
[851] Sarges R. and Witkop B., J. Am. Chem. Soc., 86 (1964) 1862.
[852] Sarges R. and Witkop B., J. Am. Chem. Soc., 87 (1965) 2011.
[853] Sarges R. and Witkop B., J. Am. Chem. Soc., 87 (1965) 2020.
[854] Sarges R. and Witkop B., J. Am. Chem. Soc., 87 (1965) 2027.
[855] Sarges R. and Witkop B., Biochemistry, 4 (1965) 2491.
[856] Sato H., Takahashi K. and Kikuchi G., Biochim. Biophys. Acta, 112 (1966) 8.
[857] Scarpa A., Abstracts of Communications presented at VIIIth FEBS Meeting
       (Amsterdam, 1972), No. 85, North-Holland, Amsterdam, 1972.
[858] Scarpa A. and Azzone G. F., Eur. J. Biochem., 12 (1970) 328.
[859] Scarpa A., Cecchetto A. and Azzone G. F., Nature, 219 (1968) 529.
[860] Scarpa A. and De Gier J., Biochim. Biophys. Acta, 241 (1971) 789.
[861] Scarpa A. and Inesi G., FEBS Lett., 22 (1972) 273.
[862] Scher A., Fette Seifen Anstrichm., 63 (1961) 617.

430

[863] Schliepake W., Junge W. and Witt H. T., Z. Naturforsch., 23b (1968) 1571.
[864] Schmitz H., Deak S. B., Crook K. E. and Hooper I. R., Antimicrob. Agents Chemother., (1963) 89.
[865] Schneeweiss F. and L'Orange R., Z. Naturforsch., 266 (1971) 624.
[866] Scholer R. P. and Simon W., Chimia, 24 (1970) 372.
[867] Scholes P., Mitchell P. and Moyle J., Eur. J. Biochem., 8 (1969) 450.
[868] Schönfeldt N., J. Am. Oil Chem. Soc., 32 (1955) 17.
[869] Schröder E. and Lübke K., The Peptides, Vol. 2, Academic Press, New York—London, 1966.
[870] Schuldiner S. and Avron M., Eur. J. Biochem., 19 (1971) 227.
[871] Schuldiner S., Rottenberg H. and Avron M., FEBS Lett., 28 (1972) 173.
[872] Schulz H., Chem. Ber., 99 (1966) 3425.
[873] Schwarzenbach G., Chimia, 27 (1973) 1.
[874] Schwesinger R. and Prinzbach H., Angew. Chem., 84 (1972) 990.
[875] Schwyzer R., Experientia, 26 (1970) 577.
[876] Schwyzer R., Tun-Kyi A., Caviezel M. and Moser P., Helv. Chim. Acta, 53 (1970) 15.
[877] See [899] p. 501.
[878] Sekeris C. E. and Schmid W., FEBS Lett., 27 (1972) 41.
[879] Selwyn M. J., Biochem. J., 130 (1972) 65P.
[880] Selwyn M. J., Dawson A. P., Stockdale M. and Gains N., Eur. J. Biochem., 14 (1970) 120.
[881] Selwyn M. J., Dunnett S. J., Philo A. P. and Dawson A. P., Biochem. J., 127 (1972) 66P.
[882] Shah D. O. and Shulman J. H., Biochim. Biophys. Acta, 135 (1967) 184.
[883] Shavit N., Allen J. M. and Starks M., Arch. Biochem. Biophys., 128 (1968) 142.
[884] Shavit N., Degani H. and San Pietro A., Biochim. Biophys. Acta, 216 (1970) 208.
[885] Shavit N., Dilley R. and San Pietro A., Biochemistry, 7 (1969) 2356.
[886] Shavit N. and San Pietro A., Biochem. Biophys. Res. Commun., 28 (1967) 277.
[887] Shavit N., Thore A., Keister D. L. and San Pietro A., Proc. Natl. Acad. Sci. U.S., 59 (1968) 917.
[888] Shaw P. D., in Antibiotics, (Gottlieb D. O. and Shaw P. D., eds), Vol. 1, Springer-Verlag, Berlin—Heidelberg—New York, 1967.
[889] Shchagina L. V., Naumov Yu. V., Shkrob A. M. and Lev A. A., in Membrane Biophysics (Russian), (Alkalene B. D., Zablotskaite, D. V. and Narushevichus E. V., eds), Part 2, Kaunas, in press.
[890] Shchori E. and Jagur-Grodzinsky J., Israel J. Chem., 10 (1972) 935.
[891] Shchori E. and Jagur-Grodzinsky J., Israel J. Chem., 10 (1972) 959.
[892] Shchori E. and Jagur-Grodzinsky J., J. Am. Chem. Soc., 94 (1972) 7957.
[893] Shchori E., Jagur-Grodzinski J., Luz Z. and Shporer M., J. Am. Chem. Soc., 93 (1971) 7133.
[894] Sheetz M. P. and Chan S. I., Biochemistry, 11 (1972) 4573.
[895] Shemyakin M. M., Aldanova N. A., Vinogradova E. I. and Feigina M. Yu., Tetrahedron Lett., (1963) 1921.
[896] Shemyakin M. M., Aldanova N. A., Vinogradova E. I. and Feigina M. Yu., Izv. Akad. Nauk S.S.S.R., Ser. Khim. (U.S.S.R.), (1966) 2143.
[897] Shemyakin M. M., Antonov V. K., Bergelson L. D., Ivanov V. T., Malenkov G. G., Ovchinnikov Yu. A. and Shkrob A. M., in The Molecular Basis of Membrane Function (Tosteson D. C., ed.), Prentice Hall, Englewood Cliffs, 1969, p. 173.
[898] Shemyakin M. M., Antonov V. K., Ovchinnikov Yu. A., Kiryushkin A. A., Ivanov V. T., Shchelokov V. I. and Shkrob A. M., Tetrahedron Lett., (1964) 47.

[899] Shemyakin M. M., Khokhlov A. S., Kolosov M. N., Bergelson L. D. and Antonov V. K., Chemistry of Antibiotics (Russian), Vol. 2, U.S.S.R. Academy of Sciences, Moscow, 1961, p. 1071.

[900] Shemyakin M. M. and Ovchinnikov Yu. A., in Recent Development in the Chemistry of Natural Carbon Compounds, Vol. 2, Hungarian Academy of Sciences, Budapest, 1967, p. 3.

[901] Shemyakin M. M., Ovchinnikov Yu. A., Antonov V. K., Kiryushkin, A. A., Ivanov V. T., Shchelokov V. I. and Shkrob A. M., Izv. Akad. Nauk S.S.S.R., Ser. Khim. (U.S.S.R.), (1963) 2233.

[902] Shemyakin M. M., Ovchinnikov Yu. A. and Ivanov V. T., Angew. Chem., 81 (1969) 523.

[903] Shemyakin M. M., Ovchinnikov Yu. A., Ivanov V. T., Antonov V. K., Shkrob A. M., Mikhaleva I. I., Evstratov A. V. and Malenkov G. G., Biochem. Biophys. Res. Commun., 29 (1967) 834.

[904] Shemyakin M. M., Ovchinnikov Yu. A., Ivanov V. T., Antonov V. K., Vinogradova E. I., Shkrob A. M., Malenkov G. G., Evstratov A. V., Ryabova I. D., Laine I. A. and Melnik E. I., J. Membrane Biol., 1 (1969) 402.

[905] Shemyakin M. M., Ovchinnikov Yu. A., Ivanov V. T. and Evstratov A. V., Nature, 213 (1967) 412.

[906] Shemyakin M. M., Ovchinnikov Yu. A., Ivanov V. T., Evstratov A. V., Mikhaleva I. I. and Ryabova I. D., Zh. Obshch. Khim. (U.S.S.R.), 42 (1972) 2320.

[907] Shemyakin M. M., Ovchinnikov Yu. A., Ivanov V. T. and Kiryushkin A. A., Izv. Akad. Nauk S.S.S.R., Ser. Khim. (U.S.S.R.), (1963) 578.

[908] Shemyakin M. M., Ovchinnikov Yu. A., Ivanov V. T. and Kiryushkin A. A., Tetrahedron, 19 (1963) 581.

[909] Shemyakin M. M., Ovchinnikov Yu. A., Ivanov V. T. and Kiryushkin A. A., Tetrahedron, 19 (1963) 995.

[910] Shemyakin M. M., Ovchinnikov Yu. A., Ivanov V. T. and Kiryushkin A. A., Tetrahedron Lett., (1963) 885.

[911] Shemyakin M. M., Ovchinnikov Yu. A., Ivanov V. T. and Kiryushkin A. A., Tetrahedron Lett., (1963) 1927.

[912] Shemyakin M. M., Ovchinnikov Yu. A., Ivanov V. T., Kiryushkin A. A. and Khalilulina K. Kh., Zh. Obshch. Khim. (U.S.S.R.), 35 (1965) 1399.

[913] Shemyakin M. M., Ovchinnikov Yu. A., Ivanov V. T., Kiryushkin A. A., Zhdanov G. L. and Ryabova I. D., Experientia, 19 (1963) 566.

[914] Shemyakin M. M., Ovchinnikov Yu. A., Kiryushkin A. A. and Ivanov V. T., Tetrahedron Lett., (1962) 301.

[915] Shemyakin M. M., Ovchinnikov Yu. A., Kiryushkin A. A. and Ivanov V. T., Izv. Akad. Nauk S.S.S.R., Ser. Khim. (U.S.S.R.), (1962) 2154.

[916] Shemyakin M. M., Ovchinnikov Yu. A., Kiryushkin A. A. and Ivanov V. T., Izv. Akad. Nauk S.S.S.R., Ser. Khim. (U.S.S.R.), (1963) 579.

[917] Shemyakin M. M., Ovchinnikov Yu. A., Kiryushkin A. A. and Ivanov V. T., Izv. Akad. Nauk S.S.S.R., Ser. Khim. (U.S.S.R.), (1963) 1148.

[918] Shemyakin M. M., Ovchinnikov Yu. A., Kiryushkin A. A. and Ivanov V. T., Izv. Akad. Nauk S.S.S.R., Ser. Khim. (U.S.S.R.), (1965) 1623.

[919] Shemyakin M. M., Shchukina L. A., Vinogradova E. I., Ravdel G. A. and Ovchinnikov Yu. A., Experientia, 22 (1966) 535.

[920] Shemyakin M. M., Vinogradova E. I., Feigina M. Yu. and Aldanova N. A., Tetrahedron Lett., (1963) 351.

[921] Shemyakin M. M., Vinogradova E. I., Feigina M. Yu. and Aldanova N. A., Zh. Obshch. Khim. (U.S.S.R.), 34 (1964) 1798.

[922] Shemyakin M. M., Vinogradova E. I., Feigina M. Yu. Aldanova N. A., Loginova N. F., Ryabova I. D. and Pavlenko I. A., Experientia, 27 (1965) 548.

432

[923] Shemyakin M. M., Vinogradova E. I., Feigina M. Yu., Aldanova N. A., Oladkina V. A. and Shchukina L. A., Dokl. Akad. Nauk S.S.S.R. (U.S.S.R.), 140 (1961) 287.

[924] Shemyakin M. M., Vinogradova E. I., Feigina M. Yu., Aldanova N. A., Ovchinnikov Yu. A. and Kiryushkin A. A., Zh. Obshch. Khim. (U.S.S.R.), 31 (1961) 1782.

[925] Shemyakin M. M., Vinogradova E. I., Feigina M. Yu., Aldanova N. A., Shvetsov Yu. B. and Fonina L. A., Zh. Obshch. Khim. (U.S.S.R.), 36 (1966) 1391.

[926] Shemyakin M. M., Vinogradova E. I., Ryabova I. D., Fonina L. A. and Sanasaryan A. A., Khim. Prir. Soedin. (U.S.S.R.), (1973) 241.

[927] Sherman L. A. and Clayton R. K., FEBS Lett., 22 (1972) 127.

[928] Shibata M., Nakazawa K., Inoue M., Terumichi J. and Miyake A., Ann. Rep. Takeda Res. Labs., 17 (1958) 19.

[929] Shields T. C., Chem. Commun., (1968) 832.

[930] Shiro M. and Koyama H., J. Chem. Soc. Ser. B, (1970) 243.

[931] Shkrob A. M. and Demin V. V., in Ion Transport Across Membranes, Abstracts of Reports on Conference of Experts from Socialistic Countries (Castle Reinhardsbrunn, GDR, May 1972), Berlin, 1972, p. 54.

[932] Shoji J., Kozuki S., Matsutani S., Kubota T., Nishimura H., Mayama M., Motokawa K., Tanaka Y., Shimaoka N. and Otsuka H., J. Antibiot., 21 (1968) 402.

[933] Sholes P. and Mitchell P., J. Bioenerg., 1 (1970) 61.

[934] Sholes P. and Mitchell P., J. Bioenerg., 1 (1970) 309.

[935] Shunnard R. F. and Callender M. E., Antimicrob. Agents Chemother., (1967) 369.

[936] Silman I., J. Gen. Physiol., 54 (1969) 265S.

[937] Simon W., Swiss. Pat. 479870 (Cl. G01n), 28 XII 1969.

[938] Simon W. and Morf W. E., in Membranes — A Series of Advances (Eisenman G., ed.), Vol. 2, Marcel Dekker, New York, 1973, p. 329.

[939] Simon W., Pioda L. A. R. and Wipf H.-K., 20 Colloquium der Gesellschaft für biologische Chemie (Mosbach/Baden, April 1969), Springer-Verlag, Heidelberg, 1969, S.356.

[940] Simon W., Wuhrmann H.-R., Vašák M., Pioda L. A. R., Dohner R. and Štefanac Z., Angew. Chem., 82 (1970) 433.

[941] Singer M. A. and Bangham A. D., Biochim. Biophys. Acta, 241 (1971) 687.

[942] Singer S. J., Ann. N.Y. Acad. Sci., 195 (1972) 16.

[943] Singer S. J. and Nicolson G. L., Science, 175 (1972) 720.

[944] Skulachev V. P., Energy Accumulation in the Cell (Russian), Nauka, Moscow, 1969.

[945] Skulachev V. P., FEBS Lett., 11 (1970) 301.

[946] Skulachev V. P., in Current Topics in Bioenergetics, Vol. 4, Academic Press, 1971, p. 127.

[947] Skulachev V. P., Usp. Sovrem. Biol. (U.S.S.R.), 71 (1971) 310.

[948] Skulachev V. P., J. Bioenerg., 3 (1972) 25.

[949] Skulachev V. P., Energy Transformations in Biological Membranes (Russian), Nauka, Moscow, 1972.

[950] Slater E. C., Eur. J. Biochem., 1 (1967) 317.

[951] Smeley R. R., Leben C., Klett G. W. and Strong F. M., Phytopathology, 42 (1952) 506.

[952] Smid J., Angew. Chem., 84 (1972) 127.

[953] Smirnova G. M., Blinova I. M., Koloditskaya T. A. and Khokhlov A. S., Antibiotiki (U.S.S.R.), 5 (1970) 387.

[954] Smith E. H. and Beyer R. E., Arch. Biochem. Biophys., 122 (1967) 614.

[955] Smith S. G. and Hanson M. P., J. Org. Chem., 36 (1971) 1931.

[956] Smith W. B., Analyst, 84 (1959) 77.

[957] Solodar J., Tetrahedron Lett., (1971) 287.

433

[958] Sotnikov P. S., Myagkov I. V., Babakov A. V., Terekhov O. P. and Dem'yanovskii O. B., in Biofizika Membran (Alkalene B. D., Zabolotskaite D. V. and Narushevichus E. V., eds), Part I, Kaunas, 1971, p. 724.
[959] Staley S. W. and Erdman J. P., J. Am. Chem. Soc., 92 (1970) 3832.
[960] Stark G. and Benz R., J. Membrane Biol., 5 (1971) 133.
[961] Stark G., Benz R., Pohl G. W. and Janko K., Biochim. Biophys. Acta, 266 (1972) 603.
[962] Stark G., Ketterer B., Benz R. and Läuger P., Biophys. J., 11 (1972) 981.
[963] Stark W. M., Knox N. G. and Westhead J. E., Antimicrob. Agents Chemother., (1967) 353.
[964] Starks C. M., J. Am. Chem. Soc., 93 (1971) 195.
[965] Stedingk von L.-V. and Baltscheffsky H., Arch. Biochem. Biophys., 117 (1966) 400.
[966] Štefanac Z. and Simon W., Chimia, 20 (1966) 436.
[967] Štefanac Z. and Simon W., Microchem. J., 12 (1967) 125.
[968] Stein W. D., Nature, 218 (1968) 570.
[969] Steinrauf L. K., Czerwinski E. W. and Pinkerton M., Biochem. Biophys. Res. Commun., 45 (1971) 1279.
[970] Steinrauf L. K., Pinkerton M. and Chamberlin J. W., Biochem. Biophys. Res. Commun., 33 (1968) 29.
[971] Stempel A., Westley J. W. and Benz W., J. Antibiot., 22 (1969) 384.
[972] Stern A., Gibbons W. A. and Craig L. C., Proc. Natl. Acad. Sci. U.S., 61 (1968) 735.
[973] Stewart D. G., Waddan D. Y. and Borrows E. T., Brit. Pat. 785229 (Oct. 23, 1957).
[974] Stillman I. M., Gilbert D. L. and Robbins M., Biochim. Biophys. Acta, 203 (1970) 338.
[975] Stoeckenius W., in Membranes of Mitochondria and Chloroplasts (Racker E., ed.), [ACS Monograph No. 165], Van Nostrand Reinhold, 1970, p. 53.
[976] Strelkina L. A., Ion Distribution in a Two Phase System in the Presence of Valinomycin (Russian), Graduate Thesis, Moscow Veterinary Academy, Shemyakin Institute for Chemistry of Natural Products, U.S.S.R. Academy of Sciences, 1971.
[977] Strichartz G. R. and Chance B., Biochim. Biophys. Acta, 256 (1972) 71.
[978] Stryer L., J. Mol. Biol., 13 (1965) 482.
[979] Studer R. O., Quitt P., Böhni F. and Vogler K., Monatsh. Chem., 96 (1965) 461.
[980] Su A. C. L. and Weiher J. F., Inorg. Chem., 7 (1968) 176.
[981] Sullivan E. A. and Hinckley A. A., J. Org. Chem., 27 (1962) 3731.
[982] Suzuki A., Kuyama S., Kodaira Y. and Tamura S., Agr. Biol. Chem., 30 (1966) 517.
[983] Suzuki A., Taguchi H. and Tamura S., Agr. Biol. Chem., 34 (1970) 813.
[984] Suzuki K., Nawata Y. and Ando K., J. Antibiot., 24 (1971) 675.
[985] Svoboda M., Hapala J. and Závada J., Tetrahedron Lett., (1972) 265.
[986] Svoboda M., Závada J. and Sicher J., Coll. Czech. Chem. Commun., 33 (1968) 1415.
[987] Swardstrom J. W., Duvall L. A. and Miller D. P., Acta Cryst., B28 (1972) 2510.
[988] Synge R. L. M., Biochem. J., 39 (1945) 355.
[989] Synge R. L. M., Cold Spring Harbor Symp. Quant. Biol., 14 (1950) 191.
[990] Szabo G., Eisenman G. and Ciani S., J. Membrane Biol., 1 (1969) 346.
[991] Szabo G., Eisenman G. and Ciani S. M., in Coral Gables Conference on Physical Principles of Biological Membranes (Snell F., Iverson G. and Lam T., eds), Gordon and Breach, New York, 1972.
[992] Szabo G., Eisenman G., Laparde R., Ciani S. and Krasne S., in Membranes — A Series of Advances (Eisenman G., ed.), Vol. 2, Marcel Dekker, New York, 1973, p. 179.

434

[993] Szwarc M., Accounts Chem. Research, 2 (1969) 87.

[994] Szwarc M., Science, 170 (1970) 23.

[995] Takaki U., Hogen Esch T. E. and Smid J., J. Am. Chem. Soc., 93 (1971) 6760.

[996] Takaki U., Hogen Esch T. E. and Smid J., J. Phys. Chem., 76 (1972) 2152.

[997] Talekar S. V., Quantum chemical studies of the electronic structure of macrocyclic antibiotic molecules (Valinomycin and its potassium complex), Ph.D. Thesis, All-India Institute of Medical Sciences, New Delhi, India, 1970.

[998] Tamura S., Kuyama S., Kodaira Y. and Higashikawa S., Agr. Biol. Chem., 28 (1964) 137.

[999] Tapley D. F., J. Biol. Chem., 222 (1956) 325.

[1000] Terry P. M. and Vidaver G. A., Biochem. Biophys. Res. Commun., 47 (1972) 539.

[1001] Thong C. M., Canet D., Granger P., Marraud M. and Neel J., Compt. Rend. Ser. C., 269 (1969) 580.

[1002] Thore A., Keister D. L., Shavit N. and San Pietro A., Biochemistry, 7 (1968) 3499.

[1003] Tinsley I. J., Haque R. and Schmedding D., Science, 174 (1971) 145.

[1004] Tirunarayan M. O. and Sirsi M., J. Indian Inst. Sci., 39AB (1957) A185.

[1005] Tirunarayan M. O. and Sirsi M., J. Indian Inst. Sci., 39AB (1957) A215.

[1006] Tirunarayan M. O. and Sirsi M., Arch. Int. Pharmacodyn., 108 (1959) 258.

[1007] Ti Tien H. and Diana A. L., Chem. Phys. Lipids, 2 (1968) 55.

[1008] Tonelli A. E., Proc. Natl. Acad. Sci. U.S., 68 (1971) 1203.

[1009] Tonelli A. E., Patel D. J., Goodman M., Naider F., Faulstich H. and Wieland Th., Biochemistry, 10 (1971) 3211.

[1010] Tosteson D. C., Fed. Proc., 27 (1968) 1269.

[1011] Tosteson D. C., Report on Biophysical Society Meeting, Baltimore, Maryland, February 1970.

[1012] Tosteson D. C., Andreoli T. E., Tieffenberg M. and Cook P., J. Gen. Physiol., 51 (1968) 373S.

[1013] Tosteson D. C., Cook P., Andreoli T. and Tieffenberg M., J. Gen. Physiol., 50 (1967) 2513.

[1014] Tosteson D. C., Gisin B., Tieffenberg M., Davis D., Gunn R. and Cook P., Report of the Presymposium The Physicochemical Basis of Ion Transport Through Biological Membranes (VIIth IUPAC Symposium on the Chemistry of Natural Products, Riga, 1970).

[1015] Tosteson D. C., Tieffenberg M. and Cook P., in Metabolism and Membrane Permeability of Erythrocytes and Thrombocytes (Deutsch E., Gerlach E. and Moser K., eds), Georg Thieme Verlag, Stuttgart, 1968, p. 424.

[1016] Tostenson D. C., Tieffenberg M., Davis D. G., Ginsburg H., Gisin B. F. and Cook P., Abstracts of papers presented at the Symposium on Molecular Mechanisms of Antibiotic Action on Protein Biosynthesis and Membranes (Granada, 1971), p. 63.

[1017] Tosteson D. C., private communication.

[1018] Travis K. and Busch D. H., Chem. Commun., (1970) 1041.

[1019] Truter M. R., Chem. Br., 7 (1971) 203.

[1020] Trutnovsky H., Z. Klin. Chem. Klin. Biochem., 9 (1971) 341.

[1021] Tsatsas A. T., Stearns R. W. and Risen W. M., J. Am. Chem. Soc., 94 (1972) 5247.

[1022] Uglestad J., Ellingsen T. and Berge A., Acta Chem. Scand., 20 (1966) 1593.

[1023] Uno T. and Miyajima K., Chem. Pharm. Bull., 11 (1963) 80.

[1024] Urry D. W., in Spectroscopic Approaches to Biomolecular Conformation (Urry D. W., ed.), American Medical Association, Chicago, 1970, p. 263.

[1025] Urry D. W., Proc. Natl. Acad. Sci. U.S., 68 (1971) 672.

[1026] Urry D. W., Proc. Natl. Acad. Sci. U.S., 68 (1971) 810.

[1027] Urry D. W., Proc. Natl. Acad. Sci. U.S., 69 (1972) 1610.

[1028] Urry D. W., Ann. N.Y. Acad. Sci., 195 (1972) 108.

[1029] Urry D. W., Biochim. Biophys. Acta, 265 (1972) 115.

[1030] Urry D. W., Glickson J. D., Mayers D. F. and Haider J., Biochemistry, 11 (1972) 487.

[1031] Urry D. W., Goodall M. C., Glickson J. D. and Mayers D. F., Proc. Natl. Acad. Sci. U.S., 68 (1971) 1907.

[1032] Urry D. W., Krivacic J. R. and Haider J., Biochem. Biophys. Res. Commun., 43 (1971) 6.

[1033] Urry D. W., Ohnishi M. and Walter R., Proc. Natl. Acad. Sci. U.S., 66 (1970) 111.

[1034] Venkatachalam M., Biopolymers, 6 (1968) 1425.

[1035] Vernon L. P., Bacteriol. Rev., 32 (1968) 243.

[1036] Villa L., Agostoni A. and Jean G., Experientia, 24 (1968) 576.

[1037] Vining L. C. and Taber W. A., Can. J. Chem., 40 (1962) 1579.

[1038] Vinogradova E. I., Sanasaryan A. A., Fonina L. A. and Ivanov V. T., Khim. Prir. Soedin. (U.S.S.R.), in press.

[1039] Vogel E., Altenbach H.-J. and Sommerfeld C.-D., Angew. Chem., 84 (1972) 986.

[1040] Waddell W. J., Biochem. Biophys. Res. Commun., 49 (1972) 127.

[1041] Walker J. L., Anal. Chem., 43 (1971) 89A.

[1042] Walker D. A. and Crofts A. R., Ann. Rev. Biochem., 39 (1970) 389.

[1043] Wallhäusser K. H., Huber G., Nessenmann G., Präve P. and Zept K., Arzneimittelforsch., 14 (1964) 356.

[1044] Warner D. T., J. Am. Oil Chem. Soc., 44 (1967) 593.

[1045] Wasserman H. H., Keggi J. J. and McKeen J. E., J. Am. Chem. Soc., 83 (1961) 4107.

[1046] Watling A. S. and Selwyn M. J., FEBS Lett., 10 (1970) 139.

[1047] Watling A. S. and Selwyn M. J., Biochem. J., 128 (1972) 86P.

[1048] Weber W. P. and Shepherd J. P., Tetrahedron Lett., (1972) 4907.

[1049] Weinkam R. J. and Jorgensen E. C., J. Am. Chem. Soc., 93 (1971) 7038.

[1050] Wenner C. E. and Hackney J. H., Biochemistry, 8 (1969) 330.

[1051] Wenner C. E., Harris E. J. and Pressman B. C., J. Biol. Chem., 242 (1967) 3454.

[1052] West I. C. and Mitchell P., Biochem. J., 127 (1972) 56P.

[1053] West I. and Mitchell P., J. Bioenerg., 3 (1972) 445.

[1054] Westley J. W., Evans R. H., Williams T. and Stempel A., Chem. Commun., (1970) 71.

[1055] Westley J. W., Evans R. H., Williams T. and Stempel A., Chem. Commun., (1970) 1467.

[1056] Whittaker V. P., in Regulation of Metabolic Processes in Mitochondria (Tager J. M., Papa S., Quagliariello E. and Slater E. C., eds), [BBA Library, Vol. 7], Elsevier, Amsterdam, 1966, p. 1.

[1057] Wittingham C. P., Progr. Biophys. Mol. Biol., 21 (1970) 127.

[1058] Wieland Th., Fortschr. Chem. Org. Naturst., 25 (1967) 214.

[1059] Wieland Th., in Jahrbuch der Max-Planck Gesellschaft zur Förderung der Wissenschaften, 1970, S.159.

[1060] Wieland Th., private communication.

[1061] Wieland Th., Naturwiss., 59 (1972) 225.

[1062] Wieland Th., Birr Ch. and von Dungen A., Liebig's Ann. Chem., 747 (1971) 207.

[1063] Wieland Th., Birr Ch., Frodl R., Lochinger W. and Stahnke G., Liebig's Ann. Chem., 757 (1972) 136.

[1064] Wieland Th., Birr Ch. and Flor F., Liebig's Ann. Chem., 727 (1969) 130.

[1065] Wieland Th., von Dungen A. and Birr Ch., FEBS Lett., 14 (1971) 299.

436

[1066] Wieland Th., Faesel J. and Konz W., in Peptides (Proc. of the IXth Eur. Peptide Symp.), (Bricas E., ed.), North-Holland, Amsterdam, 1968, p. 243.

[1067] Wieland Th., Faesel J. and Konz W., Liebig's Ann. Chem., 722 (1969) 197.

[1068] Wieland Th., Faulstich H. and Bürgermeister W., Biochem. Biophys. Res. Commun., 47 (1972) 984.

[1069] Wieland Th., Faulstich H., Bürgermeister W., Otting W., Mohle W., Shemyakin M. M., Ovchinnikov Yu. A., Ivanov V. T. and Malenkov G. G., FEBS Lett., 9 (1970) 89.

[1070] Wieland Th., Faulstich H., Jahn W., Govindan M., Puchinger H., Kopitar Z., Schamus H. and Schmitz A., Hoppe-Seyler's Z. Physiol. Chem., 353 (1972) 1337.

[1071] Wieland Th., Lapatsanis L., Faesel J. and Konz W., Liebig's Ann. Chem., 747 (1971) 194.

[1072] Wieland Th., Lewalter J. and Birr Ch., Liebig's Ann. Chem., 740 (1970) 31.

[1073] Wieland T., Lüben G., Ottenheym H., Faesel J., de Vries J. X., Konz W., Prox A. and Schmid J., Angew. Chem., 80 (1968) 209.

[1074] Wieland Th., Lüben G., Ottenheym H. and Schiefer H., Liebig's Ann. Chem., 722 (1969) 173.

[1075] Wieland Th., Penke B. and Birr Ch., Liebig's Ann. Chem., 759 (1972) 71.

[1076] Wieland Th., Rietzel C. and Seeliger A., Liebig's Ann. Chem., 759 (1972) 63.

[1077] Wieland Th. and de Vries J. X., Liebig's Ann. Chem., 700 (1966) 174.

[1078] Williams R. J. P., Q. Rev., 24 (1970) 331.

[1079] Wilson D. F. and Merz R. D., Arch. Biochem. Biophys., 119 (1967) 470.

[1080] Wilzbach K., J. Am. Chem. Soc., 79 (1957) 1013.

[1081] Wipf H.-K., Olivier A. and Simon W., Helv. Chim. Acta, 53 (1970) 1605.

[1082] Wipf H.-K., Pache W., Jordan P., Zähner H., Keller-Schierlein W. and Simon W., Biochem. Biophys. Res. Commun., 36 (1969) 387.

[1083] Wipf H.-K., Pioda L. A. R., Štefanac Z. and Simon W., Helv. Chim. Acta, 51 (1968) 377.

[1084] Wipf H.-K. and Simon W., Biochem. Biophys. Res. Commun., 34 (1969) 707.

[1085] Wise W. M., Kurey M. J. and Baum G., Clin. Chem., 16 (1970) 103.

[1086] Witonsky P. and Johnson D., Fed. Proc., 23 (1964) 266.

[1087] Witt H. T., J. Bioenerg., 3 (1972) 47.

[1088] Witt H. T., Q. Rev. Biophys., 4 (1972) 365.

[1089] Wolstenholme W. A. and Vining L. C., Tetrahedron Lett., (1966) 2785.

[1090] Wong S. M., Fischer H. P. and Cram D. J., J. Am. Chem. Soc., 93 (1971) 2235.

[1091] Wong M. K., McKinney W. J. and Popov A. I., J. Phys. Chem., 75 (1971) 56.

[1092] Wong D. T., Horug J.-S., Hamill R. L. and Lardy H. A., Biochem. Pharmacol., 20 (1971) 3169.

[1093] Wong K. H., Konizer G. and Smid J., J. Am. Chem. Soc., 92 (1970) 666.

[1094] Wraight C. A. and Crofts A. R., Eur. J. Biochem., 17 (1971) 319.

[1095] Wraight C. A. and Crofts A. R., Eur. J. Biochem., 19 (1971) 386.

[1096] Wudl F., Chem. Commun., (1972) 1229.

[1097] Wudl F. and Gaeta F., Chem. Commun., (1972) 107.

[1098] Wuepper J. L. and Popov A. I., J. Am. Chem. Soc., 92 (1970) 1493.

[1099] Wurzschmidt B., Z. Anal. Chem., 130 (1951) 105.

[1100] Wyssbrod H. R., Biochim. Biophys. Acta, 193 (1969) 361.

[1101] Yamada H., Bull. Chem. Soc. Japan, 33 (1960) 666.

[1102] Yamada H., Bull. Chem. Soc. Japan, 33 (1960) 780.

[1103] Yasin Y. M. G., Hodder O. J. R. and Powell H. M., Chem. Commun., (1966) 705.

[1104] Young J. H., Blondin G. A. and Green D. E., Proc. Natl. Acad. Sci. U.S., 68 (1971) 1364.

[1105] Young J. H., Blondin G. A., Vanderkooi G. and Green D. E., Proc. Natl. Acad. Sci. U.S., 67 (1970) 550.

[1106] Zaugg H. E., Ratajczyk J. F., Leonard J. E. and Schaefer A. D., J. Org. Chem., 37 (1972) 2249.

[1107] Zavada J., Krupička J. and Sicher J., Coll. Czech. Chem. Commun., 33 (1968) 1333.

[1108] Zook H. D. and Gumby W. L., J. Am. Chem Soc., 82 (1960) 1386.

[1109] Zutphen van H., Merola A. J., Brierley G. P. and Cornwell D. G., Arch. Biochem. Biophys., 152 (1972) 755.

[1110] Zylber E. and Penman S., Proc. Natl. Acad. Sci. U.S., 68 (1971) 2861.

[1111] Biochem. J., 126 (1972) 773.

[1112] Patel D. J., Biochemistry, 12 (1973) 667.

[1113] Patel D. J., Biochemistry, 12 (1973) 677.

[1114] Tonelli A. E., Biochemistry, 12 (1973) 689.

[1115] Karle I. L., Karle J., Wieland Th., Bürgermeister W., Faulstich H. and Witkop B., Proc. Natl. Acad. Sci. U.S., 70 (1973) 1836.

ADDITIONAL REFERENCES

This list gives papers published after the manuscript had been submitted to the publishers and also includes some older papers not mentioned in the text.

To Chapters I—III

1 Alpha S. R. and Brady A. H., Optical activity and conformation of the cation carrier X-537 A, J. Am. Chem. Soc., 95 (1973) 7043.

2 Andrasko J. and Forsen S., Pulsed NMR studies on Na binding to simple carbohydrates, Biochem. Biophys. Res. Commun., 52 (1973) 233.

3 Angyal S. J., Complex formation between sugars and metal ions, Pure Appl. Chem., 35 (1973) 131.

4 Audhya T. K. and Russell D. W., Spectrophotometric determination of enniatin A and valinomycin in fungal extracts by ion complexation, Anal. Lett., 6 (1973) 265.

5 Balasubramanian D. and Shaikh R., On the interaction of lithium salts with model amide, Biopolymers, 12 (1973) 1639.

6 Bradshaw J. S., Hui J. Y., Haymore B. L., Christensen J. J. and Izatt R. M., Macrocyclic polyether sulfide syntheses. Preparation of thia-crown-3, 4 and 5 compounds, J. Heterocycl. Chem., 10 (1973) 1.

7 Ceraso J. M. and Dye J. L., ^{23}Na NMR study of exchange rates: sodium cryptate in ethylene diamine, J. Am. Chem. Soc., 95 (1973) 4432.

8 Chastrette M. and Chastrette F., Template effect in synthesis of furan compounds, Chem. Commun., (1973) 534.

9 Cook W. J. and Bugg C. E., Calcium binding to galactose. Crystal structure of a hydrated α-galactose-calcium complex with $CaBr_2$, J. Am. Chem. Soc., 95 (1973) 6442.

10 Davies J. S., Foley M. H., Hassall C. H. and Arroyo V., The biosynthetic origin of D-isoleucine in the monamycins, Chem. Commun., (1973) 782.

11 Degani H., Friedmann H. L., Navon G. and Kosower E. M., Fluorimetric complexing constants and CD measurements for antibiotic X-537 A with univalent and bivalent cations, Chem. Commun., (1973) 431.

12 Dietrich B., Lehn J. M., Sauvage J. P. and Blanzat J., Cryptates. X. Syntheses et propriétés physique de systèmes diaza-polyoxamacrobicycliques, Tetrahedron, 29 (1973) 1629.

13 Dietrich B., Lehn J. M. and Sauvage J. P., Cryptates. XI. Complexes macro-bicycliques, formation, structure, propriétés, Tetrahedron, 29 (1973) 1647.

14 Dye J. L., Lok M. T., Tehan F. J., Coolen R. B., Papadakis N., Ceraso J. M. and de Backer M. G., Alkali metal solutions. Effect of two cyclic polyethers on solubility and spectra, Ber. Bunsenges. Phys. Chem., 75 (1971) 659.

15 Eisenman G. and Krasne S., The ion selectivity of carrier molecules, membranes and enzymes, in MTP International Review of Science, Biochemistry Series, Vol. 2 (Fox C. F., ed.), Butterworth, London, in press.

16 Evans D. F., Wellington S. L., Nadis, J. A. and Cussler E. L., Conductance of cyclic polyether-cation complexes, J. Solution Chem., 1 (1972) 499.

17 Fitzgerald P. R. and Mansfield M. E., Efficacy of monensin against bovine coccidiosis in young Holstein-Friesian calves, J. Protozool., 20 (1973) 121.

18 Flora H. B. and Gilkerson W. R., Association of alkali metal cations with triphenylphosphine oxide in tetrahydrofuran solution, J. Phys. Chem., 77 (1973) 1421.

19 Frolova V. I., Rosynov B. V. and Kusovkov A. D., Isolation and identification of the antibiotic nonactin, Antibiotiki (U.S.S.R.), (1973) 777.

20 Greene R. N. (E. I. du Pont de Nemours and Co.), Macrocyclic polyethers, Ger. Offen. 2 153 844 (Cl. C 07d), 15 VI 1972; Chem. Abstr., 77 (1972) 114450.

21 Grell E. and Funck Th., Dynamic properties and membrane activity of ion specific antibiotics, J. Supramol. Struct., (1973) 307.

22 Groth P., Crystal structure of a complex between potassium p-toluenesulfonate and 1,4,7,10,13,16-hexaoxacyclooctadecane, Acta Chem. Scand., 25 (1971) 3189.

23 Ivanov V. T., Evstratov A. V., Sumskaya L. V., Melnik E. I., Chumburidze T. S., Portnova S. L., Balashova T. A. and Ovchinnikov Yu. A., Sandwich complexes as a functional form of the enniatin ionophores, FEBS Lett., 36 (1973) 65.

24 Helgeson R. C., Timko J. M. and Cram D. J., Structural requirements for cyclic ethers to complex and lipophylize metal cations and amino acids, J. Am. Chem. Soc., 95 (1973) 3023.

25 Karagounis G. and Pandi-Agathoki J., Formation of polymeric catenanes by selective interfacial reactions, Prakt. Akad. Athenon, 45 (A-B-G) (1970) 118; Chem. Abstr., 77 (1972) 165111.

26 Kimmich G. A., Coupling between Na^+ and sugar transport in small intestine, Biochim. Biophys. Acta, 300 (1973) 31.

27 Koryta J. and Mittal M. L., Electroreduction of monovalent metal ion complexes of macrocyclic polyethers, J. Electroanal. Chem. Interfacial Electrochem., 36 (1972) App. 14.

28 Kubota T., Mayama M. and Matsutani Sh. (Shionogu and Co.), Polyetherin A 29-ethers, Japan 72 14 224 (Cl. C 07d, A 61k, A 01kn, C 12k), 27 IV 1972; Chem. Abstr., 77 (1973) 48251.

29 Kyba E. P., Siegel M. G., Sousa L. R., Sogah G. D. Y. and Cram D. J., Chiral, hinged and functionalized multiheteromacrocycles, J. Am. Chem. Soc., 95 (1973) 2691.

30 Layton A. J., Mallinson P. R., Parsons D. G. and Truter M. R., Synthesis and crystal structure of complexes between caesium thiocyanate and two isomeric macrocyclic crown polyethers, 6,7,9,10,17,18,20,21-octahydro-7R,9R,18S,20S-tetramethyl-dibenzo-[b,k]-1,4,7,10,13,16-hexaoxacyclodecin (isomer F) and 6,7,9,10,17,18,20, 21-octahydro-7R,9R,18R,20R-tetramethyldibenzo[b,k]-1,4,7,10,13,16-hexaoxacyclo-decin (isomer B), Chem. Commun., (1973) 654.

31 Layton D. and Symmons P., The effect of some uncoupling agents, ionophorous agents and inhibitors on the fluorescence of ANS bound to bovine serum albumin, FEBS Lett., 30 (1973) 325.

32 Lehn J. M., Simon J. and Wagner J., Mesomolecules. Polyaza-, polyoxa-macro-polycyclic systems, Angew. Chem. (Int. Ed.), 12 (1973) 578.

33 Lehn J. M., Simon J. and Wagner J., Molecular and cation complexes of macro-tricyclic and macrotetracyclic ligands, Angew. Chem. (Int. Ed.), 12 (1973) 579.

34 Lockhart J. C., Robson A. C., Thompson M. E., Sister D. E., Kaura C. K. and Allan A. R., Preparation of nitrogen-containing polyether crown compounds, J. Chem. Soc., Perkin Trans. (I), (1973) 577.

35 Madison V., Conformational energy and circular dichroism computed for ⌐(Gly—L-Pro)₃⌐, Biopolymers, 12 (1973) 1837.

36 Maigret B. and Pullman B., Molecular orbital study on the conformation of enniatin B, Biochem. Biophys. Res. Commun., 50 (1973) 908.

37 Mathieu F. and Weiss R., Transition metal cryptates: the crystal and molecular structure of a cobalt (II) cryptate, $[Co(C_{16}H_{32}N_2O_5)][Co(SCN)_4]$, Chem. Commun., (1973) 816.

38 Max N. L., Spherical trigonometry and the structure of valinomycin, Biopolymers, 12 (1973) 1565.

39 Mellinger M., Fischer J. and Weiss R., New cryptates containing tricyclic ligands. Crystal structure of $C_{32}H_{64}N_4O_{16} \cdot 2NaI$, Angew. Chem. (Int. Ed.), 12 (1973) 771.

40 Mercer M. and Truter M. R., Crystal structures of complexes between alkali-metal salts and cyclic polyethers. Part VI. Complex formed between dicyclohexyl-18-crown-6, isomer B (perhydrodibenzo[b,k]-1,4,7,10,13,16-hexaoxacyclooctadecin) and sodium bromide, J. Chem. Soc., Dalton Trans., (1973) 2215.

41 Mercer M. and Truter M. R., Crystal structures of complexes between alkali-metal salts and cyclic polyethers. Part VII. Complex formed between dibenzo-24-crown-8 (6,7,9,10,12,13,20,21,23,24,26,27-dodecahydrodibenzo[b,n]-1,4,7,10,13,16,19,22-octaoxacyclotetracosin) and two molecules of potassium isothiocyanate, J. Chem. Soc., Dalton Trans., (1973) 2469.

42 Metz B., Moras D. and Weiss R., Coordination des cations alkalino-terreux dans leurs complexes avec des molécules macrobicycliques. I. Structure crystalline et moléculaire du cryptate de calcium, $C_{18}H_{36}N_2O_6 \cdot CaBr_2 \cdot 3H_2O$, Acta Cryst., B29 (1972) 1377.

43 Metz B., Moras D. and Weiss R., Coordination des cations alkalino-terreux dans leurs complexes avec des molécules macrobicycliques. II. Structure crystalline et moléculaire du cryptate de barium, $C_{18}H_{36}N_2O_6 \cdot Ba(SCN)_2 \cdot H_2O$, Acta Cryst., B29 (1973) 1382.

44 Metz B., Moras D. and Weiss R., Coordination des cations alkalino-terreux dans leurs complexes avec des molécules macrobicycliques. III. Structure crystalline et moléculaire du cryptate de barium, $C_{20}H_{40}N_2O_7 \cdot Ba(SCN)_2 \cdot 2H_2O$, Acta Cryst., B29 (1973) 1388.

45 Moras D., Metz B. and Weiss R., Etude structurale des cryptates. I. Structure crystalline et moléculaire du cryptate de potassium, $C_{18}H_{36}N_2O_6 \cdot KI$, Acta Cryst., B29 (1973) 383.

46 Moras D., Metz B and Weiss R., Etude structurale des cryptates, II. Structure crystalline et moléculaire des cryptates de rubidium et de césium, $C_{18}H_{36}N_2O_6 \cdot RbSCN \cdot H_2O$ et $C_{18}H_{36}N_2O_6 \cdot CsSCN \cdot H_2O$, Acta Cryst., B29 (1973) 388.

47 Moras D. and Weiss R., Etude structurale des cryptates. III. Structure crystalline et moléculaire du cryptate de sodium $C_{18}H_{36}N_2O_6 \cdot NaI$, Acta Cryst., B29 (1973) 396.

48 Moras D. and Weiss R., Etude structurale des cryptates. IV. Structure crystalline et moléculaire du cryptate de lithium, $C_{18}H_{36}N_2O_6 \cdot LiI$, Acta Cryst., B29 (1973) 1059.

49 Moras D. and Weiss R., Etude structurale des cryptates. V. Structure crystalline et moléculaire du cryptate de thallium, $C_{18}H_{36}N_2O_6 \cdot HCOOTl \cdot H_2O$, Acta Cryst., B29 (1973) 400.

50 Newkome G. R. and Robinson J. M., Multidentate chelating agents: macrocyclic azaethers, Chem. Commun., (1973) 831.

51 Pedersen Ch. J. (E. I. du Pont de Nemours and Co.), Macrocyclic polyethers with aromatic groups and their cationic complexes, Fr. 1 440 716 (Cl. C 08g), June 1966; Chem. Abstr., 66 (1967) 46442.

52 Pedersen Ch. J., Macrocyclic polyethers. Dibenzo-18-crown-6 polyether and dicyclo-hexyl-18-crown-6 polyether, Org. Synth., 52 (1972) 66.

53 Perricaudet M. and Pullman A., SCF *ab initio*, molecular orbital study on the relative affinities of peptide and ester carbonyl groups for Na^+ and K^+ ions, FEBS Lett., 34 (1973) 222.

54 Pressman B. C., Properties of ionophores with broad range cation selectivity, Fed. Proc., 32 (1973) 1698.

55 Rothschild K. J. *et al.*, Raman spectroscopic evidence for two conformations of uncomplexed valinomycin in solid state, Science, 182 (1973) 384.

56 Shchori E. and Jagur-Grodzinski J., A conductometric study of complexation of macrocyclic polyethers with sodium salts, Israel J. Chem., 11 (1973) 243.

57 Shchori E., Jagur-Grodzinski J. and Shporer M., Kinetics of complexation of macrocyclic polyethers with sodium ions by NMR. II. Solvent effects, J. Am. Chem. Soc., 95 (1973) 3842.

58 Soloviéva N. K. and Fadeeva N. P., An organism producing nonactin, Antibiotiki (U.S.S.R.), (1973) 600.

442

59 Stahl P. and Pape H., Metabolic products of microorganisms. 110. Biosynthesis of macrotetrolides. 111. Isolation of free nonactinic acids and their function as precursors of macrotetrolides, Arch. Microbiol., 85 (1972) 239.

60 Starcher B. C. and Urry D. W., Elastin as a matrix for coacervation, Biochem. Biophys. Res. Commun., 53 (1973) 210.

61 Story P. R. and Busch P., Modern methods for the synthesis of macrocyclic compounds, Adv. Org. Chem., 8 (1972) 67.

62 Takahashi H., Hirano S., Sagawa T. and Togashi K. (Chugai Pharm. Co.), Insect sterilant, Japan Kokai 73 04 628 (Cl. 30 F25, 30 F34, 30 F922), Jan. 1973; Chem. Abstr., 79 (1973) 1368u.

63 Talekar S. V. and Sundaram K., Ion binding by macrocyclic depsipeptide antibiotics. Mutual replaceability of amide and ester groups, Adv. Exp. Med. Biol., 21 (1972) 9.

64 Titlestad K., Groth P., Dale J. and Ali M. Y., Unique conformation of the cyclic octapeptide of sarcosine and a related depsipeptide, Chem. Commun., (1973) 346.

65 Urry D. W., Cunningham W. D. and Ohnishi T., A neutral polypeptide-calcium ion complex, Biochim. Biophys. Acta, 292 (1973) 853.

66 Watts M. T., Lu M. L. and Eastman M. P., Electron spin resonance studies of Heisenberg spin exchange. The effect of macrocyclic polyethers on the spin exchange rate for ion pairs, J. Phys. Chem., 77 (1973) 625.

67 Westley J. W., Oliveto E. P., Berger J., Evans R. E., Glass R., Stempel A., Toome V. and Williams Th., Chemical transformations of antibiotic X-537 A and their effect on antibacterial activity, J. Med. Chem., 16 (1973) 397.

68 Westley J. W., Evans R. H., Williams T. and Stempel A., Pyrolytic cleavage of antibiotic X-537 A and related reactions, J. Org. Chem., 38 (1973) 3431.

69 Wiest R. and Weiss R., Cryptates with macrotricyclic ligands. The crystal and molecular structure of the silver [3]-cryptate, $C_{24}H_{48}N_4O_6 \cdot 2AgNO_3$, Chem. Commun., (1973) 678.

70 Züst Ch., Früh P. U. and Simon W., Complex formation of macrotetrolide carrier antibiotics with cations studied by microcalorimetry and vapor pressure osmometry, Helv. Chim. Acta, 56 (1973) 495.

To Chapter IV

71 Barney A. L. and Hornsberg W. (E. I. du Pont de Nemours and Co.), Vulcanization of saturated fluorinated elastomeric polymers, Ger. Offen. 1 942 675 (Cl. C 08f), Sept. 1970; Chem. Abstr., 71 (1970) 131833.

72 Barney A. L. and Hornsberg W. (E. I. du Pont de Nemours and Co.), Vulcanization of fluorinated polymers, Fr. 1 577 005 (Cl. C 08f), August 1969; Chem. Abstr., 72 (1970) 112540.

73 Berger C., Solid electrolyte device, U.S. 3 704 174 (Cl. 136-153; H 01m), Nov. 1972; Chem. Abstr., 78 (1973) 37237.

74 Del Cima F., Biggi G. and Pietra F., Origin of high ortho-para reactivity ratios in reactions of fluoronitrobenzenes with potassium tert-butoxide in tert-butyl alcohol, J. Chem. Soc., Perkin Trans.(II), (1973) 55.

75 Dockx J., Quarternary ammonium compounds in organic synthesis, Synthesis, (1973) 441.

76 Eksborg S., Lagerström P. O., Modin R. and Schill G., Ion pair chromatography of organic compounds, J. Chromatogr., 83 (1973) 99.

77 Eksborg S. and Schill G., Ion pair partition chromatography of organic ammonium compounds, Anal. Chem., 45 (1973) 2092.

78 Gokel G. W. and Cram D. J., Molecular complexation of arenediazonium and benzoyl cations by macrocyclic polyethers, Chem. Commun., (1973) 481.

79 Guthrie R. D., Jaeger D. A., Meister W. and Cram D. J., Electrophilic substitution at saturated carbon. XLVIII. High stereospecificity in a transamination reaction, J. Am. Chem. Soc., 93 (1971) 5137.

80 Harrey D. H., E1cB reactions. Stereochemistry and the counterion, J. Am. Chem. Soc., 93 (1971) 2348.

81 Helgeson R. C., Koga K., Timko J. M. and Cram D. J., Complete optical resolution by differential complexation in solution between a chiral cyclic polyether and an α-amino acid, J. Am. Chem. Soc., 95 (1973) 3021.

82 Kyba E. P., Koga K., Sousa L. R., Siegel M. G. and Cram D. J., Chiral recognition in molecular complexing, J. Am. Chem. Soc., 95 (1973) 2692.

83 Leveque A. and Rosset R., Cryptates in analytical chemistry, Analysis, 2 (1973) 218.

84 Martin J. C. and Bloch D. R., The dehydrocyclopentadienyl anion. A new aryne, J. Am. Chem. Soc., 93 (1971) 451.

85 Matsuda T. and Koida K., Effect of crown ether on the reduction of ketones by $NaBH_4$ in aromatic solvent, Bull. Chem. Soc. Japan, 46 (1973) 2253.

86 Rais J. and Seluoky P., New extraction agents for cesium. III. Complex formed between some cesium salts and dibenzo-18-crown-6, Radiochem. Radioanal. Lett., 6 (1971) 254.

87 Saraie T., Ishiguro T., Kowashima K. and Morita K., A new synthesis of nitriles, Tetrahedron Lett., (1973) 2121.

88 Stark C. M. and Owens R. M., Phase-transfer catalysis. II. Kinetic details of cyanide displacement on 1-halooctanes, J. Am. Chem. Soc., 95 (1973) 3613.

89 Vest R. D. (E. I. du Pont de Nemours and Co.), Composite plastic coatings from polysilicic acids, hydroxyalkyl vinyl ether-perfluoroolefin copolymer, and cyclic polyethers, Fr. 1 557 958 (Cl. C 09d), Febr. 1969; Chem. Abstr., 71 (1969) 82775.

90 Vest R. D. (E. I. du Pont de Nemours and Co.), Polysilicic acid (hydroxyalkyl vinyl ether-polyfluoroolefin copolymer) alkali metal compositions containing cyclic polyethers as coatings to improve the adhesion and to increase the resistance of plastics to attack by water and aqueous alkali, U.S. 3 546 318 (Cl. C 08g), Dec. 1970; Chem. Abstr., 74 (1971) 65662.

91 Wong K.-H., Complexes of alkali carbanion pairs with macrocyclic polyethers, and their kinetic behaviour in proton abstraction reaction, Diss. Abstr. Int., B33 (1972) 1062; Chem. Abstr., 78 (1973) 42390.

To Chapter V

92 Aityan S. Kh. and Chizmadjev Yu. A., The diffusion equation for ion-conducting channels in biomembranes, in The Biophysics of Membranes (Russian) (Abstr. of Communs at the U.S.S.R. Conference on Membrane Biophysics, Palanga, Sept. 1973), Kaunas Medical Institute, Kaunas, 1973, p. 30.

93 Aityan S. Kh., Markin V. S. and Chizmadjev Yu. A., Theory of the passage of alternating current through synthetic membranes in the scheme of carriers, Biofizika (U.S.S.R.), 18 (1973) 75.

94 Bakker E. P., van den Heuvel E. J., Wiechmann A. H. C. A. and van Dam K., A comparison between the effectiveness of uncouplers of oxidative phosphorylation in mitochondria and in different artificial membrane systems, Biochim. Biophys. Acta, 292 (1973) 78.

95 Carmack G. D. and Freiser H., Conductance of quaternary ammonium salt dispersions in polymeric films, Anal. Chem. 45 (1973) 1975.

96 Ciani S., Laprade R. and Eisenman G., Theory for carrier mediated zero-current conductance of bilayers extended to allow for non-equilibrium of interface reaction, spatially dependent mobility and barrier shape, J. Membrane Biol., 11 (1973) 255.

444

97 Demin V. V., Babakov A. V., Shkrob A. M. and Ovchinnikov Yu. A., Conductance of bilayer lipid membranes in the presence of valinomycin and lipophilic anions, Biofizika (U.S.S.R.), in the press.

98 Eisenman G., Szabo G., McLaughlin S. G. A. and Ciani S. M., Molecular basis for the action of macrocyclic ion carriers on passive ionic translocation across bilayer lipid membranes, J. Bioenerg., 4 (1973) 93.

99 Fielder U. and Ružička J., Selectrode — the universal ion-selective electrode. Part VII. A valinomycin-based potassium electrode with non-porous polymer membrane and solid state inner reference system, Anal. Chim. Acta, 67 (1973) 179.

100 Finer E. G., Hauser H. and Chapman D., High resolution NMR studies of alamethicin and valinomycin, Magn. Resonances Biol. Res., Rep. Int. Conf., 1969 (Franconi, Cafiero, eds), Gordon and Breach, New York, 1971, p. 337; Chem. Abstr., 78 (1973) 54406.

101 Fernandez M. S., Celis C. H. and Montal M., PMR detection of ionophore-mediated transport of praseodimium ions across phospholipid membranes, Biochim. Biophys. Acta, 323 (1973) 600.

102 Gainullin R. Z. and Nenashev V. A., Diffusion of the ionophoric antibiotics and phosphorylation uncouplers through lipid bimolecular membranes, see p. 160 of Ref. 92.

103 Gambale F., Gliozzi A. and Robello M., Determination of rate constants in carrier-mediated diffusion through lipid bilayers, Biochim. Biophys. Acta, 330 (1973) 325.

104 Goodall M. C., A synthetic transmembrane channel, Biochim. Biophys. Acta, 291 (1973) 317.

105 Goodall M. C., Action of two classes of channel forming synthetic peptides on lipid bilayers, Arch. Biochem. Biophys., 157 (1973) 514.

106 Gordon L. G. M., Conductance channels in neutral lipid bilayers. I, II, J. Membrane Biol., 12 (1973) 207 and 217.

107 Jagur-Grodzinski J., Marian S. and Vofsi D., The mechanism of a selective permeation of ions through "solvent polymeric material" membranes, Separation Sci., 8 (1973) 33.

108 Haydon D. A. and Myers V. B., Surface charge, surface dipoles, and membrane conductance, Biochim. Biophys. Acta, 307 (1973) 429.

109 Hinkle P. C., Electron transfer across membranes and energy coupling, Fed. Proc., 32 (1973) 1988.

110 Hladky S. B., The effect of stirring on the flux of carriers into black lipid membranes, Biochim. Biophys. Acta, 307 (1973) 261.

111 Hladky S. B. and Haydon D. A., Membrane conductance and surface potential, Biochim. Biophys. Acta, 318 (1973) 464.

112 Hsu M. and Chan S. I., NMR studies of the interaction of valinomycin with unsonicated lecithin bilayers, Biochemistry, 12 (1973) 3872.

113 Huang H. W., Kinetic theory of antibiotic ion carrier. III, J. Theor. Biol., 38 (1973) 191.

114 Kemp G. and Wenner C., Cation binding by valinomycin and trinactin at the air—water interface: cooperativity in cation binding by valinomycin, Biochim. Biophys. Acta, 323 (1973) 161.

115 Kuo K.-H. and Bruner L. J., Uncoupler antagonism of valinomycin-induced bilayer membrane conductance, Biochem. Biophys. Res. Commun., 52 (1973) 1079.

116 Lambert P. A. and Hammond S. M., Potassium flux. First indications of membrane damage in microorganisms, Biochem. Biophys. Res. Commun., 54 (1973) 796.

117 Läuger P., Ion transport through pores: a rate-theory analysis, Biochim. Biophys. Acta, 34 (1973) 423.

445

118 Läuger P. and Neumcke B., Theoretical analysis of ion conductance in lipid bilayer membranes, in Membranes — A Series of Advances (Eisenman G., ed.), Vol. 2, Marcel Dekker, New York, 1973, p. 1.

119 Leonard F., Nelson J. and Wade C. W. R., Calcium transport through CCl_4 solution of vitamin D_3, Calcified Tissue Res., 12 (1973) 113.

120 Markin V. S., Current—voltage characteristics and impedance of artificial phospholipid membranes taking into account non-stirred aqueous layers, Dokl. Acad. Nauk S.S.S.R. (U.S.S.R.), 202 (1973) 703.

121 Markin V. S. and Liberman E. A., Theory of the relaxation current in clamp experiments on the membrane with ion carrier, Biofizika (U.S.S.R.), 18 (1973) 453.

122 Mauro A., Nanavati R. P. and Heyer E., Time-variant conductance of bilayer membranes treated with monazomycin and alamethicin, Proc. Natl. Acad. Sci. U.S., 69 (1972) 3742.

123 McLaughlin S.G.A., Salicylates and phospholipid bilayer membranes, Nature, 243 (1973) 234.

124 Montal M. and Mueller P., Formation of bimolecular membranes from lipid monolayers and a study of their electrical properties, Proc. Natl. Acad. Sci. U.S., 69 (1972) 3561.

125 Niedrach L. W. (General Electric Co.), Ion-specific electrode, Ger. Offen. 2 251 287 (Cl.G 01n), Apr. 1973; Chem. Abstr., 78 (1973) 168254.

126 Pick J., Toth K., Pungor E., Vasák M. and Simon W., A. potassium-selective silicon-rubber membrane electrode based on a neutral carrier, Anal. Chim. Acta, 64 (1973) 481.

127 Ryba O., Kuizakova E. and Petranek J., Potassium polymeric membrane electrodes based on neutral carriers, Coll. Czech. Chem. Commun., 38 (1973) 497.

128 Shamoo A. E. and Albers R. W., Na^+ selective ionophoric material derived from electrical organ and kidney membrane, Proc. Natl. Acad. Sci. U.S., 70 (1973) 1191.

129 Shkrob A. M., Demin V. V. and Ovchinnikov Yu. A., The effect of potential jump decreasing modifiers on the electric properties of bilayer lipid membranes containing synthetic analogs of valinomycin, Biofizika (U.S.S.R.), in the press.

130 Shkrob A. M., Demin V. V. and Ovchinnikov Yu. A., The compounds decreasing electric potential boundary jump: a new class of membrane modifiers, Biofizika (U.S.S.R.), in the press.

131 Shlyachter T. A., Shchagina L. V. and Lev A. A., Study of the interface properties in the system heptanic solution of the valinomycin/aqueous solution of alkali metal salts, see p. 662 of Ref. 92.

132 Sokolova A. E., Malev V. V. and Lev A. A., The relative contribution of the molecular and micellar forms to the penetration of bulk membranes by valinomycin, see p. 581 of Ref. 92.

133 Stark G., Rectification phenomena in carrier-mediated ion transport, Biochim. Biophys. Acta, 298 (1973) 323.

134 Ting-Beall H. P., Tosteson M. T., Gisin B. F. and Tosteson D. C., The effect of peptide PV on the ionic permeability of lipid bilayer membranes, unpublished observations.

135 Tosteson D. C., Ion transport across thin lipid bilayer membranes, in Alfred Benzon Symposium V. Transport mechanisms in epithelia (Ussing H. H. and Thorn N. A., eds), Munksgaard, Copenhagen, 1973, p. 346.

136 Wu S. H. and McConnel H. M., Lateral phase separation and perpendicular transport in membranes, Biochem. Biophys. Res. Commun., 55 (1973) 484.

137 Wuhrmann H.-E., Morf W. and Simon W., Modellberechnung der EMK und der Ionenselektivität von Membranelectrodenmessketten, Helv. Chim. Acta, 56 (1973) 1011.

446

138 Wuhrmann P., Thoma A. P. and Simon W., Calcium carrier properties of neutral synthetic ligands in bulk membranes, Chimia, 27 (1973) 637.

To Chapter VI

139 Adamyan S. Ya., Martirosov S. M. and Simonyan A. L., Membrane potentials of muscle in the presence of phosphorylation uncouplers, Biofizika (U.S.S.R.), 18 (1973) 163.
140 Aksenova O. P. and Antonov V. F., Action of some antibiotics on the electrical properties of giant neurons of medical leeches, see p. 36 of Ref. 92.
141 Asghar S. S., Levin E. and Harold F. M., Accumulation of neutral amino acids by S. faecalis. Energy coupling by a proton-motive force, J. Biol. Chem., 248 (1973) 5225.
142 Astashkin E. I. and Antonov V. F., Valinomycin-induced coupled ion transport in erythrocytes, see p. 46 of Ref. 92.
143 Bamberger E. S., Rottenberg H. and Avron M., Internal pH, Δ pH, and the kinetics of electron transport in chloroplasts, Eur. J. Biochem., 34 (1973) 557.
144 Beatie D. S. and Ibrahim N. G., Optimal conditions for amino acid incorporation by isolated rat liver mitochondria. Stimulation by valinomycin and other agents, Biochemistry, 12 (1973) 176.
145 Breitbart H. and Herzberg M., Membrane-mediated inhibition of protein synthesis by valinomycin in reticulocytes, FEBS Lett., 32 (1973) 15.
146 Brierley G. P., Jurkowitz M. and Scott K. M., Ion transport in heart mitochondria. XXVII. The relation of mercurial-dependent ATPase activity to ion movements, Arch. Biochem. Biophys., 159 (1973) 742.
147 Bromm B., Valinomycin as a potassium carrier through Ranvier's node membrane, in Proc. Eur. Biophys. Congr. (Broda E., ed.), Wiener Med. Acad., Vienna, 1971; Chem. Abstr., 77 (1972) 15673.
148 Cereijo-Santalo R., The effects of non-electrolytes on the DNP-activated ATPase of rat liver mitochondria, Arch. Biochem. Biophys., 148 (1972) 22.
149 Cereijo-Santalo R., The effects of non-electrolytes on the DNP-activated ATPase of rat liver mitochondria, Arch. Biochem. Biophys., 150 (1972) 542.
150 Cockrell R. S., The influence of permeant anions upon mitochondrial energy conservation by means of a K^+ gradient, Biochem. Biophys. Res. Commun., 46 (1973) 1991.
151 Cockrell R. S., Energy-linked ion translocation in submitochondrial particles, J. Biol. Chem., 248 (1973) 6828.
152 Cogdell R. J. and Crofts A. R., Effect of antimycin A and 1,10-phenanthroline on rapid hydrogen ion-uptake by chromatophores from R. spheroides, Photosynthesis, Two Centuries after its Discovery, J. Priestley, Proc. Int. Congr. Photosynth. Res. II. (Forti, Giorgio, eds), The Hague, Netherlands, 2 (1972) 977; Chem. Abstr., 78 (1973) 119652.
153 Dalmasso A. P. and Lelchuk R., Effect of alkali cations on the interaction between detergents and erythrocyte membrane, Experientia, 29 (1973) 288.
154 Deol B. S., Bermingham M. A. C., Still J. L., Haydon D. A. and Gale E. F., The action of serratamolide on ion movement in lipid bilayers and biomembranes, Biochim. Biophys. Acta, 330 (1973) 192.
155 Dorigo P., Gaion R. M., Toth E. and Fassina G., Effect of ionoforous antibiotics (nigericin, gramicidin, valinomycin) on cyclic AMP synthesis in fat adipose tissue, Biochem. Pharm., 22 (1973) 1949.
156 Entman M. L., Allen J. C., Bornet E. P., Gillette P. C., Wallick E. T. and Schwartz A., Mechanisms of calcium accumulation and transport in cardiac relaxing system (sarcoplasmic reticulum membranes): effects of verapamil, D-600, X-537 A and A23187, J. Mol. Cell. Cardiol., 4 (1972) 681.

157 Erecinska M., Wilson D. F., Dutton P. L. and Chance B., Kinetic interactions at site II during energy coupling reactions, Fed. Proc., 32 (1973) 1981.

158 Ernster L., Juntii K. and Asami K., Mechanisms of energy conservation in the mitochondrial membrane, J. Bioenerg., 4 (1973) 149.

159 Evans E. H. and Crofts A. R., The relationship between delayed fluorescence and the pH gradient in chloroplasts, Biochim. Biophys. Acta, 292 (1973) 130.

160 Ganser A. L. and Forte J. G., Ionophoretic stimulation of K^+ ATPase of oxyntic cell microsomes, Biochem. Biophys. Res. Commun., 54 (1973) 690.

161 Gemba M. and Ueda J., Repression of antimicin A on mitochondrial ATPase of rat kidney cortex, Jap. J. Pharmacol., 21 (1971) 839.

162 Gibb L. E. and Eddy A. A., An electrogenic sodium pump as a possible factor leading to the concentration of amino acids by mouse ascites-tumour cells with reversed sodium ion concentration gradient, Biochem. J., 129 (1972) 979.

163 Gómez-Puyou A., Sandoval F., Chávez E., Freites D. and Gómez-Puyou M. T., Dependency of the ATPase and 32P-ATP exchange reaction of mitochondria on K^+ and electron transport, Arch. Biochem. Biophys., 153 (1972) 215.

164 Granger M. and Harris E. J., Use of a continuous assay system to show the stimulation of phosphoenolpyruvate production in intact mitochondria by valinomycin and by Mn^{2+}, J. Bioenerg., 2 (1971) 151.

165 Grinius L. L. Kadzyauskas Yu. P., Lokene R. and Markevichute I., Membrane potential and pH gradient in the transport of penetrating ions in mitochondria, see p. 210 of Ref. 92.

166 Heldt H. W., Werdan K., Milovancev M. and Geller G., Alkalization of the chloroplast stroma, caused by light-dependent proton flux into the thylakoid space, Biochim. Biophys. Acta, 314 (1973) 224.

167 Hirata H., Altendorf K. and Harold F. M., Role of an electrical potential in the coupling of metabolic energy to active transport by membrane vesicles of *E. coli*, Proc. Natl. Acad. Sci. U.S., 70 (1973) 1804.

168 Kashket E. R. and Wilson T. H., Proton-coupled accumulation of galactoside in *S. lactis* 7962, Proc. Natl. Acad. Sci. U.S., 70 (1973) 2866.

169 Larkum A. W. D. and Bonner W. D., Light-induced absorbance changes of cytochromes and other pigments in pea chloroplast fragments. II. Effect of inhibitors, uncouplers and ionophores, Arch. Biochem. Biophys., 153 (1972) 249.

170 Lee P.-F. and Lam K.-W., Effect of valinomycin on intraocular pressure, Ann. Ophthalmol., 5 (1973) 33 and 38.

171 Leung J. and Eisenberg R. S., The effects of the antibiotics gramicidin A, amphotericin B and nystatin on the electrical properties of frog skeletal muscle, Biochim. Biophys. Acta, 298 (1973) 718.

172 Levy I. V., Cohen I. A. and Inesi G., Contractile effects of a calcium ionophore, Nature, 242 (1973) 461.

173 Lin D. C. and Kun E., Mode of action of the antibiotic X-537 A on mitochondrial glutamate oxidation, Biochem. Biophys. Res. Commun., 50 (1973) 820.

174 Lin D. C. and Kun E., Inhibition of the oxidation of glutamate and isocitrate in liver mitochondria at a specific $NADP^+$-reducing site, Proc. Natl. Acad. Sci. U.S., 70 (1973) 3450.

175 Lombardi F., Reeves I. P. and Kaback H. R., Mechanism of active transport in isolated bacterial membrane vesicles. XIII. Valinomycin-induced rubidium transport, J. Biol. Chem., 248 (1973) 3551.

176 Martirosov S. M., Alikhanyan M. A. and Petrosyan L. S., Transport systems of glycolizing bacteria, see p. 78 of Ref. 92.

177 Massari S., Balboni E. F. and Azzone G., Distribution of permeant cations in rat liver mitochondria under steady-state conditions, Biochim. Biophys. Acta, 238 (1972) 16.

178 Meissner G. and Fleischer S., Ca^{2+} uptake in reconstituted sarcoplasmic reticulum vesicles, Biochem. Biophys. Res. Commun., 52 (1973) 913.
179 Miller R. N., Smith E. E. and Hunter F. E., Halothane-induced alterations in energy-dependent, and energy-independent membrane carrier functions in isolated rat liver mitochondria with some electron microscopic correlations, Cell Biol. Toxicity Anesth., Proc. Res. Symp. 1970 (Fink and Raymond, eds), Williams and Wilkins, Baltimore, 1972, p. 93; Chem. Abstr., 77 (1972) 122500.
180 Mitchell P., Performance and conservation of osmotic work by proton coupled solute porter systems, J. Bioenerg., 4 (1973) 63.
181 Morel F. M., A study of passive potassium efflux from human red blood cells using ion-specific electrodes, J. Membrane Biol., 12 (1973) 69.
182 Muravieva T. I., Ryabova I. D., Oreshnikova N. A. and Novikova M. A., The effect of valinomycin on the respiration and K^+ transport in the yeast mitochondria, Biokhimiya (U.S.S.R.), 38 (1973) 845.
183 Muravieva T. I., Ryabova I. D., Oreshnikova N. A. and Novikova M. A., The effect of valinomycin on yeasts, Biokhimiya (U.S.S.R.), 42 (1973) 83.
184 Nicholls D. G. and Lindberg O., Inhibited respiration and ATPase activity of rat liver mitochondria under conditions of matrix condensation, FEBS Lett., 25 (1972) 61.
185 Niven D. F., Jeacocke R. E. and Hamilton W. A., The membrane potential as the driving force for the accumulation of lysine by S. aureus, FEBS Lett., 29 (1973) 248.
186 Niven D. F. and Hamilton W. A., Valinomycin-induced amino acid uptake by S. aureus, FEBS Lett., 37 (1973) 244.
187 Papa S., Guerrieri F., Simone S. and Lorusso M., Action of local anaesthetics on passive and energy-linked ion translocation in the inner mitochondria membrane, J. Bioenerg., 3 (1973) 553.
188 Papa S., Guerrieri F., Simone S., Lorusso M. and Larosa D., Mechanism of respiration-driven proton translocation in the inner mitochondrial membrane, Biochim. Biophys. Acta, 292 (1973) 20.
189 Papa S., Storey B. T., Lorusso M., Lee C. P. and Chance B., Energy-linked swelling of EDTA submitochondrial particles, Biochem. Biophys. Res. Commun., 52 (1973) 1395.
190 Poole D. T., Butler T. C. and Williams M. E., Intracellular pH and glycolysis in Ehrlich ascites tumor cells. Effect of membrane-active drugs in potassium depletion, J. Natl. Cancer Inst., 49 (1972) 1659.
191 Portis A. R. and McCarty R. E., On the pH-dependence of the light-induced hydrogen ion gradient in spinach chloroplasts, Arch. Biochem. Biophys., 156 (1973) 621.
192 Postma P. W., Visser A. S. and van Dam K., The movement of cations connected with metabolism in Azotobacter vinelandii, Biochim. Biophys. Acta, 298 (1973) 341.
193 Prince W. T., Rasmussen H. and Berridge M. J., The role of calcium in fly salivary gland secretion analyzed with the ionophore A 23187, Biochim. Biophys. Acta, 329 (1973) 98.
194 Scarpa A., Baldassare J. and Inesi G., Effect of calcium ionophores on a fragmented sarcoplasmic reticulum, J. Gen. Physiol., 60 (1972) 735.
195 Schuldiner S., Rottenberg H. and Avron M., Simulation of ATP synthesis by a membrane potential in chloroplasts, Eur. J. Biochem., 39 (1973) 455.
196 Seshachalam D., Frahm D. H. and Ferraro F. M., Cation reversal of inhibition of growth by valinomycin in Streptococcus pyogenes and Clostridium sporogenes, Antimicrob. Agents Chemother., 3 (1973) 63.
197 Terry P. M. and Vidaver G. A., The effect of gramicidin and Na^+-accumulation of glycine by pigeon red cells. A test of the cation gradient hypothesis, Biochim. Biophys. Acta, 323 (1973) 441.

198 Uribe E., ATP synthesis driven by a K^+ valinomycin-induced charge imbalance across chloroplast grana membranes, FEBS Lett., 36 (1973) 143.

199 Uribe E. F. and Li B. C. Y., Stimulation and inhibition of membrane dependent ATP synthesis in chloroplasts by artificially induced K^+ gradients, J. Bioenerg., 4 (1973) 435.

200 Utsumi K., Pereirae J., Mustafa M. G. and Oda T., Changes of proton gradient in mitochondria at various energy states, Acta Med. Okayama, 25 (1971) 493.

201 Vainio H., Baltscheffsky M., Baltscheffsky H. and Azzi A., Energy dependent changes in membranes of Rh. rubrum chromatophores as measured by 8-anilino-1-naphthalene sulfonic acid, Eur. J. Biochem., 30 (1972) 301.

202 Van der Neut-Kok E. C. M., De Gier J., Middlebeek E. J. and Van Deenen L. L. M., Valinomycin-induced potassium and rubidium permeability of intact cells of Acholeplasma laidlawii B, Biochim. Biophys. Acta, 332 (1974) 97.

203 Visser A. S. and Postma P. W., Permeability of Azotobacter vinelandii to cations and anions, Biochim. Biophys. Acta, 298 (1973) 333.

204 Walsh C. T., Abeles R. H. and Kaback H. R., Mechanisms of active transport in isolated bacterial membrane vesicles. X. Inactivation of D-lactate dehydrogenase and D-lactate dehydrogenase-coupled transport in E. coli membrane vesicles by an acetylenic substrate, J. Biol. Chem., 247 (1973) 7858.

205 Weiss L., Cellular adhesion in tissue culture. XI. Some effects of ouabain, Exp. Cell Res., 71 (1972) 281.

206 Wilson R. H., Dever J., Harper W. and Fry R., Effect of valinomycin on respiration and volume changes in plant mitochondria, Plant Cell. Physiol., 13 (1973) 1103.

207 Witt H. T. and Zickler A., Electrical evidence for the field indicating absorption change in bioenergetic membranes, FEBS Lett., 37 (1973) 307.

208 Wong D. T., Wilkinson J. R., Hamill R. L. and Horug J.-S., Effect of antibiotic ionophore A 23187 on oxidative phosphorylation and Ca transport of liver mitochondria, Arch. Biochem. Biophys., 156 (1973) 578.

SUBJECT INDEX

Acetate 324, 325, 333, 337, 339, 381, 387, 392
Adrenaline 234
Alamethicin, association 55, 95, 319
—, biosynthesis 42
—, CD curves 44, 175
—, effects on ANS fluorescence 95
—, — ion transport through bilayers 316—321
—, — lipid packing in liposomes 329
—, — liposome surface charge 329
—, interaction with metal ions 44, 316
—, methyl ester 44
—, NMR spectra 44
—, proposed conformation 174
—, structure 44
—, surfactant properties 45
Alkali metal ions, crystallographic radii 224
—, crystallosolvates 212, 213
—, effects on complexone-induced inhibition of bacterial growth 397, 398
—, far infrared spectra of solvated 84, 85
—, ion exchanger selectivity toward 246, 247
—, ion pairing 84, 92
—, solvation energies 222
Alkaline earth metal ions, effects on complexone-induced transport 297
—, antibiotic A23187 as ionophore specific toward 404, 405
—, relative specificity of complexones toward 239, 240
Amanitine 333
α-Aminoisobutyric acid 42, 174
Amylose 215
Angolide 23
8-Anilino-1-naphthalene sulfonate, as counter-ion in salt extraction 105
—, association with complexes 94, 95
—, complexone effects on fluorescence 94, 95, 328
—, effect on the nonactin-induced bilayer conductance 300, 301
—, transport through mitochondrial membrane 402, 403
Anions (lipophilic), as uncouplers in chromatophores 369
—, — in submitochondrial particles 394
—, complexation effects on spectra of 92—95

—, effect on nigericin-induced aerobic ion transport in submitochondrial particles 394
—, — valinomycin-induced K^+ permeability of bilayers 300
—, in the spectrometric study of complexes 91
—, in complexone detection by thin-layer chromatography 94
—, in complexone-induced salt extraction 104—106
—, synergism in uncoupling action with nigericin 369
ANS, see 8-anilino-1-naphthalene sulfonate
Antamanide, analogs, complexation with Na^+ and K^+ 38—41
—, —, conformational analysis 172—174
—, —, structure 38—41
—, antitoxic activity 33
—, CD curves 166
—, complexation effects on ultraviolet spectra 85
Antamanide, complexes, limiting mobilities 97
—, —, stability constants 37
—, —, Stokes' radii 97
—, conformation 168, 217
—, effects on cation permeability of bilayers 379
—, — electroconductivity of salt solutions 95
—, — mitochondrial respiration and K^+ transport 378, 379
—, — salt extraction 37, 103
—, infrared spectra 167
—, Li^+ complex 171
—, Na^+ complex, conformation 168—172, 235
—, —, infrared spectra 167
—, —, mass spectrometry 109
—, —, NMR spectrum 169
—, —, ORD curves 166
—, ORD curves 166
—, perhydro, antimicrobial activity 42
—, —, far infrared spectra of complexes 85
—, —, membrane activity 174
—, [Phe^4, Val^6]-Na^+ complex 171
—, —, ^{13}C-NMR spectra 86, 170
—, —, ^{1}H-NMR spectra 167
—, rearrangements 34

464

Valinomycin—*continued*

—, effects on K⁺ self-diffusion across water/
decane interface 271

—, — Na⁺-amino acid symport in pigeon
erythrocytes 401

—, — proton carrier-activated mitochondrial
ATPase and respiration 386, 387

—, — proton-β-galactoside symport in *S.
aureus* 401

—, — — and K⁺ uptake by respiring bacteria
399

—, — — in mitochondria 377—387

—, — — in submitochondrial particles 394

—, — — in sonic particles from *M. denitrificans* 395

—, — respiratory carrier kinetics 401

—, — reversed electron transport in mitochondria 402

—, — salt extraction by organic solvents
103, 105, 108, 274

—, — secondary blackening of lipid films
278, 279

—, in cation-sensitive membrane electrodes
100, 221, 265, 267, 270

—, infrared spectra 84

—, isolation 1

—, isotope labeling 3, 4

—, low angle X-ray scattering 102

—, micelle formation 274

—, NMR spectra 130

—, ORD curves 119

—, sensitization of bacterial membranes to
342

—, solubility 2

—, structure 2

—, synthesis 4

—, water/lipid decane solution partition coefficient 280

—, water/liposome partition coefficient 323

Verramycin 47